Three Decades of Progress in Control Sciences

Xiaoming Hu, Ulf Jonsson, Bo Wahlberg, and
Bijoy K. Ghosh (Eds.)

Three Decades of Progress in Control Sciences

Dedicated to Chris Byrnes and Anders Lindquist

Springer

Prof. Dr. Xiaoming Hu
Optimization and Systems Theory
School of Engineering Sciences
KTH – Royal Institute of technology
Sweden
E-mail: hu@math.kth.se

Prof. Dr. Ulf Jonsson
Optimization and Systems Theory
School of Engineering Sciences
KTH – Royal Institute of technology
Sweden
E-mail: ulfj@math.kth.se

Prof. Dr. Bo Wahlberg
Automatic control
School of Electrical Engineering
KTH – Royal Institute of Technology
Sweden
E-mail: bo.wahlberg@s3.kth.se

Prof. Dr. Bijoy K. Ghosh
Mathematics and Statistics Department
Texas Tech University
Lubbock, Texas
USA
E-mail: bijoy.ghosh@ttu.edu

ISBN 978-3-642-11277-5 e-ISBN 978-3-642-11278-2

DOI 10.1007/978-3-642-11278-2

Library of Congress Control Number: 2010935850

Cover Design: Erich Kirchner, Heidelberg

Printed on acid-free paper

9 8 7 6 5 4 3 2 1

springer.com

Dedicated to Christopher I. Byrnes and Anders Lindquist
for their lifelong contributions in Systems and Control Theory

Christopher I. Byrnes

Anders Lindquist

Preface

In this edited collection we commemorate the 60th birthday of Prof. Christopher Byrnes and the retirement of Prof. Anders Lindquist from the Chair of Optimization and Systems Theory at KTH. These papers were presented in part at a 2009 workshop in KTH, Stockholm, honoring the lifetime contributions of Professors Byrnes and Lindquist in various fields of applied mathematics.

Outstanding in their fields of research, Byrnes and Lindquist have made significant advances in systems & control and left an indelible mark on a long list of colleagues and PhD students. As co editors of this collection, we have tried to showcase parts of this exciting interaction and congratulate both Byrnes and Lindquist for their years of successful research and a shining career.

About a quarter of a century ago, Anders Lindquist came to KTH to provide new leadership for the Division of Optimization and Systems Theory. In 1985 Chris spent his sabbatical leave at KTH, and the two of them organized the 7th International Symposium on the Mathematical Theory of Networks and Systems (MTNS 85) at KTH that showcased both the field and a thriving academic division at the university and highlighted the start of a long lasting collaboration between the two. Chris Byrnes was recruited recently as a Distinguished Visiting Professor at KTH to continue what has now become a very successful research program, some results from which will be mentioned below.

Chris Byrnes's career began as a PhD student of Marshall Stone, from whom he learned that a good approach to doing research has to begin with an understanding of what makes the problem hard and must ultimately bring the right mixture of applied and pure mathematics techniques to bear on the problem. What is characteristic of his contributions is the unanticipated application of seemingly unrelated branches of pure mathematics. This was exhibited early in his career with the application of techniques from algebraic geometry to solve some long-standing open problems, such as pole-placement by output feedback, in classical linear control systems. In characteristic form, he made this seem understandable and inevitable because "the Laplace transform turns the analysis of linear differential systems into the algebra of rational functions." In collaboration with Alberto Isidori, he helped transform modern nonlinear control systems using nonlinear dynamics and the geometry of manifolds, developing natural analogs of classical notions such as zeros (zero dynamics), minimum phase systems, instantaneous gain and the steady-state response of a system in a nonlinear setting. Together with J. C. Willems, they further enhanced these concepts in terms of their relationship with passive (positive real) systems, i.e., nonlinear systems which dissipate energy. These enhancements of classical control were then used to develop feedback design methods for asymptotic stabilization, asymptotic tracking and disturbance rejection of nonlinear control systems, conceptualized in seemingly familiar terms drawn from classical automatic control.

After receiving his PhD degree at KTH, in 1972 Anders Lindquist went to the Center for Mathematical Systems Theory at the University of Florida as a post doc with R. E. Kalman, followed by a visiting research position at Brown University. He became a full professor at the University of Kentucky in 1980 before returning to KTH in 1983. He has delivered fundamental contributions to the field of systems, signals and control for almost four decades, especially in the areas of stochastic control, modeling, estimation and filtering, and, more recently, feedback and robust control. Anders has produced seminal work in the area of stochastic systems theory, often with a veritable sense for the underlying geometry of the problems. His contributions to filtering and estimation include the very first development of fast filtering algorithms for Kalman filtering and a rigourous proof of the separation principles for stochastic control systems. With Bill Gragg he wrote a widely cited paper on the partial realization problem that has gained considerable attention in the numerical linear algebra community. Together with Giorgio Picci (and coworkers) he developed a comprehensive geometric theory for Markovian representations that provides coordinate-free representations of stochastic systems, and that turned out to be an excellent tool for understanding the principles of the subspace algorithms for system identification developed later.

Anders and Chris published their first joint paper in 1982 and have most recently published two joint articles in 2009 and numerous papers in between. Both Anders and Chris are grateful to have each found a research soul mate who gets excited about the same things. This has played a profound role in their mutual careers. As evidence of their successful collaboration, Anders and Chris, together with coworkers, have worked on the partial realization theory and developed a comprehensive geometric theory of the moment problem for rational measures. A major initial step was the final proof of a conjecture by Tryphon Georgiou on the rational covariance extension problem, formulated in the 1970s by Kalman, and left open for 20 years. This is now the basis of a progressive area of research, which has provided entirely new paradigms based on analytic interpolation and mathematical tools for solving key problems in robust control, spectral estimation, systems identification, and many other engineering problems.

Xiaoming Hu, Ulf Jönsson and Bo Wahlberg,
Kungliga Tekniska Högskolan,
Stockholm, Sweden.

Bijoy K. Ghosh,
Texas Tech University,
Lubbock, Texas, USA.

Christopher I. Byrnes

Christopher I. Byrnes received his doctorate in 1975 from University of Massachusetts under Marshall Stone. He has served on the faculty of the University of Utah, Harvard University, Arizona State University and Washington University in St. Louis, where he served as dean of engineering and the The Edward H. and Florence G. Skinner Professor of Systems Science and Mathematics. The author of more than 250 technical papers and books, Chris received an Honorary Doctorate of Technology from the Royal Institute of Technology (KTH) in Stockholm in 1998 and in 2002 was named a Foreign Member of the Royal Swedish Academy of Engineering Sciences. He is a Fellow of the IEEE, two time winner of The George Axelby Prize and the recipient of the Hendrik W. Bode Prize. In 2005 he was awarded the Reid Prize from SIAM for his contributions to Control Theory and Differential Equations and in 2009 was named an inaugural Fellow of SIAM. He held hold the Giovanni Prodi Chair in Nonlinear Analysis at the University of Wuerzburg in the summer of 2009 and is spending the 2009-2012 academic years as Distinguished Visiting Professor at KTH.

Dissertation Students of Christopher I. Byrnes

1. D. Delchamps, "The Geometry of Spaces of Linear Systems with an Application to the Identification Problem", Ph. D. , Harvard University, 1982.
2. P. K. Stevens, "Algebro-Geometric Methods for Linear Multivariable Feedback Systems", Ph. D. , Harvard University, 1982.
3. B. K. Ghosh, "Simultaneous Pole Assignability of Multi-Mode Linear Dynamical Systems", Ph. D. , Harvard University, 1983.
4. A. Bloch, "Least Squares Estimation and Completely Integrable Hamiltonian Systems", Ph. D. , Harvard University, 1985.
5. B. Martensson (co-directed with K. J. Åström), "Adaptive Stabilization", Ph. D. , Lund Institute of Technology, 1986.
6. P. Baltas (co-directed with P. E. Russell), "Optimal Control of a PV-Powered Pumping System", Ph. D. , Arizona State University, 1987.
7. X. Hu, "Robust Stabilization of Nonlinear Control Systems", Ph. D. , Arizona State University, 1989.
8. S. Pinzoni, "Stabilization and Control of Linear Time-Varying Systems", Ph. D. , Arizona State University, 1989.
9. X. Wang, "Additive Inverse Eigenvalue Problems and Pole-Placement of Linear Systems", Ph. D. , Arizona State University, 1989.
10. J. Rosenthal, "Geometric Methods for Feedback Stabilization of Multivariable Linear Systems", Ph. D. , Arizona State University, 1990.
11. X. Zhu, "Adaptive Stabilization of Multivariable Systems", Ph. D. , Arizona State University, 1991.
12. D. Gupta, "Global Analysis of Splitting Subspaces", Ph. D. , Arizona State University, 1993.
13. W. Lin, "Synthesis of Discrete-Time Nonlinear Control Systems", D. Sc. , Washington University, 1993.
14. J. Roltgen, "Inner-Loop Outer-Loop Control of Nonlinear Systems", D. Sc. , Washington University, 1995.
15. R. Eberhardt, "Optimal Trajectories for Infinite Horizon Problems for Nonlinear Systems", D. Sc. , Washington University, 1996.
16. S. Pandian, "Observers for Nonlinear Systems", D. Sc. , Washington University, 1996.
17. J. Ramsey, "Nonlinear Robust Output Regulation for Parameterized Systems Near a Codimension One Bifurcation", Ph. D. , Washington University, December 2000.
18. F. Celani (co-directed with A. Isidori), "Omega-limit Sets of Nonlinear Systems That Are Semiglobally Practically Stabilized", D. Sc. , Washington University, 2003.
19. N. McGregor (co-directed with A. Isidori), "Semiglobal and Global Output Regulation for Classes of Nonlinear Systems", D. Sc. , Washington University, 2007.
20. B. Whitehead, "Adaptive Output Regulation: Model Reference and Internal Model Techniques", D. Sc. , Washington University, 2009.

Anders Lindquist

Anders Lindquist received his doctorate in 1972 from the Royal Institute of Technology (KTH), Stockholm, Sweden, after which he held visiting positions at the University of Florida and Brown University. In 1974 he joined the faculty at the University of Kentucky, where in 1980 he became a Professor of Mathematics. In 1982 he was appointed to the Chair of Optimization and Systems Theory at KTH, and from 2000 to 2009 he was the Head of the Mathematics Department at the same university. Presently, he is the Director of the Strategic Research Center for Industrial and Applied Mathematics (CIAM) at KTH. He was elected a Member of the Royal Swedish Academy of Engineering Sciences in 1996 and a Foreign Member of the Russian Academy of Natural Sciences in 1997. He is a Fellow of the IEEE and an Honorary Member the Hungarian Operations Research Society. He was awarded the 2009 Reid Prize from SIAM and the 2003 George S. Axelby Outstanding Paper Award of the IEEE Control Systems Society. He is also receiving an Honorary Doctorate (*Doctor Scientiarum Honoris Causa*) from Technion, Haifa, Israel (conferred in June 2010).

Dissertation Students of Anders Lindquist

1. Michele Pavon, "Duality Theory, Stochastic Realization and Invariant Directions for Linear Discrete Time Stochastic Systems", Ph. D. , University of Kentucky, 1979.
2. David Miller, "The Optimal Impulse Control of Jump Stochastic Processes", Ph. D. , University of Kentucky, 1979.
3. Faris Badawi, "Structures and Algorithms in Stochastic Realization Theory and the Smoothing Problem", Ph. D. , University of Kentucky, 1981.
4. Carl Engblom (co-directed with P. O. Lindberg), "Aspects on Relaxations in Optimal Control Theory", Ph. D. , Royal Institute of Technology, 1984.
5. Andrea Gombani, "Stochastic Model Reduction", Ph. D. , Royal Institute of Technology, 1986.
6. Anders Rantzer, "Parametric Uncertainty and Feedback Complexity in Linear Control Systems", Ph. D. , Royal Institute of Technology, 1991.
7. Martin Hagström, "The Positive Real Region and the Dynamics of Fast Kalman Filtering in Some Low Dimensional Cases", TeknL, Royal Institute of Technology, 1993.
8. Yishao Zhou, "On the Dynamical Behavior of the Discrete-Time Matrix Riccati Equation and Related Filtering Algorithms", Ph. D. , Royal Institute of Technology, 1992.
9. Jan-Åke Sand, "Four Papers in Stochastic Realization Theory", Ph. D. , Royal Institute of Technology, 1994.
10. Jöran Petersson (codirected with K. Holmström), "Algorithms for Fitting Two Classes of Exponential Sums to Empirical Data", TeknL, Royal Institute of Technology, 1998.
11. Jore Mari, "Rational Modeling of Time Series and Applications of Geometric Control", Ph. D. , Royal Institute of Technology, 1998.
12. Magnus Egerstedt (co-directed with X. Hu), "Motion Planning and Control of Mobile Robots", Ph. D. , Royal Institute of Technology, 2000.
13. Mattias Nordin (co-directed with Per-Olof Gutman), "Nonlinear Backlash Compensation of Speed Controlled Elastic System", Ph. D. , Royal Institute of Technology, 2000.
14. Camilla Landén (co-directed with Tomas Björk), "On the Term Structure of Forwards, Futures and Interest Rates", Ph. D. , Royal Institute of Technology, 2001.
15. Per Enqvist, "Spectral Estimation by Geometric, Topological and Optimization Methods", Ph. D. , Royal Institute of Technology, 2001.
16. Claudio Altafini (co-directed with X. Hu), "Geometric Control Methods for Nonlinear Systems and Robotic Applications", Ph. D. , Royal Institute of Technology, 2001.
17. Anders Dahlén, "Identification of Stochastic Systems: Subspace Methods and Covariance Extension", Ph. D. , Royal Institute of Technology, 2001.
18. Henrik Rehbinder (co-directed with X. Hu), "State Estimation and Limited Communication Control for Nonlinear Robotic Systems", Ph. D. , Royal Institute of Technology, 2001.

19. Ryozo Nagamune, "Robust Control with Complexity Constraint: A Nevanlinna-Pick Interpolation Approach", Ph. D. , Royal Institute of Technology, 2002.
20. Anders Blomqvist, "A Convex Optimization Approach to Complexity Constrained Analytic Interpolation with Applications to ARMA Estimation and Robust Control", Ph. D. , Royal Institute of Technology, 2005.
21. Gianantonio Bortolin (co-directed with Per-Olof Gutman), "Modeling and Grey-Box Identification of Curl and Twist in Paperboard Manufacturing", Ph. D. , Royal Institute of Technology, 2006.
22. Christelle Gaillemard (codirected with Per-Olof Gutman), "Modeling the Moisture Content of Multi-Ply Paperboard in the Paper Machine Drying Section", TeknL, Royal Institute of Technology, 2006.
23. Giovanna Fanizza, "Modeling and Model Reduction by Analytic Interpolation and Optimization", Ph. D. , Royal Institute of Technology, 2008.
24. Johan Karlsson, "Inverse Problems in Analytic Interpolation for Robust Control and Spectral Estimation", Ph. D. , Royal Institute of Technology, 2008.
25. Yohei Kuriowa, "A Parametrization of Positive Real Residue Interpolants with McMillan Constraint", Ph. D. , Royal Institute of Technology, 2009.

Acknowledgement

The editors of this manuscript would like to thank Mr. Mervyn P. B. Ekanayake for his tireless efforts in formatting this collection. One of the co editor was supported by the National Science Foundation under Grant No. 0523983 and 0425749. Any opinions, findings, and conclusions or recommendations expressed in this material are those of the author(s) and do not necessarily reflect the views of the National Science Foundation. The KTH workshop was supported in part by the Swedish Research Council Conference Grant No. 2009-1099.

Contents

1

Information Acquisition in the Exploration of Random Fields*

J. Baillieul and D. Baronov

Intelligent Mechatronics Lab (IML), Boston University, Boston, MA 02215, USA

Summary. An information-like metric that characterizes the complexity of functions on compact planar domains is presented. Combined with some recently introduced control laws for level following and gradient climbing, it is shown how the metric can be used in designing reconnaissance strategies for sensor-enabled mobile robots. Reconnaissance of unknown scalar potential fields—describing physical quantities such as temperature, RF field strength, chemical species concentration, and so forth—may be thought of as an empirical approach to determining critical point geometries and other important topological features. It is hoped that this will be of interest to Professors Byrnes and Lindquist on the occasion of the career milestones that this volume celebrates.

1.1 Appreciation

When a distinguished scientist passes a certain age or achievement milestone, it is nowadays standard practice to publish a collection of scholarly articles reflect on the work of that scholar. More often than not, the authors who contribute to such volumes struggle to write something that is original and significant on the one hand, while being somehow related to the honoree on the other. The task is doubly challenging when there are two who are being honored at the same time. Some years ago, I had the honor of collaborating with Byrnes in an attempt to apply differential topology to models of electric power grids. (See [6] and the references cited therein.) At the same time, Lindquist, with whom I have not collaborated, has done definitive work in stochastic systems with particular emphasis on covariance methods and the study of moments. (See [18] and the references cited therein.) By happenstance, my own current research has led to the study of *random polynomials* and *random potential fields*. Hence, in providing a brief summary of some ongoing work, I hope that I will have succeeded in showing yet another direction in which the work of Byrnes and Lindquist can be seen as providing inspiration.

*The authors gratefully acknowledge support from ODDR&E MURI07 Program Grant Number FA9550-07-1-0528, and the National Science Foundation ITR Program Grant Number DMI-0330171.

X. Hu et al. (Eds.): Three Decades of Progress in Control Sciences, pp. 1–17, 2010.

1.2 Decision-Making in the Performance of Search and Reconnaissance

The study of optimal decision-making has spawned an enormous body of technical literature spanning large disciplinary segments of control theory, operations research, and statistics—and other branches of applied mathematics as well. Roots of the theory of optimal decisions can be found in the early theory of games and economic decisions, the pioneers of which included Von Neumann, ([24]), Nash, Kuhn, and Tucker, ([17]) with more recent advances chronicled in the work of Raiffa, Schlaifer, and Pratt ([21]). Recently, interest has shifted to studying ways that groups and individuals actually make decisions in various settings and how these decisions compare with what would be optimal in some sense. (The session entitled Mixed Robot/Human Team Decision Dynamics at the 2008 IEEE Conference on Decisions and Control describes some of this research. See [19],[25],[8],[2],[11],[23].) Contemporary work on decision modeling has drawn inspiration from cognitive and social psychology where researchers have been working to understand experimentally observed dynamics of human decision-making in instance where subjects systematically fail to make optimal decisions. (See [13] and the references therein.) This research has also shown that human decision-making behaviors can change a great deal depending on factors such as level of boredom, reward rate, and social context.

To understand these issues in the context of common yet important human activities, we have begun to study how humans approach search and reconnaissance problems. Such problems are of interest in a variety of practical settings, and they lend themselves to being abstracted as computer games where realistic choices need to be made. Some prior work has been reported on distributed algorithms for optimal random search of building interiors. (See [1],[7],[10],[12],[15],[16],[20], and [22] for recent results and a discussion relating search methods to models arising in statistical mechanics.) The present paper introduces a new class of search (or more precisely, *reconnaissance*) problems in which there are time vs accuracy trade offs. The goal of the research is to define and characterize information-like metrics that will permit quantifying the relative importance of speed versus accuracy in simulated search and reconnaissance tasks. This metric that will be described in what follows is a refinement of an earlier version that we presented in [2].

1.3 Formal Models of Information-Gathering during Reconnaissance

The search problems being studied involve estimating important characteristic features of smooth functions $f : \mathbb{R}^m \to \mathbb{R}$ on compact, connected, and simply connected domains $D \subset \mathbb{R}^m$ where $m = 1$ or 2. The types of features we have in mind include values of the function argument at which the function has a zero (especially in the case $m = 1$) or values where the function achieves a maximum or minimum. It may also be of interest to estimate how much the function varies over a domain that is of interest. In order to rule out uninteresting pathologies, it is assumed that for all functions under consideration, the inverse image $f^{-1}(y)$ of any point in the range has only a finite number of connected components, and if $m = 2$, the connected components of $f^{-1}(y)$ are almost surely (with respect to an appropriate measure) *simple* curves of two types. One type consists of closed curves, and the other type is made up of simple curves whose beginning and ending points lie on the boundary of D. In the present paper, we shall emphasize the case $m = 2$, and note that functions of the type we shall study arise in modeling physical fields—thermal, RF, chemical species concentrations and so forth. The goal of the work is to understand how to use sensor-enabled mobile agents to acquire knowledge of an unknown field as efficiently as possible.

1.3.1 Acquiring Empirical Information about Smooth Functions on Bounded Domains

In [2], the valuation of a search strategy was approached by means of an information-based measure of *complexity* of functions. Let $D \subset \mathbb{R}^m$ be a compact, connected, simply connected domain with $m = 1$ or 2, and let $f : \mathbb{R}^m \to \mathbb{R}$. Then $f(D)$ is a compact connected subset of $\mathbb{R}$ which we write as $[a,b]$. At the outset, we fix a finite partition of this interval:

$$a = x_0 < x_1 < \cdots < x_n = b.$$

For each x_j, $j=1,\ldots,n$, we denote the set of connected components of $f^{-1}([x_{j-1},x_j])$ by

$$cc\,[f^{-1}([x_{j-1},x_j])].$$

For any such partition, we obtain a corresponding partition

$$\mathscr{V} = \cup_{j=1}^{n} cc[f^{-1}([x_{j-1},x_j])]$$

of D. We define the *complexity* of f with respect to $\mathscr{V} = \{V_1,\ldots,V_N\}$ as

$$H(f,\mathscr{V}) = -\sum_{j=1}^{N} \frac{\mu(V_j)}{\mu(D)} \log_2 \frac{\mu(V_j)}{\mu(D)}, \tag{1.1}$$

where μ is Lebesgue measure on $\mathbb{R}^m$. We shall also refer to (1.1) as the *partition entropy* of f with respect to $\mathscr{V}$.

As pointed out in [2], the properties of this measure of *function complexity* are directly analogous to corresponding properties of Shannon's *entropy*:

1. If the closed interval $[a,b]$ in fact contains only a single element (i.e. if f is a constant), then we adopt the convention that $\mathscr{H}(f,\mathscr{V}) = 0$. The trivial partition of $[a,b]$ with the two elements $\{a,b\}$ also has $\mathscr{H}(f,\mathscr{V}) = 0$.
2. If the connected components of inverse images of all cells $[x_{j-1},x_j]$ in the range partition have identical measure $\mu(V_j)$, then $\mathscr{H}(f,\mathscr{V}) = \log_2 N$ (where N is the number of elements in the partition $\mathscr{V}$.
3. If $\mu(V_i) \neq \mu(V_j)$ for some pair of cells $V_i, V_j \in \mathscr{V}$, then $\mathscr{H}(f,\mathscr{V}) < \log_2 N$.

We wish to use this concept of function complexity to provide guideposts in a strategy for seeking out important characteristics of unknown functions. As in [2], there are two important features of the domain partition $\mathscr{V}$ associated with the function f. First, because all search strategies under consideration will discover only connected components of sets in $\mathscr{V}$, it is important to recall that by construction, the elements $V_j \in \mathscr{V}$ are connected subsets of $\mathbb{R}^m$. Second, we shall assume that each search problem under consideration will be posed with respect to a fixed partition $\{x_i\}$ of the range $[a,b]$ and corresponding fixed partition $\mathscr{V}$ of $D \subset \mathbb{R}^m$. We define a *search chain* to be a sequence of nested subsets

$$S_1 \subset S_2 \subset \cdots \subset S_n = \{x_i\}_{i=1}^n$$

such that the cardinality of S_k is k. A search chain is thus a maximal ascending path in the lattice of subsets of $\{x_1,\ldots,x_n\}$. A *search sequence* is then defined to be a corresponding set of elements $V_{i_1},\ldots V_{i_n} \in \mathscr{V}$ such that $V_{i_j} \subset f^{-1}([x_{j-1},x_j])$ for $j = 1,\ldots,n$.

It is in terms of these constructions that we pursue the discussion of reconnaissance strategies. Given a smooth function f mapping a compact connected domain $D \subset \mathbb{R}^m$ onto an interval $[a,b]$, together with a partition $a = x_0 < x_1 < \cdots < b$ as above, we let $\mathscr{S}$ denote the set of all search chains. That is to say $\mathscr{S}$ is the set of all maximal ascending chains in the lattice of subsets of $\{x_1,\ldots,x_n\}$. We let $\mathscr{W}$ denote the set of all search sequences corresponding to elements of $\mathscr{S}$.

We next apply our complexity measure to compare search sequences. Let $\mathscr{V}_\alpha \in \mathscr{W}$ be a search sequence—i.e. a set of elements of $\mathscr{V}$ (subsets of D) corresponding to a search chain as defined above. The search sequence $\mathscr{V}_\alpha$ is said to be *monotone* if the elements can be ordered $\bar{V}_1,\ldots,\bar{V}_n \in \mathscr{V}_\alpha$ such that for $k = 1,\ldots,n$, $\cup_{j=1}^k \bar{V}_j$ is connected.

Now to each set

$$S_k = \{x_{i_1},\ldots,x_{i_k}\}$$

in a search chain, there is an associated partition $\mathscr{V}_k$ of D consisting of all connected components of $\{f^{-1}([x_{j_{i-1}},x_{j_i}]) : i = 1,\ldots,k+1\}$, where we adopt the conventions

1. $x_{i_1} < \cdots < x_{i_k}$,
2. $x_{i_0} = x_0 = a$, and
3. $x_{i_{k+1}} = x_n = b$.

The notation is cumbersome, but the meaning is simple: in order to define $\mathscr{V}_k$, we consider S_k together with the endpoints $x_0 = a$ and $x_n = b$. To this partition there is (as defined above) an associated complexity measure given by the *partition entropy*:

$$H(f,\mathscr{V}_k) = -\sum_{V_\alpha \in \mathscr{V}_k} \frac{\mu(V_\alpha)}{\mu(D)} \log_2 \frac{\mu(V_\alpha)}{\mu(D)}.$$

For each partition of $[a,b]$ and for each search chain $S_1 \subset \cdots \subset S_n$, there is a corresponding increasing chain of partition entropies. The stepwise refining of domain partitions leading successively from $\mathscr{V}_k$ to $\mathscr{V}_{k+1}$ defines the reconnaissance process,

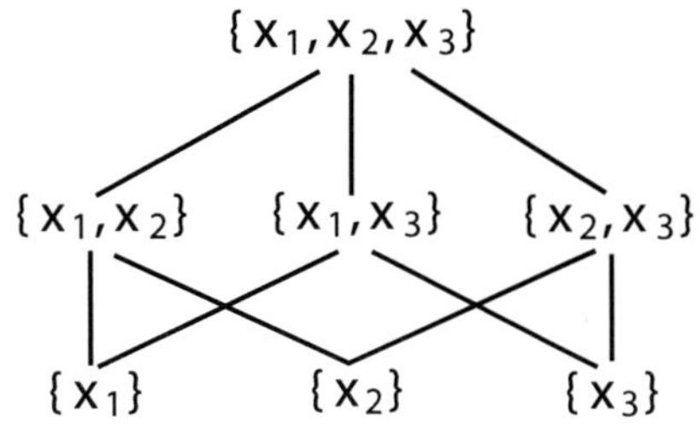

Fig. 1.1. The lattice of subsets of $\{x_1,x_2,x_3\}$.

and the changes in partition entropy going from $\mathscr{V}_k$ to $\mathscr{V}_{k+1}$ measures the efficiency of the reconnaissance effort at that step.

Let $f: D \to [a,b]$ as above, and let $P=\{x_i\}_{i=1}^n$ be a random partition of $[a,b]$. Let $S_1 \subset \cdots \subset S_{n-1}$ and $\overline{S}_1 \subset \cdots \subset \overline{S}_{n-1}$ be two search chains with associated partitions $\mathscr{V}_1 \subset \cdots \subseteq \mathscr{V}_n$ and $\overline{\mathscr{V}}_1 \subset \cdots \subset \overline{\mathscr{V}}_n$ of D. We say that $\overline{S}$ *dominates* S, and write $\overline{S} \succeq S$, if $H(f,\overline{\mathscr{V}}_k) \geq H(f,\mathscr{V}_k)$ for all k, $1 \leq k \leq n$.

With $P=\{x_i\}_{i=1}^n$ a random partition of $[a,b]$, the relation "$\succeq$" defines a *quasi-order* on the set of all search sequences on P. It is clear that this relation is both reflexive and transitive. That it does not have the antisymmetry property is illustrated by the following.

Example 1.3.1. Let $D=[a,b]=[0,1]$ and $f: D \to [a,b]$ be given by $f(x)=x$. Consider the partition $\{x_0,x_1,x_2,x_3,x_4\}$ where $x_k = k/4$. The lattice of subsets of $\{x_1,x_2,x_3\}$ is depicted in Fig. 1.1. Consider the search sequences

$$\begin{aligned} \mathscr{S}&: \ S_1=\{x_1\} \subset S_2=\{x_1,x_2\} \subset S_3=\{x_1,x_2,x_3\}, \\ \overline{\mathscr{S}}&: \ \overline{S}_1=\{x_1\} \subset \overline{S}_2=\{x_1,x_3\} \subset \overline{S}_3=\{x_1,x_2,x_3\}. \end{aligned}$$

The partition entropies corresponding to $\mathscr{S}$ are

$$\begin{aligned} &H(f,\mathscr{V}_1)=2-\tfrac{3}{4}\log_2(3) \approx 0.811278 \\ &H(f,\mathscr{V}_2)=3/2 \ \ H(f,\mathscr{V}_3)=2. \end{aligned}$$

Since $H(f,\overline{\mathscr{V}}_k)=H(f,\mathscr{V}_k)$ for $k=1,2,3$, $\mathscr{S} \succeq \overline{\mathscr{S}}$ and $\overline{\mathscr{S}} \succeq \mathscr{S}$ but because $\mathscr{S} \neq \overline{\mathscr{S}}$ the relation does not have the anti-symmetry property and thus fails to be a partial order on the set of search sequences. From Fig. 1.1 it is easy to see determine that among the six distinct ascending paths in the subset lattice, the search sequences $\{x_2\} \subset \{x_1,x_2\} \subset \{x_1,x_2,x_3\}$ and $\{x_2\} \subset \{x_2,x_3\} \subset \{x_1,x_2,x_3\}$ are *dominating* with respect to the quasi-ordering. Each of these is related to the binary subdivision search that is discussed below.

Remark 1.3.1. Let N be a positive integer. Suppose that the compact domain D has area A_D. The maximum possible entropy of a partition of D into N cells is $\log_2 N$. This partition entropy is achieve by any partition of D into N cells of area all equal to A_D/N. We omit the proof, but note that a proof using the *log sum inequality* very much along the lines of the proof of Proposition 1.4.1 below can be carried out.

*Remark 1.3.2. Binary subdivision*is a particular sequential partition refinement procedure that at various stages achieves the maximum possible partition entropy. Given any set D, partition it into two subsets V_1 and V_2 of equal area. Then, in either order, subdivide each of these into two smaller subsets, each of which has area equal to one-fourth that of D. Because we are conducting our discussion under the assumption that robots need to actually be in motion to carry out subdivisions, the subdivision of V_1 and V_2 does not occur simultaneously. To continue the process, we partition each of the cells previously obtained in two smaller subsets—each having the same area. Continuing with successive partition refinements of this type where we step-wise divide cells in half, we find that each time in the process at which there are 2^k subsets for some integer k, the maximum possible partition entropy of $\log_2(2^k) = k$ is achieved.

Remark 1.3.3. For functions that map domains D to $[a,b]$ in more complex ways than in Example 1.3.1, it is generally not so straightforward to find a dominating search strategy. This is easily illustrated by reworking the above example with the same $D = [a,b] = [0,1]$ and $f : D \to [a,b]$ given by $f(x) = x^2$. For this function and the same partition as in Example 1.3.1, there is a unique dominating search sequence: $\{x_1\} \subset \{x_1,x_2\} \subset \{x_1,x_2,x_3\}$, and it may be observed that all search sequences have distinct information entropy patterns.

1.3.2 A Reconnaissance Strategy for Two Dimensional Domains

While the information-like complexity metric $H(f,\mathscr{V})$ and the notion of partition entropy provide useful guides in sample-based exploration of unknown functions, there are important aspects of search and exploration that are not directly captured. In the exploration of two-dimensional domains—as described in [2], for instance—contour-following control laws enable search agents to map connected components of level sets of functions, but the complete search protocol must include the additional capability of discovering all connected components of the level sets. This remark is illustrated in Fig. 1.2 where the inverse images of mesh points in the range $a = x_0 < x_1 \cdots < x_n = b$ need not be connected sets.

Corresponding to such a range partition, the following strategy for mapping values of f may be based on the *level-curve-following* control law for mobile robots that was proposed in [3]. Start at the lowest level in the range, x_0, and choose an arbitrary point ξ_0 in the domain such that $f(\xi_0) = x_0$. Starting at ξ_0, follow the curve $f(\zeta) \equiv x_0$ until the path either returns to ξ_0 or intersects the boundary of the domain. (One of these two must occur.) Denote this point on the curve ζ_0. Starting at ζ_0, follow an ascending curve ([5]) until either $f(\zeta) = x_1$ or until no further ascent is possible. If it happens that the search agent has arrived at $\zeta = \xi_1$ such that $f(\xi_1) = x_1$, the next step in the search process is to follow the curve $f(\zeta) \equiv x_1$ until the path either returns to ξ_1 or intersects the boundary of the domain. Label this "stopping point" on the curve ζ_1. Starting at ζ_1, again follow an ascending path until either $f(\zeta) = x_2$ or until no further ascent is possible. By repeating this strategy of alternating between a process of step-wise ascent between mesh points x_k and x_{k+1}, followed by tracing

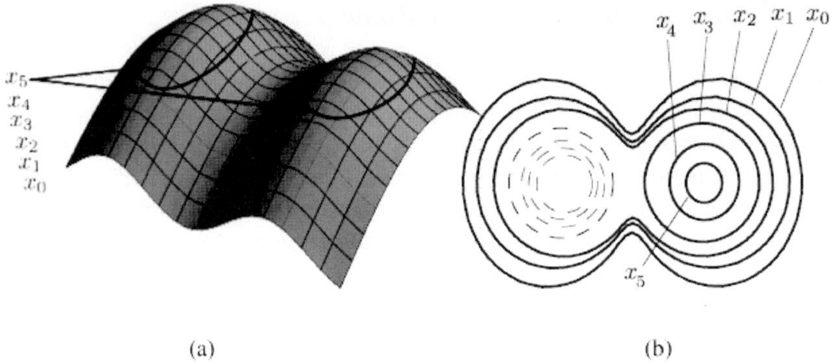

Fig. 1.2. The level sets corresponding to a partition of the range of a function in 2-*d* are typically not connected. This is illustrated by the surface plot (a) and the contour plot (b).

the level curve $f(\zeta) \equiv x_{k+1}$, we have specified a protocol by which an agent can trace and record the locations of points on connected components of level sets of f corresponding to the given partition of the range.

In this way, a monotone search sequence (as defined above) can be mapped. The monotone sequence is associated with contours such as those depicted by the thick (as opposed to dashed) curves in Fig. 1.2(b). It is clear that this ascend-and-trace protocol will be effective in identifying monotone search sequences, but in order to map all components of the level sets, it must be enhanced in some way. Non-monotone search sequences, which are important because they provide information on the numbers and locations of critical points of f, must be treated differently.

This is stated more precisely as the following proposition whose proof is omitted.

Proposition 1.3.1. *Let $\mathscr{V}_\alpha \in \mathscr{W}$ be a (not necessarily monotone) search sequence: $\mathscr{V}_\alpha = \{V_1,\dots,V_n\}$. The number of connected components of $\cup_{j=1}^n V_j$ is a lower bound on the number of relative extrema of f.*

We say that a function $f : D \subset \mathbb{R}^2 \to \mathbb{R}$ is *locally radially symmetric* on a subset $V \subset D$ if there is a point $(x^*,y^*) \in V$ such that for all $(x,y) \in V$, f depends on (x,y) only as a function of $\sqrt{(x-x^*)^2+(y-y^*)^2}$. We conclude the section by noting the following geometric feature of monotone search sequences.

Proposition 1.3.2. *Let $\mathscr{V}_\alpha$ be a monotone search sequence whose elements are labeled such that $V_j \subset f^{-1}([x_{j-1},x_j])$. Let $\partial \overline{V}_j$ denote the boundary of V_j that is the preimage of $f^{-1}(x_j)$, and suppose that on the set of points enclosed by $\partial \overline{V}_0$, f is locally radially symmetric. Then if $i<k$, the arc length of $\partial \overline{V}_i$ is great than the arc length of $\partial \overline{V}_k$. In other words, the boundaries of the sets in the domain partition of a monotone search sequence are a nested set of simple closed curves.*

1.4 Monotone Functions in the Plane

Let f be a smooth function defined on a compact domain $D \subset \mathbb{R}^2$ as in the previous section with $f(D) = [a,b]$. If for every partition $a = x_0 < x_1 < \cdots < x_n = b$ all associated search sequences are monotone, the function itself is said to be *monotone*. Monotone functions are unimodal—i.e. a monotone function has a unique maximum in its domain. We examine several monotone functions and the corresponding monotone search sequences associated with uniform partitions of the range.

Example 1.4.1. (*Cone-like Potential Fields*) Consider a right circular cone in $\mathbb{R}^3$ whose base has radius r and whose height is h. Assume the base lies in the x,y-plane and is centered at the origin. The function f maps the domain $\{(x,y) : x^2 + y^2 \le r\}$ onto $[0,h]$ by $f(x,y) = h(1-(1/h)\sqrt{x^2+y^2})$. That is, f maps the point (x,y) onto the point on the surface of the cone lying above (x,y). Partition the range $[0,h]$ into n subintervals of uniform length h/n. The corresponding partition of the domain, $f^{-1}([(k-1)h/n, kh/n])$, consists of annular regions whose outer boundary is a circle of radius $(n+1-k)r/n$ and inner boundary a circle of radius $(n-k)r/n$. The area of this annulus is

$$Area_k = \pi \frac{2(n-k)+1}{n^2} r^2,$$

and the normalized area is $A_k = Area_k/(\pi r^2) = 2(n-k)+1)/n^2$. The partition entropy, as defined in the previous section, is given by

$$H(f,\mathscr{V}) = -\sum_{k=1}^{n} A_k \log_2 A_k.$$

The dependence of this entropy on the number of cells in the partition is shown in Figure 1.3. The discrete values of this entropy are shown as small circular dots in the plot. Using standard data fitting techniques we have found that the dependence of this partition entropy is well approximated for the given range of values depicted by

$$H(n) = 1.45421 \log_e(2.31129\, n + 4.99357) - 1.54152.$$

This function was found by fitting the values of $H(f,\mathscr{V})$ for n between 3 and 25. The plot illustrates the goodness of fit in the range $n = 3$ to 45.

Example 1.4.2. (*Hemispherical Potential Fields*) Next consider the unit hemisphere as defining a potential field over the unit disk. We partition the range $[0,1]$ into n subintervals of equal length. The corresponding partition of the domain is into annular regions $\{(x,y) : \sqrt{1-k^2/n^2} \le \sqrt{x^2+y^2} \le \sqrt{1-(k-1)^2/n^2}\}$. The areas of such regions are given by $Area_k = \pi(2k-1)/n^2$, so that the normalized areas are $A_k = (2k-1)/n^2$. It is interesting to note that these values, as k ranges from 1 to n are the same as the values of the previous example (cone-like potentials) listed in reverse order. Hence, the partition entropies are the same in both cases.

Example 1.4.3. (*Gaussian Potential Fields*) A unimodal Gaussian function has the form $f(x,y) = \exp(-(x^2+y^2)/c^2)$. The range of interest is $[0,1]$. If we subdivide

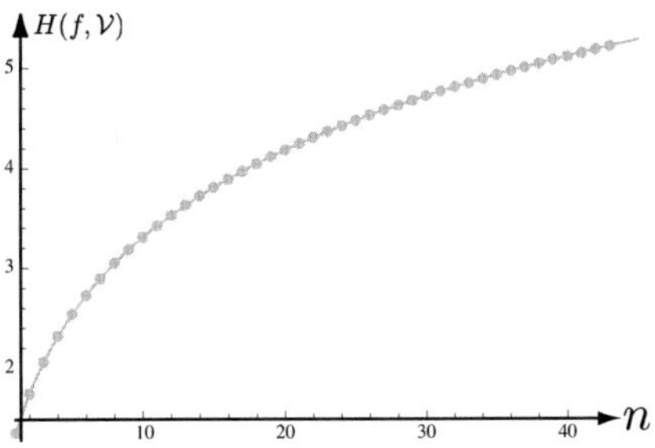

Fig. 1.3. The partition entropy of a cone-like potential field as a function of the number n of cells in the uniform of the range $[0,h]$.

this into subintervals of equal length $1/n$, we obtain a corresponding set of concentric annular regions $\{(x,y) : c^2[\log n - \log k] \le x^2 + y^2 \le c^2[\log n - \log(k-1)\}$. Unlike the previous two cases, the first of these regions has infinite area. A natural approach to pass to consideration of a finite domain is to restrict our attention to the second through the n-th regions. The normalized areas of these are

$$A_k = \frac{\log k - \log(k-1)}{\log n}.$$

As in the preceding examples, for moderate values of n we can approximate the partition entropy by writing

$$H(n) = -\sum_{k=1}^{n} A_k \log_2 A_k \approx 1.14174 \log(1.44092\, n + 0.838691).$$

Thus, in each of the Examples 1.4.1-1.4.3, the partition entropy has an approximately logarithmic dependence on the number of cells in the range partition. Figure 1.4 compares the partition entropies of this and the preceding examples as a function of the number n of uniform subintervals in the range partition.

A somewhat different comparison of these functions—in terms of the relative sizes of cells in the domain partition—illustrates the way that the partition entropy encodes qualitative features of the field. As noted, the cells in the monotone search sequence associated with a uniform range partition and the cone potential have the same normalized areas as those for the hemisphere potential. If the concentric annular cells are ordered from the outer boundary of the domain inwards, the cone potential's cell areas are linearly decreasing, whereas the hemisphere potential's cell areas are linearly increasing. See Figure 1.5. We also see from this figure that the cell areas of the Gaussian potential decrease in area in a nonlinear fashion as a function of their place in the ordering from outermost to innermost.

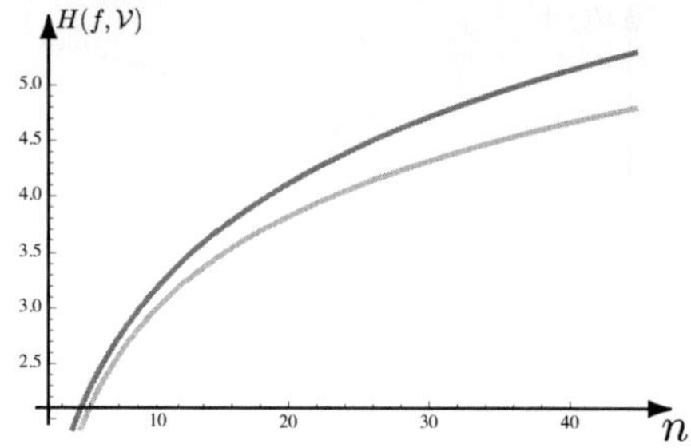

Fig. 1.4. A comparison of the partition entropies of the cone-like and hemispherical potentials (upper) and the Gaussian potential (lower) as a function of the number n of cells in the partitions.

1.4.1 Maximally Complex Symmetric Monotone Functions

Examples 1.4.1 through 1.4.3 are special cases of a more general class of functions on planar domains that can be constructed in terms of continuous scalar functions. We define the class $\mathscr{H}$ of continuous, non-negative functions defined on the unit interval that satisfy (*i*) $h(1)=0$, and (*ii*) h is monotonically decreasing on $[0,1]$. To each $h\in\mathscr{H}$, there is an associated function f defined on the compact domain $D=\{(x,y):x^2+y^2\leq 1\}$ defined by $f(x,y)=h(\sqrt{x^2+y^2})$. As in the previous examples, partition the range $[0,h(0)]$ into n equal subintervals. This partition determines an associated partition of D into n concentric annular regions, the k-th of which has normalized area $h^{-1}(\frac{k-1}{n})^2-h^{-1}(\frac{k}{n})^2$. The partition entropy is

$$H(h)=-\sum_{k=1}^{n}\big(h^{-1}(\frac{k-1}{n})^2-h^{-1}(\frac{k}{n})^2\big)\log_2\big[h^{-1}(\frac{k-1}{n})^2-h^{-1}(\frac{k}{n})^2\big].$$

Let $h^*(x)=1-x^2$. Restricted to the unit interval $[0,1]$, h^* is in the class $\mathscr{H}$, and on this interval, $h^{*-1}(x)=\sqrt{1-x}$.

Proposition 1.4.1. *For all $h\in\mathscr{H}$, $H(h)\leq H(h^*)$.*

Proof. Let $h\in\mathscr{H}$, and let $a_k=h^{-1}(\frac{k-1}{n})^2-h^{-1}(\frac{k}{n})^2$, and let $b_k=h^{*-1}(\frac{k-1}{n})^2-h^{*-1}(\frac{k}{n})^2=1/n$. Note that $H(h^*)=-\sum_{k=1}^{n}(1/n)\log(1/n)=\log(n)$. The well-known *log sum inequality* (See [14].) states that

$$\sum_{i=1}^{n}a_i\log\frac{a_i}{b_i}\geq(\sum_{i=1}^{n}a_i)\log\frac{\sum_{i=1}^{n}a_i}{\sum_{i=1}^{n}b_i}$$

with equality holding if and only if a_i/b_i =const. Plugging in our values of b_i, this inequality is easily seen to reduce to

$$-\sum_{i=1}^{n}a_i\log a_i\leq\log n.$$

The inequality is valid for logarithms of any base ≥ 1, and this proves the proposition. □

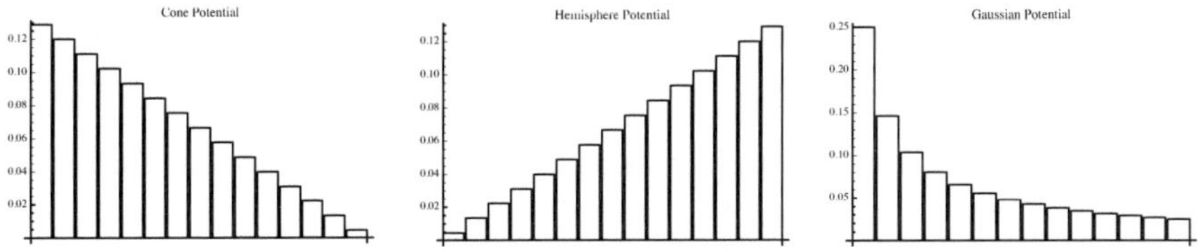

Fig. 1.5. In the case of the three potential functions considered in Examples 1.4.1,1.4.2 and 1.4.3 respectively, the monotone search sequence associated with a uniform partition of the range defines a partition of the domain that is made up of concentric annular regions. The area of each region depends on its position in the sequential order in which the outermost is first and the innermost (disk) is last. This dependence in each of the three cases is displayed above. The dependence is linear decreasing for the cone, linear increasing for the hemisphere, and nonlinear decreasing for the Gaussian.

1.5 Models of Robot-Assisted Reconnaissance of Potential Fields

The gradient-climbing and level-following control laws reported in [5] and [3] can be used in concert with the partition entropy metric to design efficient reconnaissance strategies. The premise is that there is a sensor guided mobile robot that is able to determine the value of an unknown potential field at its present location. The unknown potential field is our abstraction of an unknown terrain, unknown concentration of a chemical species, an unknown thermal field etc. The search strategy is essentially what was described in Section 1.3.2, but the distinction here is that the potential field, f, is not known a priori. This means in particular that the maximum and minimum values of f are not known. Nor do we know whether f is monotone or not.

Many reconnaissance strategies are possible, and a broad survey will be given elsewhere. The strategy we describe here is somewhat conservative in that it methodically accumulates small increments of information regarding the level sets of the potential field, while at the same time looking for characteristic changes that indicate whether the field is non-monotone (multimodal). The exploration begins at an arbitrarily chosen initial point (x_0, y_0) at which the field value $L = f(x_0, y_0)$ is measured. Using an isoline-following control law (e.g [3]), a connected contour of points in the domain that achieve this level of the field is determined. (Assume for the moment that the contour is completely contained within the domain that is of interest—i.e. it does not intersect the boundary.) Depending on what is being measured, it is possible to make *ad hoc* but reasonable assumptions regarding the range of f. For toxic chemicals, for instance, there are published values of concentrations that are known to produce health hazards ([9]). Such values can be taken to define the upper limit T of the range of interest. Given T, the range $[L, T]$ can be partitioned, and the reconnaissance strategy of Section 1.3.2 can be executed.

As the ascend-and-trace reconnaissance protocol is executed, a sequence of domain partitions is successively refined, and each time a new level contour is mapped, a cell in the domain partition is subdivided. As discussed in Section 1.3.1, we obtain a search chain with corresponding increasing chain of partition entropies. The ascend-and-trace strategy is associated with the particular search chain $S_1 \subset S_2 \subset \cdots$ where $S_k = \{x_1, \ldots, x_k\}$ is defined in terms of the range partition $L < x_1 < \cdots < x_n = T$. This chain is in turn associated with a sequence of partitions of the domain D as follows:

$\mathscr{V}_1 = \{V_1, \bar{V}_2\}$ where V_1 is the set of points enclosed between the contours of level L and level $\underline{x_1}$; $\bar{V}_2$ is the complement of V_1 in D—i.e. $\bar{V}_2 = D - V_1$.

$\mathscr{V}_2 = \{V_1, V_2, \bar{V}_3\}$ where V_1 remains the same, V_2 is the set of points enclosed between the mapped contours corresponding to range levels x_1 and x_2, and $\bar{V}_3 = D - (V_1 \cup V_2)$.

The k-th partition refinement is given by $\mathscr{V}_k = \{V_1, \ldots, V_k, \bar{V}_{k+1}\}$ where $V_1, \ldots, V_{k-1}$ are cells defined for $\mathscr{V}_{k-1}$, and V_k is the cell enclosed between the mapped contours corresponding to range levels x_{k-1} and x_k. $\bar{V}_{k+1} = D - (\cup_j = 1^k V_j)$.

To each partition $\mathscr{V}_k$ we have an associated partition entropy

$$H(f, \mathscr{V}_k) = -\frac{\mu(\bar{V}_{k+1})}{\mu(D)} \log_2 \frac{\mu(\bar{V}_{k+1})}{\mu(D)} - \sum_{j=1}^{k} \frac{\mu(V_j)}{\mu(D)} \log_2 \frac{\mu(V_j)}{\mu(D)}.$$

The stepwise change in going from $H(f, \mathscr{V}_k)$ to $H(f, \mathscr{V}_{k+1})$ indicates how effectively the reconnaissance strategy is increasing our knowledge about the potential field (function) f.

The following notation will be useful in our effort to characterize the *entropy rate* $\Delta H_k = H(f, \mathscr{V}_{k+1}) - H(f, \mathscr{V}_k)$ determined by the given partition refinement. For each m-element set of positive numbers $p_1, \ldots, p_m$ satisfying $p_1 + \cdots + p_m = 1$, define

$$H_m(p_1, \ldots, p_m) = -\sum_{j=1}^{m} p_j \log p_j.$$

Then we have the following.

Proposition 1.5.1. *Given* $p_1, \ldots, p_m$, *such that* $p_j > 0$; $\Sigma_{j=1}^{n} p_j = 1$,

$$\begin{aligned} H_{k+1}(p_1, \ldots, p_k, 1 - p_1 - \cdots - p_k) &= H_k(p_1, \ldots, p_{k-1}, 1 - \Sigma_{j=1}^{k-1} p_j) \\ &\quad + (1 - \Sigma_{j=1}^{k-1} p_j) H_2\Big(\tfrac{p_k}{1 - \Sigma_{j=1}^{k-1} p_j}, \tfrac{1 - \Sigma_{j=1}^{k} p_j}{1 - \Sigma_{j=1}^{k-1} p_j}\Big). \end{aligned}$$

In particular,

$$\Delta H_k = \left[1 - \frac{\Sigma_{j-1}^{k-1} \mu(V_j))}{\mu(D)}\right] H_2\left(\frac{\mu(V_k)}{\mu(D) - \Sigma_{j-1}^{k-1} \mu(V_j)}, \frac{\mu(D) - \Sigma_{j-1}^{k} \mu(V_j)}{\mu(D) - \Sigma_{j-1}^{k-1} \mu(V_j)}\right).$$

Proof. The terms making up

$$(1-\sum_{j=1}^{k-1} p_j)H_2(\frac{p_k}{1-\sum_{j=1}^{k-1} p_j}, \frac{1-\sum_{j=1}^{k} p_j}{1-\sum_{j=1}^{k-1} p_j})$$

may be rearranged by simple algebra to yield

$$A_k + B_k + C_k + D_k + E_k,$$

where $A_k = -p_k \log p_k$, $B_k = -(1-\sum_{j=1}^{k} p_j)\log(1-\sum_{j=1}^{k} p_j)$, $C_k = p_k \log(1-\sum_{j=1}^{k-1} p_j)$, $D_k = (1-\sum_{j=1}^{k-1} p_j)\log(1-\sum_{j=1}^{k-1} p_j)$, and $E_k = -p_k \log(1-\sum_{j=1}^{k-1} p_j)$. Defined in this way, C_k and E_k cancel each other, and the remaining terms provide the appropriate adjustment when added to $H_k(p_1,\ldots,p_{k-1},1-\sum_{j=1}^{k-1} p_j)$ to give the desired result. The remainder of the proposition follows by replacing p_j with $\mu(V_j)/\mu(D)$. □

The proposition sheds light on the rate at which a reconnaissance protocol can be expected to increase the partition entropy. To further illustrate this, we examine some radially-symmetric monotone fields associated with the scalar function class $\mathscr{H}$ introduced in Section 1.4.1. Consider the functions displayed in the following table. The corresponding function $f : D \to \mathbb{R}^2$ are depicted in Figure 1.6 and the corresponding sequence of partition entropies (based on a 20 interval uniform partition of the range) are depicted in Figure 1.7.

Table 1.1. Functions on $[0,1]$ (first row) and their inverses (second row) that determine radially symmetric functions on the unit disk as in Section 1.4.1.

$h_1(x) = 1 - x^5$	$h_2(x) = (1-x)^5$	$h_3(x) = 1 - x^{\frac{1}{5}}$	$h_4(x) = (1-x)^{\frac{1}{5}}$
$h_1^{-1}(x) = (1-x)^{\frac{1}{5}}$	$h_2^{-1}(x) = 1 - x^{\frac{1}{5}}$	$h_3^{-1}(x) = (1-x)^5$	$h_4^{-1}(x) = 1 - x^5$

1.6 Non-simple Reconnaissance Strategies and Non-monotone Fields

While a complete understanding of the relationship between the geometric and topological characteristics of $f : D \to \mathbb{R}^2$ and the associated partition entropies is not presently at hand, certain qualitative aspects of the relationship are revealed in the examples of the previous section. First, we note that the rates at which partition entropies increase (the entropy rates ΔH_k) in the simple reconnaissance protocol under investigation is fairly regular, and inflection points that appear in the plots in Figure 1.7 depend on the curvature characteristics of the surfaces determined by f. Less

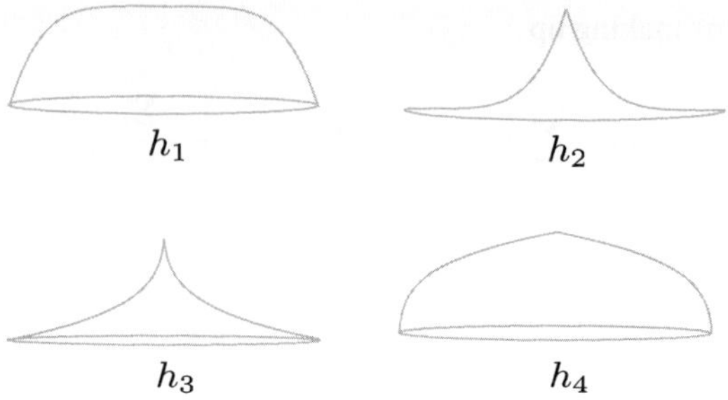

Fig. 1.6. The functions $h_k(\cdot)$ listed in Table 1.1 define radially symmetric functions on the unit circle in the way described in Section 1.4.1. The figures are the silhouettes of the surfaces defined by these functions $f_k(x,y) = h_k(\sqrt{x^2+y^2})$ for each function appearing in the table.

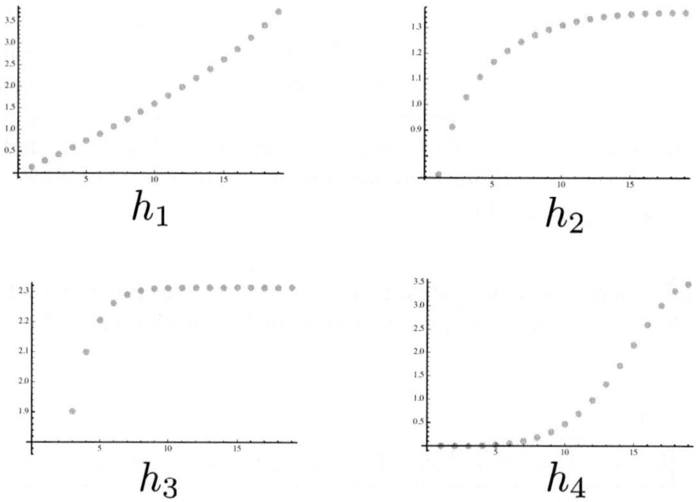

Fig. 1.7. The figures display the monotonic increase in partition entropy and partitions go through n successive refinements corresponding to the simple search chain and uniform twenty interval partition of the range of the monotone fields associated with the functions in the table and depicted in Fig. 1.6.

localized features of the field f may be revealed as well. It is clear from well-known properties of the binary entropy function $H_2(p, 1-p)$ and from the expression for ΔH_k in Proposition 1.5.1 that the maximum possible change in the partition entropy at the k-th search step will be achieved if the newly identified cell V_k in the domain partition has measure $\mu(V_k) = (1/2)(\mu(D) - \sum_{j=1}^{k-1} \mu(V_j))$. The simple reconnaissance protocol being employed determines a monotone search sequence, and for reasonably regular functions f, the successively determined cells V_k in the partition typically have areas that do not vary a great deal from step to step. Exceptions to very regular changes in the areas of partition cells—and corresponding regular changes in partition entropies—can occur in the case that a subinterval in the range partition encloses a critical value corresponding to an index 1 critical point of the function. A correspondingly large value of ΔH_k at the k-th step would be associated with going from a relatively long level curve (corresponding to $f^{-1}(x_{k-1})$) to a relatively shorter level curve contained in $f^{-1}(x_k)$ and defining the outer boundary of the next cell V_k. While a large increase in the value of the partition entropy could be due solely to the geometry of a single monotone peak of the function f, large increases are also characteristic of successive level curves enclosing different numbers of extrema of f. The geometry of this is illustrated in Figure 1.2 where the level curve corresponding to x_2 encloses two local maxima (and one index 1 critical point), whereas the traced curve corresponding to x_3 encloses only a single local maximum.

These remarks are more heuristic than precise. Nevertheless, the concept of partition entropy shows promise of providing a useful guide for reconnaissance of scalar fields in 2-d domains. An important factor in the design of reconnaissance strategies for sensor-enabled mobile robots is the trade-off of speed and accuracy. In cases where neither speed nor energy expenditures are important considerations, a raster scan of the domain of interest will be no worse than any other approach to experimental determination of the unknown field. When time and energy are major design criteria, however, it becomes important to identify the most important qualitative features of the field as early as possible in the process with details of the level contours being filled in as time and energy reserves permit.

Current research is aimed at designing enhancements to the trace-and-ascend reconnaissance protocol described in this paper. Hybrid reconnaissance protocols that balance competing objectives of speed and accuracy are currently under study. The protocols involve switching back and forth between an *exploitation* strategy and and *exploration* strategy. The exploitation phase executes trace-and-ascend as we have outlined in this paper. If run to completion, the exploitation phase would provide a detailed contour map of a single monotone feature in the potential field. That is, it would completely map a single mountain peak. In the absence of indications that there are multiple maxima in the domain of interest, a pure search-and-ascend can be designed to be generally more efficient than a raster scan. The exploration phase of our hybrid protocol can be triggered by either a sharp inflection in the cumulative partition entropy (i.e. by possible detection of additional extrema of the field) or by a noticeable flattening of the cumulative partition entropy—indicating that the ascend-and-trace protocol is yielding relatively little new information about the field. The switch to the exploration phase involves having the mobile robots cease their methodical trace-and-ascend mapping activity and go off in search of new points of rising gradients in parts of the domain that have not already been mapped. Preliminary results on such hybrid protocols have appeared in [2], and further details are to appear.

References

1. Baillieul, J., Grace, J.: The Fastest Random Search of a Class of Building Interiors. In: Proceedings of the 17-th International Symposium on Mathematical Theory of Networks and Systems, Kyoto, Japan, July 24-28, 2006, pp. 2222–2226 (2006)
2. Baronov, D., Baillieul, J.: Search Decisions for Teams of Automata. In: Proceedings of the 47th IEEE Conference on Decision and Control, Cancun, Mexico, Dec. 9-11, 2008, pp. 1133–1138 (2008), doi:10.1109/CDC.2008.4739365
3. Baronov, D., Baillieul, J.: Reactive Exploration Through Following Isolines in a Potential Field. In: Proceedings of the 2007 Automatic Control Conference, New York, NY, July11-13, ThA01.1, pp. 2141–2146 (2007), doi:10.1109/ACC.2007.4282460
4. Baronov, D., Anderson, S.B., Bailieul, J.: Tracking a nanosize magnetic particle using a magnetic force microscope. In: Proceedings of the 46-th IEEE Conference on Decision and Control, New Orleans, December 12-14, 2007, pp. 2445–2450. ThPI20.20 (2007), doi:10.1109/CDC.2007.4434192
5. Baronov, D., Baillieul, J.: Autonomous vehicle control for ascending/descending along a potential field with two applications. In: Proceedings of the 2008 American Control Conference, Seattle, Washington, Seattle, Washington, June 11-13, 2008, WeBI01.7, pp. 678–683 (2008), doi:10.1109/ACC.2008.4586571
6. Baillieul, J., Byrnes, C.I.: The singularity theory of the load flow equations for a 3-node electrical power system. Systems and Control Letters 2(6), 330–340 (1983)
7. Boyd, S., Diaconis, P., Xiao, L.: Fastest Mixing Markov Chain on a Graph. SIAM Review 46(4), 667–689 (2004)
8. Cao, M., Stewart, A., Leonard, N.E.: Integrating human and robot decision-making dynamics with feedback: Models and convergence analysis. In: Proceedings of the 47th IEEE Conference on Decision and Control, Cancun, Mexico, Dec. 9-11, 2008, pp. 1127–1132 (2008), doi:10.1109/CDC.2008.4739103
9. California Office of Environmental Health Hazard Assessments (OEHHA): The Air Toxics Hot Spots Program Guidance Manual for Preparation of Health Risk Assessment (2003), available online at `http://www.oehha.ca.gov/air/hot_spots/HRAguidefinal.html`
10. Caputo, P., Martinelli, F.: Relaxation Time of Anisotropic Simple Exclusion Processes and Quantum Heisenberg Models. Preprint, arXiv:math (2002)
11. Castanon, D.A., Ahner, D.K.: Team task allocation and routing in risky environments under human guidance. In: Proceedings of the 47th IEEE Conference on Decision and Control, Cancun, Mexico, Dec. 9-11, 2008, pp. 1139–1144 (2008), doi:10.1109/CDC.2008.4739148
12. Chin, W.-P., Ntafos, S.: Optimum watchman routes. In: Proceedings of the Second Annual Symposium on Computational Geometry, Yorktown Heights, New York, United States, pp. 24–33. ACM (1986), `http://doi.acm.org/10.1145/10515.10518`
13. Bogacz, R., Brown, E., Moehlis, J., Holmes, P., Cohen, J.D.: The physics of optimal decision-making: A formal analysis of models of performance in two-alternative forced-choice tasks. Psychological Review 113(4), 700–765 (2006)
14. Cover, T.M., Thomas, J.A.: Elements of Information Theory. John Wiley & Sons, New York (1991)
15. Ganguli, A., Cortes, J., Bullo, F.: Distributed deployment of asynchronous guards in art galleries. In: American Control Conference, Minneapolis, MN, June 2006, pp. 1416–1421 (2006), doi:10.1109/ACC.2006.1656416

16. Grace, J., Baillieul, J.: Stochastic Strategies for Autonomous Robotic Surveillance. In: Proceedings of the 2005 IEEE Conf. on Decision and Control/Europ. Control Conf., Seville, Spain, December 13, Paper TuA03.5, pp. 2200–2205 (2005)
17. Kuhn, H.W., Tucker, K.W.: Contributions to the Theory of Games, **I**. In: Annals of Mathematics Studies, 24, Princeton University Press, Princeton (1950)
18. Byrnes, C.I., Gusev, S.V., Lindquist, A.: From Finite Covariance Windows to Modeling Filters: A Convex Optimization Approach. SIAM Review 63(4), 645–675 (2001)
19. Nedic, A., Tomlin, D., Holmes, P., Prentice, D.A., Cohen, J.D.: A simple decision task in a social context: Experiments, a model, and preliminary analyses of behavioral data. In: Proceedings of the 47th IEEE Conference on Decision and Control, Cancun, Mexico, Dec. 9-11, 2008, pp. 1115–1120 (2008), doi:10.1109/CDC.2008.4739153
20. O'Rourke, J.: Galleries Need Fewer Mobile Watchmen. Geometriae Dedicata 14, 273–283 (1983)
21. Pratt, J.W., Raiffa, H., Schlaifer, R.: Introduction to Statistical Decision Theory. MIT Press, Cambridge (1995)
22. Rosenthal, J.: Convergence Rates of Markov Chains. SIAM Review 37, 387–405 (1994)
23. Savla, K., Temple, T., Frazzoli, E.: Human-in-the-loop vehicle routing policies for dynamic environments. In: Proceedings of the 47th IEEE Conference on Decision and Control, Cancun, Mexico, Dec. 9-11, 2008, pp. 1145–1150 (2008), doi:10.1109/CDC.2008.4739443
24. Von Neumann, J., Morgenstern, O.: Theory of Games and Economic Behavior. Princeton University Press, Princeton (1947)
25. Vu, L., Morgansen, K.A.: Modeling and analysis of dynamic decision making in sequential two-choice tasks. In: Proceedings of the 47th IEEE Conference on Decision and Control, Cancun, Mexico, Dec. 9-11, 2008, pp. 1121–1126 (2008), doi:10.1109/CDC.2008.4739374

2

A Computational Comparison of Alternatives to Including Uncertainty in Structured Population Models*,†

H.T. Banks, Jimena L. Davis, and Shuhua Hu

Center for Research in Scientific Computation, Center for Quantitative Sciences in Biomedicine, North Carolina State University, Raleigh, NC 27695-8212, USA

Summary. Two conceptually different approaches to incorporate growth uncertainty into size-structured population models have recently been investigated. One entails imposing a probabilistic structure on all the possible growth rates across the entire population, which results in a growth rate distribution model. The other involves formulating growth as a Markov stochastic diffusion process, which leads to a Fokker-Planck model. Numerical computations verify that a Fokker-Planck model and a growth rate distribution model can, with properly chosen parameters, yield quite similar time dependent population densities. The relationship between the two models is based on the theoretical analysis in [7].

2.1 Introduction

Class and size-structured population models, which have been extensively investigated for some time, have proved useful in modeling the dynamics of a wide variety of populations. Applications are diverse and include populations ranging from cells to whole organisms in animal, plant and marine species [1, 3, 5, 7, 8, 9, 12, 14, 17, 18, 19, 20, 21, 22, 24]. One of the intrinsic assumptions in standard size-structured population models is that all individuals of the same size have the same size-dependent growth rate. This does not allow for differences due to inherent genetic differences, chronic disease or disability, underlying local environmental variability, etc. This means that if there is no reproduction involved then the variability in size at any time is totally determined by the variability in initial size. Such models are termed *cryptodeterministic* [16] and embody the fundamental feature that uncertainty or stochastic variability enters the population only through that in the initial data. However, the

*This research was supported in part (HTB and SH) by grant number R01AI071915-07 from the National Institute of Allergy and Infectious Diseases, in part (HTB and SH) by the Air Force Office of Scientific Research under grant number FA9550-09-1-0226 and in part (JLD) by the US Department of Energy Computational Science Graduate Fellowship under grant DE-FG02-97ER25308.

†On the occasion of the 2009 Festschrift in honor of Chris Byrnes and Anders Lindquist.

X. Hu et al. (Eds.): Three Decades of Progress in Control Sciences, pp. 19–33, 2010.

experimental data in [7] for the early growth of shrimp reveals that shrimp exhibit a great deal of variability in size as time evolves even though all the shrimp begin with similar size. It was also reported in [5, 9] that experimental size-structured field data on mosquitofish population (no reproduction involved) exhibits both dispersion and bimodality in size as time progresses even though the initial population density is unimodal. Hence, standard size-structured population models such as that first proposed by Sinko and Streifer [24] are inadequate to describe the dynamics of these populations. For these situations we need to incorporate some type of uncertainty or variability into the growth process so that the variability in size is not only determined by the variability in initial size but also by the variability in individual growth.

We consider here two conceptually different approaches to incorporating the growth uncertainty into a size-structured population model. One entails imposing a probabilistic structure on the set of possible growth rates permissible in the entire population while the other involves formulating growth as a stochastic diffusion process. In [7] these are referred to as probabilistic formulations and stochastic formulations, respectively. Because we are only interested in modeling growth uncertainty in this paper, for simplicity, we will not consider either reproduction and mortality rates in our formulations.

2.1.1 Probabilistic Formulation

The probabilistic formulation is motivated by the observation that genetic differences or non-lethal infections of some chronic disease can have an effect on individual growth. For example, in many marine species such as mosquitofish, females grow faster than males, which means that individuals with the same size may have different growth rates. The probabilistic formulation is constructed based on the assumption that each individual does grow according to a deterministic growth model $\frac{dx}{dt} = g(x,t)$ as posited in the Sinko-Streifer formulation, but that different individuals may have different size-dependent growth rates. Based on this underlying assumption, one partitions the entire population into (possibly a continuum of) subpopulations where individuals in each subpopulation have the same size-dependent growth rate, and then assigns a probability distribution to this partition of possible growth rates in the population.

The growth process for individuals in a subpopulation with growth rate g is assumed to be described by the dynamics

$$\frac{dx(t;g)}{dt} = g(x(t;g),t), \qquad g \in \mathcal{G}, \tag{2.1}$$

where $\mathcal{G}$ is a collection of admissible growth rates. Model (2.1) combined with the probability distribution imposed on $\mathcal{G}$ will be called the *probabilistic growth model* in this paper. Hence, we can see that for the probabilistic formulation, the growth uncertainty is introduced into the entire population by the variability of growth rates among subpopulations. In the literature, it is common to assume that growth rate is a nonnegative function, that is, no loss in size occurs. However, individuals may experience loss in size due to disease or some other involuntary factors. Hence, we will permit these situations in this formulation, but for simplicity we assume that growth rate in each subpopulation is either a nonnegative function or a negative function, that is, the size of each individual is either nondecreasing or decreasing continuously in its growth period.

With this assumption of a family of admissible growth rates and an associated probability distribution, one thus obtains a generalization of Sinko-Streifer model, called the growth rate distribution (GRD) model, which has been formulated and studied in [2, 4, 5, 9, 10]. The model consists of solving

$$\begin{aligned} &v_t(x,t;g) + (g(x,t)v(x,t;g))_x = 0, \qquad x \in (0,L), \quad t > 0, \\ &g(0,t)v(0,t;g) = 0 \text{ if } g \geq 0 \text{ or } g(L,t)v(L,t;g) = 0 \text{ if } g < 0, \\ &v(x,0;g) = v_0(x;g), \end{aligned} \tag{2.2}$$

for a given $g \in \mathscr{G}$ and then "summing" (with respect to the probability) the corresponding solutions over all $g \in \mathscr{G}$. Thus if $v(x;t;g)$ is the population density of individuals with size x at time t having growth rate g, the expectation of the total population density for size x at time t is given by

$$u(x,t) = \int_{g \in \mathscr{G}} v(x,t;g)d\mathscr{P}(g), \tag{2.3}$$

where $\mathscr{P}$ is a probability measure on $\mathscr{G}$. Thus, this probabilistic formulation involves a *stationary probabilistic structure* on a family of *deterministic dynamical systems*, and $\mathscr{P}$ is the fundamental "parameter" that is to be estimated by either parametric or nonparametric methods (which depends on the prior information known about the form for $\mathscr{P}$). As detailed in [5, 10], the growth rate distribution model is sufficiently rich to exhibit a number of phenomena of interest, for example, dispersion and development of two modes from one.

Observe that if all the subpopulations have nonnegative growth rates, then we need to set $g(L,t)v(L,t;g) = 0$ for each $g \in \mathscr{G}$ in order to provide a conservation law for the GRD model. Specifically if L denotes the maximum attainable size of individuals in a life time, then it is reasonable to set $g(L,t) = 0$ (as commonly done in the literature). However, if we just consider the model in a short time period, then we may choose L sufficiently large so that $u(L,t)$ is negligible or zero if possible. We observe that if there exist some subpopulations whose growth rates are negative, then we can not provide a conservation law for these subpopulations as $g(0,t) < 0$. Hence, in this case, once the size of an individual is decreased to below the minimum size, then that individual will be removed from the system. In other words, we exclude those individuals whose size go below the minimum size. This effectively serves as a sink for these subpopulations.

2.1.2 Stochastic Formulation

A stochastic formulation may be motivated by the acknowledgment that environmental or emotional fluctuations can have a significant influence on the individual growth. For example, the growth rate of shrimp are affected by several environmental factors [3] such as temperature, dissolved oxygen level and salinity. The stochastic formulation is constructed under the assumption that movement from one size class to another can be described by a stochastic diffusion process [1, 13, 16, 22]. Let $\{X(t) : t \geq 0\}$ be a Markov diffusion process with $X(t)$ representing size at time t (i.e., each process realization corresponds to the size trajectory of an individual). Then $X(t)$ is described by the Ito stochastic differential equation (we refer to this equation as the *stochastic growth model*)

$$dX(t) = g(X(t),t)dt + \sigma(X(t),t)dW(t), \tag{2.4}$$

where $W(t)$ is the standard Wiener process [1, 16]. Here $g(x,t)$ denotes the average growth rate (the first moment of rate of change in size) of individuals with size x at time t, and is given by

$$g(x,t) = \lim_{\Delta t \to 0+} \frac{1}{\Delta t} \mathrm{E}\{\Delta X(t) | X(t) = x\}. \tag{2.5}$$

For application purposes, we assume that g is a nonnegative function here. The function $\sigma(x,t)$ represents the variability in the growth rate of individuals (the second moment of rate of change in size) and is given by

$$\sigma^2(x,t) = \lim_{\Delta t \to 0+} \frac{1}{\Delta t} \mathrm{E}\left\{[\Delta X(t)]^2 | X(t) = x\right\}. \tag{2.6}$$

Hence, the growth process of each individual is stochastic, and each individual grows according to stochastic growth model (2.4). Thus, for this formulation the growth uncertainty is introduced into the entire population by the stochastic growth of each individual. In addition, individuals with the same size at the same time have the same uncertainty in growth, and individuals also have the possibility of reducing their size during a growth period.

With this assumption on the growth process, we obtain the Fokker-Planck (FP) or forward Kolmogorov model for the population density u, which was carefully derived in [22] among numerous other places and subsequently studied in many references (e.g., [1, 13, 16]). The equation and appropriate boundary conditions are given by

$$\begin{aligned} &u_t(x,t) + (g(x,t)u(x,t))_x = \tfrac{1}{2}(\sigma^2(x,t)u(x,t))_{xx}, \quad x \in (0,L), \quad t > 0, \\ &g(0,t)u(0,t) - \tfrac{1}{2}(\sigma^2(x,t)u(x,t))_x|_{x=0} = 0, \\ &g(L,t)u(L,t) - \tfrac{1}{2}(\sigma^2(x,t)u(x,t))_x|_{x=L} = 0, \\ &u(x,0) = u_0(x). \end{aligned} \tag{2.7}$$

Here L is the maximum size that individuals may attain in any given time period. Observe that the boundary conditions in (2.7) provide a conservation law for the FP model. Because both mortality and reproduction rates are assumed zero, the total number of in the population is a constant given by $\int_0^L u_0(x)dx$. In addition, we observe that with the zero-flux boundary condition at zero (minimum size) one can equivalently set $X(t) = 0$ if $X(t) \leq 0$ for the stochastic growth model (2.4) in the sense that both are used to keep individuals in the system. This means that if the size of an individual is decreased to the minimum size, it remains in the system with the possibility to once again increase its size.

The discussions in Sections 2.1.1 and 2.1.2 indicate that these probabilistic and stochastic formulations are *conceptually* quite different. However, the analysis in [7] reveals that in some cases the size distribution (the probability density function of $X(t)$) obtained from the stochastic growth model is exactly the same as that obtained from the probabilistic growth model. For example, if we consider the two models

$$\begin{aligned} &\text{stochastic formulation: } dX(t) = b_0(X(t)+c_0)dt + \sqrt{2t}\sigma_0(X(t)+c_0)dW(t) \\ &\text{probabilistic formulation: } \tfrac{dx(t;b)}{dt} = (b-\sigma_0^2 t)(x(t;b)+c_0), \quad b \in \mathbb{R} \text{ with } B \sim \mathcal{N}(b_0,\sigma_0^2), \end{aligned} \tag{2.8}$$

and assume their initial size distributions are the same, then we obtain at each time t the same size distribution from these two distinct formulations. Here b_0, σ_0 and c_0 are positive constants (for application purposes), and B is a normal random variable with b a realization of B. Moreover, by using the same analysis as in [7] we can show that if we compare

$$\begin{aligned} &\text{stochastic formulation: } dX(t) = (b_0+\sigma_0^2 t)(X(t)+c_0)dt + \sqrt{2t}\sigma_0(X(t)+c_0)dW(t) \\ &\text{probabilistic formulation: } \tfrac{dx(t;b)}{dt} = b(x(t;b)+c_0), \quad b \in \mathbb{R} \text{ with } B \sim \mathcal{N}(b_0,\sigma_0^2), \end{aligned} \tag{2.9}$$

with the same initial size distributions, then we can also obtain at each time t the same size distribution for these two formulations. In addition, we see that both the stochastic growth models and the probabilistic growth models in (2.8) and (2.9) reduce to the same deterministic growth model $\dot{x} = b_0(x+c_0)$ when there is no uncertainty or variability in growth (i.e., $\sigma_0 = 0$) even though both models in (2.9) *do not* satisfy the mean growth dynamics

$$\frac{d\mathrm{E}(X(t))}{dt} = b_0(\mathrm{E}(X(t))+c_0) \tag{2.10}$$

while both models in (2.8) *do*.

As remarked in [7], if in the probabilistic formulation we impose a normal distribution $\mathcal{N}(b_0,\sigma_0^2)$ for B, this is not completely reasonable in applications because the intrinsic growth rate b can be negative which results in the size having non-negligible probability of being negative in a finite time period when σ_0 is sufficiently large relative to b_0. A standard approach in practice to remedy this problem is to impose a truncated normal distribution $\mathcal{N}_{[\underline{b},\bar{b}]}(b_0,\sigma_0^2)$ instead of a normal distribution; that is, we restrict B in some reasonable range $[\underline{b},\bar{b}]$. We observe that the stochastic formulation also can lead to the size having non negligible probability of being negative when σ_0 is sufficiently large relative to b_0. This is because $W(t) \sim \mathcal{N}(0,t)$ for any fixed t and hence decreases in size are possible. One way to remedy this situation is to set $X(t) = 0$ if $X(t) \leq 0$. Thus, if σ_0 is sufficiently large relative to b_0, then we may obtain different size distributions for these two formulations *after* we have made these different modifications to each. The same anomalies hold for the solutions of the FP models and the GRD models themselves because we impose zero-flux boundary conditions in the FP model and put constraints on B in the GRD model. In this paper, we present some computational examples using the models in (2.8) and (2.9) to investigate how the solutions to the modified FP models and the modified GRD models change as we vary the values of σ_0 and $\underline{b}$.

The remainder of this paper is organized as follows. In Section 2.2 we outline the numerical scheme we use to numerically solve the Fokker-Planck model. In Section 2.3 we present computational examples using (2.8) and (2.9) to investigate the influence of the values of σ_0 and $\underline{b}$ on the solutions to the FP model and the GRD model. Finally, we conclude the paper in Section 2.4 with some conclusions and further remarks.

2.2 Numerical Scheme to Solve the FP Model

For the computational results presented here, we used the finite difference scheme developed by Chang and Cooper in [15] to numerically solve the FP model (2.7). This scheme provides numerical solutions which preserve some of the more important intrinsic properties of the FP model. In particular, the solution is non-negative, is particle conserving in the absence of sources or sinks, and gives exact representations of the analytic solution upon equilibration.

In the following exposition, we assume that all the model parameters are sufficiently smooth to allow implementation of this scheme. For convenience, the following notation will be used in this section:

$$d(x,t) = \sigma^2(x,t),\ F(x,t) = g(x,t)u(x,t) - \tfrac{1}{2}(d(x,t)u(x,t))_x,\ h(x,t) = g(x,t) - \tfrac{1}{2}d_x(x,t).$$

Hence, we can rewrite F as

$$F(x,t) = h(x,t)u(x,t) - \frac{1}{2}d(x,t)u_x(x,t).$$

Let $\Delta x = L/n$ and $\Delta t = T/l$ be the spatial and time mesh sizes, respectively, where T is the maximum time considered in the simulations. The mesh points are given by $x_j = j\Delta x$, $j = 0,1,2,\ldots,n$, and $t_k = k\Delta t$, $k = 0,1,2,\ldots,l$. We denote by u_j^k the finite difference approximation of $u(x_j,t_k)$, and we let $u_j^0 = u_0(x_j)$, $j = 0,1,2,\ldots,n$. The mid point between two space mesh points is given by $x_{j+\frac{1}{2}} = \frac{x_j+x_{j+1}}{2}$, and $h^k_{j+\frac{1}{2}} = g(x_{j+\frac{1}{2}},t_k) - \frac{1}{2}d_x(x_{j+\frac{1}{2}},t_k)$. The scheme to solve the FP model (2.7) is given by

$$\frac{u_j^{k+1} - u_j^k}{\Delta t} + \frac{F^{k+1}_{j+\frac{1}{2}} - F^{k+1}_{j-\frac{1}{2}}}{\Delta x} = 0,\ j = 0,1,2,\ldots,n,\ k = 0,1,2,\ldots,l-1. \tag{2.11}$$

Here $F^{k+1}_{j+\frac{1}{2}}$, $j = 0,1,2,\dots,n-1$ are defined by

$$
\begin{aligned}
F^{k+1}_{j+\frac{1}{2}} &= h^{k+1}_{j+\frac{1}{2}} u^{k+1}_{j+\frac{1}{2}} - \tfrac{1}{2} d^{k+1}_{j+\frac{1}{2}} \frac{u^{k+1}_{j+1} - u^{k+1}_{j}}{\Delta x} \\
&= h^{k+1}_{j+\frac{1}{2}} \left[\delta^{k+1}_j u^{k+1}_{j+1} + (1-\delta^{k+1}_j) u^{k+1}_j \right] - \tfrac{1}{2} d^{k+1}_{j+\frac{1}{2}} \frac{u^{k+1}_{j+1} - u^{k+1}_{j}}{\Delta x} \\
&= \left[\delta^{k+1}_j h^{k+1}_{j+\frac{1}{2}} - \tfrac{1}{2\Delta x} d^{k+1}_{j+\frac{1}{2}} \right] u^{k+1}_{j+1} + \left[(1-\delta^{k+1}_j) h^{k+1}_{j+\frac{1}{2}} + \tfrac{1}{2\Delta x} d^{k+1}_{j+\frac{1}{2}} \right] u^{k+1}_j ,
\end{aligned}
\tag{2.12}
$$

where $\delta^{k+1}_j = \frac{1}{\tau^{k+1}_j} - \frac{1}{\exp(\tau^{k+1}_j)-1}$ with $\tau^{k+1}_j = \frac{2h^{k+1}_{j+\frac{1}{2}}\Delta x}{d^{k+1}_{j+\frac{1}{2}}}$. Note that if $h^{k+1}_{j+\frac{1}{2}} = 0$, then we do not need to figure out the value of $u^{k+1}_{j+\frac{1}{2}}$. Hence, we do not need to worry about δ^{k+1}_j in this case.

Define $f(\tau) = \frac{1}{\tau} - \frac{1}{\exp(\tau)-1}$. By a Taylor series expansion, we know that $\exp(\tau) + \exp(-\tau) > 2 + \tau^2$. Hence, $f'(\tau) < 0$. Thus, f is monotonically decreasing. Note that $\lim_{\tau\to-\infty} f(\tau) = 1$ and $\lim_{\tau\to\infty} f(\tau) = 0$. Hence, $0 \le \delta^{k+1}_j \le 1$ for $j = 0,1,2,\dots,n-1$, $k = 0,1,2,\dots,l-1$. Thus, we can see that when this choice for $u^{k+1}_{j+\frac{1}{2}}$ is used in a first derivative, the scheme continuously shifts from a backward difference ($\delta^{k+1}_j = 0$) to a centered difference ($\delta^{k+1}_j = \frac{1}{2}$) to a forward difference ($\delta^{k+1}_j = 1$).

To preserve the conservation law, we use $F^{k+1}_{-\frac{1}{2}} = 0$ and $F^{k+1}_{n+\frac{1}{2}} = 0$ to approximate boundary conditions $F(0,t_{k+1}) = 0$ and $F(L,t_{k+1}) = 0$ in the FP model, respectively. To the order of accuracy of the difference scheme, these numerical boundary conditions are consistent with the boundary conditions in the FP model. Note that scheme (2.11) can also be written as the following tridiagonal system

$$
-a^{k+1}_{1,j} u^{k+1}_{j+1} + a^{k+1}_{0,j} u^{k+1}_j - a^{k+1}_{-1,j} u^{k+1}_{j-1} = u^k_j, \; j = 0,1,2,\dots,n, \; k = 0,1,2,\dots,l-1.
$$

By (2.11), we have for $j = 1,2,\dots,n-1$,

$$
\begin{aligned}
a^{k+1}_{1,j} &= \tfrac{\Delta t}{\Delta x} \left[\tfrac{1}{2\Delta x} d^{k+1}_{j+\frac{1}{2}} - \delta^{k+1}_j h^{k+1}_{j+\frac{1}{2}} \right] = \tfrac{\Delta t}{\Delta x} \left[\frac{h^{k+1}_{j+\frac{1}{2}}}{\exp(\tau^{k+1}_j)-1} \right], \\
a^{k+1}_{0,j} &= 1 + \tfrac{\Delta t}{\Delta x} \left[(1-\delta^{k+1}_j) h^{k+1}_{j+\frac{1}{2}} - \delta^{k+1}_{j-1} h^{k+1}_{j-\frac{1}{2}} \right] + \tfrac{\Delta t}{2\Delta x^2} \left(d^{k+1}_{j+\frac{1}{2}} + d^{k+1}_{j-\frac{1}{2}} \right) \\
&= 1 + \tfrac{\Delta t}{\Delta x} \left[\frac{\exp(\tau^{k+1}_j)}{\exp(\tau^{k+1}_j)-1} h^{k+1}_{j+\frac{1}{2}} + \frac{1}{\exp(\tau^{k+1}_{j-1})-1} h^{k+1}_{j-\frac{1}{2}} \right], \\
a^{k+1}_{-1,j} &= \tfrac{\Delta t}{\Delta x} \left[(1-\delta^{k+1}_{j-1}) h^{k+1}_{j-\frac{1}{2}} + \tfrac{1}{2\Delta x} d^{k+1}_{j-\frac{1}{2}} \right] = \tfrac{\Delta t}{\Delta x} \left[\frac{\exp(\tau^{k+1}_{j-1})}{\exp(\tau^{k+1}_{j-1})-1} h^{k+1}_{j-\frac{1}{2}} \right].
\end{aligned}
$$

By (2.11) with $j=0$ and boundary condition $F^{k+1}_{-\frac{1}{2}}=0$, we find that

$$
\begin{aligned}
a^{k+1}_{1,0} &= \tfrac{\Delta t}{\Delta x}\left[\tfrac{1}{2\Delta x}d^{k+1}_{\frac{1}{2}} - \delta^{k+1}_0 h^{k+1}_{\frac{1}{2}}\right] = \tfrac{\Delta t}{\Delta x}\left[\frac{h^{k+1}_{\frac{1}{2}}}{\exp(\tau^{k+1}_0)-1}\right],\\
a^{k+1}_{0,0} &= 1+\tfrac{\Delta t}{\Delta x}\left[(1-\delta^{k+1}_0)h^{k+1}_{\frac{1}{2}} + \tfrac{1}{2\Delta x}d^{k+1}_{\frac{1}{2}}\right] = 1+\tfrac{\Delta t}{\Delta x}\left[\frac{\exp(\tau^{k+1}_0)}{\exp(\tau^{k+1}_0)-1}h^{k+1}_{\frac{1}{2}}\right],\\
a^{k+1}_{-1,0} &= 0.
\end{aligned}
$$

By (2.11) with $j=n$ and boundary condition $F^{k+1}_{n+\frac{1}{2}}=0$, we find that

$$
\begin{aligned}
a^{k+1}_{1,n} &= 0,\\
a^{k+1}_{0,n} &= 1+\tfrac{\Delta t}{\Delta x}\left[\tfrac{1}{2\Delta x}d^{k+1}_{n-\frac{1}{2}} - \delta^{k+1}_{n-1} h^{k+1}_{n-\frac{1}{2}}\right] = 1+\tfrac{\Delta t}{\Delta x}\left[\frac{h^{k+1}_{n-\frac{1}{2}}}{\exp(\tau^{k+1}_{n-1})-1}\right],\\
a^{k+1}_{-1,n} &= \tfrac{\Delta t}{\Delta x}\left[(1-\delta^{k+1}_{n-1})h^{k+1}_{n-\frac{1}{2}} + \tfrac{1}{2\Delta x}d^{k+1}_{n-\frac{1}{2}}\right] = \tfrac{\Delta t}{\Delta x}\left[\frac{\exp(\tau^{k+1}_{n-1})}{\exp(\tau^{k+1}_{n-1})-1}h^{k+1}_{n-\frac{1}{2}}\right].
\end{aligned}
$$

It is obvious that if we set $\Delta t < \frac{1}{\|h_x\|_\infty}$ and $\frac{\Delta t}{\Delta x} < \frac{1}{2\|h\|_\infty}$, then $a^{k+1}_{-1,j}$, $a^{k+1}_{0,j}$ and $a^{k+1}_{1,j}$ satisfy the following conditions

$$
\begin{cases}
a^{k+1}_{-1,j},\ a^{k+1}_{0,j},\ a^{k+1}_{1,j},\ u^0_j \geq 0, \\
a^{k+1}_{0,j} \geq a^{k+1}_{-1,j} + a^{k+1}_{1,j}.
\end{cases}
\quad j=0,1,2,\ldots,n,\ k=0,1,2,\ldots,l-1, \tag{2.13}
$$

which guarantee that $u^{k+1}_j \geq 0$, $j=0,1,2,\ldots,n$, $k=0,1,2,\ldots,l-1$ (see [15, 23]).

2.3 Numerical Results

For all the examples given in this section, the maximum time is set at $T=10$. The initial condition in the FP model is given by $u_0(x)=100\exp(-100(x-0.4)^2)$, and initial conditions in the GRD model are given by $v_0(x;b)=100\exp(-100(x-0.4)^2)$ for $b\in[\underline{b},\bar{b}]$. We set $c_0=0.1$, $b_0=0.045$, and $\sigma_0=rb_0$, where r is a positive constant. We use $\Delta x=10^{-3}$ and $\Delta t=10^{-3}$ in the finite difference scheme to numerically solve the FP model.

Section 2.3.1 details results for an example where model parameters in the FP and the GRD models are chosen based on (2.8), and Section 2.3.2 contains results comparing the FP and the GRD models in (2.9). In these two examples, we vary the values of r and $\underline{b}$ to illustrate their effect on the solutions to the FP and the GRD models.

2.3.1 Example 1

Model parameters in the FP and the GRD models in this example are chosen based on (2.8) and are given by

$$\text{FP model: } g(x) = b_0(x+c_0), \quad \sigma(x,t) = \sqrt{2t}\sigma_0(x+c_0)$$

$$\text{GRD model: } g(x,t;b) = (b-\sigma_0^2 t)(x+c_0), \text{ where } b \in [\underline{b}, \bar{b}] \text{ with } B \sim \mathcal{N}_{[\underline{b},\bar{b}]}(b_0,\sigma_0^2). \tag{2.14}$$

We choose $\underline{b} = b_0 - 3\sigma_0$ and $\bar{b} = b_0 + 3\sigma_0$. Let $r_0 = \frac{-3+\sqrt{4b_0T+9}}{2b_0T}$ (≈ 0.3182). It is easy to show that if $r < r_0$, then $g(x,t;b) = (b-\sigma_0^2 t)(x+c) > 0$ in $\{(x,t)|(x,t) \in [0,L]\times[0,T]\}$ for all $b \in [\underline{b},\bar{b}]$. Here we just consider the case for $r < r_0$, i.e., the growth rate of each subpopulation is positive. To conserve the total number of the population in the system, we must choose L sufficiently large so that $v(L,t;b)$ is negligible for any $t \in [0,T]$ and $b \in [\underline{b},\bar{b}]$. For this example we chose $L = 6$.

We observe that with this choice of $g(x,t) = (b-\sigma_0^2 t)(x+c_0)$ in the GRD model, we can analytically solve (2.2) by the method of characteristics, and the solution is given by

$$v(x,t;b) = \begin{cases} v_0(\omega(x,t);b)\exp\left(-bt+\frac{1}{2}\sigma_0^2t^2\right) & \text{if } \omega(x,t) \geq 0 \\ 0 & \text{if } \omega(x,t) < 0, \end{cases} \tag{2.15}$$

where $\omega(x,t) = -c_0 + (x+c_0)\exp(-bt+\frac{1}{2}\sigma_0^2t^2)$. Hence, by (2.3) we have

$$u(x,t) = \int_{\underline{b}}^{\bar{b}} v(x,t;b)\frac{\frac{1}{\sigma_0}\phi\left(\frac{b-b_0}{\sigma_0}\right)}{\Phi\left(\frac{\bar{b}-b_0}{\sigma_0}\right)-\Phi\left(\frac{\underline{b}-b_0}{\sigma_0}\right)}db, \tag{2.16}$$

where ϕ is the probability density function of the standard normal distribution, and Φ is its corresponding cumulative distribution function. In the simulations, the trapezoidal rule with $\Delta b = (\bar{b}-\underline{b})/128$ was used to calculate the integral in (2.16).

Snapshots of the numerical solution of the Fokker-Planck equation and the solution of the GRD model at $t = T$ with $r = 0.1$ (left) and $r = 0.3$ (right) are graphed in Figure 2.1.

These results, along with other snapshots (not depicted here) demonstrate that we do indeed obtain quite similar (in fact indistinguishable in these graphs) population densities for these two models and parameter values. This is because $\mathcal{N}_{[\underline{b},\bar{b}]}(b_0,\sigma_0^2)$ is a good approximation of $\mathcal{N}(b_0,\sigma_0^2)$ (for this setup of $\underline{b}$ and $\bar{b}$) and σ_0 is chosen sufficiently small so that the size distributions obtained in (2.8) are good approximations of size distributions obtained computationally with the GRD models and the FP models. Note that the population density $u(x,t)$ is just the product of the total number of the population and the probability density function.

2.3.2 Example 2

We consider model parameters in the FP and GRD models of (2.9). That is, we compare models with

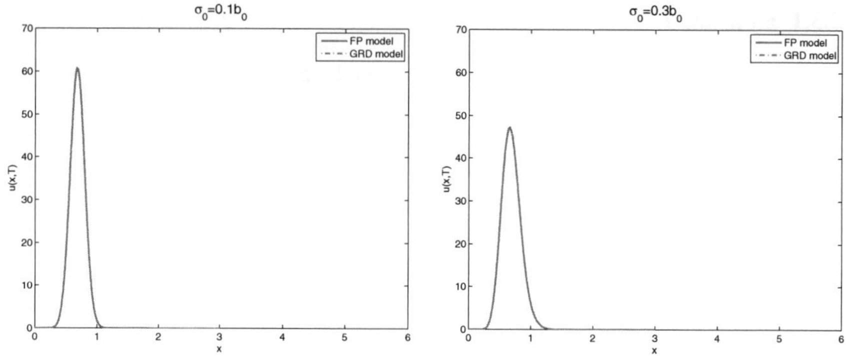

Fig. 2.1. Numerical solutions $u(x,T)$ to the FP model and the GRD model with model parameters chosen as (2.14), where $\underline{b} = b_0 - 3\sigma_0$ and $\bar{b} = b_0 + 3\sigma_0$.

$$\text{FP model: } g(x,t) = (b_0 + \sigma_0^2 t)(x + c_0), \quad \sigma(x,t) = \sqrt{2t}\sigma_0(x + c_0)$$
$$\text{GRD model: } g(x;b) = b(x + c_0), \text{ where } b \in [\underline{b}, \bar{b}] \text{ with } B \sim \mathcal{N}_{[\underline{b},\bar{b}]}(b_0, \sigma_0^2). \tag{2.17}$$

Because the growth rate g in the GRD model is a positive function if $b > 0$, we need to choose L sufficiently large so that $v(L,t;b)$ is negligible for any $t \in [0,T]$ in any subpopulation with positive intrinsic growth rate b. Doing so will conserve the total number in the population. Here we again chose $L = 6$.

With this choice of $g(x) = b(x + c_0)$ in the GRD model, we can again analytically solve (2.2) by the method of characteristics, and the solutions for subpopulations with nonnegative b (the boundary condition in (2.2) is $v(0,t;b) = 0$ in this case) is given by

$$v(x,t;b) = \begin{cases} v_0(\omega(x,t);b)\exp(-bt) & \text{if } \omega(x,t) \geq 0 \\ 0 & \text{if } \omega(x,t) < 0. \end{cases} \tag{2.18}$$

The solution for subpopulations with negative b (the boundary condition in (2.2) is $v(L,t;b) = 0$ in this case) is given by

$$v(x,t;b) = \begin{cases} v_0(\omega(x,t);b)\exp(-bt) & \text{if } \omega(x,t) \leq L \\ 0 & \text{if } \omega(x,t) > L, \end{cases} \tag{2.19}$$

where $\omega(x,t) = -c_0 + (x + c_0)\exp(-bt)$. We use these with (2.16) to calculate $u(x,t)$.

The numerical solutions of the Fokker-Planck equation and the corresponding solutions of the GRD model at $t = T$ with $r = 0.1, 0.3, 0.7, 0.9, 1.3$ and 1.5 are depicted in Figure 2.2, where $\underline{b} = \max\{b_0 - 3\sigma_0, 10^{-6}\}$ and $\bar{b} = b_0 + 3\sigma_0$. Let $r_0 = \frac{b_0 - 10^{-6}}{3b_0}$ (≈ 0.3333). It is easy to see that if $r \leq r_0$, then $\mathcal{N}_{[\underline{b},\bar{b}]}(b_0, \sigma_0^2)$ is a good approximation of $\mathcal{N}(b_0, \sigma_0^2)$ as $\underline{b} = b_0 - 3\sigma_0$ in these cases. Figure 2.2 reveals that we obtained quite similar population densities for these two models for $r = 0.1$ and 0.3, again because for these cases the size distributions obtained with (2.9) are good approximations of size distributions obtained by both the FP and GRD models. However, when

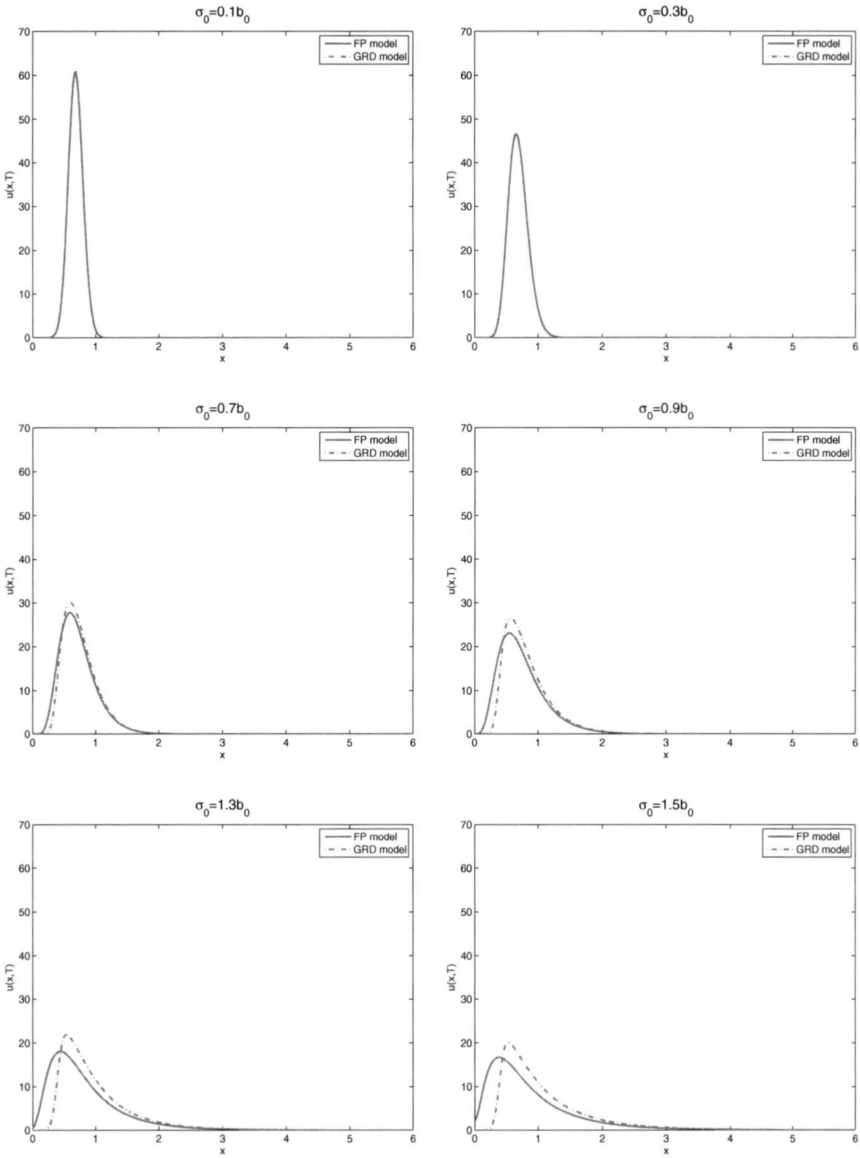

Fig. 2.2. Numerical solutions $u(x,T)$ to the FP model and the GRD model with model parameters chosen as in (2.17), with $\underline{b} = \max\{b_0 - 3\sigma_0, 10^{-6}\}$ and $\bar{b} = b_0 + 3\sigma_0$.

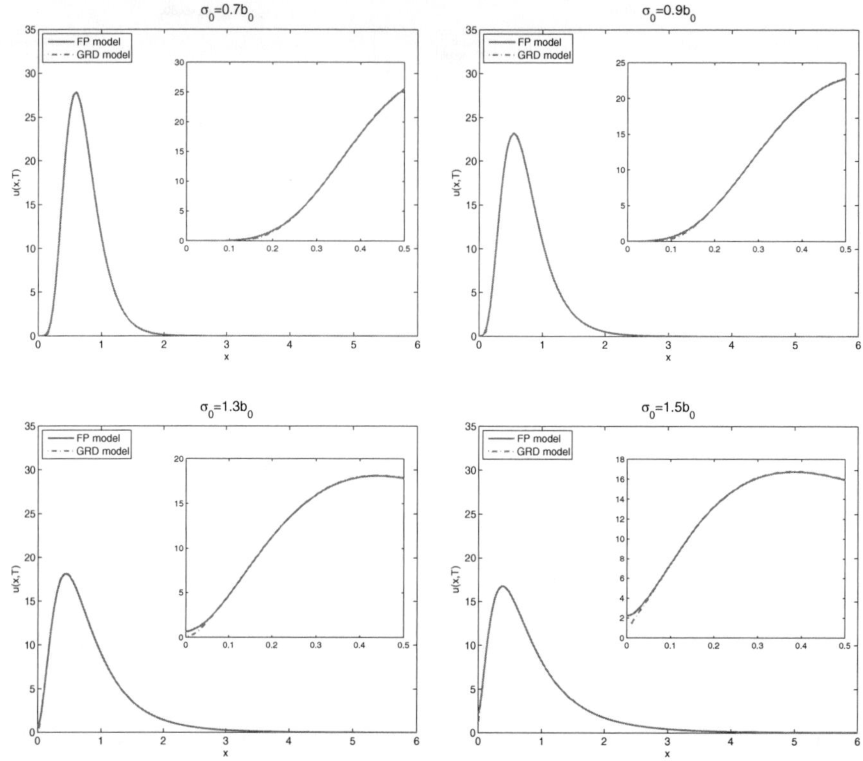

Fig. 2.3. Numerical solutions $u(x,T)$ to the FP model and the GRD model with model parameters chosen as in (2.17), where $\underline{b} = b_0 - 3\sigma_0$ and $\bar{b} = b_0 + 3\sigma_0$. The embedded plots are enlarged snapshots of the plot in the region $[0, 0.5]$.

$r > r_0$, the two solutions begin to diverge further as r increases. The reason is that $\mathcal{N}_{[\underline{b},\bar{b}]}(b_0, \sigma_0^2)$ is no longer a good approximation of $\mathcal{N}(b_0, \sigma_0^2)$ because $\underline{b} = 10^{-6}$. This is greater than $b_0 - 3\sigma_0$ in these cases, which means the size distributions obtained with (2.9) are no longer a good approximation of size distributions obtained by the GRD model. Indeed, for the FP model with the case $r > r_0$, there exist some non-negligible fraction of individuals whose size is decreased, while in the GRD model the size of each individual always increases as b is always positive.

Figure 2.3 illustrates the numerical solutions of the FP model and the solutions of the GRD model at $t = T$ with $r = 0.7$, 0.9, 1.3 and 1.5, where $\underline{b} = b_0 - 3\sigma_0$ and $\bar{b} = b_0 + 3\sigma_0$. With this choice of $\underline{b}$, we see that if $r > 1/3$, then there also exist some subpopulations in the GRD model with negative growth rates. Thus individuals in these subpopulations continue to lose weight and they will be removed from the population once their size is less than zero (the minimum size). If this situation occurs, then the total number in the population is no longer conserved, and this difficulty becomes worse as r becomes larger. However, for the FP model the total number of population is always conserved because of the zero-flux boundary conditions. In the FP model, once the size of individuals is decreased to the minimum size, they either stay there or they may increase their size in future time increments. From Figure 2.3 we can see that these two models yield pretty much similar solutions for $r = 0.7$ and 0.9. This is because in these cases r is not sufficiently large, which results in the size having negligible probability being negative in the given time period. Thus most of individuals in the GRD model remain in the system. However, we can also see that for the cases $r = 1.3$ and $r = 1.5$, the solutions to the FP models and the GRD models diverge (at the left part of the lower figures). This is because the size has non-negligible probability of being negative in these cases and these individuals with negative size in the GRD models are removed from the system.

2.4 Concluding Remarks

The computational results in this paper illustrate that, as predicted based on the analysis in [7], the Fokker-Planck model and the growth rate distribution model can, with properly chosen parameters in the individual growth dynamics, yield quite similar population densities. This implies that if one formulation is much more computationally difficult than the other, then we can use the easier one to compute solutions if we can find the corresponding equivalent forms. For example, the computational time needed to solve the Fokker-Planck model is usually much longer than that for growth rate distribution model for both examples given in Section 2.3. This is especially true when the initial population density is a sharp pulse, because then we need to employ a very fine mesh size to have a reasonably accurate solution to the FP model. In this case we can equivalently use the growth rate distribution model to compute the solution for the Fokker-Planck model when σ_0 is relatively small compared to b_0.

In closing we note that the arguments of [7, 11] guarantee equivalent size distributions at any time t for the two formulations discussed in this paper. Moreover, while the GRD formulation is not defined in terms of a stochastic process, one can argue that there does exist an equivalent underlying stochastic process satisfying a random differential equation (but not a stochastic differential equation for a Markov process). It can be argued that while the corresponding stochastic processes have the same size distribution at any time t, they are *not* the same stochastic process. This can be seen, for example, by computing the covariances of the respective processes which are different [11].

References

1. Allen, L.J.S.: An Introduction to Stochastic Processes with Applications to Biology. Prentice Hall, New Jersey (2003)
2. Banks, H.T., Bihari, K.L.: Modelling and estimating uncertainty in parameter estimation. Inverse Problems 17, 95–111 (2001)
3. Banks, H.T., Bokil, V.A., Hu, S., Dhar, A.K., Bullis, R.A., Browdy, C.L., Allnutt, F.C.T.: Modeling shrimp biomass and viral infection for production of biological countermeasures, CRSC-TR05-45, NCSU, December, 2005. Mathematical Biosciences and Engineering 3, 635–660 (2006)
4. Banks, H.T., Bortz, D.M., Pinter, G.A., Potter, L.K.: Modeling and imaging techniques with potential for application in bioterrorism, CRSC-TR03-02, NCSU, January, 2003. In: Banks, H.T., Castillo-Chavez, C. (eds.) Bioterrorism: Mathematical Modeling Applications in Homeland Security. Frontiers in Applied Math, vol. FR28, pp. 129–154. SIAM, Philadelphia (2003)
5. Banks, H.T., Botsford, L.W., Kappel, F., Wang, C.: Modeling and estimation in size structured population models, LCDS-CCS Report 87-13, Brown University. In: Proceedings 2nd Course on Mathematical Ecology, Trieste, December 8-12, 1986, pp. 521–541. World Press, Singapore (1988)
6. Banks, H.T., Davis, J.L.: Quantifying uncertainty in the estimation of probability distributions, CRSC-TR07-21, December, 2007. Math. Biosci. Engr. 5, 647–667 (2008)
7. Banks, H.T., Davis, J.L., Ernstberger, S.L., Hu, S., Artimovich, E., Dhar, A.K., Browdy, C.L.: A comparison of probabilistic and stochastic formulations in modeling growth uncertainty and variability, CRSC-TR08-03, NCSU, February, 2008. Journal of Biological Dynamics 3, 130–148 (2009)
8. Banks, H.T., Davis, J.L., Ernstberger, S.L., Hu, S., Artimovich, E., Dhar, A.K.: Experimental design and estimation of growth rate distributions in size-structured shrimp populations, CRSC-TR08-20, NCSU, November 2008. Inverse Problems (to appear)
9. Banks, H.T., Fitzpatrick, B.G., Potter, L.K., Zhang, Y.: Estimation of probability distributions for individual parameters using aggregate population data, CRSC-TR98-6, NCSU, January, 1998. In: McEneaney, W., Yin, G., Zhang, Q. (eds.) Stochastic Analysis, Control, Optimization and Applications, pp. 353–371. Birkhäuser, Boston (1998)
10. Banks, H.T., Fitzpatrick, B.G.: Estimation of growth rate distributions in size structured population models. Quart. Appl. Math. 49, 215–235 (1991)
11. Banks, H.T., Hu, S.: An equivalence between nonlinear stochastic Markov processes and probabilistic structures on deterministic systems (in preparation)
12. Banks, H.T., Tran, H.T.: Mathematical and Experimental Modeling of Physical and Biological Processes. CRC Press, Boca Raton (2009)
13. Banks, H.T., Tran, H.T., Woodward, D.E.: Estimation of variable coefficients in the Fokker-Planck equations using moving node finite elements. SIAM J. Numer. Anal. 30, 1574–1602 (1993)
14. Bell, G., Anderson, E.: Cell growth and division I. A mathematical model with applications to cell volume distributions in mammalian suspension cultures. Biophysical Journal 7, 329–351 (1967)
15. Chang, J.S., Cooper, G.: A practical difference scheme for Fokker-Planck equations. J. Comp. Phy. 6, 1–16 (1970)
16. Gard, T.C.: Introduction to Stochastic Differential Equations. Marcel Dekker, New York (1988)
17. Gyllenberg, M., Webb, G.F.: A nonlinear structured population model of tumor growth with quiescence. J. Math. Biol. 28, 671–694 (1990)

18. Kot, M.: Elements of Mathematical Ecology. Cambridge University Press, Cambridge (2001)
19. Luzyanina, T., Roose, D., Bocharov, G.: Distributed parameter identification for a label-structured cell population dynamics model using CFSE histogram time-series data. J. Math. Biol. (to appear)
20. Luzyanina, T., Roose, D., Schenkel, T., Sester, M., Ehl, S., Meyerhans, A., Bocharov, G.: Numerical modelling of label-structured cell population growth using CFSE distribution data. Theoretical Biology and Medical Modelling 4, 1–26 (2007)
21. Metz, J.A.J., Diekmann, O. (eds.): The Dynamics of Physiologically Structured Populations. Lecture Notes in Biomathematics. Springer, Berlin (1986)
22. Okubo, A.: Diffusion and Ecological Problems: Mathematical Models. Lecture Notes in Biomathematics, vol. 10. Springer, Berlin (1980)
23. Richtmyer, R.D., Morton, K.W.: Difference Methods for Initial-value Problems. Wiley, New York (1967)
24. Sinko, J., Streifer, W.: A new model for age-size structure of a population. Ecology 48, 910–918 (1967)

22. [illegible]
23. [illegible] (2007)

3

Sorting: The Gauss Thermostat, the Toda Lattice and Double Bracket Equations*,†

Anthony M. Bloch[1,‡] and Alberto G. Rojo[2,§]

[1] Department of Mathematics, University of Michigan, Ann Arbor, MI 48109, USA.
[2] Department of Physics, Oakland University, Rochester, MI 48309, USA.

Summary. In this paper we consider certain equations that have gradient like behavior and which sort numbers in an analog fashion. Two kinds of equation that have been discussed earlier that achieve this are the Toda lattice equations and the double bracket equations. The Toda lattice equations are Hamiltonian and can be shown to be a special type of double bracket equation. The double bracket equations themselves are gradient (and hence the Toda lattice has a dual Hamiltonian/gradient form). Here we compare these systems to a system that arises from imposing a constant kinetic energy constraint on a one dimensional forced system. This is a nonlinear nonholonomic constraint on these oscillators and the dynamics are consistent with Gauss's law of least constraint. Dynamics of this sort are of interest in nonequilibrium molecular dynamics. This system is neither Hamiltonian nor gradient.

3.1 Introduction

In this paper we consider certain equations that have gradient like (asymptotic) behavior and which sort numbers in an analog fashion. Two kinds of equation that have been discussed earlier that achieve this are the Toda lattice equations and the double bracket equations (see [32], [16] and [6]). The Toda lattice equations are Hamiltonian and can be shown to be a special type of double bracket equation. The double bracket equations themselves are gradient (and hence the Toda lattice has a dual Hamiltonian/gradient form). Here we compare these systems to a system that arises from imposing a constant kinetic energy constraint on a one dimensional forced system. This is a nonlinear nonholonomic constraint on these oscillators and the dynamics are consistent with Gauss's law of least constraint. Dynamics of this sort are of interest in nonequilibrium molecular dynamics. This system is neither Hamiltonian nor gradient.

---

*We would like to thank Roger Brockett for useful remarks.
†In honor of Professors Chris Byrnes and Anders Lindquist.
‡Research partially supported by the National Science Foundation.
§Research partially supported by the Research Corporation.

X. Hu et al. (Eds.): Three Decades of Progress in Control Sciences, pp. 35–48, 2010.

Nonholonomic mechanics is the study of systems subject to nonintegrable constraints on their velocities. The classical study of such systems (see e.g. [5] and references therein) is concerned with constraints that are linear in their velocities. Nonlinear nonholonomic constraints essentially do not arise in classical mechanics but are however of interest in the study of nonequilibrium or constant temperature dynamics which model the interaction of system with a bath (see e.g. [24], [20], [18], [31], [21]). In this setting the dynamics be derived using the classical Gauss's principle of least constraint. In this paper we analyze some simple examples of such systems and show that the dynamics gives rise to a generalization of another very interesting class of dynamical systems, gradient flows and in particular double bracket flows. Double bracket flows on matrices (see [16], [3],[6], [7]) arise as the gradient flows on orbits of certain Lie groups with respect to the so called normal metric. It was shown in [3] and [6] that in the tridiagonal matrix setting the Toda lattice flow (see [22]), an integrable Hamiltonian flow, may be written in double bracket form. This elucidates its dynamics and scattering behavior. Double bracket flows have also been show to give a very interesting kind of dissipation in classical mechanical systems (see [13] and also [26]).

The study of the first author of the Toda lattice and gradient flows goes back to interesting years at Harvard working with Chris Byrnes and Roger Brockett and he continues to find much inspiration from those and continuing contacts. Chris set a remarkable standard and example for the understanding of pure mathematics and for how to apply it to interesting applied problems. Chris also helped me enormously in my understanding of Morse theory and critical point theory of how to apply it to the Total Least Squares problem discussed below. The first author also enjoyed very much a visit in 1985 to the Royal Institute of Technology with Chris Byrnes and Anders Lindquist which included learning about identification and realization from Anders.

3.2 The Toda Lattice and Double Bracket Equations

An important and beautiful mechanical system that describes the interaction of particles on the line (i.e., in one dimension) is the Toda lattice. We shall describe the nonperiodic finite Toda lattice following the treatment of [27].

This is a key example in integrable systems theory.

The model consists of n particles moving freely on the x-axis and interacting under an exponential potential. Denoting the position of the kth particle by x_k, the Hamiltonian is given by

$$H(x,y) = \frac{1}{2}\sum_{k=1}^{n} y_k^2 + \sum_{k=1}^{n-1} e^{(x_k - x_{k+1})}.$$

The associated Hamiltonian equations are

$$\dot{x}_k = \frac{\partial H}{\partial y_k} = y_k\,, \tag{3.1}$$

$$\dot{y}_k = -\frac{\partial H}{\partial x_k} = e^{x_{k-1}-x_k} - e^{x_k-x_{k+1}}\,, \tag{3.2}$$

where we use the convention $e^{x_0-x_1} = e^{x_n-x_{n+1}} = 0$, which corresponds to formally setting $x_0 = -\infty$ and $x_{n+1} = +\infty$.

This system of equations has an extraordinarily rich structure. Part of this is revealed by Flaschka's ([22]) change of variables given by

$$a_k = \frac{1}{2}e^{(x_k-x_{k+1})/2} \quad \text{and} \quad b_k = -\frac{1}{2}y_k\,. \tag{3.3}$$

In these new variables, the equations of motion then become

$$\dot{a}_k = a_k(b_{k+1}-b_k)\,, \quad k = 1,\ldots,n-1\,, \tag{3.4}$$

$$\dot{b}_k = 2(a_k^2 - a_{k-1}^2)\,, \quad k = 1,\ldots,n\,, \tag{3.5}$$

with the boundary conditions $a_0 = a_n = 0$. This system may be written in the following Lax pair representation:

$$\frac{d}{dt}L = [B,L] = BL - LB, \tag{3.6}$$

where

$$L = \begin{pmatrix} b_1 & a_1 & 0 & \cdots & 0 \\ a_1 & b_2 & a_2 & \cdots & 0 \\ & & \ddots & & \\ & & & b_{n-1} & a_{n-1} \\ 0 & & & a_{n-1} & b_n \end{pmatrix}, \quad B = \begin{pmatrix} 0 & a_1 & 0 & \cdots & 0 \\ -a_1 & 0 & a_2 & \cdots & 0 \\ & & \ddots & & \\ & & & 0 & a_{n-1} \\ 0 & & & -a_{n-1} & 0 \end{pmatrix}.$$

If $O(t)$ is the orthogonal matrix solving the equation

$$\frac{d}{dt}O = BO\,, \; O(0) = \text{Identity}\,,$$

then from (3.6) we have

$$\frac{d}{dt}(O^{-1}LO) = 0\,.$$

Thus, $O^{-1}LO = L(0)$; i.e., $L(t)$ is related to $L(0)$ by a similarity transformation, and thus the eigenvalues of L, which are real and distinct, are preserved along the flow. This is enough to show that in fact this system is explicitly solvable or integrable.

There is, however, much more structure in this example. For instance, if N is the matrix $\text{diag}[1,2,\ldots,n]$, the Toda flow (3.6) may be written in the following double bracket form:

$$\dot{L} = [L,[L,N]]\,. \tag{3.7}$$

This was shown in [3] and analyzed further in [6], [7], and [10]. This double bracket equation restricted to a level set of the integrals described above is in fact the gradient flow of the function TrLN with respect to the so-called normal metric; see [6]. Double bracket flows are derived in [16].

From this observation it is easy to show that the flow tends asymptotically to a diagonal matrix with the eigenvalues of $L(0)$ on the diagonal and ordered according to magnitude, recovering the observation of Moser, [32], and [19].

A very important feature of the tridiagonal aperiodic Toda lattice flow is that it can be solved explicitly as follows: Let the initial data be given by $L(0) = L_0$. Given a matrix A, use the Gram–Schmidt process on the columns of A to factorize A as $A = k(A)u(A)$, where $k(A)$ is orthogonal and $u(A)$ is upper triangular. Then the explicit solution of the Toda flow is given by

$$L(t) = k(\exp(tL_0))L_0k^T(\exp(tL_0)). \tag{3.8}$$

The reader can check this explicitly or refer for example to [32].

Four-Dimensional Toda

Here we simulate the Toda lattice in four dimensions. The Hamiltonian is

$$H(a,b) = a_1^2 + a_2^2 + b_1^2 + b_2^2 + b_1b_2. \tag{3.9}$$

and one has the equations of motion

$$\begin{aligned} \dot{a}_1 &= -a_1(b_1 - b_2) \quad & \dot{b}_1 &= 2a_1^2, \\ \dot{a}_2 &= -a_2(b_1 + 2b_2) \quad & \dot{b}_2 &= -2(a_1^2 - a_2^2). \end{aligned} \tag{3.10}$$

(setting $b_1 + b_2 + b_3 = 0$, for convenience, which we may do since the trace is preserved along the flow). In particular, TraceLN is, in this case, equal to b_2 and can be checked to decrease along the flow.

Figure 3.1 exhibits the asymptotic behavior of the Toda flow. It is also of interest to note that the Toda flow may be written as a different double bracket flow on the space of rank one projection matrices. The idea is to represent the flow in the variables $\lambda = (\lambda_1, \lambda_2, \ldots, \lambda_n)$ and $\mathbf{r} = (r_1, r_2, \ldots, r_n)$ where the λ_i are the (conserved) eigenvalues of L and r_i, $\sum_i r_i^2 = 1$ are the top components of the normalized eigenvectors of L (see [27] and [19]). Then one can show (see [3],[4], [10]) that the flow may be written as

$$\dot{P} = [P, [P, \Lambda]] \tag{3.11}$$

where $P = \mathbf{r}\mathbf{r}^T$ and $\Lambda = \mathrm{diag}(\lambda)$.

This flow is a flow on a simplex (see [3]). The Toda flow in its original variables can also be mapped to a flow convex polytope (see [10], [7]).

More generally one can consider the gradient flow on the space of Grassmannians of the function TrΛP where P is a projection matrix representing the projection onto a k-plane in n-space (in the real or complex setting). It also useful to replace the diagonal matrix Λ by a general symmetric matrix C. In this case the function TrCP is of the form of a function that represents the Total Least Squares distance function and has an elegant critical point structure (see [17], [2],[3], [10]). In this case the double

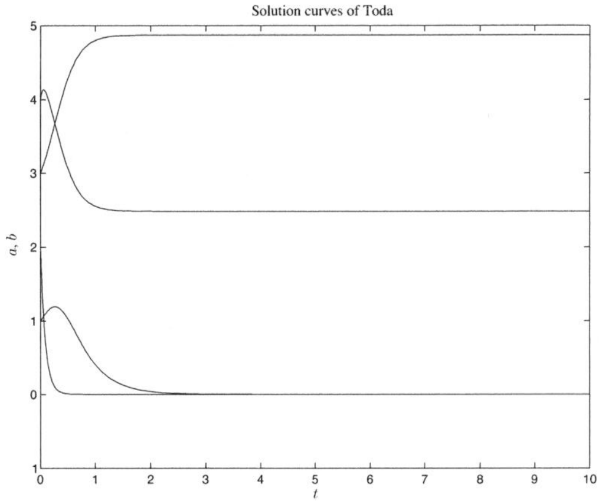

Fig. 3.1. Asymptotic behavior of the solutions of the four-dimensional Toda lattice.

bracket equation can determine the minimum of the this function. The critical point structure in the infinite setting is also interesting (see [8]). The role of the momentum map is all these setting is of great interest and discussed in the above references,.

As we shall see below the thermostat flow may be regarded as a flow of rank two matrices rather like the flows of Moser in [28].

3.3 Dynamics of Particles with Constant Kinetic Energy Constraint

3.3.1 Nonholonomic Constraints

The standard setting for nonholonomic systems (see e.g. [5]) is the following: one has n coordinates $q_i(t)$ and m (linear in the) velocity-dependent constraints of the form

$$\sum_{i=1}^{n} a_i^{(j)}(\mathbf{q})\dot{q}_i = 0, \quad j = 1, \cdots, m. \tag{3.12}$$

The general form of the equations can be written using the *unconstrained* Lagrangian $L(q_i, \dot{q}_i)$:

$$\frac{d}{dt}\frac{\partial L}{\partial \dot{q}_i} - \frac{\partial L}{\partial q_i} = F_i, \tag{3.13}$$

with F_i the virtual forces necessary to impose the constraints (3.12).

Suppose the m velocity constraints, are represented by the equation

$$A(q)\dot{q} = 0. \tag{3.14}$$

Here $A(q)$ is an $m \times n$ matrix and $\dot{q}$ is a column vector. Let λ be a row vector whose elements are called "Lagrange multipliers." The equations we obtain are thus

$$\frac{d}{dt}\left(\frac{\partial L}{\partial \dot{q}}\right) - \frac{\partial L}{\partial q} = \lambda A(q), \quad A(q)\dot{q} = 0. \tag{3.15}$$

In the current setting we are interested in a nonlinear constraint, the constraint of constant kinetic energy. This again may be implemented using Lagrange multipliers, by differentiating the constraint and enforcing the system to lie on the resultant hypersurface defined by this constraint. This is equivalent to Gauss's principle of least constraint.

In the linear setting (see [5]), the system energy is preserved. This is not true in the nonlinear setting.

3.3.2 Constraint in the Case of Equal Masses

The simplest setting is the case of N particles with equal mass. In this case the constraint of kinetic energy correspond to the norm of the velocity being constant under the flow.

Consider an N dimensional vector $\mathbf{V} = (\dot{x}_1, \cdots, \dot{x}_N)$ and an N dimensional force $\mathbf{F} = (f_1, \cdots, f_N)$. The constraint of constant kinetic energy is imposed by a "time dependent viscosity feedback" $\eta(t)$

$$\dot{\mathbf{V}} = \mathbf{F} - \eta(t)\mathbf{V}.$$

The crucial ingredient is that the viscosity term can be positive or negative. The condition that the norm of $\mathbf{V}$ is constant (or constant kinetic energy) means:

$$\dot{\mathbf{V}} \cdot \mathbf{V} = 0 \Rightarrow \eta(t) = \frac{\mathbf{F} \cdot \mathbf{V}}{\mathbf{V} \cdot \mathbf{V}} \tag{3.16}$$

The equation of motion is therefore:

$$\dot{\mathbf{V}} = \mathbf{F} - \frac{\mathbf{F} \cdot \mathbf{V}}{\mathbf{V} \cdot \mathbf{V}}\mathbf{V}.$$

3.4 Correlations Induced by the Constraint in the Case of Constant Force

Consider the case of N particles in one dimension subject to a constant gravitational force $f = mg$. In the absence of the constraint the particles move independently and the kinetic energy fluctuates. We now show that the constraint induces correlations and that the long time behavior corresponds to *all* particles moving with the same velocity, regardless of the initial conditions.

The equation of motion of the n-th particle is

$$\dot{v}_n = g - \frac{\sum_{m=1}^{N} g v_m}{\mathbf{V}^2} v_n. \tag{3.17}$$

Of course $\mathbf{V}^2 = \sum v_n(t)^2$ is preserved by the dynamics.

Define

$$u_q = \frac{1}{N}\sum_n v_n e^{iqn}, \tag{3.18}$$

with $q = \frac{2\pi}{N}k,\ k = 0, 1, \cdots, (N-1)$. Also define a (constant) mean quadratic velocity as $v_M^2 = \frac{\mathbf{v}^2}{N}$.

Replace these two transformations in (3.17) to obtain

$$\dot{u}_q(t) = g\delta_{q,0} - \frac{gu_0(t)}{v_M^2}u_q(t). \tag{3.19}$$

From this equation, the equation of motion for u_0 is

$$\dot{u}_0 = g\left(1 - \frac{u_0^2}{v_M^2}\right). \tag{3.20}$$

with solution (and long time limit) given by:

$$u_0(t) = v_M \tanh(gt/v_M) \to v_M.$$

The solution for $u_q(t)$ for $q > 0$ is given by

$$u_q(t) = \frac{u_q(0)}{\cosh(gt/v_m)}$$

In the long time limit $u_q(t) \to 0$.

Substituting in (3.18) we see that the long time solution is

$$v_n(t \to \infty) = v_M.$$

This means that in this particular example, at long times, the constraint enforces all particles to move *with the same velocity* v_M.

In the absence of the constraint, the velocities are of course independent, and the total energy is conserved.

In the constrained case the long time behavior for each $x_n(t)$ is a linear increase, meaning that, although the kinetic energy is constant, the potential energy is linearly decreasing:

$$\dot{U}_n = -mgv_M.$$

3.4.1 Breaking of Equipartition for Particles of Different Mass

Consider now the case different masses.

The equation of motion of the n-th particle is

$$M_n\dot{v}_n = M_n g - \frac{\sum_{m=1}^N M_m g v_m}{\sum M_n v_n^2} M_n v_n. \tag{3.21}$$

Of course the kinetic energy K, with $2K = \sum M_n v_n(t)^2$ is preserved by the dynamics.

Define the momentum modes $P_q(t)$

$$P_q(t) = \frac{1}{N}\sum_n M_n v_n(t) e^{iqn}, \tag{3.22}$$

and a (time independent) "mass mode"

$$M_q = \frac{1}{N}\sum_n M_n e^{iqn}, \tag{3.23}$$

with $q = \frac{2\pi}{N}k,\ k = 0, 1, \cdots, (N-1)$. Also define a (constant) mean square velocity as

$$v_M^2 = \frac{\sum_n M_n v_n^2}{\sum_n M_n}.$$

Replace these two transformations in (3.21) to obtain

$$\dot{P}_q(t) = M_q g - \frac{g}{M_0 v_M^2} P_0(t) P_q(t). \tag{3.24}$$

From this equation, the equation of motion for P_0 is

$$\dot{P}_0 = M_0 g \left[1 - \left(\frac{P_0}{M_0 v_M}\right)^2\right]. \tag{3.25}$$

with solution (and long time limit) given by $P_0(t) = M_0 v_M \tanh(gt/v_M) \to M_0 v_M$.

In this long time limit, the equation for P_q for $q \neq 0$ is

$$\dot{P}_q(t) = M_q g - \frac{g}{v_M} P_q(t), \tag{3.26}$$

with obvious solution

$$P_q(t) = M_q v_M + [P_q(0) - M_q v_M] e^{-gt/v_M} \to M_q v_M.$$

Substituting these in (3.22) and (3.23) we see that the long time solution is $v_n(t \to \infty) = v_M$.

This means that in this particular example, at long times, the constraint again enforces all particles to move *with the same velocity* v_M. However, large mass particles get more kinetic energy than low mass ones, breaking the equipartition theorem.

3.4.2 Three Particles in One Dimension and the Evolution as a Rotation

Since, for particles of equal mass, the motion is always in a sphere of radius $|\mathbf{V}_0|$, for 3 particles we can formulate the dynamics as a rotation:

$$\dot{\mathbf{V}} = \vec{\Omega} \times \mathbf{V},$$

with

$$\Omega_i = \frac{1}{\mathbf{V}_0^2} \varepsilon_{ijk} v_j f_k.$$

Explicitly,

$$\begin{aligned}
\dot{v}_1 &= \Omega_2 v_3 - \Omega_3 v_2 \\
&= \frac{1}{\mathbf{V}_0^2} \left[(v_3 f_1 - v_1 f_3) v_3 - (v_1 f_2 - v_2 f_1) v_2 \right] \\
&= \frac{1}{\mathbf{V}_0^2} \left[(v_1^2 + v_2^2 + v_3^2) f_1 - (f_1 v_1 + f_2 v_2 + f_3 v_3) v_1 \right] \\
&\equiv f_1 - \frac{\sum f_i v_i}{\mathbf{V}_0^2} v_1
\end{aligned} \tag{3.27}$$

3-Particle Case as a Double Bracket Equation

Note that in fact Ω we have $\Omega = \frac{1}{\mathbf{V}_0^2} \mathbf{V} \times \mathbf{F}$. Hence

$$\dot{\mathbf{V}} = -\frac{1}{\mathbf{V}_0^2} \mathbf{V} \times (\mathbf{V} \times \mathbf{F}).$$

Now using the standard map from 3-vectors to matrices in $so(3)$ (see e.g. [25]), denoted by $\mathbf{V} \to \hat{\mathbf{V}}$ this equation may be rewritten in the form

$$\dot{\mathbf{V}} = -\frac{1}{\mathbf{V}_0^2} [\hat{\mathbf{V}}, [\hat{\mathbf{V}}, \hat{\mathbf{F}}]],$$

This is the classic double bracket form and links nonlinear nonholonomic mechanics (second order!) to double bracket flows. Note also that this tells us precisely what the equilibria (steady state solutions) should be: when $\hat{\mathbf{V}}$ and $\hat{\mathbf{F}}$ commute. See also [13] for its use as a nonlinear dissipative mechanism.

N-Particle Case

For N particles in one dimension, the extension of the discussion above is immediate. The dynamics in general is given by the skew matrix $\mathbf{O}$:

$$\dot{\mathbf{V}} = \mathbf{O}\mathbf{V},$$

with

$$O_{ij} = \frac{f_i v_j - v_i f_j}{\mathbf{V}_0^2},$$

and formal solution

$$\mathbf{V}(t) = T e^{\int_0^t dt' \mathbf{O}(t')} \mathbf{V}_0,$$

and T the time ordering operator.

3.4.3 Stability and Generalized Double Bracket Form

Note that this equation can be reformulated in the following way:

O_{ij} is the rank two matrix

$$\mathbf{O} = \frac{\mathbf{F}\mathbf{V}^T - \mathbf{V}\mathbf{F}^T}{\mathbf{V}_0^2}$$

Hence the flow may be written:

$$\dot{\mathbf{V}} = \frac{\mathbf{F}\mathbf{V}^T - \mathbf{V}\mathbf{F}^T}{\mathbf{V}_0^2}\mathbf{V} = \frac{\mathbf{F}\otimes\mathbf{V} - \mathbf{V}\otimes\mathbf{F}}{\mathbf{V}_0^2}\mathbf{V} \tag{3.28}$$

(Note that this is effectively a generalization of the double bracket form above to the N-vector setting.)

Now consider the derivative of $\mathbf{V}\cdot\mathbf{F}$ in the case $\mathbf{F}$ is constant.

We have

$$\frac{d}{dt}(\mathbf{V}\cdot\mathbf{F}) = \mathbf{F}\cdot\dot{\mathbf{V}} = \mathbf{F}\cdot\mathbf{O}\mathbf{V} = \mathbf{F}\cdot\frac{\mathbf{F}\mathbf{V}^T - \mathbf{V}\mathbf{F}^T}{\mathbf{V}_0^2}\mathbf{V}.$$

But the numerator here just equals $||\mathbf{V}||^2||\mathbf{F}||^2 - ||\mathbf{V}\cdot\mathbf{F}||^2$ which is sign definite. Hence $\mathbf{V}\cdot\mathbf{F}$ changes monotonically along the flow. Note that this is similar to what happens in the double bracket flow (see [16] and [6]). Note also that it has the right equilibrium structure: when $\mathbf{V}$ and $\mathbf{F}$ are parallel one gets a dynamic equilibrium.

Note these flows are not Hamiltonian and in this setting one expects this kind of asymptotic behavior (see. e.g [18]).

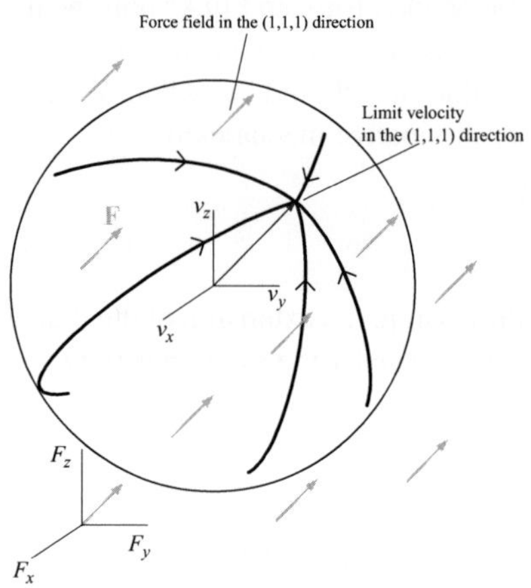

Fig. 3.2. Flow in constant force case

3.5 General Case of Constant Forces

Now we consider the general case of constant forces. The physical situation can be viewed as that of n charged particles in an electric field with equal masses but different charges. We show in this case that the particle velocities get sorted according to the original charges.

The equation of motion of the n-th particle is then of the form

$$\dot{v}_n = f_n - \frac{\sum_{m=1}^{N} f_m v_m}{\mathbf{V}^2} v_n. \tag{3.29}$$

where $\mathbf{V}^2 = \sum v_n(t)^2$ is preserved by the dynamics and we assume the f_i are distinct.

Rewrite this as

$$f_n \dot{v}_n = f_n^2 - \frac{\sum_{m=1}^{N} f_m v_m}{\mathbf{V}^2} f_n v_n. \tag{3.30}$$

Then one does a Fourier analysis as before where we define

$$P_q = \frac{1}{N} \sum_n f_n v_n e^{iqn}, \tag{3.31}$$

with $q = \frac{2\pi}{N} k,\ k = 0, 1, \cdots, (N-1)$

and

$$F_q = \frac{1}{N} \sum_n f_n^2 e^{iqn}. \tag{3.32}$$

We find

$$\dot{P}_q(t) = F_q - \frac{1}{\mathbf{V}^2} P_0(t) P_q(t). \tag{3.33}$$

Thus the equation of motion for P_0 is

$$\dot{P}_0 = F_0 \left[1 - \frac{P_0^2}{F_0 \mathbf{V}^2} \right]. \tag{3.34}$$

with solution (and long time limit) given by:

$$P_0(t) = \sqrt{F_0}\mathbf{V} \tanh(F_0 t / \mathbf{V}^{3/2}) \to \sqrt{F_0}\mathbf{V}_0.$$

In this long time limit, the equation for P_q for $q \neq 0$ is

$$\dot{P}_q(t) = F_q - \frac{\sqrt{F_0}}{\mathbf{V}_0} P_q(t), \tag{3.35}$$

This implies $P_q \to F_q$ and $v_n \to f_n$. Thus sorting occurs as illustrated by figure 3.3 where we consider the 4×4 case where the the f_i are monotonic.

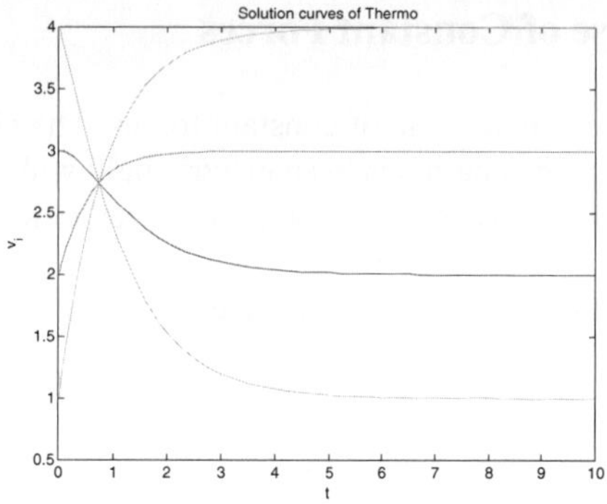

Fig. 3.3. 4 by 4 sorting for the thermostat

3.6 Symmetric Bracket Equation for Constant Forces

We now show that in the constant force setting the flow may be described by a symmetric bracket. We note that a similar result also applies in the case of a harmonic potential, which gives rise to very interesting dynamics (see [30]).

Note that this is a flow on rank two matrices – this is related in form to integrable systems which are rank two perturbations as discussed in [28]. This includes a special class of rigid body flows.

The equation of motion for **V** becomes

$$\dot{\mathbf{V}} = \frac{1}{\mathbf{V}_0^2}\left[\mathbf{V}\otimes\mathbf{F} - \mathbf{F}\otimes\mathbf{V}\right]\mathbf{V},$$

or, re-scaling the time

$$\dot{\mathbf{V}} = \left[\mathbf{V}\otimes\mathbf{F} - \mathbf{F}\otimes\mathbf{V}\right]\mathbf{V} \equiv \mathbf{L}\mathbf{V}.$$

Now consider the evolution of the operator **L** defined above

$$\begin{aligned}\dot{\mathbf{L}} &= \dot{\mathbf{V}}\otimes\mathbf{F} - \mathbf{F}\otimes\dot{\mathbf{V}}\\ &= (\left[\mathbf{V}\otimes\mathbf{F} - \mathbf{F}\otimes\mathbf{V}\right]\mathbf{V})\otimes\mathbf{F} - \mathbf{F}\otimes(\left[\mathbf{V}\otimes\mathbf{F} - \mathbf{F}\otimes\mathbf{V}\right]\mathbf{V})\\ &= (\mathbf{V}\otimes\mathbf{F})(\mathbf{V}\otimes\mathbf{F}) - (\mathbf{F}\otimes\mathbf{V})(\mathbf{F}\otimes\mathbf{V}),\end{aligned} \tag{3.36}$$

where we have used $\left[(\mathbf{a}\otimes\mathbf{b})\mathbf{c}\right]\otimes\mathbf{d} = (\mathbf{a}\otimes\mathbf{b})(\mathbf{c}\otimes\mathbf{d})$, $\mathbf{a}\otimes\left[(\mathbf{b}\otimes\mathbf{c})\mathbf{d}\right] = (\mathbf{a}\otimes\mathbf{c})(\mathbf{d}\otimes\mathbf{b})$.

Now we can show that, in terms of the operator **B**, defined as

$$\mathbf{B} = \frac{1}{2}(\mathbf{V}\otimes\mathbf{F} + \mathbf{F}\otimes\mathbf{V}),$$

equation (3.36) can be written as

$$\dot{\mathbf{L}} = \mathbf{B}\mathbf{L} + \mathbf{L}\mathbf{B}. \tag{3.37}$$

In summary, the equation of motion can be cast into an *anticommutator* form

$$\dot{\mathbf{L}} = \{\mathbf{B}, \mathbf{L}\}. \tag{3.38}$$

3.7 Conclusion

We have analyzed some nonlinear nonholonomic flows that arise in the nonequilibrium thermodynamics setting and described the structure and solutions of these flows in special cases, yielding double bracket and symmetric bracket flows. These flows are compared with the Toda lattice flow and the sorting property is examined.

References

1. Arnold, V., Kozlov, I.V.V., Neishtadt, A.I.: Dynamical Systems III. Encyclopedia of Math., vol. 3. Springer, Heidelberg (1988)
2. Bloch, A.M.: A completely integrable Hamiltonian system associated with line fitting in complex vector spaces. Bull. Amer. Math. Soc. 12, 250–254 (1985)
3. Bloch, A.M.: Steepest descent, linear programming and Hamiltonian flows. Contemp. Math. Amer. Math. Soc. 114, 77–88 (1990)
4. Bloch, A.M.: The Kahler structure of the total least squares problem, Brockett's steepest descent equations and constrained flow. In: Realization and Modeling in Systems Theory, pp. 83–88. Birkhauser, Boston (1990)
5. Bloch, A.M., Baillieul, J., Crouch, P., Marsden, J.E.: Nonholonomic Mechanics and Control. Springer, Heidelberg (2003)
6. Bloch, A.M., Brockett, R.W., Ratiu, T.: A new formulation of the generalized Toda lattice equations and their fixed-point analysis via the moment map. Bulletin of the AMS 23, 447–456 (1990)
7. Bloch, A., Brockett, M.R., Ratiu, T.S.: Completely integrable gradient flows. Comm. Math. Phys. 147, 57–74 (1992)
8. Bloch, A.M., Byrnes, C.I.: An infinite-dimensional variational problem arising in estimation theory. In: Fliess, M., Hazewinkel, M. (eds.) Algebraic and Geometric Methods in Nonlinear Control Theory, pp. 487–498. D. Reidel Publishing Co., Dordrecht (1986)
9. Bloch, A.M., Crouch, P.E.: Nonholonomic and vakonomic control systems on Riemannian manifolds. Fields Institute Communications 1, 25 (1993)
10. Bloch, A.M., Flaschka, H., Ratiu, T.S.: A convexity theorem for isospectral manifolds of Jacobi matrices in a compact Lie algebra. Duke Math. J. 61, 41–65 (1990)
11. Bloch, A.M., Iserles, A.: The optimality of double bracket flows. The International Journal of Mathematics and Mathematical Sciences 62, 3301–3319 (2004)
12. Bloch, A., Krishnaprasad, M.P.S., Marsden, J.E., Murray, R.: Nonholonomic mechanical systems with symmetry. Arch. Rat. Mech. An. 136, 21–99 (1996)
13. Bloch, A., Krishnaprasad, M.P.S., Marsden, J.E., Ratiu, T.S.: The Euler–Poincaré equations and double bracket dissipation. Comm. Math. Phys. 175, 1–42 (1996)
14. Bloch, A., Marsden, M.J.E., Zenkov, D.: Nonholonomic Dynamics. Notices AMS 52, 324–333 (1996)
15. Bloch, A.M., Rojo, A.G.: Quantization of a nonholonomic system. Phys. Rev. Letters 101, 030404 (2008)
16. Brockett, R.W.: Dynamical systems that sort lists and solve linear programming problems. In: Proc. 27th IEEE Conf. and Control. See also: Linear Algebra and Its Appl. 146, 79–91 (1991)
17. Byrnes, C.I., Willems, J.C.: Least squares estimation, linear programming and momentum: A geometric parametrization of local minima. IMA Journal of Mathematical Control and Information 3, 103–118 (1986)

18. Dettman, C.P., Morris, G.P.: Proof of Lyapunov exponent pairing for systems at constant kinetic energy. Physical Review E, 2495–2598
19. Deift, P., Nanda, T., Tomei, C.: Differential equations for the symmetric eigenvalue problem. SIAM J. on Numerical Analysis 20, 1–22 (1983)
20. Evans, D.J., Hoover, W.G., Failor, B.H., Moran, B., Ladd, A.J.C.: Nonequilibrium thermodynamics via Gauss's principle of least constraint. Phys, Rev. A 28, 1016–1021 (1983)
21. Ezra, G., Wiggins, S.: Impenetrable barriers in phase space for deterministic thermostats. J. Phys. A, Math. and Theor. 42, 042001 (2009)
22. Flaschka, H.: The Toda Lattice. Phys. Rev. B 9, 1924–1925 (1974)
23. Helmke, U., Moore, J.: Optimization and Dyamical Systems. Springer, New York (1994)
24. Hoover, W.G.: Computational Statistical Mechancis. Elsevier, Amsterdam (1991)
25. Marsden, J.E., Ratiu, T.S. (eds.): Introduction to Mechanics and Symmetry. Springer, Heidelberg (1999), Texts in Applied Mathematics, 17 (First Edition 1994, Second Edition, 1999)
26. Morrison, P.: A paradigm for joined Hamiltonian and dissipative systems. Physica D 18, 410–419 (1986)
27. Moser, J.: Finitely many mass points on the line under the influence of an exponential potential — an integrable system. Springer Lecture Notes in Physics 38, 467–497 (1974)
28. Moser, J.: Geometry of quadrics and spectral theory, *The Chern Symposium*, pp. 147–188. Springer, New York (1980)
29. Neimark, J.I., Fufaev, N.A.: Dynamics of Nonholonomic Systems. Translations of Mathematical Monographs, AMS 33 (1972)
30. Rojo, A.G., Bloch, A.M.: Nonholonomic double bracket equations and the Gauss thermostat. Phys. Rev. E. E 80, 025601(R), 2009 (to appear)
31. Sergi, A.: Phase space flow for non-Hamiltonian systems with constraints. Phys. Rev. E 72, 031104 (2005)
32. Symes, W.W.: The QR algorithm and scattering for the nonperiodic Toda lattice. Physica D 4, 275–280 (1982)

4

Rational Functions and Flows with Periodic Solutions*

R.W. Brockett

School of Engineering and Applied Sciences, Harvard University, USA

Summary. The geometry of the space of real, proper, rational functions of a fixed degree and without common factors has been of interest in system theory for some time because of the central role transfer functions play in modeling linear time invariant systems. The $2n$-dimensional manifold of real proper rational functions of degree n can also be identified with the product of the set of $(2n-1)$-dimensional manifold of n-by-n real nonsingular Hankel matrices and the real line. The distinct possibilities for the signature of a nonsingular n-by-n Hankel matrix serves to characterize the distinct connected components of the correspond set of rational functions and, at the same time, serve to decompose the space into connected components. In this paper we consider the construction of the de Rham cohomology of the n-by-n real nonsingular Hankel matrices of signature $n-2$ as a further step in the quest for more useful parameterizations of various families of rational functions.

4.1 Introduction

In our collaboration with Byrnes [1] the focus is on the development of testable conditions for establishing the existence of periodic solutions of differential equations. These conditions involve the identification of an monotone increasing angle-like quantity and an invariant set with the topology of a disk cross a circle. This basic setup is useful more generally for establishing qualitative properties trajectories even if their initial conditions do not lie on a periodic orbit.

The concept of angle playing a role in this work can be thought of as a natural generalization of familiar ideas involving the ambiguities associated with the formula

$$\frac{d}{dt}\tan^{-1}\frac{y}{x}=\frac{\dot{y}x-\dot{x}y}{x^2+y^2}\,;\,x^2+y^2>0$$

and its differential version

$$d\left(\tan^{-1}\frac{y}{x}\right)=\left(\frac{x}{x^2+y^2}\right)dy-\left(\frac{y}{x^2+y^2}\right)dx$$

*This work was supported in part by the US Army Research Office under grant DAAG 55 97 1 0114.

X. Hu et al. (Eds.): Three Decades of Progress in Control Sciences, pp. 49–57, 2010.

The language used comes from differential geometry where objects of the form $\sum \alpha_i(x)dx_i$ are called one-forms. They are said to be *closed* if there is equality of the mixed partials

$$\frac{\partial \alpha_i(x)}{\partial x_j} = \frac{\partial \alpha_j(x)}{\partial x_i}$$

A closed one-form defined in a set X_0 is said to be *exact* if there is an everywhere defined smooth function on X_0 such that the one-form is its differential. Poincare's lemma asserts that a closed one-form on a contractable set is exact but on sets such as the punctured plane $\{(x,y)|\, x^2+y^2>0\}$ (think $\tan^{-1}(y/x)$ as above) there may not be any such function. In this way closed, but not exact, one-forms bear witness to "holes" in the space and are said to represent a de Rham cohomology class in H^1. In this paper we describe a method to construct such one-forms for certain kinds of spaces of interest in system theory. The method involves linear constant coefficient differential equations and one might think that something like the standard procedures for constructing Liapunov functions would be available but this does not seem to be the case.

One of our application areas involves rational functions. The geometry of the space of rational functions, and the closely related theory of nonsingular Hankel matrices, has been of interest in system theory for some time [2-6]. The system theoretic motivation comes from realization theory, and the related partial realization problem discussed in [3-5]. Although it is known that certain connected components of the space of all nonsingular Hankel matrices have a geometry that permits the existence of closed but not exact one-forms, it seems that the explicit construction of a representative of H^1 for these spaces has not been reported. However, to use the method of the solid torus to investigate the existence of periodic solutions it is desirable to know explicitly a suitable representation of the cohomology class and it is for this reason that we give a construction. It may be noted that rational functions play a role in various examples of completely integrable systems [6-8] with and without periodic solutions, providing additional motivation for this work.

Example 4.1.1. On the space of two-by-two matrices with determinant 1 we have in the notation

$$F = \begin{bmatrix} \alpha & \beta \\ \gamma & \delta \end{bmatrix}$$

a closed but not exact one-form

$$d\tan^{-1}\left(\frac{\beta-\gamma}{\alpha+\delta}\right) = \frac{(\alpha+\delta)(d\beta - d\gamma)-(\beta-\gamma)(d\alpha+d\delta)}{(\alpha+\delta)^2+(\beta-\gamma)^2}$$

Note that because $\alpha\delta-\beta\gamma=1$ the denominator can be written as $\alpha^2+\delta^2+\beta^2+\gamma^2+2$ and hence is never zero. When integrated along the closed path

$$F(\theta) = \begin{bmatrix} \cos\theta & \sin\theta \\ -\sin\theta & \cos\theta \end{bmatrix}$$

with θ increasing from 0 to 2π it evaluates to 2π, confirming the fact that the form is not exact. Closely related is the three-dimensional manifold of two-by-two symmetric matrices with negative determinant; $\alpha\delta - \beta^2 < 0$. In the notion just used, if replace γ by β an appropriate one-form is

$$d\,\tan^{-1}\left(\frac{2\beta}{\alpha-\delta}\right) = 2\frac{d\,\beta\,(\alpha-\delta)-(d\,\alpha-d\,\delta)\beta}{(\alpha-\delta)^2+4\beta^2}$$

The denominator can not vanish because if $\beta = 0$ then $\alpha\delta < 0$ and $(\alpha-\delta)^2 > 0$. When integrated along the closed path defined above the result is the same.

Remark 4.1.1. Consider a differential equation on the space of symmetric matrices with negative determinant, $\dot{F} = f(F)$, adopting the notation

$$\frac{d}{dt}\begin{bmatrix} a & b \\ b & c \end{bmatrix} = \begin{bmatrix} g_1(a,b,c) & g_2(a,b,c) \\ g_2(a,b,c) & g_3(a,b,c) \end{bmatrix}$$

Imposing a condition such as requiring the g's to vanish when $ac - b^2$ will constrain the solutions stay in the given space. The condition

$$(a-c)g_2 - b(g_1 - g_3) > 0$$

implies that the angle $\tan^{-1}(2b/(a-c)$ is advancing along all solutions. With this, and some condition that prevents solutions from going to infinity, one can expect to be able to prove the existence of a periodic solution. It is an elaboration of this idea that motivates our search for one-forms representing nontrivial cohomology classes.

Example 4.1.2. Consider the space of three-by-three Hankel matrices parametrized as

$$H(a,b,c,d,e) = \begin{bmatrix} a & b & c \\ b & c & d \\ c & d & e \end{bmatrix}$$

We restrict attention to $\mathscr{H}(2,1)$, the manifold consisting of nonsingular three-by-three Hankel matrices of signature $(2,1)$. We will show that the one-form

$$\omega = d\left(\tan^{-1}\frac{b(c+\sqrt{c^2+d^2+e^2})-d(a+\sqrt{a^2+b^2+c^2})}{c(c+\sqrt{c^2+d^2+e^2})-e(a+\sqrt{a^2+b^2+c^2})}\right)$$

is closed but not exact on this space. The proof requires the verification of a number of concrete items

1. Show that it is defined for all $H \in \mathscr{H}(2,1)$
2. Show that it is not exact.
3. Evaluate the least period.

The rest of this paper is devoted to aspects of verifying these properties.

4.2 Group Actions on Hankel Matrices

We collect here a few facts about Hankel matrices over the real field that will play a role.

Remark 4.2.1. The set of n-by-n Hankel matrix with a fixed signature is a connected set. By virtue of a theorem of Frobenius, we know that the pattern of signs (including the zeros) of the principal minors of H determine its signature. The set of all nonsingular two-by-two Hankel matrices have three connected components and the assignment of a particular matrix to one of these connected components can be done on the basis of the signs of its eigenvalues.

It is an old observation that the Hankel matrices and the binary forms of degree $2p$, i.e., forms homogeneous of degree $2p$ in two variables,

$$\phi(x,y) = a_0 x^{2p} + a_1 x^{2p-1} y + \cdots a_{p-1} x^1 y^{2p-1} + a_{2p} y^{2p}$$

are closely related. These can be represented as quadratic form using a Hankel matrix by introducing the vector of monomals

$$\begin{bmatrix} x \\ y \end{bmatrix}^{[p]} = \begin{bmatrix} x^p \\ x^{p-1}y \\ \vdots \\ xy^{p-1} \\ y^p \end{bmatrix}$$

Explicitly,

$$\phi(x,y) = \left\langle \begin{bmatrix} x \\ y \end{bmatrix}^{[p]}, H \begin{bmatrix} x \\ y \end{bmatrix}^{[p]} \right\rangle$$

with

$$H = \begin{bmatrix} a_0 & a_1/2 & \cdots & a_{p-1}/(p-2) & a_p/(p-1) \\ a_1/2 & a_2/3 & \cdots & a_p/(p-1) & a_{p+1}/(p-2) \\ \cdots & \cdots & \cdots & \cdots & \cdots \\ a_{p-1}/(p-2) & a_p/(p-1) & \cdots & a_{2p-2}/3 & a_{2p-1}/2 \\ a_p/(p-1) & a_{p+1}/(p-1) & \cdots & a_{2p-1}/2 & a_{2p} \end{bmatrix}$$

If we let the special linear group in two dimensions act on (x,y) via

$$\begin{bmatrix} x \\ y \end{bmatrix} \mapsto \begin{bmatrix} a & b \\ c & d \end{bmatrix} \begin{bmatrix} x \\ y \end{bmatrix}$$

the corresponding change in the coefficients of the binary form $\phi(x,y)$ defines an action on Hankel matrices. We can describe this concretely in terms of the three-parameter Lie group group of n-by-n matrices generated by matices of the form

$$F(\alpha,\beta,\gamma) = \exp(\alpha\tau^+ + \beta\tau^- + \gamma h)$$

with

$$\tau^+ = \begin{bmatrix} 0 & n-1 & 0 & \dots & 0 \\ 0 & 0 & n-2 & \dots & 0 \\ \dots & \dots & \dots & \dots & \dots \\ 0 & 0 & 0 & \dots & 1 \\ 0 & 0 & 0 & \dots & 0 \end{bmatrix} \; ; \; \tau^- = \begin{bmatrix} 0 & 0 & \dots & 0 & 0 \\ 1 & 0 & \dots & 0 & 0 \\ 0 & 2 & \dots & 0 & 0 \\ \dots & \dots & \dots & \dots & \dots \\ 0 & 0 & \dots & n-1 & 0 \end{bmatrix}$$

and $h = \mathrm{diag(n, n-2, ..., -n+2, -n)}$. This group is isomorphic to the special linear group in two dimensions and plays an important role in earlier work on properties of Hankel matrices [3,[5]. It is not difficult to see that if H is a n-by-n Hankel matrix and if L is a linear combination of τ^+, τ^-, h then

$$\dot{H} = LH + HL^T$$

defines a flow on the space of Hankel matrices of a fixed signature. Moreover, the particular linear combination

$$L = \begin{bmatrix} 0 & n-1 & 0 & 0 & \dots & 0 \\ -1 & 0 & n-2 & 0 & \dots & 0 \\ 0 & -2 & 0 & n-3 & \dots & 0 \\ \dots & \dots & \dots & \dots & \dots & \dots \\ 0 & 0 & 0 & 0 & \dots & 1 \\ 0 & 0 & 0 & 0 & \dots & 0 \end{bmatrix}$$

generates a periodic solution. This will be used below.

If $n = 3$ the eigenvalues of L are $2i, 0, -2i$. More generally, the eigenvalues of L are purely imaginary and range in evenly spaced steps from ni to $-ni$; 0 will be an eigenvalue if n is odd but not if n is even. Similarly, the eigenvalues of the operator $\tilde{L} = L(\cdot) + (\cdot)L^T$ are purely imaginary and range in equally spaced steps from $(2n-1)i$ to $-(2n-1)i$

In terms of the variables

$$\begin{bmatrix} \mu_1 \\ \mu_2 \\ \mu_3 \\ \mu_4 \\ \mu_5 \end{bmatrix} = \begin{bmatrix} 1 & 0 & 0 & 0 & -1 \\ 0 & 2 & 0 & 2 & 0 \\ 1 & 0 & -6 & 0 & 1 \\ 0 & -4 & 0 & 4 & 0 \\ 1 & 0 & 2 & 0 & 1 \end{bmatrix} \begin{bmatrix} a \\ b \\ c \\ d \\ e \end{bmatrix}$$

these equations decouple as

$$\frac{d}{dt} \begin{bmatrix} \mu_1 \\ \mu_2 \\ \mu_3 \\ \mu_4 \\ \mu_5 \end{bmatrix} = \begin{bmatrix} 0 & 2 & 0 & 0 & 0 \\ -2 & 0 & 0 & 0 & 0 \\ 0 & 0 & 0 & 4 & 0 \\ 0 & 0 & -4 & 0 & 0 \\ 0 & 0 & 0 & 0 & 0 \end{bmatrix} \begin{bmatrix} \mu_1 \\ \mu_2 \\ \mu_3 \\ \mu_4 \\ \mu_5 \end{bmatrix}$$

The following equations relate μ and the Hankel parameters:

$$\begin{bmatrix} a \\ b \\ c \\ d \\ e \end{bmatrix} = \begin{bmatrix} 1/2 & 0 & 1/8 & 0 & 3/8 \\ 0 & 1/4 & 0 & -1/8 & 0 \\ 0 & 0 & -1/8 & 0 & 1/8 \\ 0 & 1/4 & 0 & 1/8 & 0 \\ -1/2 & 0 & 1/8 & 0 & 3/8 \end{bmatrix} \begin{bmatrix} \mu_1 \\ \mu_2 \\ \mu_3 \\ \mu_4 \\ \mu_5 \end{bmatrix}$$

If the initial conditions for μ are $\mu_1 = 12$ and $\mu_5 = 16$ with $\mu_2 = \mu_3 = \mu_4 = 0$ then $a(0) = 12, b(0) = 0, c(0) = 2, d(0) = 0, e(0) = 0$ and the solution is

$$H(t) = \begin{bmatrix} 6+6\cos 2t & -3\sin 2t & 2 \\ -3\sin 2t & 2 & -3\sin 2t \\ 2 & -3\sin 2t & 6-6\cos 2t \end{bmatrix}$$

The path Γ defined by letting t range from 0 to π defines a closed curve in Hank(2,1). We will show that this path is not contractable in Hank(2,1) by constructing a one-form on Hank(2,1) which integrates to 2π along this path.

4.3 One-Forms and Differential Equations

The flow defined above puts the matter of finding a suitable one-form in the following setting. We have a real linear constant coefficient differential equation $\dot{x} = Ax$ with the eigenvalues of A rationally related and lying on the imaginary axis. Their geometric multiplicity is one. It happens that there are some inequalities $\phi_k(x) > 0$ that define a connected region $X_0 \subset \mathbb{R}^n$, which is invariant under the flow. In our case this comes about because the differential equation can be written as $\dot{H} = AH + HA^T$ and thus the signature of H is preserved. Matters being so, one can look for a pair of functions, $\psi(x)$, $\chi(x)$ such that $\psi^2 + \chi^2 > 0$ on X_0 and the one-form $d\left(\tan^{-1}(\psi/\chi)\right)$ is not exact. The problem we are now faced with is that of finding such a ψ and χ.

4.4 Getting to the One-Form

For the three-dimensional Hankel matrices we adopt the notation.

$$H = \begin{bmatrix} a & b & c \\ b & c & d \\ c & d & e \end{bmatrix}$$

with the determinant being $\det = ace + b^2c + cd^2 - eb^2 - eb^2 - c^3$. Of course the columns of this matrix must be linearly independent and, in particular, the first and third column are independent. Moreover, in the connected component of the Hankel matrices characterized as $H(2,1)$ the first and third columns cannot take on certain values because they would imply that H has the wrong signature. Specifically, this excludes the possibility that the minor

$$H_2 = \begin{bmatrix} a & b \\ b & c \end{bmatrix}$$

might be negative definite. Thus, neither $[a,b,c]$ or $[c,d,e]$ can take on the values $[-1,0,0]$ because this would imply that H has two negative eigenvalues.

We now normalize the first and third columns of H to get

$$\xi_1 = \frac{1}{\sqrt{a^2+b^2+c^2}}\begin{bmatrix} a \\ b \\ c \end{bmatrix} \ ; \ \xi_2 = \frac{1}{\sqrt{c^2+d^2+e^2}}\begin{bmatrix} c \\ d \\ e \end{bmatrix}$$

These unit vectors can then be projected stereo-graphically, using the point $[-1,0,0]$ as the pole. The projections of these two vectors, expressed in terms of $n_1 = \sqrt{a^2+b^2+c^2}$ and $n_2 = \sqrt{c^2+d^2+e^2}$ are

$$\chi_1 = \frac{1}{n_1}\begin{bmatrix} \frac{b}{1+a/n_1} \\ \frac{c}{1+a/n_1} \end{bmatrix} \ ; \ \chi_2 = \frac{1}{n_2}\begin{bmatrix} \frac{d}{1+c/n_2} \\ \frac{e}{1+c/tn_2} \end{bmatrix}$$

Linear independence in the original space implies that these two vectors cannot coincide and so the difference between them is nonzero. After some algebraic manipulations this statement is seen to be equivalent to saying that the two-dimensional vector

$$\eta = \begin{bmatrix} b(c+n_2)-d(a+n_1) \\ c(c+n_2)-e(a+n_1) \end{bmatrix}$$

is nonzero. From this we see that

$$\omega = d\left(\tan^{-1}\frac{b(c+\sqrt{c^2+d^2+e^2})-d(a+\sqrt{a^2+b^2+c^2})}{c(c+\sqrt{c^2+d^2+e^2})-e(a+\sqrt{a^2+b^2+c^2})}\right)$$

is everywhere defined on $Hank(2,1)$. (Of course we make no such claim for the other connected components of the Hankel matrices.)

It remains to determine if this one-form is exact or not. We show that it is not by displaying a particular closed path such that the line integral along this closed path integrates to something nonzero. The path will be the integral curve of $\dot{H} = LH + HL^T$ described above. In matrix form, the initial condition is

$$H(0) = \begin{bmatrix} 12 & 0 & 2 \\ 0 & 2 & 0 \\ 2 & 0 & 0 \end{bmatrix}$$

and thus is in Hank(2,1). The $\det H(0) = -8$ and is constant along this path, confirming that this is a loop in $\mathscr{H}(2,1)$. The normalization factors needed for χ_i are

$$n_1 = \sqrt{49+72\cos 2t+27\cos^2 2t} \ ; \ n_2 = \sqrt{49-72\cos 2t+27\cos^2 2t}$$

and the formula for the angle is

$$\tan\theta = \frac{3\sin 2t\,(4+6\cos 2t+n_1-n_2)}{4+2n_2-(6-6\cos 2t)(6\cos 2t+n_1)}$$

Figure 4.1 shows the graph of this function. As t advances from 0 to π the inverse tangent advances by 2π. Thus the path is not contractable.

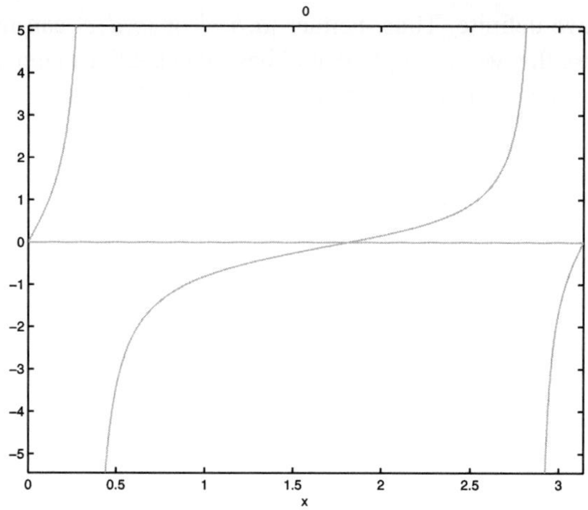

Fig. 4.1. The graph of the ratio defining $\tan\theta$ showing that as t advances from 0 to π the angle θ increases by 2π.

4.5 Generalizations

Of course any path that is homotopic to the one used above to evaluate the integral will result in the same value of the integral. In particular, any closed path generated by solving $\dot{H} = LH + HL^T$ with an initial condition in Hank(2,1) will give the same value and hence will not be contractable. From the connectedness of $\mathscr{H}(2,1)$ we see that this means that for all initial conditions in this space the integral will have the same value. It is, of course, natural to ask if there is a simpler, every defined on Hank(2,1), one-form that represents this cohomology class. Also, because we have described an analogous path in Hankel matrices of all dimensions, one would like to know if such a path is also not contractable in the higher dimensional cases.

4.6 Other Aproaches

Graeme Segal [9] used with good effect a reformulation of the common factor condition for a rational function q/p as the condition that the complex polynomial $p(s)+iq(s)$ should not have any roots that appear together with their complex conjugates. That is, $p(s)+iq(s)$ and $p(s)-iq(s)$ should not have common factors. Thus there is a corresponding complex rational function without common factors,

$$f(s) = \frac{p(s)+iq(s)}{p(s)-iq(s)}$$

One can obtain a corresponding Hankel matrix by subtracting off the value at infinity and dividing to get

$$g(s) = \frac{2iq(s)}{p(s) - iq(s)} = h_1 s^{-1} + h_2 s^{-2} + h_3 s^{-3} \cdots$$

This complex Hankel matrix will then be of rank n if and only there are no common factors. Such a reformulation over the complex field is potentially useful for a variety of reasons and it has been suggested that the differential of ln det H is a candidate for a useful one-form.

References

1. Byrnes, C.I., Brockett, R.W.: Nonlinear Oscillations and Vector Fields Paired with a Closed One-Form (submitted for publication)
2. Brockett, R.W.: Some Geometric Questions in the Theory of Linear Systems. IEEE AC 29, 449–455 (1976)
3. Brockett, R.: The Geometry of the Partial Realization Problem. In: Proceedings of the 1978 IEEE Conference on Decision and Control, pp. 1048–1052. IEEE, New York (1978)
4. Byrnes, C.I., Lindquist, A.: On the Partial Stochastic Realization Problem. Linear Algebra Appl. 50, 277–319 (1997)
5. Manthey, W., Helmke, U., Hinrichsen, D.: Topological aspects of the partial realization problem. Mathematics of Control, Signals, and Systems 5(2), 117–149 (1992)
6. Krishnaprasad, P.S.: Symplectic Mechanics and and rational functions. Ricerche di Automatica 10, 107–135 (1979)
7. Atiyah, M., Hitchin, N.: The Geometery abd Dynamics of Magnetic Monopoles. Princeton University Press, Princeton (1988)
8. Brockett, R.W.: A Rational Flow for the Toda Lattice Equations. In: Helmke, U., et al. (eds.) Operators, Systems and Linear Algebra, pp. 33–44. B.G. Teubner, Stuttgart (1997)
9. Segal, G.: On the Topology of spaces of Rational Functions. Acta Mathematica 143(1), 39–72 (1979)

5

Dynamic Programming or Direct Comparison?*,†

Xi-Ren Cao

Hong Kong University of Science and Technology, Clear Water Bay, Kowloon, Hong Kong

Summary. The standard approach to stochastic control is dynamic programming. In our recent research, we proposed an alternative approach based on direct comparison of the performance of any two policies. This approach has a number of advantages: the results may be derived in a simple and intuitive way; the approach applies to different optimization problems, including finite and infinite horizon, discounting and average performance, discrete time discrete states and continuous time and continuous stats, etc., in the same way; and it may be generalized to some non-standard problems where dynamic programming fails. This approach also links stochastic control to perturbation analysis, reinforcement learning and other research subjects in optimization, which may stimulate new research directions.

5.1 Introduction

Control, or performance optimization, of stochastic systems is a multi-disciplinary subject that has attracted wide attention from many research communities. The standard approach to stochastic control is dynamic programming [2, 3, 9]. The approach is particularly suitable for finite-horizon problems; it works backwards in time. The problem with infinite horizon can be treated as the limiting cases of the finite-horizon problem when time going to infinity, and the long-run average cost problem can be treated as the limiting case of the problems with discounted costs. In the approach, the Hamilton-Jacobi-Bellman (HJB) equation for the optimal policies is first established with the dynamic programming principles, and a verification theorem is then proved which verifies that the solution to the HJB equation indeed provides the value function, from which an optimal control process can be constructed. The HJB equations are usually differential equations, and the concept of viscosity solution is introduced when the value functions are not differentiable [9].

In this paper, we review another approach to stochastic control, called the *direct-comparison* approach. The idea of this approach is very simple: searching for an optimal policy, we always start with a comparison of the performance of any two

*Supported in part by a grant from Hong Kong UGC.

†Tribute to Chris Byrnes and Anders Lindquist.

X. Hu et al. (Eds.): Three Decades of Progress in Control Sciences, pp. 59–76, 2010.

policies. The underlying philosophy is that one can only compare two policies at a time, and performance optimization stems from such comparisons. Therefore, one can always start with a formula that gives the difference of the performance of two policies. Not surprisingly, it has been shown that from this performance difference formula, many results of dynamic programming can be easily derived and intuitively explained, some new results are obtained, and in addition, this approach can also solve some problems that go beyond the scope of dynamic programming.

Compared with dynamic programming, the direct comparison approach has the following advantages:

1. Many results become intuitively clear and the derivation and proof become simpler, because they are based on a direct comparison of the performance of any two policies. In particular, it is clear that under some minor conditions, a policy is optimal, if and only if its value function (it is called a "potential" in the direct-comparison approach) satisfies the HJB optimality equations almost everywhere, i.e, the value function is allowed to be non-differentiable at a set with a zero Lebesgue measure; and in such cases, the verification theorem is almost obvious and no viscosity solutions are needed.
2. The approach applies to different problems with, finite- and infinite horizons, discounted and long-run average performance, continuous and jump diffusions, in the same way. Discounting is not needed when dealing with long-run average performance. Furthermore, this approach can be easily extended to different problems, including the impulse control [7] used in financial engineering [1, 14].
3. Under the same framework of direct comparison, this approach links stochastic control to other research areas in performance optimization, including perturbation analysis (PA) [10, 4, 5] and reinforcement learning (RL) [15, 5], that are mainly for systems with discrete time and discrete state spaces (DTDS). Therefore, the ideas and methods in these areas may stimulate new research directions in stochastic control, e.g., sample-path-based reinforcement learning, gradient-based optimization with PA, and event-based optimization, which are active research topics mainly in DTDS communities. The direct comparison approach provides a unified framework for a number of disciplines, including stochastic control, PA, Markov decision processes (MDP), and RL.
4. The approach provides some new insights to the area of stochastic control and can also solve some problems that go beyond the scope of dynamic programming. For example, for ergodic systems, the approach is based on the fact: the performance difference of any two policies can be decomposed into the product of two factors, each of them is determined by only one policy. That is, the effect of each policy on the difference can be separated. This decomposition property clearly illustrates why the optimality condition exists and how they can be found. This insight, in the DTDS case, leads to the event-based approach in which policy depends on event rather than state [5]. Another example is our on-going research in the gain-risk multi-objective optimization; the direct comparison approach may easily obtain the efficient frontier for a wide class of problem.

In this paper, we survey the main ideas and results of the direct comparison approach and discuss some future research directions.

1. We illustrate, in Section 5.2, the main ideas of the direct comparison approach with the discrete-time and finite-state model for Markov systems. We show that the HJB optimality equation and policy iteration are direct consequences of the performance difference formula.
2. We further illustrate the power of the direct comparison approach in Section 5.3. In fact, with this approach we may develop a simple, intuitively clear, and coherent theory for MDP that covers bias and nth bias, Blackwell optimality, and multi-chain processes for long-run average performance in a unified way; the results are equivalent to, but simpler and more direct than, Veinott's n-discount theory [16, 13], and discounting is not needed.
3. We show, in Section 5.4, this simple approach can be applied to stochastic control problems of continuous-time continuous-state (CTCS) systems. The results can be simply derived and intuitively explained. We also show, in Section 5.5, this simple approach can be extended to impulse control.
4. We briefly discuss the new methods stimulated by this direct comparison approach and the new problems they may solve; we also discuss the possible future research topics. These include event-based optimization [5], gradient-based learning and optimization, and gain-risk multi-objective optimization, etc.

5.2 Direct Comparison Illustrated

We illustrate the main idea by considering the optimization problem of a discrete-time and finite state system with the long run average performance.

Consider an irreducible and aperiodic Markov chain $\mathbf{X} = \{X_l : l \geq 0\}$ on a finite state space $\mathscr{S} = \{1, 2, \cdots, M\}$ with transition probability matrix $P = [p(j|i)] \in [0,1]^{M\times M}$. Let $\pi = (\pi(1), \ldots, \pi(M))$ be the (row) vector representing its steady-state probabilities, and $f = (f(1), f(2), \cdots, f(M))^T$ be the performance (column) vector, where "T" represents transpose. We use (P, f) to represent this Markov chain. We have $Pe = e$, where $e = (1, 1, \cdots, 1)^T$ is an M-dimensional vector whose all components equal 1, and $\pi = \pi P$. The performance measure is the long-run average defined as

$$\eta = \sum_{i=1}^{M} \pi(i) f(i) = \pi f = \lim_{L\to\infty} \frac{1}{L} \sum_{l=0}^{L-1} f(X_l), \quad w.p.1. \tag{5.1}$$

The last equation holds sample path-wisely with probability one (w.p.1).

The performance *potential* vector g of a Markov chain (P, f) is defined as a solution to the *Poisson equation*

$$(I - P)g + \eta e = f. \tag{5.2}$$

The solution to this equation is only up to an additive constant; i.e., if g is a solution, then $g + ce$ is also a solution for any constant c.

Now, we consider two Markov chains (P,f) and (P',f') defined on the same state space $\mathscr{S}$. We use prime " $'$ " to denote the values associated with (P',f'). Thus, $\eta' = \pi' f'$ is the long-run average performance of the Markov chain (P',f'). Multiplying both sides of (5.2) with π' on the left yields

$$\eta' - \eta = \pi'\{[P'g + f'] - [Pg + f]\}. \tag{5.3}$$

We call it the *performance difference formula*.

To know the exact value of the performance difference from (5.3), one needs to know π' and g. On the other hand, if π' is known, one can get η' directly by $\pi' f'$; thus, in terms of obtaining the exact value of $\eta' - \eta$, (5.3) is no better than using $\eta' - \eta = \pi' f' - \pi f$ directly. Furthermore, it is impossible to calculate π' for all the policies since the policy space is usually very large. Fortunately, since $\pi' > 0$ (componentwisely), (5.3) may help us to determine which Markov process, (P,f) or (P',f'), is better without solving for π'. This leads to the following discussion.

For two M-dimensional vectors a and b, we define $a = b$, $a \le b$, and $a < b$ if $a(i) = b(i)$, $a(i) \le b(i)$, or $a(i) < b(i)$ for all $i = 1, 2 \cdots, M$, respectively; and we define $a \preceq b$ if $a \le b$ and $a(i) < b(i)$ for at least one i. The relations $>$, $\ge$, and $\succeq$ are defined similarly. From (5.3) and the fact $\pi' > 0$, the following lemma follows directly.

Lemma 5.2.1.

a) If $Pg + f \preceq$ (or $\succeq$) $P'g + f'$, then $\eta <$ (or $>$) η'.
b) If $Pg + f \le$ (or $\ge$) $P'g + f'$, then $\eta \le$ (or $\ge$) η'.

In the lemma, we use only the potentials with one Markov chain, i.e., g.

In an MDP, at any transition instant $n \ge 0$ of a Markov chain $\mathbf{X} = \{X_n, n \ge 0\}$, we take an action chosen from an action space $\mathscr{A}$. The actions that are available when the state is $X_n = i \in \mathscr{S}$ form a nonempty subset $A(i) \subseteq \mathscr{A}$. A stationary policy is a mapping $d : \mathscr{S} \to \mathscr{A}$, i.e., for any state i, d specifies an action $d(i) \in A(i)$. Let $\mathscr{D}$ be the policy space. If action α is taken at state i, then the state transition probabilities at state i are denoted as $p^{\alpha}(j|i)$, $j = 1, 2, \cdots, M$, and the cost is denoted as $f(i,\alpha)$. With a policy d, the Markov process evolves according to the transition matrix $P^d = [p^{d(i)}(j|i)]_{i=1}^{M}|_{j=1}^{M}$, and the cost function is $f^d := (f(1,d(1)), \cdots, f(M,d(M)))^T$. For simplicity, we assume that the number of actions is finite, and all the policies are ergodic (i.e., the Markov chains they generate are ergodic). A Markov chain with (P,f) is also said to be under policy $d = (P,f)$.

We use the superscript $*^d$ to denote the quantities associated with policy d. Thus, the steady-state probability corresponding to policy d is denoted as a vector $\pi^d = (\pi^d(1), \cdots, \pi^d(M))$. The long-run average performance corresponding to policy d is

$$\eta^d = \lim_{L\to\infty} \frac{1}{L} E\{\sum_{l=0}^{L-1} f[X_l, d(X_l)]\}, \quad w.p.1.$$

For ergodic chains, this limit exists with probability one (w.p.1) and does not depend on the initial state. we wish to minimize η^d over the policy space $\mathscr{D}$, i.e., to obtain $\min_{d\in\mathscr{D}} \eta^d$.

For policy d, the Poisson equation (5.2) becomes

$$(I-P^d)g^d + \eta^d e = f^d. \tag{5.4}$$

The following optimality theorem follows almost immediately from Lemma 5.2.1.b). (The "only if" part can be proved easily by construction, see [5].)

Theorem 5.2.1. A policy $\widehat{d}$ is optimal if and only if

$$P^{\widehat{d}} g^{\widehat{d}} + f^{\widehat{d}} \le P^d g^{\widehat{d}} + f^d \tag{5.5}$$

for all $d \in \mathscr{D}$.

From (5.4), we have

$$\eta^d e + g^d = f^d + P^d g^d. \tag{5.6}$$

Then Theorem 5.2.1 becomes: A policy $\widehat{d}$ is optimal if and only if

$$\eta^{\widehat{d}} e + g^{\widehat{d}} = \min_{d \in \mathscr{D}} \{P^d g^{\widehat{d}} + f^d\}. \tag{5.7}$$

The minimum is taken component-wisely. This fact is very important because it means that the minimization is taken on the actions space $\mathscr{A}(i)$, $i = 1, 2, \cdots, M$, rather than on the policy space, and the former is much smaller than the latter. (5.7) is the Hamilton-Jacobi-Bellman (HJB) equation. g^d is equivalent to the "differential" or "relative cost vector" in [2], or the "bias" in [13].

Policy iteration algorithms for finding an optimal policy can be easily developed by combining Lemma 5.2.1 and Theorem 5.2.1. Roughly speaking, the algorithm works as follows. It starts with any policy d_0 at step 0. At the kth step with policy d_k, $k = 0, 1, \cdots$, we set the policy for the next step (the $(k+1)$th step) as $d_{k+1} \in arg\{\min[P^d g^{d_k} + f^d]\}$ component-wisely, with g^{d_k} being the potential vector of (P^{d_k}, f^{d_k}). Lemma 5.2.1 implies that performance usually improves at each iteration. Theorem 5.2.1 shows that the minimum is reached when no performance improvement can be achieved. We shall not state the details here because they are standard.

The core of this approach is the performance difference formula (5.3), in which the performance difference $\eta' - \eta$ is decomposed into two factors: the first one is π', which reflects the contribution of policy (P', f') to the difference, and the second one is $\{(P'-P)g + (f'-f)\}$, which reflects the contribution of (P, f) to the difference and it indicates that this contribution is through its potential g. Furthermore, we know $\pi' > 0$ for any ergodic policies. Because of this decomposition, by analyzing one policy (P, f) to obtain its potential g and using only the structure parameters P' and f', we may find a policy better than (P, f), if such a policy exists, without analyzing any other policies. This decomposition is the foundation of the optimization theory; it leads to the optimality equation and policy iteration algorithms, etc.

Finally, the direct-comparison approach is closely related to perturbation analysis. Suppose the policies depend on a continuous parameter, denoted as (P_θ, f_θ). We use subscript $*_\theta$ to denote its quantities. Setting $P' = P_{\theta+d\theta}$ and $P = P_\theta$ in (5.3), we can easily derive the *performance derivative formula*:

$$\frac{d\eta_\theta}{d\theta} = \pi_\theta\{\frac{dP_\theta}{d\theta}g_\theta + \frac{df_\theta}{d\theta}\}. \tag{5.8}$$

Because (P_θ, f_θ) are known, performance derivatives depend only on local information π_θ and g_θ. Furthermore, if we have π and g at a policy, we may get the derivative for any parameters at this policy easily.

It has been shown that for problems with discounted performance and finite horizon, we may derive the corresponding performance difference formulas easily, and the direct comparison approach applies in a similar way [5],

5.3 A Complete Theory of Markov Decision Processes

The direct-comparison approach can be used to develop a complete theory for Markov decision processes with the general multi-chain model (for the definition of multi-chain, see e.g., [13]). For multi-chain Markov processes, the long-run average cost for a policy $d = (P, f) \in \mathscr{D}$, also called the *0th bias*, depends on the initial state and is defined as a vector $\eta^d := g_0^d$ with components

$$g_0^d(i) := \eta^d(i) = \lim_{L\to\infty}\frac{1}{L}E\Big\{\sum_{l=0}^{L-1} f^d(X_l)\Big|X_0 = i\Big\}, \qquad i \in \mathscr{S}.$$

The *bias* or the *1st bias* is denoted as $g_1^d := g^d$, its ith component is

$$g_1^d(i) := g^d(i) = \sum_{l=0}^{\infty} E\left[f^d(X_l) - \eta^d(i)|X_0 = i\right].$$

The *nth bias*, $n > 1$, is defined as a vector g_n^d whose ith component is

$$g_n^d(i) = -\sum_{l=0}^{\infty} E[g_{n-1}^d(X_l)|X_0 = i], \qquad n > 1.$$

In the above equations, $g_1^d \equiv g^d$ satisfies

$$(I - P^d)g^d + \eta^d = f^d,$$

in which η^d is a vector, and g_n^d, $n > 1$, satisfy [5, 6]

$$(I - P^d)g_{n+1}^d = -g_n^d.$$

A policy $\widehat{d}$ is said to be *gain (0th bias) optimal* if

$$g_0^{\widehat{d}} \le g_0^d, \qquad \text{for all } d \in \mathscr{D}.$$

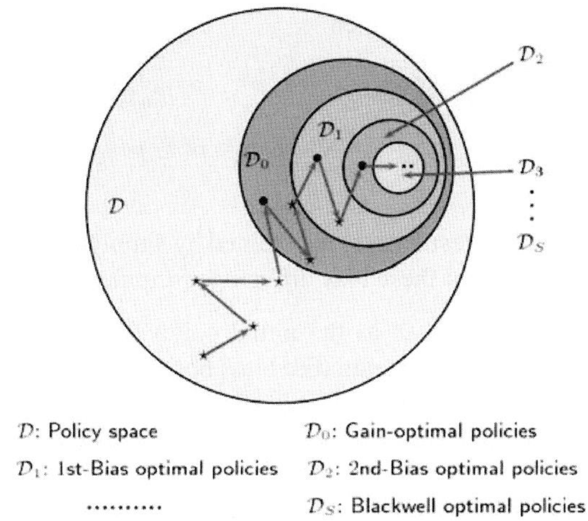

Fig. 5.1. Policy Iteration for *n*th-Bias and Blackwell Optimal Policies

Let $\mathscr{D}_0$ be the set of all gain-optimal policies. A policy $\widehat{d}$ is said to be *nth-bias optimal*, $n > 0$, if $\widehat{d} \in \mathscr{D}_{n-1}$ and

$$g_n^{\widehat{d}} \leq g_n^d, \qquad \text{for all } d \in \mathscr{D}_{n-1}, \quad n > 0.$$

Let $\mathscr{D}_n$ be the set of all nth-bias optimal policies in $\mathscr{D}_{n-1}, n > 0$. We have $\mathscr{D}_n \subseteq \mathscr{D}_{n-1}$, $n \geq 0$, $\mathscr{D}_{-1} \equiv \mathscr{D}$.

The sets $\mathscr{D}, \mathscr{D}_0, \mathscr{D}_1, \ldots$, are illustrated in Figure 5.1. Our goal is to find an nth bias optimal policy in $\mathscr{D}_n$, $n = 0, 1, \ldots$. In the direct-comparison approach, we start with the difference formulas for the n-biases of any two $(n-1)$th bias optimal policies, $n = 0, 1, \ldots$; these formulas can be easily derived. For any two policies $d, h \in \mathscr{D}$, we have [5, 6]

$$\begin{aligned} g_0^h - g_0^d = {} & (P^h)^* \left[(f^h + P^h g_1^d) - (f^d + P^d g_1^d) \right] \\ & + \left[\left(P^h\right)^* - I \right] g_0^d, \end{aligned} \tag{5.9}$$

where for any policy P, we define

$$P^* = \lim_{L \to \infty} \frac{1}{L} \sum_{l=0}^{L-1} P^l. \tag{5.10}$$

If $g_0^h = g_0^d$, then

$$\begin{aligned} g_1^h - g_1^d = {} & (P^h)^* (P^h - P^d) g_2^d \\ & + \sum_{k=0}^{\infty} \left\{ (P^h)^k \left[(f^h + P^h g_1^d) - (f^d + P^d g_1^d) \right] \right\}. \end{aligned} \tag{5.11}$$

If $g_n^h = g_n^d$ for a particular $n \geq 1$, then

$$g_{n+1}^h - g_{n+1}^d = (P^h)^*(P^h - P^d)g_{n+2}^d + \sum_{k=0}^{\infty} \left\{ (P^h)^k (P^h - P^d) g_{n+1}^d \right\}. \quad (5.12)$$

Indeed, all the following results can be obtained by simply exploring and manipulating the special structures of these bias difference formulas. For details, see [5, 6].

1. Choose any policy $d_0 \in \mathscr{D}$ as the initial policy. Applying the policy iteration algorithm, we may obtain a gain (0th bias) optimal policy $\widehat{d_0} \in \mathscr{D}_0$.
2. Staring from any nth bias optimal policy $\widehat{d_n} \in \mathscr{D}_n$, $n = 0, 1 \ldots$, applying a similar policy iteration algorithm, we may obtain an $(n+1)$th bias optimal policy $\widehat{d_{n+1}} \in \mathscr{D}_{n+1}$.
3. If a policy is an Mth bias optimal, with M being the number of states, it is also an nth bias optimal for all $n > M$; i.e., $\mathscr{D}_M = \mathscr{D}_{M+1} = \mathscr{D}_{M+2} = \ldots$.
4. An Mth bias optimal policy is a *Blackwell* optimal policy.
5. The optimality equations for nth bias optimal polices, both necessary and sufficient, can be derived from the bias difference formulas (5.9) to (5.12).

The direct comparison approach provides a unified approach to all these MDP-types of optimization problems; and the basic principle behind this approach is surprisingly simple and clear: all these results can be derived simply by a comparison of the performance, or of the bias or nth bias, of any two policies. These results are equivalent to, and simpler than, Veinott's n-discount theory [16], and discounting is not used in the derivation.

5.4 Stochastic Control

In this section, we extend the direct comparison approach to the control of continuous-time and continuous-state (CTCS) systems. The basic principle is the same as that for the DTDS systems, and the major challenge is that in CTCS systems transition probabilities cannot be represented by matrices and should be represented by continuous time operator in continuous state spaces. The main part of this section devotes to the introduction of mathematic notations.

Consider the n-dimensional space of real numbers denoted as $\mathscr{R}^n$. Let $\mathscr{B}^n$ be the σ-field of $\mathscr{R}^n$ containing all the Lebesgue measurable sets. For technical simplicity, we assume that the functions considered in this paper are bounded, and let $\mathscr{C}$ be the space of all the bounded Lebesgue measurable functions on $\mathscr{R}^n$.

In general, an *operator* $\mathbb{T}$ is defined as a mapping $\mathscr{C}_I(\mathbb{T}) \to \mathscr{C}_O(\mathbb{T})$, or $\mathscr{C}_I \to \mathscr{C}_O$ for short, such that for any $h \in \mathscr{C}_I$, we have $\mathbb{T}h \in \mathscr{C}_O$, where $\mathscr{C}_I$ and $\mathscr{C}_O$ are the input and output spaces of $\mathbb{T}$. We assume that $\mathscr{C}_I \subseteq \mathscr{C}$. In a more precise way, we may set $\mathbb{T} \triangleq \{\mathbb{T}_x, x \in \mathscr{R}^n\}$, with $\mathbb{T}_x$ being a mapping from $h \in \mathscr{C}$ to $\mathbb{T}_x h \in \mathscr{R}$. We denote $(\mathbb{T}h)(x) \triangleq \mathbb{T}_x h$.

Now, we consider a CTCS Markov process $\mathbf{X} = \{X(t), t \in [0, \infty)\}$ with state space $\mathscr{S} = \mathscr{R}^n$. We consider *time-homogeneous* systems and let $P_t(B|x)$ be the probability that $X(t)$ lies in a set $B \in \mathscr{B}^n$ given that $X(0) = x$. For any given $x \in \mathscr{R}^n$, $P_t(B|x)$ is a probability measure on $\mathscr{B}^n$, and for any $B \in \mathscr{B}^n$, it is a Lebesgue measurable function. Define a *transition operator* $\mathbb{P}$: $h \to \mathbb{P}h$, $h \in \mathscr{C}$, as follows

$$\begin{aligned}(\mathbb{P}_t h)(x) &\triangleq \int_{R^n} h(y) P_t(dy|x) \\ &= E\{h[X(t)]|X(0) = x\}.\end{aligned} \tag{5.13}$$

For any transition operator $\mathbb{P}$, we have $(\mathbb{P}e)(x) = 1$ for all $x \in \mathscr{R}^n$. Thus, we can write $\mathbb{P}e = e$. Define the n-dimensional identity function I:

$$I(B|x) \triangleq \begin{cases} 1 & if\ x \in B, \\ 0 & otherwise. \end{cases} \tag{5.14}$$

The corresponding operator $\mathbb{I}$ is the *identity operator*: $(\mathbb{I}h)(x) = h(x)$, $x \in \mathscr{R}^n$, for any function $h \in \mathscr{C}_I(\mathbb{I}) \equiv \mathscr{C}$; and we have $P_{t=0}(B|x) = I(B|x)$ for any $x \in \mathscr{R}^n$, i.e., $\mathbb{P}_0 = \mathbb{I}$.

The product of two transition functions $P_{t_1}(B|x)$ and $P_{t_2}(B|x)$, $t_1 \geq 0$, $t_2 \geq 0$, is

$$(P_{t_1} * P_{t_2})(B|x) \triangleq \int_{\mathscr{R}^n} P_{t_2}(B|y) P_{t_1}(dy|x), \quad x \in \mathscr{R}^n,\ B \in \mathscr{B}^n.$$

By definition, we may prove $(P_{t_1} * P_{t_2})(B|x) = P_{t_1+t_2}(B|x)$, and for any three transition functions, we have $(P_{t_1} * P_{t_2}) * P_{t_3} = P_{t_1} * (P_{t_2} * P_{t_3}) = P_{t_1+t_2+t_3}$. Define $P_t^{*k} \triangleq (P_t^{*(k-1)}) * P_t = P_t * (P_t^{*(k-1)}) = P_{kt}$. In operator forms, we have $(\mathbb{P}_{t_1}\mathbb{P}_{t_2})h(x) = \mathbb{P}_{t_1+t_2}h(x)$, for any function $h \in \mathscr{C}$. We denote it as $\mathbb{P}_{t_1}\mathbb{P}_{t_2} = \mathbb{P}_{t_1+t_2}$.

Next, for any probability measure $\nu(B)$, $B \in \mathscr{B}^n$, we define an operator
$n\boldsymbol{\nu}: \mathscr{C} \to \mathscr{R}$ with

$$n\boldsymbol{\nu} h \triangleq \int_{\mathscr{R}^n} h(y)\nu(dy) \triangleq \nu * h, \quad h \in \mathscr{C}, \tag{5.15}$$

which is the mean of h under measure ν. We have
$n\boldsymbol{\nu} e = 1$.

For any transition operator $\mathbb{P}_t$ and a probability measure ν, By definition, define
$n\boldsymbol{\nu}\mathbb{P}_t : \mathscr{C} \to \mathscr{R}$

$$(n\boldsymbol{\nu}\mathbb{P}_t)h \triangleq n\boldsymbol{\nu}(\mathbb{P}_t h). \tag{5.16}$$

Correspondingly, we define a measure, denoted as $\nu * P$, by

$$(\nu * P_t)(B) \triangleq \int_{\mathscr{R}^n} \nu(dx) P_t(B|x), \qquad B \in \mathscr{B}^n.$$

In many cases, we need to change the orders of limits, expectations, and integrations etc., which are guaranteed under some technical conditions [7]. For simplicity, we will not present them in this paper; instead, we will use the notation $\rightleftharpoons$ to indicate that the order changes are involved in the equality.

5.4.1 The Infinitesimal Generator

An *infinitesimal generator* of a Markov process $\mathbf{X} = \{X(t), t \in [0,\infty)\}$ with transition function $P_t(B|x)$, $B \in \mathscr{B}^n$, $x \in \mathscr{R}^n$, is defined as an operator $\mathbb{A}$:

$$
\begin{aligned}
(\mathbb{A}h)(x) &\triangleq \lim_{\tau\to 0}\frac{1}{\tau}\{E[h(X(\tau))|X(0)=x] - h(x)\} \\
&= \frac{\partial}{\partial\tau}\Big\{E[h(X(\tau))\Big|X(0)=x]\Big\}_{\tau=0} \\
&= \lim_{\tau\to 0}\Big\{\frac{\mathbb{P}_\tau - \mathbb{I}}{\tau}\Big\}h, \qquad h \in \mathscr{C}_I(\mathbb{A}).
\end{aligned} \tag{5.17}
$$

$\mathscr{C}_I(\mathbb{A})$ is a subset of $\mathscr{C}$ for which the limit exists. We may write

$$
\mathbb{A} \triangleq \lim_{\tau\to 0}\frac{\mathbb{P}_\tau - \mathbb{I}}{\tau} \equiv \frac{\partial \mathbb{P}_t}{\partial t}\Big|_{t=0}. \tag{5.18}
$$

By definition, we have

$$
\mathbb{A}e = 0.
$$

From (5.17), we have

$$
\begin{aligned}
&\mathbb{P}_t\mathbb{A}h \\
&= \int_{\mathscr{R}^n} P_t(dz|x)\frac{\partial}{\partial\tau}\Big\{E[h(X(\tau))\Big|X(0)=z]\Big\}_{\tau=0} \\
&\rightleftharpoons \frac{\partial}{\partial\tau}\Big\{\int_{\mathscr{R}^n} h(y)\Big[\int_{\mathscr{R}^n} P_\tau(dy|z)P_t(dz|x)\Big]\Big\}_{\tau=0} && (5.19) \\
&= \frac{\partial}{\partial t}\Big\{E[h(X(t))\Big|X(0)=x]\Big\} && (5.20) \\
&= \frac{\partial}{\partial\tau}\Big\{E[h(X(t+\tau))\Big|X(0)=x]\Big\}_{\tau=0}, && (5.21)
\end{aligned}
$$

in which (5.19) holds because $\mathbb{P}_{t+\tau} = \mathbb{P}_t\mathbb{P}_\tau$. From (5.21), we may write

$$
\mathbb{P}_t\mathbb{A} = \lim_{\tau\to 0}\frac{\mathbb{P}_{t+\tau} - \mathbb{P}_t}{\tau} := \frac{\partial \mathbb{P}_t}{\partial t}. \tag{5.22}
$$

Next, from $\mathbb{P}_t h(x) = E[h(X(t))|X(0)=x]$, we have

$$
\mathbb{P}_t h(X(\tau)) = E[h(X(t+\tau))|X(\tau)].
$$

Thus, replacing h by $\mathbb{P}_t h$ in (5.17), we have

$$
\begin{aligned}
&\mathbb{A}(\mathbb{P}_t h) \\
&= \lim_{\tau\to 0}\frac{1}{\tau}\Big\{E\Big[E[h(X(t+\tau))|X(\tau)]\Big|X(0)=x\Big] \\
&\quad -E[h(X(t))|X(0)=x]\Big\} \\
&= \frac{\partial}{\partial t}\Big\{E[h(X(t))\Big|X(0)=x]\Big\} = \Big(\frac{\partial \mathbb{P}_t}{\partial t}\Big)h.
\end{aligned}
$$

Combining with (5.22), we have the Kolmogorov forward and backward equations:

$$\frac{\partial \mathbb{P}_t}{\partial t} = \mathbb{P}_t \mathbb{A} = \mathbb{A} \mathbb{P}_t. \tag{5.23}$$

5.4.2 The Steady-State Probability

A probability measure $\pi(B)$, $B \in \mathscr{B}^n$, is called a *steady-state probability measure* of $\mathbf{X}$ if its corresponding operator defined via (5.15), denoted as $\boldsymbol{\pi}$, satisfies (cf. (5.16))

$$\boldsymbol{\pi} \mathbb{A} = 0. \tag{5.24}$$

By definition, this means $(\boldsymbol{\pi}\mathbb{A})h = 0$, or $\int_{\mathscr{R}^n} (\mathbb{A}h)(x)\pi(dx) = 0$, for all $h \in \mathscr{C}_I(\boldsymbol{\pi}\mathbb{A}) \subseteq \mathscr{C}$.

A Markov process $\mathbf{X} = \{X(t), t \in [0,\infty)\}$ on $\mathscr{R}^n$ (and its transition function P_t) is said to be (weakly) *ergodic* if there exists a probability measure π on $\mathscr{R}^n$ such that for all $B \in \mathscr{B}^n$ and $x \in \mathscr{R}^n$,

$$\lim_{t\to\infty} P_t(B|x) = e(x)\pi(B). \tag{5.25}$$

If $X(t)$ is ergodic, then for any fixed $x \in \mathscr{R}^n$, we have

$$\begin{aligned} &\lim_{t\to\infty} E[h(X(t))\big|X(0)=x] = \lim_{t\to\infty} \mathbb{P}_t h(x) \\ &\rightleftharpoons \int_{\mathscr{R}^n} h(y) \lim_{t\to\infty} P_t(dy|x) = \Big\{\int_{\mathscr{R}^n} h(y)\pi(dy)\Big\} e(x). \end{aligned} \tag{5.26}$$

Under some conditions, (5.26) holds. Thus, for an ergodic process, we have

$$\lim_{t\to\infty} \mathbb{P}_t \rightleftharpoons e\boldsymbol{\pi}. \tag{5.27}$$

From (5.21) and (5.27), we have for any $h \in \mathscr{C}_I(\mathbb{A})$,

$$(e\boldsymbol{\pi})\mathbb{A}h \rightleftharpoons \lim_{t\to\infty}(\mathbb{P}_t\mathbb{A}h) = \lim_{t\to\infty} \frac{\partial}{\partial t}\Big\{E[h(X(t))\big|X(0)=x]\Big\} = 0.$$

Thus, (5.24) holds and π in (5.25) is indeed the steady-state measure.

5.4.3 The Long-Run Average Performance

To study the sample path average, we denote (cf. (5.10))

$$\mathbb{P}^* \triangleq \lim_{T\to\infty} \frac{1}{T}\int_0^T \mathbb{P}_t dt,$$

which means

$$\mathbb{P}^* h \triangleq \lim_{T\to\infty} \frac{1}{T}\int_0^T (\mathbb{P}_t h) dt, \quad h \in \mathscr{C}. \tag{5.28}$$

We call $\mathbb{P}^*$ the *sample path average operator*, or simply the *average operator*. Define

$$P^*(B|x) \triangleq \lim_{T\to\infty} \frac{1}{T}\int_0^T P_t(B|x)dt, \quad x \in \mathscr{R}^n,\ B \in \mathscr{B}^n. \tag{5.29}$$

Next, we assume that $\lim_{t\to\infty} P_t(B|x)$ exists (not necessary ergodic, i.e., may not equal $e\pi$). Then we have $\lim_{t\to\infty} P_t(B|x) = P^*(B|x)$, $\lim_{t\to\infty} \mathbb{P}_t h = \int_{\mathscr{R}^n} h(y)P^*(dy|x) = \mathbb{P}^* h$, and

$$\mathbb{P}^* h \rightleftharpoons \int_{\mathscr{R}^n} h(y)P^*(dy|x), \qquad h \in \mathscr{C}. \tag{5.30}$$

Also, from (5.28), we have

$$(\mathbb{P}^*\mathbb{A})h = \mathbb{P}^*(\mathbb{A}h) = \lim_{T\to\infty} \frac{1}{T}\int_0^T \mathbb{P}_t(\mathbb{A}h)dt.$$

From (5.20), we have $\lim_{t\to\infty} \mathbb{P}_t(\mathbb{A}h) = 0$. Thus, $\mathbb{P}^*\mathbb{A} = 0$. Finally, from (5.23), we have

$$(\mathbb{A}\mathbb{P}^*) = \mathbb{P}^*\mathbb{A} = 0. \tag{5.31}$$

If P_t is ergodic, then $\mathbb{P}^* = e\boldsymbol{\pi}$, and $\mathbb{P}^*\mathbb{P}^* = \mathbb{P}^*$.

If $h \in \mathscr{C}_I(\mathbb{A})$, $\mathbb{A}h \in \mathscr{C}$, and $\lim_{t\to\infty} \mathbb{P}_t h = \mathbb{P}^* h$, we have the *Dynkin formula* [14]:

$$\lim_{T\to\infty} E\left\{ \int_0^T [\mathbb{A}h(X(\tau))]d\tau \middle| X(0) = x \right\} = (\mathbb{P}^*h)(x) - h(x). \tag{5.32}$$

Let $f(x)$ be a cost function. The long-run average performance is defined as (assuming it exists)

$$\eta(x) \triangleq \lim_{T\to\infty} \frac{1}{T} E\{\int_0^T f(X(t))dt \Big| X(0) = x\}.$$

From (5.28) and (5.30), we have

$$\begin{aligned} \eta(x) &\rightleftharpoons \lim_{T\to\infty} \frac{1}{T}\{\int_0^T (\mathbb{P}_t f)(x)dt\} \\ &= (\mathbb{P}^* f)(x) = \int_{\mathscr{R}^n} f(y)P^*(dy|x). \end{aligned} \tag{5.33}$$

and from (5.33),

$$\mathbb{P}^*\eta = \eta.$$

For ergodic systems, we have

$$\eta(x) = [(e\boldsymbol{\pi})f](x) = (\boldsymbol{\pi}f)e(x),$$

with $\boldsymbol{\pi}f = \int_{\mathscr{R}^n} f(x)\pi(dx)$. We set $\underline{\eta} := \boldsymbol{\pi}f$ be a constant. Then, we have $\eta(x) = \underline{\eta}e(x)$.

5.4.4 Performance Potentials and Difference Formulas

With the infinitesimal generator $\mathbb{A}$, we may define the *Poisson equation*:

$$-\mathbb{A}g(x) + \eta(x) = f(x). \tag{5.34}$$

Any solution $g(x)$ to the Poisson equation is called a *performance potential* function. The solution to the Poisson equation is only up to an additive term, i.e., if $g(x)$ is a solution to (5.34), then so is $g(x) + cr(x)$, with $\mathbb{A}r(x) = 0$, for any constant c. For any solution g, by (5.31), $\mathbb{A}(\mathbb{P}^*g) = 0$. Thus, $g' = g - \mathbb{P}^*g$ is also a solution with $\mathbb{P}^*g' = 0$. Therefore, there is a solution g such that $\mathbb{P}^*g = 0$. Next, from (5.34), we have

$$-\mathbb{P}_t\mathbb{A}g = \mathbb{P}_t[f(x) - \eta(x)].$$

By Dynkin's formula (5.32) and $\mathbb{P}^*g = 0$, we get

$$g(x) = \lim_{T\to\infty}\int_0^T \mathbb{P}_t[f(X(t)) - \eta(X(t))]dt \tag{5.35}$$

$$\rightleftharpoons \lim_{T\to\infty} E\left\{\int_0^T [f(X(t)) - \eta(X(t))]dt \middle| X(0) = x\right\}. \tag{5.36}$$

This is the sample-path based expression for the potentials.

For ergodic processes, we can write the Poisson equation as follows:

$$-\mathbb{A}g(x) + \underline{\eta}e(x) = f(x). \tag{5.37}$$

If $\{g(x), \underline{\eta}\}$ is a solution to (5.37), then we have $\underline{\eta} = \boldsymbol{\pi}f$.

Now, we consider two ergodic Markov processes $\mathbf{X} = \{X(t),\ t \in [0,\infty)\}$ and $\mathbf{X}' = \{X'(t), t \in [0,\infty)\}$ on the same state space $\mathscr{R}^n$. We use superscript "′" to denote the quantities associated with process $\mathbf{X}'$. Thus, $f'(x)$ is the cost function of $\mathbf{X}'$, π' is its steady-state probability measure, $\mathbb{A}'$ is its infinitesimal generator, and $\mathbb{P}'^*$ is its average operator, with $\boldsymbol{\pi}'\mathbb{A}' = 0$. We can easily derive the following *performance difference formula*

$$\underline{\eta}' - \underline{\eta} = \boldsymbol{\pi}'\{(f' + \mathbb{A}'g) - (f + \mathbb{A}g)\}. \tag{5.38}$$

Proof. Left-multiplying both sides of the Poisson equation (5.37) with $\boldsymbol{\pi}'$, we get

$$-\boldsymbol{\pi}'(\mathbb{A}g) + \underline{\eta} = \boldsymbol{\pi}'f.$$

Therefore,

$$\begin{aligned}\underline{\eta}' - \underline{\eta} &= \boldsymbol{\pi}'f' - \underline{\eta} = (\boldsymbol{\pi}'f - \underline{\eta}) + \boldsymbol{\pi}'(f' - f)\\ &= \{(\boldsymbol{\pi}'\mathbb{A}')g - \boldsymbol{\pi}'(\mathbb{A}g)\} + \boldsymbol{\pi}'(f' - f)\\ &= \boldsymbol{\pi}'\{(f' + \mathbb{A}'g) - (f + \mathbb{A}g)\},\end{aligned}$$

in which we used $\pi'\mathbb{A}' = 0$. □

Equation (5.38) keeps the same form if g is replaced by $g + cr$ with $\mathbb{A}r = 0$.

5.4.5 Policy Iteration and Optimality Conditions

With the performance difference formula, we may develop the policy iteration and optimization theory for CTCS systems by simply translating the corresponding results for the discrete-time case discussed in Section 5.2.

First, we modify the definition of the relations $=$, $\leq$, $<$, and $\preceq$ for two functions on $\mathscr{R}^n$. Given a probability measure ν on $\mathscr{R}^n$, for two functions $h(x)$ and $h'(x)$, $x \in \mathscr{R}^n$, we define $h' =_\nu h$, $h' \leq_\nu h$, and $h' <_\nu h$, respectively, if $h'(x) = h(x)$, $h'(x) \leq h(x)$, and $h'(x) < h(x)$, respectively, for all $x \in \mathscr{R}^n$ except on a set H with $\nu(H) = 0$. We further define $h'(x) \preceq_\nu h(x)$ if $h'(x) \leq_\nu h(x)$ and $h'(x) < h(x)$ on a set H with $\nu(H) > 0$. Similar definitions are used for the relations $>_\nu$, $\succeq_\nu$, and $\geq_\nu$.

Let $(\mathbb{A}, f)$, and $(\mathbb{A}', f')$ be the infinitesimal generators and cost functions of two ergodic Markov processes with the same state space $\mathscr{S} = \mathscr{R}^n$, and η, g, π and η', g', π' be their corresponding long-run average performance functions, performance potential functions, and steady-state probability measures, respectively. The following lemma follows directly from (5.38).

Lemma 5.4.1.

a) If $f' + \mathbb{A}'g \preceq_{\pi'} f + \mathbb{A}g$ (or $f' + \mathbb{A}'g \succeq_{\pi'} f + \mathbb{A}g$), then $\underline{\eta}' < \underline{\eta}$ (or $\underline{\eta}' > \underline{\eta}$).
b) If $f' + \mathbb{A}'g \leq_{\pi'} f + \mathbb{A}g$ (or $f' + \mathbb{A}'g \geq_{\pi'} f + \mathbb{A}g$), then $\underline{\eta}' \leq \underline{\eta}$ (or $\underline{\eta}' \geq \underline{\eta}$).

The difficulty in verifying the condition $\preceq_{\pi'}$ (or $\succeq_{\pi'}$) lies in the fact that we may not know π', so we may not know which sets have positive measures under π'. Fortunately, in many cases (e.g., for diffusion processes) we can show that $\pi'(B) > 0$ if and only if B is a subset of $\mathscr{R}^n$ with a positive Lebesgue measure.

In a control problem, when the system state is $x \in \mathscr{R}^n$, we may take an *action*, denoted as $u(x)$, which determines the infinitesimal generator at x, $\mathbb{A}_x^{u(x)}$, and the cost $f(x, u(x))$, at x. The function $u(x)$, $x \in \mathscr{R}^n$, is called a *policy*. We may also refer to a pair $(\mathbb{A}^u, f^u)$ as a policy, where $(\mathbb{A}h)(x) = \mathbb{A}_x^{u(x)} h$ and $f^u(x) = f(x, u(x))$. A policy is said to be ergodic, if the Markov process it generates is ergodic. We use superscript $*^u$ to denote the quantities associated with policy u; e.g., π^u and η^u are the steady-state probability measure and long-run average performance of policy u, respectively. The goal is to find a policy $\widehat{u} \in U$ with the best performance $\eta^{\widehat{u}} = \min_{u \in U} \eta^u$, where U denotes the policy space.

Theorem 5.4.1. *Suppose that for a Markov system all the policies are ergodic. A policy $\widehat{u}(x)$ is optimal, if and only if*

$$f^{\widehat{u}} + \mathbb{A}^{\widehat{u}} g^{\widehat{u}} \leq_{\pi^u} f^u + \mathbb{A}^u g^{\widehat{u}}, \tag{5.39}$$

for all policies u.

Note that in the theorem, we use the assumption that $\mathbb{A}h(x)$ depends only the on action taken at x, $u(x)$, which can be chosen independently to each other.

We say that two policies u and u' have the same *support* if for any set $B \in \mathscr{B}^n$, $\pi^u(B) > 0$ if and only if $\pi^{u'}(B) > 0$ (i.e., π^u and $\pi^{u'}$ are equivalent). We assume that all the policies in the policy space have the same support. Because in many problems with continuous state spaces, $\pi^u(B) > 0$ if B is a subset of $\mathscr{S}$ with a positive Lebesgue measure, the assumption essentially requires that $\mathscr{S}$ is the same for all policies, except for a set with a zero Lebesgue measure. In control problems and in particular in financial applications, the noise is usually the Brownian motion, which is supported by the entire state space $\mathscr{R}^n$, then $\mathscr{S} = \mathscr{R}^n$ and the assumption holds.

If all the policies have the same support, we may drop the subscript π^u in the relationship notations such as $\leq$ and $\preceq$, etc, and we may understand them as under the Lebesgue measure and say that the relations hold almost everywhere (a.e.).

Theorem 5.4.2. *Suppose that for a Markov system all the policies are ergodic and have the same support. A policy $\widehat{u}(x)$ is optimal, if and only if*

$$f^{\widehat{u}} + \mathbb{A}^{\widehat{u}} g^{\widehat{u}} \leq f^u + \mathbb{A}^u g^{\widehat{u}}, \qquad a.e., \tag{5.40}$$

for all policies u.

From Theorem 5.4.2, policy $\widehat{u}$ is optimal if and only if the optimality equation

$$\min_{u \in U}\{\mathbb{A}^u g^{\widehat{u}} + f^u\} = \mathbb{A}^{\widehat{u}} g^{\widehat{u}} + f^{\widehat{u}} = \eta^{\widehat{u}}. \tag{5.41}$$

holds a.e. We assume that the policy space is, in a sense, compact and the functions have some sort of continuity and so the minimum can be reached.

With the performance difference formula (5.38), policy iteration algorithms can be designed. Roughly speaking, we may start with any policy u_0. At the kth step with policy u_k, $k = 0, 1, \cdots$, we set

$$u_{k+1}(x) = \arg\{\min_{u \in U}[\mathbb{A}^u g^{u_k}(x) + f^u(x)]\},$$

with g^{u_k} being the potential function of $(\mathbb{A}^{u_k}, f^{u_k})$. If at some x, $u_k(x)$ attains the maximum, we set $u_{k+1}(x) = u_k(x)$. The iteration stops if u_{k+1} and u_k differ only on a set with a zero Lebesgue measure. Denote $\eta_k = \eta^{u_k}$. When the iteration stops, we have $\eta_{k+1} = \eta_k$. Lemma 5.4.1 implies that performance improves at each step. Theorem 5.4.2 shows that the minimum is reached when no performance improvement can be further achieved. If the policy space is finite, the policy iteration will stop in a finite number of steps. However, if the action space is not finite, the iteration scheme may not stop at a finite number of steps, although the sequence of the performance η_k is increasing and hence converges. We may prove that under some conditions the iteration does stop (see, e.g., [11]).

In control problems, we apply a feedback control law $u(x) = (u_\alpha(x), u_\sigma(x), u_\gamma(x)) \in U$ to a stochastic system; its state process $X(t)$ is described as a controlled Levy process

$$\begin{aligned} dX(t) &= \alpha(X(t), u_\alpha[X(t)])dt + \sigma(X(t), u_\sigma[X(t)])dW(t) \\ &\quad + \int_{\mathscr{R}^l} \gamma(X(t-), u_\gamma[X(t-)], z)N(dt, dz), \end{aligned}$$

in which $X(t) \in \mathscr{R}^n$ is the state process, $W(t) \in \mathscr{R}^m$ is a Browian motion, $N(t,z)$ denotes an l-dimensional jump process, and α, σ, and γ represent three coefficient matrices with the proper dimensions. The probability that $N_j(dt,dz_j)$ jumps in $[t,t+dt)$ with a size in $[z_j,z_j+dz_j)$ is $\nu_j(dz_j)dt$. At time t, with probability $\nu_j(dz_j)dt$, the process $X(t)$ jumps from $X(t-)=x$ to $X(t)=x+\gamma^{(j)}(x,u_\gamma(x),z)$.

Let $\mathbb{A}^u$ be the infinitesimal generator of $X(t)$ with control law $u(x)$. For any function h with continuous second order derivatives, we have [7, 14]

$$\mathbb{A}^u h(x) = \sum_{i=1}^{n} \alpha_i(x,u_\alpha(x))\frac{\partial h}{\partial x_i}(x) + \frac{1}{2}\sum_{i,j=1}^{n}(\sigma\sigma^T)_{ij}(x,u_\sigma(x))\frac{\partial^2 h}{\partial x_i \partial x_j}(x)$$
$$+ \int_{\mathscr{R}} \sum_{j=1}^{l}\{h(x+\gamma^{(j)}(x,u_\gamma(x),z)) - h(x)\}\nu_j(dz_j). \qquad (5.42)$$

Therefore, the HJB equations for the performance potentials of the optimal policy $\widehat{u}$ is Equation (5.41), with $\mathbb{A}^u$ specified by (5.42). Although $\mathbb{A}^u$ contains differentials, the HJB equations is required to hold only almost everywhere; i.e., we may allow the potential (value) function $g^{\widehat{u}}$ not differential at a set of zero Lebesgue measure. In such cases, the concept of viscosity solution is not needed.

It has been verified that the same approach works well for control problems with finite-horizon and discounted criteria. This indeed provides a unified approach to these problems and discounting is not needed for problems with long-run average performance.

5.5 Impulse Control

The impulse stochastic control is motivated by the portfolio management problem in which one has to determine when to buy or sell which stock in order to obtain the maximum profit. Let us model the stock values as an n-dimensional Levy process

$$dX(t) = \alpha(X(t))dt + \sigma(X(t))dW(t) + \int_{\mathscr{R}^l} \gamma(X(t-),z)N(dt,dz).$$

The standard way of modeling the control actions (selling and buying) as n dimensional jump (cadlag) processes $L(t) = (L_1(t),\cdots,L_n(t))^T$ (for buying) and $M(t) = (M_1(t),\cdots,M_n(t))^T$ (for selling). The stochastic process with the controls is

$$dX(t) = \alpha(X(t))dt + \sigma(X(t))dW(t) + \int_{\mathscr{R}^l} \gamma(X(t-),z)N(dt,dz) + dL(t) - dM(t).$$

The goal is to determine the jump instants and the jump heights to obtain the maximum profit (e.g., average growth rate).

The standard approach is dynamic programming, which requires viscosity solution and other deep mathematics [1]. We can show that we may simply apply the direct comparison approach to obtain the HJB equation; the approach is simple and intuitive.

To apply the direct comparison approach, we first propose a composite model for Markov processes [8]. The state space of a composite Markov process consists of two parts, J and $\overline{J}$. When the process is in $\overline{J}$, it evolves like a continuous-time Levy process; and once the process enters J, it makes a jump instantly according to a transition function like a direct-time Markov chain. The composite Markov process provides a new model for impulse stochastic control problem, with the instant jumps in J modeling the impulse control feature (e.g., selling or buying stocks in the portfolio management problem).

With this model, we may develop a direct-comparison based approach to the impulse stochastic control problem. The derivation and results look simpler than dynamic programming [2] and enjoys the other advantages as the direct-comparison approach. In particular, this work puts the impulse stochastic control problem in the same framework as the other research areas in control and optimization, and therefore, stimulates new research directions.

5.6 New Approaches

So far, we have assumed the Markov property for the systems to be controlled. It is well known that the Markov model suffers from the following disadvantages:

1. The state space and the policy space are too large for most problems.
2. The MDP theory requires that the actions taken at different states can be chosen independently.
3. The model does not utilize any special feature of the system.

As we discussed, the essential feature used in the direct comparison approach is the decomposition nature of the difference formula (5.3) and (5.38). Under some conditions, this decomposition may hold without the Markov property. A new formulation, called the *event-based optimization* [5], is developed along this direction for DTDS systems. In the event-based approach, actions depend on events, rather than on states. The events are defined as a set of state transitions. It is shown that under some conditions the difference formula for the performance of two event-based policies also enjoys the decomposition property as in (5.3). Therefore, optimality equations and policy iteration can be derived for event-based optimization. Events capture the special features of the system structure. Because the number of events is usually much smaller than that of states, the computation is reduced.

The direct comparison approach links the stochastic control problem to other optimization approaches in DTDS systems. Therefore, it is natural to expect that methods similar to those in areas such as PA (cf. (5.8)) and RL can be developed for stochastic control. These methods may provide numerical solutions to the stochastic control problem.

Furthermore, recently, we are applying the direct comparison approach to multi-objective optimization problems such as the gain-risk management, the efficient frontiers can be obtained in a simple and intuitive way.

5.7 Discussion and Conclusion

In this paper, we reviewed the main ideas and results in the direct comparison approach to stochastic control. The work is a part of our effort in developing a sensitivity-based unified approach to the area of control and optimization of stochastic systems [5]. The sensitivity-based approach is based on a simple philosophic-like view: the most fundamental action in optimization is a direct comparison of the performance of any two policies. In other words, in general, whether we may develop efficient optimization methods for a particular problem relies on the structure of the performance difference of any two policies. We have verified this philosophical view in many problems.

References

1. Akian, M., Sulem, A., Taksar, M.: Dynamic Optimization of Long Term Growth Rate for a Portfolio with Transaction Costs and Logarithmic Utility. Mathematical Finance 11, 153–188 (2001)
2. Bertsekas, D.P.: Dynamic Programming and Optimal Control, vol. I, II. Athena Scientific, Belmont (2007)
3. Brockett, R.: Stochastic Control. Preprint (2009)
4. Cao, X.R.: Realization Probabilities - The Dynamics of Queueing Systems. Springer, New York (1994)
5. Cao, X.R.: Stochastic Learning and Optimization - a Sensitivity-Based Approach. Springer, Heidelberg (2007)
6. Cao, X.R., Zhang, J.: The nth-Order Bias Optimality for Multi-chain Markov Decision Processes. IEEE Transactions on Automatic Control 53, 496–508 (2008)
7. Cao, X.R.: Stochastic Control via Direct Comparison. Submitted to IEEE Transaction on Automatic Control (2009)
8. Cao, X.R.: Singular Stochastic Control and Composite Markov Processes. Manuscript to be submitted (2009)
9. Fleming, W.H., Soner, H.M.: Controlled Markov Processes and Viscosity Solutions, 2nd edn. Springer, Heidelberg (2006)
10. Ho, Y.C., Cao, X.R.: Perturbation Analysis of Discrete-Event Dynamic Systems. Kluwer Academic Publisher, Boston (1991)
11. Meyn, S.P.: The Policy Iteration Algorithm for Average Reward Markov Decision Processes with General State Space. IEEE Transactions on Automatic Control 42, 1663–1680 (1997)
12. Muthuraman, K., Zha, H.: Simulation-based portfolio optimization for large portfolios with transaction costs. Mathematical Finance 18, 115–134 (2008)
13. Puterman, M.L.: Markov Decision Processes: Discrete Stochastic Dynamic Programming. Wiley, Chichester (1994)
14. Oksendal, B., Sulem, A.: Applied Stochastic Control of Jump Diffusions. Springer, Heidelberg (2007)
15. Sutton, R.S., Barto, A.G.: Reinforcement Learning: An Introduction. MIT Press, Cambridge (1998)
16. Veinott, A.F.: Discrete Dynamic Programming with Sensitive Discount Optimality Criteria. The Annals of Mathematical Statistics 40(5), 1635–1660 (1969)

6

A Maximum Entropy Solution of the Covariance Selection Problem for Reciprocal Processes

Francesca Carli[1], Augusto Ferrante[1], Michele Pavon[2], and Giorgio Picci[1]

[1] Department of Information Engineering, University of Padova, via Gradenigo 6/B, Padova, Italy

[2] Department of Pure and Applied Mathematics, University of Padova, Italy

Summary. Stationary *reciprocal processes* defined on a finite interval of the integer line can be seen as a special class of Markov random fields restricted to one dimension. Non stationary reciprocal processes have been extensively studied in the past especially by Krener, Levy, Frezza and co-workers. However the specialization of the non-stationary theory to the stationary case does not seem to have been pursued in sufficient depth in the litarature. Stationary reciprocal processes (and reciprocal stochastic models) are potentially useful for describing signals which naturally live in a finite region of the time (or space) line and estimation or identification of these models starting from observed data is a completely open problem which can in principle lead to many interesting applications in signal and image processing. In this paper we discuss the analog of the covariance extension problem for stationary reciprocal processes which is motivated by maximum likelihood identification. As in the usual stationary setting on the integer line, the covariance extension problem is a basic conceptual and practical step in solving the identification problem. We show that the maximum entropy principle leads to a complete solution of the problem.

6.1 Introduction: Stationary Reciprocal Processes

For an introduction to circulant matrices we refer the reader to the monograph [5]. Here we shall just recall the definition. A *block-circulant* matrix with N blocks, is a finite block-Toeplitz matrix whose entries are permuted cyclically. It looks like

$$\mathbf{M}_N = \begin{bmatrix} M_0 & M_{N-1} & \dots & \dots & M_1 \\ M_1 & M_0 & M_{N-1} & \dots & \dots \\ \vdots & & \ddots & & \vdots \\ \vdots & & & \ddots & M_{N-1} \\ M_{N-1} & M_{N-2} & \dots & M_1 & M_0 \end{bmatrix}.$$

where $M_k \in \mathbb{R}^{m\times m}$ say. It will be denoted $\mathbf{M}_N = \text{Circ}\{M_0, M_1, \dots, M_{N-1}\}$. Nonsingular block circulant matrices of a fixed size form a group. These matrices play an important role in the second-order description of *stationary processes defined on a finite interval.*

X. Hu et al. (Eds.): Three Decades of Progress in Control Sciences, pp. 77–93, 2010.

A m-dimensional stochastic process on a finite interval $[1,N]$, is just an ordered collection of (zero-mean) random m-vectors $\mathbf{y} := \{\mathbf{y}(k),\, k = 1,2,\dots,N\}$ which will be written as a column vector with N, m-dimensional components. We shall say that $\mathbf{y}$ is *stationary* if the covariances $\mathbb{E}\mathbf{y}(k)\mathbf{y}(j)^\top$ depend only on the difference of the arguments, namely

$$\mathbb{E}\mathbf{y}(k)\mathbf{y}(j)^\top = R_{k-j}, \qquad k,j = 1,\dots,N$$

in which case the covariance matrix of $\mathbf{y}$ has a symmetric block-Toeplitz structure; i.e.

$$\mathbf{R}_N := \mathbb{E}\mathbf{y}\mathbf{y}^\top = \begin{bmatrix} R_0 & R_1^\top & \dots & R_{N-1}^\top \\ R_1 & R_0 & R_1^\top & \dots \\ \dots & & & \dots \\ R_{N-1} & \dots & R_1 & R_0 \end{bmatrix}$$

Processes $\mathbf{y}$ which have a positive definite covariance $\mathbf{R}_N$ are called of *full rank* (or *minimal*). The processes that we shall deal with in this paper will normally be of full rank.

Now let us consider a process $\mathbf{y}$ on the integer line $\mathbb{Z}$ which is periodic of period N; i.e. a process satisfying $\mathbf{y}(k+nN) := \mathbf{y}(k)$ (almost surely) for arbitrary $n \in \mathbb{Z}$. In particular, $\mathbf{y}(0) = \mathbf{y}(N), \mathbf{y}(-1) = \mathbf{y}(N-1), \dots$ etc. We can think of $\mathbf{y}$ as a process on the *discrete group* $\mathbb{Z}_N \equiv \{1,2,\dots,N\}$ with arithmetics mod N. Clearly its covariance function[1] must also be periodic of period N; i.e. $R(k+N) = R(k)$ for arbitrary $k \in \mathbb{Z}$. Hence we may also consider the covariance sequence as a function on the discrete group $\mathbb{Z}_N \equiv [0, N-1]$ with arithmetics mod N. In particular we have $R(N) = R(0)$ etc. But more must be true. Just to fix the ideas assume that N is an even number and consider the midpoint $k = N/2$ of the interval $[1,N]$; for $\tau = 0,1,\dots,N/2$ we have $R(N/2+\tau) = \mathbb{E}\mathbf{y}(t+\tau+N/2)\mathbf{y}(t+N)^\top = R(N/2-\tau)^\top$ which we describe by saying that the covariance function must be *symmetric with respect to the mid-point* $\tau = N/2$ *of the interval*. In particular, for $\tau = N/2-1, N/2-2,\dots,0$, it must happen that

$$\begin{aligned} R(N-1) &= R(N/2+N/2-1) = R(N/2-N/2+1)^\top = R(1)^\top \\ R(N-2) &= R(N/2+N/2-2) = R(N/2-N/2+2)^\top = R(2)^\top \\ &\dots = \dots \text{ etc.} \end{aligned}$$

Hence the $mN \times mN$ covariance matrix of a periodic process of period N must be a **symmetric block circulant** matrix with N blocks; i.e. of the form

[1] For typographical reasons we shall occasionally switch notation from R_k to $R(k)$.

$$\begin{bmatrix} R_0 & R_1^\top & \dots & R_\tau^\top & \dots & R_\tau & \dots & R_1 \\ R_1 & R_0 & R_1^\top & \ddots & R_\tau^\top & \dots & \ddots & \vdots \\ \vdots & & \ddots & \ddots & & \ddots & & R_\tau \\ R_\tau & \dots & R_1 & R_0 & R_1^\top & \dots & \ddots & \\ \vdots & R_\tau & \dots & & R_0 & & \dots & R_\tau^\top \\ R_\tau^\top & & \ddots & & & & & \vdots \\ \vdots & \ddots & & \ddots & & \ddots & \ddots & R_1^\top \\ R_1^\top & \dots & R_\tau^\top & \dots & R_\tau & & R_1 & R_0 \end{bmatrix}$$

that is,

$$\mathbf{R}_N = \mathrm{Circ}\{R_0, R_1, \dots, R_\tau, \dots, R_{N/2}, \dots, R_\tau^\top, \dots, R_1^\top\}, \tag{6.1}$$

with the proviso that, for N odd (contrary to what we have assumed so far), $R_{(N+1)/2} = R_{(N-1)/2}^\top$. One can easily derive the following characterization.

Proposition 6.1.1. *A stationary process* $\mathbf{y}$ *on the interval* $[1,N]$ *is the restriction to* $[1,N]$ *of a stationary process on* $\mathbb{Z}$ *which is periodic of period N, if and only if its covariance matrix is a symmetric block-circulant matrix.*

When all the middle entries between R_τ and $R_\tau^\top$ in the listing (6.1) are zero, $\mathbf{R}_N$ is called a **banded block circulant** of bandwidth τ. Such a matrix has the following structure

$$\begin{bmatrix} R_0 & R_1^\top & \dots & R_\tau^\top & 0 & \dots & 0 & R_\tau & \dots & R_1 \\ R_1 & R_0 & R_1^\top & \ddots & R_\tau^\top & 0 & & 0 & \ddots & \vdots \\ \vdots & & \ddots & & & \ddots & & \vdots & & R_\tau \\ R_\tau & \dots & R_1 & R_0 & R_1^\top & \dots & R_\tau^\top & \ddots & & 0 \\ 0 & R_\tau & & \dots & R_0 & \dots & & R_\tau^\top & & \vdots \\ \vdots & & & \dots & & \ddots & & & & 0 \\ 0 & & & \dots & & \dots & & & & R_\tau^\top \\ R_\tau^\top & & \ddots & & & & & & & \vdots \\ \vdots & \ddots & & \ddots & & \ddots & & & \ddots & R_1^\top \\ R_1^\top & \dots & R_\tau^\top & 0 & \dots & 0 & R_\tau & \dots & R_1 & R_0 \end{bmatrix} \tag{6.2}$$

6.2 Reciprocal Processes

In this section we shall describe a class of stationary processes which are a natural generalization of the reciprocal processes introduced in [13] and discussed in [12], [16]. See also [9]. In a sense they are an acausal "symmetric" generalization of AR processes.

Definition 6.2.1. *Let $N > 2n$. A (stationary) reciprocal process of index n on $[1,N]$, is a zero-mean m-dimensional process $\mathbf{y}$ which can be described by a linear model of the following form*

$$\sum_{k=-n}^{n} F_k \mathbf{y}(t-k) = \mathbf{d}(t), \qquad t \in [1,N] \tag{6.3}$$

where the F_k's are $m \times m$ matrices with F_0 normalized to the identity ($F_0 = I$) and

1. *the model is associated to the cyclic boundary conditions:*

$$\begin{aligned} \mathbf{y}(-k) = \mathbf{y}(N-k); \qquad k = 0,1,\ldots,n-1 \\ \mathbf{y}(N+k) = \mathbf{y}(k); \qquad k = 1,2,\ldots,n. \end{aligned} \tag{6.4}$$

2. *The process $\{\mathbf{d}(t)\}$ is stationary finitely correlated of bandwidth n; i.e.*[2]

$$\mathbb{E}\mathbf{d}(t)\mathbf{d}(s)^\top = 0 \quad \text{for} \quad |t-s| \geq n, \qquad t,s \in [1,N] \tag{6.5}$$

   and has positive definite variance matrix $\mathbb{E}\mathbf{d}(t)\mathbf{d}(t)^\top := \Delta > 0$.
3. *The following orthogonality condition holds*

$$\mathbb{E}\mathbf{y}(t)\mathbf{d}(s)^\top = \Delta\,\delta(t-s), \qquad t \neq s \in [1,N], \tag{6.6}$$

   where δ is the Kronecker function.

Example: for $n = 1$ the process is just called *reciprocal* in the literature; in this case there are only two cyclic boundary conditions: $\mathbf{y}(0) = \mathbf{y}(N)$ and $\mathbf{y}(N+1) = \mathbf{y}(1)$.

Because of condition (6.6) the sum of the two terms in the second member of the relation

$$\mathbf{y}(t) = -\sum_{k=-n,\,k\neq 0}^{n} F_k \mathbf{y}(t-k) + \mathbf{d}(t), \qquad t \in [1,N] \tag{6.7}$$

is an orthogonal sum. Hence $\mathbf{d}(t)$ has the interpretation of estimation error of $\mathbf{y}(t)$ given the complementary history of the process, namely

$$\mathbf{d}(t) = \mathbf{y}(t) - \mathbb{E}\left[\mathbf{y}(t) \mid \mathbf{y}(s)\, s \neq t\right].$$

In the same spirit of Masani's definition [14], $\mathbf{d}$ is called the (unnormalized) *conjugate process* of $\mathbf{y}$.

Let $\mathbf{y}$ denote the mN-dimensional vector obtained by stacking the random vectors $\{\mathbf{y}(1), \ldots, \mathbf{y}(N)\}$ in the sequence. Introducing the N-block circulant matrix of bandwidth n,

$$\mathbf{F}_N := \text{Circ}\{I\; F_1\; \ldots\; F_n\; 0 \ldots 0\; F_{-n}\; \ldots\; F_{-1}\}, \tag{6.8}$$

and given a finitely correlated process $\mathbf{d}$ as in condition 1) above, the model (6.3) with the boundary conditions (6.4) can be written in matrix form as

[2] This, as we shall see later, is equivalent to $\mathbf{d}$ admitting a representation by a Moving Average (M.A.) model of order n.

$$\mathbf{F}_N \mathbf{y} = \mathbf{d}. \tag{6.9}$$

From this, multiplying both members from the right by $\mathbf{y}^\top$ and taking expectations we get

$$\mathbf{F}_N \mathbf{R}_N = \mathbf{F}_N \mathbb{E}\mathbf{y}\mathbf{y}^\top = \mathbb{E}\mathbf{d}\mathbf{y}^\top = \text{diag}\{\Delta, \dots, \Delta\} \tag{6.10}$$

in virtue of the orthogonality relation (6.6). Note that our assumption that $\Delta > 0$ (strictly positive definite) implies that $\mathbf{F}_N$, and hence $\mathbf{R}_N$ are invertible and hence the process $\mathbf{y}$ must be of full rank. In fact, the model (6.3) with the boundary conditions (6.4), defines uniquely the vector $\mathbf{y}$ as a solution of the linear equation (6.9).

Solving (6.10) we can express the inverse as

$$\mathbf{R}_N^{-1} = \text{diag}\{\Delta^{-1}, \dots, \Delta^{-1}\}\mathbf{F}_N := \mathbf{M}_N. \tag{6.11}$$

so that ($\mathbf{F}_N$ and) $\mathbf{M}_N$ is nonsingular and positive definite. If we normalize the conjugate process by setting $\mathbf{e}(t) := \Delta^{-1}\mathbf{d}(t)$ so that $\text{Var}\,\mathbf{e}(t) = \Delta^{-1}$, the model (6.3) can be rewritten

$$\sum_{k=-n}^{n} M_k \mathbf{y}(t-k) = \mathbf{e}(t), \qquad t \in \mathbb{Z}_N \tag{6.12}$$

for which the orthogonality relation (6.6) is replaced by

$$\mathbb{E}\mathbf{y}\mathbf{e}^\top = I. \tag{6.13}$$

Definition 6.2.2. *We shall say that the model* (6.3) *is* self-adjoint *if*

$$\Delta^{-1}F_{-k} = [\Delta^{-1}F_k]^\top \qquad k = 1, 2, \dots, n \tag{6.14}$$

equivalently, $M_k := \Delta^{-1}F_k$, $k = -n, \dots, n$ must form a center-symmetric sequence; i.e.

$$M_{-k} = M_k^\top, \qquad k = 1, \dots, n. \tag{6.15}$$

Hence a reciprocal model is self-adjoint if and only if $\mathbf{M}_N$ is a *symmetric* positive definite block-circulant matrix, banded of bandwith n with $M_0 = \Delta^{-1}$. Note that by convention the transposes are coefficients of "future" samples and lie immediately above the main diagonal. From this we obtain the following fundamental characterization of reciprocal processes on the discrete group $\mathbb{Z}_N$.

Theorem 6.2.1. *A nonsingular $mN \times mN$-dimensional matrix $\mathbf{R}_N$ is the covariance matrix of a reciprocal process of index n on the discrete group $\mathbb{Z}_N$ if and only if its inverse is a positive-definite symmetric block-circulant matrix which is banded of bandwidth n.*

Proof. That the condition is necessary follows from the discussion above and Proposition 6.1.1. Conversely, assume that $\mathbf{M}_N := \mathbf{R}_N^{-1}$ has the properties of the theorem. Pick a finitely correlated process $\mathbf{e}$ with covariance matrix $\mathbf{M}_N$ (we can construct such a, say Gaussian, process on a suitable probability space) and define $\mathbf{y}$ by the equation (6.12) with boundary conditions (6.4). Then $\mathbf{y}$ is uniquely defined on the interval $[1, N]$ by the equation $\mathbf{M}_N\mathbf{y} = \mathbf{e}$. The covariance of $\mathbf{y}$ is in fact $\mathbf{R}_N$ since

$$\mathbf{M}_N \mathbb{E}\mathbf{y}\mathbf{e}^\top = \mathbb{E}\mathbf{e}\mathbf{e}^\top = \mathbf{M}_N$$

and hence $\mathbb{E}\mathbf{y}\mathbf{e}^\top = \mathbf{I}_N$ which in turn implies $\mathbf{M}_N \mathbb{E}\mathbf{y}\mathbf{y}^\top = \mathbb{E}\mathbf{e}\mathbf{y}^\top = \mathbf{I}_N$. Hence $\mathbf{y}$ is reciprocal of index n. Since $\mathbf{e}$ has a symmetric block-circulant covariance matrix, it can be seen as is the restriction of a periodic process to the interval $[1, N]$ (Proposition 6.1.1) and since the covarince of $\mathbf{y}$ has the same properties, the same must be true for $\mathbf{y}$. Because of this property process $\mathbf{y}$ can equivalently be imagined as being defined on $\mathbb{Z}_N$. □

From now on we shall consider only self-adjoint models so that reciprocal processes may automatically be imagined as being defined on the discrete unit circle. Note that the whole model is captured by the matrix $\mathbf{M}_N$. For, rewriting (6.12) in vector form as $\mathbf{e} = \mathbf{M}_N\mathbf{y}$ and multiplying from the right by $\mathbf{e}^\top$ and using (6.13) we obtain

$$\mathrm{Var}\{\mathbf{e}\} = \mathbf{M}_N\mathbf{R}_N\mathbf{M}_N^\top = \mathbf{M}_N^\top = \mathbf{M}_N$$

so that the matrix $\mathbf{M}_N$ is in fact the covariance matrix of the normalized conjugate process $\mathbf{e}$. Hence the second order statistics of both $\mathbf{y}$ and $\mathbf{e}$ are encapsulated in the covariance $\mathbf{M}_N$. Note also that this result makes the stochastic realization problem for reciprocal processes of index n conceptually trivial. In fact given the covariance matrix $\mathbf{R}_N$ (the external description of the process), assuming it is in fact the covariance matrix of such a process, the model matrix $\mathbf{M}_N$ can be computed by simply inverting $\mathbf{R}_N$. This is the simplest answer one could hope for. This observation in turn leads to the following

Problem. *Characterize the covariance matrix of a reciprocal process of index n. In other words, when does a (full rank) symmetric block-circulant covariance matrix have a symmetric banded block-circulant inverse of bandwidth n.*

We note that a full rank reciprocal process of index n can always be represented as a linear memoryless function of a reciprocal process of index 1. This reciprocal process will however not have full rank in general. To see that this is the case introduce the vectors

$$\mathbf{y}_t^+ := \begin{bmatrix} \mathbf{y}(t) \\ \vdots \\ \mathbf{y}(t+n-1) \end{bmatrix}, \quad \mathbf{y}_t^- := \begin{bmatrix} \mathbf{y}(t-n+1) \\ \vdots \\ \mathbf{y}(t) \end{bmatrix}, \tag{6.16}$$

and letting $\mathbf{x}(t)^\top := [(\mathbf{y}_t^-)^\top \ (\mathbf{y}_t^+)^\top]$ we find the representation

$$\mathbf{x}(t) = \begin{bmatrix} F_- & 0 \\ 0 & 0 \end{bmatrix} \mathbf{x}(t-1) + \begin{bmatrix} 0 & 0 \\ 0 & F_+ \end{bmatrix} \mathbf{x}(t+1) + \tilde{\mathbf{d}}(t) \tag{6.17}$$

$$\mathbf{y}(t) = [0 \dots 0 \ 1/2 \ 1/2 \ 0 \dots 0] \, \mathbf{x}(t) \tag{6.18}$$

where F_- and F_+ are block-companion matrices and $\tilde{\mathbf{d}}(t) := [0 \dots 0 \mathbf{d}(t)^\top \mathbf{d}(t)^\top 0 \dots 0]^\top$ has a singular covariance matrix. This model is in general non-minimal [16].

6.3 Identification

Assume that T independent samples of the process $\mathbf{y}$ are available[3] and let us denote the sample values by $\underline{y} := \left(y^{(1)}, .., y^{(T)}\right)$. We want to solve the following

Problem. *Given the observations $\underline{y}$ of a reciprocal process* $\mathbf{y}$ *of (known) index n, estimate the parameters $\{M_k\}$ of the underlying reciprocal model $\mathbf{M}_N\mathbf{y} = \mathbf{e}$.*

In an attempt to get asymptotically efficient estimates we shall consider maximum likelihood estimation. Under the assumption of a Gaussian distribution for $\mathbf{y}$, the density can be parametrized by the model parameters $(M_0, \dots, M_n)$ as

$$p_{(M_0,\dots,M_n)}(y) = \frac{1}{\sqrt{(2\pi)^N \det\left(\mathbf{M}_N^{-1}\right)}} \exp\left(-\frac{1}{2} y^\top \mathbf{M}_N y\right),$$

where $y \in \mathbb{R}^{mN}$. Taking logarithms and neglecting terms which do not depend on the parameters, one can rewrite this expression as

$$\log p_{(M_0,\dots,M_n)}(y) = \log\det\left(\mathbf{M}_N^{-1}\right) - \frac{1}{2}\,\mathrm{Trace}\left\{\mathbf{M}_N y y^\top\right\} \tag{6.19}$$

$$= \log\det\left(\mathbf{M}_N^{-1}\right) - \frac{1}{2}\sum_{k=0}^{n} \mathrm{Trace}\left\{M_k\,\varphi_k(y)\right\} \tag{6.20}$$

where the φ_k's are certain quadratic functions of y. Assuming that the T sample measurements are independent, the negative log-likelihood function depending on the $n+1$ matrix parameters $\{M_k\,;\, k = 0, 1, \dots, n\}$ can be written

$$L(M_0, \dots, M_n) = \log\det(\mathbf{M}_N) - \sum_{k=0}^{n} \mathrm{Trace}\left\{M_k T_k\left(\underline{y}\right)\right\} + C \tag{6.21}$$

where each matrix-valued statistic $T_k(\underline{y})$ has the structure of sample estimate of the lag k covariance. For example T_0 and T_1 are given by:

$$T_0\left(\underline{y}\right) = \frac{1}{T}\sum_{t=1}^{T}\left\{\sum_{k=0}^{N} y^{(t)}(k)\left[y^{(t)}(k)\right]^\top\right\}$$

$$T_1\left(\underline{y}\right) = \frac{2}{T}\sum_{t=1}^{T}\left\{\frac{1}{N}\sum_{j=1}^{N} y^{(t)}(k)\left[y^{(t)}(k-1)\right]^\top\right\} + \frac{2}{T}\sum_{t=1}^{T} y^{(t)}(0)\left[y^{(t)}(N)\right]^\top$$

etc.

From exponential class theory [1] we see that the T_k are (matrix-valued) sufficient statistics, hence we have the well-known characterization that the statistics

[3] For example a "movie" consisting of T images of the same texture.

$T_0, T_1, \dots, T_n$ (suitably normalized) are Maximum Likelihood estimators of their expected values, namely

$$\hat{\Sigma}_0 := \frac{1}{N} T_0 = \text{M.L. Estimator of } \mathbb{E}\,\mathbf{y}(k)\mathbf{y}(k)^\top$$

$$\hat{\Sigma}_n := \frac{1}{N} T_1 = \text{M.L. Estimator of } \mathbb{E}\,\mathbf{y}(k+n)\mathbf{y}(k)^\top$$

In other words, by writing the likelihood function in the form (6.21) we directly get the M.L. estimates of the entries in the main and upper diagonal blocks of the covariance matrix of the process $\mathbf{R}_N$ up to lag n.

Theorem 6.3.1. *Given the estimates defined above, the ML estimates of* $(M_0, M_1, \dots, M_n)$ *are obtained by solving the following* **block-circulant band extension problem:**

Complete the estimated covariances $\hat{\Sigma}_0 \dots, \hat{\Sigma}_n$ *with a sequence* $\Sigma_{n+1}, \Sigma_{n+2}, \dots$ *in such a way as to form a symmetric block-circulant positive definite matrix* Σ_N *which has a* **banded inverse of bandwith** *n.*

The inverse Σ_N^{-1} *will then be the maximum likelihood estimate of* $\mathbf{M}_N$.

General covariance extension problems, of which our is a special case, are discussed in the seminal paper by A. P. DEMPSTER, [6]. In particular, statement (a) in [6, p. 160] can be rephrased in our setting as follows.

Proposition 6.3.1. *If there is any positive definite symmetric matrix* Σ_N *which agrees with the data* $\hat{\Sigma}_0 \dots, \hat{\Sigma}_n$ *in the main and upper diagonal blocks up to lag n, then there exists exactly one such matrix with the additional property that* Σ_N^{-1} *has a banded inverse of bandwith n.*

Such a matrix Σ_N is called a (symmetric) *positive extension* of the data $\hat{\Sigma}_0 \dots, \hat{\Sigma}_n$. It is clear that a necessary condition for the existence of an extension is that the Toeplitz matrix

$$\begin{bmatrix} \hat{\Sigma}_0 & \dots & \hat{\Sigma}_n^\top \\ \dots & \dots & \\ \hat{\Sigma}_n & \dots & \hat{\Sigma}_0 \end{bmatrix}$$

be positive definite.

The circulant band extension problem of Theorem 6.3.1 looks similar to classical band extension problems studied in the literature, [7, 10], which are all solvable by factorization techniques. However the banded algebra framework on which all those papers rely does not apply here. Circulant band extension seems to be a new (and harder) extension problem. Unfortunately the problem is very nonlinear and it is hard to see what is going on by elementary means. Below we give a scalar example.

Example. Let $m = 1, N = 7, n = 2$ and assume we are assigned covariance estimates $\hat{\sigma}_0, \hat{\sigma}_1, \hat{\sigma}_2$ forming a positive definite Toeplitz matrix. The three unknown coefficients in the reciprocal model (6.12) of order 2 are scalars, denoted m_0, m_1, m_2. Equation $\mathbf{M}_N \mathbf{R}_N = \mathbf{I}_N$ leads to

$$\begin{bmatrix} m_0 & m_1 & m_2 & 0 & 0 & 0 & m_2 & m_1 \\ m_1 & m_0 & m_1 & m_2 & 0 & 0 & 0 & m_2 \\ m_2 & m_1 & m_0 & m_1 & m_2 & 0 & 0 & 0 \\ 0 & m_2 & m_1 & m_0 & m_1 & m_2 & 0 & 0 \\ 0 & 0 & m_2 & m_1 & m_0 & m_1 & m_2 & 0 \\ 0 & 0 & 0 & m_2 & m_1 & m_0 & m_1 & m_2 \\ m_2 & 0 & 0 & 0 & m_2 & m_1 & m_0 & m_1 \\ m_1 & m_2 & 0 & 0 & 0 & m_2 & m_1 & m_0 \end{bmatrix} \begin{bmatrix} \hat{\sigma}_0 \\ \hat{\sigma}_1 \\ \hat{\sigma}_2 \\ x_3 \\ x_4 \\ x_3 \\ \hat{\sigma}_2 \\ \hat{\sigma}_1 \end{bmatrix} = \begin{bmatrix} 1 \\ 0 \\ \vdots \\ 0 \\ 0 \end{bmatrix}$$

where $x_3 := r_3 = r_5$ and $x_4 := r_5$ are the unknown extended covariance lags. Rearranging and eliminating the last three redundant equations one obtains

$$\begin{aligned} m_0\hat{\sigma}_0 + 2m_1\hat{\sigma}_1 + 2m_2\hat{\sigma}_2 &= 1 \\ m_0\hat{\sigma}_1 + m_1(\hat{\sigma}_1 + \hat{\sigma}_2) + m_2(\hat{\sigma}_1 + x_3) &= 0 \\ m_0\hat{\sigma}_2 + m_1(\hat{\sigma}_1 + x_3) + m_2(\hat{\sigma}_0 + x_4) &= 0 \\ m_0 x_3 + m_1(\hat{\sigma}_2 + x_4) + m_2(\hat{\sigma}_1 + x_3) &= 0 \\ m_0 x_4 + 2m_1 x_3 + 2m_2\hat{\sigma}_2 &= 0 \end{aligned}$$

which is a system of five quadratic equations in five unknowns whose solution looks already non-trivial. It may be checked that under positivity of the Toeplitz matrix of $\{\hat{\sigma}_0, \hat{\sigma}_1, \hat{\sigma}_2\}$ it has a unique positive definite solution (i.e. making $\mathbf{M}_N$ positive definite).

6.3.1 Algorithms for Circulant Band Extension

In the literature one can find a couple of ways to approach the circulant band extension problem, none of which so far seems to be really satisfactory. One is based on a result of B. Levy [11] which in the present setting implies that for $N \to \infty$ the problem becomes one of band extension for infinite positive definite symmetric block-Toeplitz matrices for which satisfactory algorithms exist. For N finite the approximation in some cases may be poor. Another route is to adapt the general idea of Dempster's algorithm [6] to the present setting. Even if in our case we deal with circulant matrices and the calculations for inverting circulant matrices can be done efficiently by FFT, the algorithm is computationally very demanding as it requires iterative inversion of large matrices.

A key observation in this respect reveals to be statement (b) in [6, p. 160], which reads as follows

Proposition 6.3.2. *Among all covariance extensions of the data $\hat{\Sigma}_0 \ldots, \hat{\Sigma}_n$, the one with a banded inverse of bandwith n has maximum Entropy.*

This statement will be the guideline for the developments which follow and, as we shall see, it will in fact lead to a new convex optimization procedure for computing the band extension. Note that both propositions 6.3.1 and 6.3.2 in Dempster's paper refer to general covariance matrices and it is not clear whether they should hold verbatim for block-circulant covariance matrices. That this is indeed the case will be proven in the next sections.

6.4 Maximum Entropy on the Discrete Circle

Let U denote the "block circulant shift" matrix

$$U = \begin{bmatrix} 0 & I_m & 0 & \dots & 0 \\ 0 & 0 & I_m & \dots & 0 \\ \vdots & \vdots & & \ddots & \vdots \\ 0 & 0 & 0 & \dots & I_m \\ I_m & 0 & 0 & \dots & 0 \end{bmatrix},$$

where I_m denotes the $m \times m$ identity matrix. Clearly, $U^\top U = UU^\top = I_{mN}$; i.e. U is orthogonal. Note that a matrix C with $N \times N$ blocks is block-circulant if and only if it commutes with U, namely if and only if it satisfies

$$U^\top CU = C. \tag{6.22}$$

Recall that the *differential entropy* $H(p)$ of a probability density function p on $\mathbb{R}^n$ is defined by

$$H(p) = -\int_{\mathbb{R}^n} \log(p(x))p(x)dx.$$

In the case of a zero-mean Gaussian distribution p with covariance matrix Σ, we get

$$H(p) = \frac{1}{2}\log(\det\Sigma)) + \frac{1}{2}n(1+\log(2\pi)). \tag{6.23}$$

Let $\mathfrak{S}_N$ denote the vector space of *symmetric* matrices with $N \times N$ square blocks of dimension $m \times m$. Let $\mathbf{T}_n \in \mathfrak{S}_{n+1}$ denote the matrix of *boundary data*:

$$\mathbf{T}_n = \begin{bmatrix} \Sigma_0 & \Sigma_1^\top & \dots & \Sigma_n^\top \\ \Sigma_1 & \dots & & \dots \\ \dots & \dots & & \dots \\ \Sigma_n & \dots & & \Sigma_0 \end{bmatrix}$$

and let E_n denote the $N \times (n+1)$ block matrix

$$E_n = \begin{bmatrix} I_m & 0 & \dots & 0 \\ 0 & I_m & \dots & 0 \\ 0 & 0 & \dots & \dots \\ \dots & & 0 & I_m \\ 0 & 0 & \dots & 0 \end{bmatrix}.$$

Consider the following *maximum entropy* problem (MEP) on the discrete circle:

Problem 6.4.1.

$$\min\{-\operatorname{tr}\log\Sigma \mid \Sigma \in \mathfrak{S},\ \Sigma > 0\} \tag{6.24}$$

subject to

$$E_n^\top \Sigma E_n = \mathbf{T}_n, \tag{6.25}$$

$$U^\top \Sigma U = \Sigma. \tag{6.26}$$

Recalling that $\operatorname{tr}\log\Sigma = \log\det\Sigma$ and (6.23), we see that the above problem indeed amounts to finding the maximum entropy Gaussian distribution with block-circulant covariance, whose first $n+1$ blocks are precisely $\Sigma_0,\dots,\Sigma_n$. The circulant structure is equivalent to requiring this distribution to be stationary on the discrete circle $\mathbb{Z}_N$. We observe that in this problem we are minimizing a strictly convex function on the intersection of a convex cone (minus the zero matrix) with a linear manifold. Hence we are dealing with a convex optimization problem.

The first question to be addressed is feasibility of (MEP), namely the existence of a positive definite, symmetric matrix Σ satisfying (6.25)-(6.26). Obviously, $\mathbf{T}_n$ positive definite is a necessary condition for the existence of such a Σ. In general it turns out that feasibility holds for N large enough. However since the details of the proof are complicated we shall just proceed assuming it holds, leaving the statement of precise conditions to a future publication.

6.5 Variational Analysis

We shall introduce a suitable set of "Lagrange multipliers" for our constrained optimization problem. Consider the linear map $A:\mathfrak{S}_{n+1}\times\mathfrak{S}_N\to\mathfrak{S}_N$ defined by

$$A(\Lambda,\Theta)=E_n\Lambda E_n^\top+U\Theta U^\top-\Theta,\quad (\Lambda,\Theta)\in\mathfrak{S}_{n+1}\times\mathfrak{S}_N.$$

and define the set

$$\mathscr{L}_+:=\{(\Lambda,\Theta)\in(\mathfrak{S}_{n+1}\times\mathfrak{S}_N)\mid(\Lambda,\Theta)\in(\ker(A))^\perp,\left(E_n\Lambda E_n^\top+U\Theta U^\top-\Theta\right)>0\}.$$

Observe that $\mathscr{L}_+$ is an open, convex subset of $(\ker(A))^\perp$. For each $(\Lambda,\Theta)\in\mathscr{L}_+$, we consider the unconstrained minimization of the *Lagrangian function*

$$\begin{aligned}L(\Sigma,\Lambda,\Theta) &:= -\operatorname{tr}\log\Sigma+\operatorname{tr}\left(\Lambda\left(E_n^\top\Sigma E_n-\mathbf{T}_n\right)\right)+\operatorname{tr}\left(\Theta\left(U^\top\Sigma U-\Sigma\right)\right)\\ &= -\operatorname{tr}\log\Sigma+\operatorname{tr}\left(E_n\Lambda E_n^\top\Sigma\right)-\operatorname{tr}(\Lambda\mathbf{T}_n)+\operatorname{tr}\left(U\Theta U^\top\right)\\ &\quad -\operatorname{tr}(\Theta\Sigma)\end{aligned}$$

over $\mathfrak{S}_{N,+}:=\{\Sigma\in\mathfrak{S}_N,\ \Sigma>0\}$. For $\delta\Sigma\in\mathfrak{S}_N$, we get

$$\delta L(\Sigma,\Lambda,\Theta;\delta\Sigma)=-\operatorname{tr}\left(\Sigma^{-1}\delta\Sigma\right)+\operatorname{tr}\left(E_n\Lambda E_n^\top\delta\Sigma\right)+\operatorname{tr}\left(\left(U\Theta U^\top-\Theta\right)\delta\Sigma\right).$$

We conclude that $\delta L(\Sigma,\Lambda,\Theta;\delta\Sigma)=0,\ \forall\delta\Sigma\in\mathfrak{S}_N$ if and only if

$$\Sigma^{-1}=E_n\Lambda E_n^\top+U\Theta U^\top-\Theta.$$

Thus, for each fixed pair $(\Lambda,\Theta)\in\mathscr{L}_+$, the unique Σ^o minimizing the Lagrangian is given by

$$\Sigma^o=\left(E_n\Lambda E_n^\top+U\Theta U^\top-\Theta\right)^{-1}. \tag{6.27}$$

Consider next $L(\Sigma^o,\Lambda,\Theta)$. We get

$$L(\Sigma^o,\Lambda,\Theta) = -\operatorname{tr}\log\left(\left(E_n\Lambda E_n^\top + U\Theta U^\top - \Theta\right)^{-1}\right)$$
$$+\operatorname{tr}\left[\left(E_n\Lambda E_n^\top + U\Theta U^\top - \Theta\right)\left(E_n\Lambda E_n^\top + U\Theta U^\top - \Theta\right)^{-1}\right] - \operatorname{tr}(\Lambda\mathbf{T}_n) \quad (6.28)$$
$$= \operatorname{tr}\log\left(E_n\Lambda E_n^\top + U\Theta U^\top - \Theta\right) + \operatorname{tr}I_{mN} - \operatorname{tr}(\Lambda\mathbf{T}_n).$$

This is a strictly concave function on $\mathscr{L}_+$ whose maximization is the *dual problem* of (MEP). We can equivalently consider the convex problem

$$\min\{J(\Lambda,\Theta),(\Lambda,\Theta)\in\mathscr{L}_+\}, \tag{6.29}$$

where J (henceforth called dual function) is given by

$$J(\Lambda,\Theta) = \operatorname{tr}(\Lambda\mathbf{T}_n) - \operatorname{tr}\log\left(E_n\Lambda E_n^\top + U\Theta U^\top - \Theta\right). \tag{6.30}$$

6.5.1 Existence for the Dual Problem

The minimization of the strictly convex function $J(\Lambda,\Theta)$ on the convex set $\mathscr{L}_+$ is a challenging problem as $\mathscr{L}_+$ is an *open* and *unbounded* subset of $(\ker(A))^\perp$. Nevertheless, the following existence result in the Byrnes-Lindquist spirit [2], [8] can be established.

Theorem 6.5.1. *The function J admits a unique minimum point $(\bar{\Lambda},\bar{\Theta})$ in $\mathscr{L}_+$.*

In order to prove this theorem, we need first to derive a number of auxiliary results. Let $\mathfrak{C}_N$ denote the vector subspace of block-circulant matrices in $\mathfrak{S}_N$. We proceed to characterize the orthogonal complement of $\mathfrak{C}_N$ in $\mathfrak{S}_N$.

Lemma 6.5.1. *Let $M\in\mathfrak{S}_N$. Then $M\in(\mathfrak{C}_N)^\perp$ if and only if it can be expressed as*

$$M = UNU^\top - N \tag{6.31}$$

for some $N\in\mathfrak{S}_N$.

Proof. By (6.22), $\mathfrak{C}_N$ is the kernel of the linear map from $\mathfrak{S}_N$ to $\mathfrak{S}_N$ given by $M\mapsto U^\top MU - M$. Hence, its orthogonal complement is the range of the adjoint map. Since

$$\operatorname{tr}\left((U^\top MU - M)N\right) = \langle U^\top MU - M, N\rangle = \langle M, UNU^\top - N\rangle,$$

the conclusion follows. □

Next we show that, as expected, feasibility of the primal problem (MEP) implies that the dual function J is bounded below.

Lemma 6.5.2. *Assume that there exists $\bar{\Sigma} \in \mathfrak{S}_{N,+}$ satisfying (6.25)-(6.26). Then, for any pair $(\Lambda,\Theta) \in \mathscr{L}_+$, we have*

$$J(\Lambda,\Theta) \geq mN + \operatorname{tr}\log\bar{\Sigma}. \tag{6.32}$$

Proof. By (6.25), $\operatorname{tr}(\Lambda\mathbf{T}_n) = \operatorname{tr}(\Lambda E_n^\top \bar{\Sigma} E_n) = \operatorname{tr}(E_n\Lambda E_n^\top \bar{\Sigma})$. Using this fact and Lemma 6.5.1, we can now rewrite the dual function J as follows

$$\begin{aligned} J(\Lambda,\Theta) &= \operatorname{tr}(\Lambda\mathbf{T}_n) - \operatorname{tr}\log\left(E_n\Lambda E_n^\top + U\Theta U^\top - \Theta\right) \\ &= \operatorname{tr}\left[\left(E_n\Lambda E_n^\top + U\Theta U^\top - \Theta\right)\bar{\Sigma}\right] - \operatorname{tr}\log\left(E_n\Lambda E_n^\top + U\Theta U^\top - \Theta\right). \end{aligned}$$

Define $M(\Lambda,\Theta) = \left(E_n\Lambda E_n^\top + U\Theta U^\top - \Theta\right)$ which is positive definite for (Λ,Θ) in $\mathscr{L}_+$. Then

$$J(\Lambda,\Theta) = \operatorname{tr}\left(M(\Lambda,\Theta)\bar{\Sigma}\right) - \operatorname{tr}\log M(\Lambda,\Theta).$$

As a function of M, this is a strictly convex function on $\mathfrak{S}_{N,+}$, whose unique minimum occurs at $M = \bar{\Sigma}^{-1}$ where the minimum value is $\operatorname{tr}(I_{mN}) + \operatorname{tr}\log\bar{\Sigma}$. □

Lemma 6.5.3. *Let $(\Lambda_k,\Theta_k), n \geq 1$ be a sequence of pairs in $\mathscr{L}_+$ such that $\|(\Lambda_k,\Theta_k)\| \to \infty$. Then also $\|A(\Lambda_k,\Theta_k)\| \to \infty$. It then follows that $\|(\Lambda_k,\Theta_k)\| \to \infty$ implies that $J(\Lambda_k,\Theta_k) \to \infty$*

Proof. Notice that A is a linear operator between finite-dimensional linear spaces. Denote by σ_m the smallest singular value of the restriction of A to $(\ker A)^\perp$ (the orthogonal complement of $\ker A$). Clearly, $\sigma_m > 0$, so that, since each element of the sequence (Λ_k,Θ_k) is in $(\ker A)^\perp$, $\|A(\Lambda_k,\Theta_k)\| \geq \sigma_m \|(\Lambda_k,\Theta_k)\| \to \infty$.

Assume now that $\|A(\Lambda_k,\Theta_k)\| = \|\left(E_n\Lambda_k E_n^\top + U\Theta_k U^\top - \Theta_k\right)\| \to \infty$. Since these are all positive definite matrices and all matrix norms are equivalent, it follows that $\operatorname{tr}\left(E_n\Lambda E_n^\top + U\Theta U^\top - \Theta\right) \to \infty$. As a consequence, $\operatorname{tr}\left(\left(E_n\Lambda E_n^\top + U\Theta U^\top - \Theta\right)\bar{\Sigma}\right) \to \infty$ and, finally, $J(\Lambda_k,\Theta_k) \to \infty$. □

We show next that the dual function tends to infinity also when approaching the boundary of $\mathscr{L}_+$, namely

$$\begin{aligned} \partial\mathscr{L}_+ := \{(\Lambda,\Theta) \in (\mathfrak{S}_{n+1} \times \mathfrak{S}_N) | (\Lambda,\Theta) \in (\ker(A))^\perp, \left(E_n\Lambda E_n^\top + U\Theta U^\top - \Theta\right) \geq 0, \\ \det\left(E_n\Lambda E_n^\top + U\Theta U^\top - \Theta\right) = 0\}. \end{aligned}$$

Lemma 6.5.4. *Consider a sequence $(\Lambda_k,\Theta_k), k \geq 1$ in $\mathscr{L}_+$ such that the matrix $\lim_k \left(E_n\Lambda_k E_n^\top + U\Theta_k U^\top - \Theta_k \right)$ is singular. Assume also that the sequence (Λ_k,Θ_k) is bounded. Then, $J(\Lambda_k,\Theta_k) \to \infty$.*

Proof. Simply write

$$J(\Lambda_k,\Theta_k) = -\log\det\left(E_n\Lambda_k E_n^\top + U\Theta_k U^\top - \Theta_k\right) + \operatorname{tr}(\Lambda_k\mathbf{T}_k).$$

Since $\operatorname{tr}(\Lambda_k\mathbf{T}_k)$ is bounded, the conclusion follows. □

Proof of Theorem 6.5.1. Observe that the function J is a continuous, bounded below (Lemma 6.5.2) function that tends to infinity both when $\|(\Lambda,\Theta)\|$ tends to infinity (Lemma 6.5.3) and when it tends to the boundary $\partial\mathscr{L}_+$ with $\|(\Lambda,\Theta)\|$ remaining bounded (Lemma 6.5.4). It follows that J is *inf-compact* on $\mathscr{L}_+$, namely it has compact sublevel sets. By Weierstrass' theorem, it admits at least one minimum point. Since J is strictly convex, the minimum point is unique. □

6.6 Reconciliation with Dempster's Covariance Selection

Let $(\bar\Lambda,\bar\Theta)$ be the unique minimum point of J in $\mathscr{L}_+$ (Theorem 6.5.1). Then $\Sigma^o \in \mathfrak{S}_{N,+}$ given by

$$\Sigma^o = \left(E_n\bar\Lambda E_n^\top + U\bar\Theta U^\top - \bar\Theta\right)^{-1} \tag{6.33}$$

satisfies (6.25) and (6.26). Hence, it is the unique solution of the primal problem (MEP). Since it satisfies (6.26), Σ^o is in particular a block-circulant matrix. Then so is $(\Sigma^o)^{-1} = \left(E_n\bar\Lambda E_n^\top + U\bar\Theta U^\top - \bar\Theta\right)$. Let $\pi_{\mathfrak{C}_N}$ denote the orthogonal projection onto the linear subspace of symmetric, block-circulant matrices $\mathfrak{C}_N$. It follows that, in force of Lemma 6.5.1,

$$(\Sigma^o)^{-1} = \pi_{\mathfrak{C}_N}((\Sigma^o)^{-1}) = \pi_{\mathfrak{C}_N}\left(E_n\bar\Lambda E_n^\top + U\bar\Theta U^\top - \bar\Theta\right) = \pi_{\mathfrak{C}_N}\left(E_n\bar\Lambda E_n^\top\right). \tag{6.34}$$

Theorem 6.6.1. *Let Σ^o be the maximum Entropy covariance given by (6.33). Then $(\Sigma^o)^{-1}$ is a symmetric block-circulant matrix which is banded of bandwith n. Hence the solution of (MEP) may be viewed as a Gaussian stationary reciprocal process of index n defined on $\mathbb{Z}_N$.*

Proof. Let

$$\Pi_{\bar\Lambda} := \pi_{\mathfrak{C}_N}\left(E_n\bar\Lambda E_n^\top\right) = \begin{bmatrix} \Pi_0 & \Pi_1^\top & \Pi_2^\top & \dots & \Pi_1 \\ \Pi_1 & \Pi_0 & \Pi_1^\top & \dots & \Pi_2 \\ \vdots & \ddots & \ddots & \ddots & \vdots \\ \Pi_2^\top & \dots & \Pi_1 & \Pi_0 & \Pi_1^\top \\ \Pi_1^\top & \Pi_2^\top & \dots & \Pi_1 & \Pi_0 \end{bmatrix}$$

be the orthogonal projection of $\left(E_n\bar\Lambda E_n^\top\right)$ onto $\mathfrak{C}_N$. Since $\Pi_{\bar\Lambda}$ is symmetric and block-circulant, it is characterized by the orthogonality condition

$$\operatorname{tr}\left[\left(E_n\bar\Lambda E_n^\top - \Pi_{\bar\Lambda}\right)C\right] = \langle E_n\bar\Lambda E_n^\top - \Pi_{\bar\Lambda}, C\rangle = 0, \quad \forall C \in \mathfrak{C}_N. \tag{6.35}$$

Next observe that, if we write $C = \operatorname{Circ}\left[C_0, C_1, C_2, \dots, C_2^\top, C_1^\top\right]$ and

$$\bar\Lambda = \begin{bmatrix} \bar\Lambda_{00} & \bar\Lambda_{01} & \dots\dots & \bar\Lambda_{0n} \\ \bar\Lambda_{10}^\top & \bar\Lambda_{11} & \dots & \bar\Lambda_{1n} \\ \dots & & \dots & \dots \\ \bar\Lambda_{n0}^\top & \bar\Lambda_{n1}^\top & \dots & \bar\Lambda_{nn} \end{bmatrix}, \qquad \bar\Lambda_{k,j} = \bar\Lambda_{j,k}^\top$$

then

$$\begin{aligned}\operatorname{tr}\left[E_n\bar{\Lambda}E_n^\top C\right] = \operatorname{tr}\left[\bar{\Lambda}E_n^\top CE_n\right] = \operatorname{tr}\big[&(\bar{\Lambda}_{00}+\bar{\Lambda}_{11}+\ldots+\bar{\Lambda}_{nn})C_0\\ &+(\bar{\Lambda}_{01}+\bar{\Lambda}_{12}+\ldots+\bar{\Lambda}_{n-1,n})C_1+\ldots+\bar{\Lambda}_{0n}C_n\\ &+(\bar{\Lambda}_{10}+\bar{\Lambda}_{21}+\ldots,\bar{\Lambda}_{n,n-1})C_1^\top+\ldots+\bar{\Lambda}_{n0}C_n^\top\Big].\end{aligned}$$

On the other hand, recalling that the product of two block-circulant matrices is block-circulant, we have that $\operatorname{tr}[\Pi_{\bar{\Lambda}}C]$ is simply N times the trace of the first block row of $\Pi_{\bar{\Lambda}}$ times the first block column of C. We get

$$\operatorname{tr}[\Pi_{\bar{\Lambda}}C] = N\operatorname{tr}\left[\Pi_0C_0+\Pi_1^\top C_1+\Pi_2^\top C_2+\ldots+\Pi_2C_2^\top+\Pi_1C_1^\top\right].$$

Hence, the orthogonality condition (6.35), reads

$$\begin{aligned}\operatorname{tr}\left[\left(E_n\bar{\Lambda}E_n^\top-\Pi_{\bar{\Lambda}}\right)C\right] = \quad &\operatorname{tr}\big[\left((\bar{\Lambda}_{00}+\bar{\Lambda}_{11}+\ldots+\bar{\Lambda}_{nn})-N\Pi_0\right)C_0+\\ &+\left((\bar{\Lambda}_{01}+\bar{\Lambda}_{12}+\ldots+\bar{\Lambda}_{n-1,n})-N\Pi_1^\top\right)C_1\\ &+\left((\bar{\Lambda}_{10}+\bar{\Lambda}_{21}+\ldots,\bar{\Lambda}_{n,n-1})-N\Pi_1\right)C_1^\top\\ &+\ldots(\bar{\Lambda}_{0n}-N\Pi_1^\top)C_n+(\bar{\Lambda}_{n0}-N\Pi_1)C_n^\top)\big]\\ &+N\Pi_{n+1}^\top C_{n+1}+N\Pi_{n+1}C_{n+1}^\top\\ &+N\Pi_{n+2}^\top C_{n+2}+N\Pi_{n+2}C_{n+2}^\top\\ &+\ldots\\ &=0.\end{aligned} \tag{6.36}$$

Since this must hold true forall $C\in\mathfrak{C}_N$, we conclude that

$$\begin{aligned}\Pi_0 &= \frac{1}{N}(\bar{\Lambda}_{00}+\bar{\Lambda}_{11}+\ldots+\bar{\Lambda}_{nn}),\\ \Pi_1 &= \frac{1}{N}(\bar{\Lambda}_{01}+\bar{\Lambda}_{12}+\ldots+\bar{\Lambda}_{n-1,n})^\top,\\ &\ldots\\ \Pi_n &= \frac{1}{N}\bar{\Lambda}_{0n}^\top,\end{aligned}$$

while from the last equation we get $\Pi_i=0$, forall i in the interval $n+1\le i\le N-n-1$. From this it is clear that the inverse of the covariance matrix solving the primal problem (MEP), namely $\Pi_{\bar{\Lambda}}=(\Sigma^o)^{-1}$ has a circulant block-banded structure of bandwith n. □

The above can be seen as a specialization of the classical covariance selection result of Dempster [6], namely Proposition 6.3.1, to the block-circulant case. In fact the results of this section specialize also the maximum Entropy characterization of Dempster (Proposition 6.3.2) to the block-circulant setting. Apparently none of these two results follows from the characterizations of Dempster's paper which deals with a very unstructured setting. In particular the proof that the solution, Σ^o, to our primal problem (MEP) has a block-circulant banded inverse (Theorem 6.6.1), uses in an essential way both the characterization of the MEP solution provided by our variational analysis and cleverly exploits the block-circulant structure.

Finally, we anticipate that the results of this section lead to an efficient iterative algorithm for the explicit solution of the MEP which is guaranteed to converge to a unique minimum. This solves the variational problem and hence the circulant band extension problem which subsumes maximum likelihood identification of reciprocal processes. This algorithm, which will not be described here for reasons of space limitations, compares very favourably with the best techniques available so far.

6.7 Conclusions

Band extension problems for block-circulant matrices of the type discussed in this paper occur in particular in applications to image modeling and simulation. For reasons of space we shall not provide details but rather refer to the literature. See [3, 4] and [15] for examples.

References

1. Barndorff-Nilsen, O.E.: Information and Exponential families in Statistica Theory. Wiley, New York (1978)
2. Byrnes, C., Lindquist, A.: Interior point solutions of variational problems and global inverse function theorems. International Journal of Robust and Nonlinear Control (special issue in honor of V.A.Yakubovich on the occation of his 80th birthday) 17, 463–481 (2007)
3. Chiuso, A., Ferrante, A., Picci, G.: Reciprocal realization and modeling of textured images. In: Proceedings of the 44rd IEEE Conference on Decision and Control, Seville, Spain (December 2005)
4. Chiuso, A., Picci, G.: Some identification techniques in computer vision (invited paper). In: Proceedings of the 47th IEEE Conference on Decision and Control, Cancun, Mexico (December 2008)
5. Davis, P.: Circulant Matrices. John Wiley & Sons, Chichester (1979)
6. Dempster, A.P.: Covariance selection. Biometrics 28, 157–175 (1972)
7. Dym, H., Gohberg, I.: Extension of band matrices with band inverses. Linear Algebra and Applications 36, 1–24 (1981)
8. Ferrante, A., Pavon, M., Ramponi, F.: Further results on the Byrnes-Georgiou-Lindquist generalized moment problem. In: Ferrante, A., Chiuso, A., Pinzoni, S. (eds.) Modeling, Estimation and Control: Festschrift in honor of Giorgio Picci on the occasion of his sixty-fifth Birthday, pp. 73–83. Springer, Heidelberg (2007)
9. Frezza, R.: Models of Higher-order and Mixed-order Gaussian Reciprocal Processes with Application to the Smoothing Problem. PhD thesis, Applied Mathematics Program, U.C.Davis (1990)

10. Gohberg, I., Goldberg, S., Kaashoek, M.: Classes of Linear Operators vol. II. Birkhäuser, Boston (1994)
11. Levy, B.C.: Regular and reciprocal multivariate stationary Gaussian reciprocal processes over **Z** are necessarily Markov. J. Math. Systems, Estimation and Control 2, 133–154 (1992)
12. Levy, B.C., Ferrante, A.: Characterization of stationary discrete-time Gaussian reciprocal processes over a finite interval. SIAM J. Matrix Anal. Appl. 24, 334–355 (2002)
13. Levy, B.C., Frezza, R., Krener, A.J.: Modeling and estimation of discrete-time Gaussian reciprocal processes. IEEE Trans. Automatic Control 35(9), 1013–1023 (1990)
14. Masani, P.: The prediction theory of multivariate stochastic proceses, iii. Acta Mathematica 104, 141–162 (1960)
15. Picci, G., Carli, F.: Modelling and simulation of images by reciprocal processes. In: Proc. Tenth International Conference on Computer Modeling and Simulation UKSIM 2008, pp. 513–518 (2008)
16. Sand, J.A.: Reciprocal realizations on the circle. SIAM J. Control and Optimization 34, 507–520 (1996)

7

Cumulative Distribution Estimation via Control Theoretic Smoothing Splines

Janelle K. Charles[1], Shan Sun[2], and Clyde F. Martin[1]

[1] Department of Mathematics and Statistics, Texas Tech University, Lubbock, TX 79409-1042, USA
[2] Department of Mathematics, University of Texas at Arlington, Arlington, TX 76019, USA

Summary. In this paper, we explore the relationship between control theory and statistics. Specifically, we consider the use of cubic monotone control theoretic smoothing splines in estimating the cumulative distribution function (CDF) defined on a finite interval $[0,T]$. The spline construction is obtained by imposing an infinite dimensional, non-negativity constraint on the derivative of the optimal curve. The main theorem of this paper states that the optimal curve $y(t)$ is a piecewise polynomial of known degree with $y(0) = 0$ and $y(T) = 1$. The solution is determined through dynamic programming which takes advantage of a finite reparametrization of the problem.

7.1 Introduction

Probability distribution estimation has been a wide studied topic in statistics for many years. Methods of such distribution estimation include kernel estimation [9] and nonparametric estimation from quantized samples [6]. The goal of this paper is to approximate the cumulative distribution function (CDF) when given the empirical CDF. Our aim is to show that control theoretic splines have some favorable properties over traditional smoothing splines of statistics. Control theoretic smoothing splines were developed in control theory mostly in the area of trajectory planning [7]. In this paper, we will examine smoothing spline construction where the optimal curve preserves monotonicity. This property translates to the non-negativity constraint on the first derivative of the spline. In this case, we have a nonlinear constraint which is very difficult to handle directly; however, we show that this infinite dimensional problem can be translated and solved in a finite setting following the dynamic programming algorithm as illustrated in [3] and [4].

Interpolating splines were developed as a tool in approximation in numerical analysis, where the errors were assumed to be insignificant or nonexistent. However, in most statistical applications, where noisy data is apparent, interpolation gives very little insight into the underlying distribution function from which the data is sampled. The use of smoothing splines for statistics was not explored until it was determined that you could balance the trade off between the deviation from the data points and

X. Hu et al. (Eds.): Three Decades of Progress in Control Sciences, pp. 95–104, 2010.

smoothness of the spline. As such, the construction of the smoothing spline is so that the errors between the spline and the data possess favorable statistical properties, for instances that the variance of residuals is small. Significant work in the application of smoothing spline approximation has been done in [2], [10], [11], and [12].

Monotone interpolating splines have been studied extensively in literature. In this paper, we will focus on the use of monotone smoothing splines in CDF estimation on a specified interval [0,T], that is, we do not require exact interpolation of the data. However, we do require that the spline $y(t)$ has end conditions $y(0) = 0$ and $y(T) = 1$. Moreover, following [3], we show that our approximation can be implemented numerically with a dynamic programming algorithm in the case of second order systems. The outline for this paper is as follows: In section 2 we will discuss the smoothing spline construction and describe some properties that the optimal solution possesses. In Section 3, we illustrate the dynamic programming algorithm used in solving the cubic monotone smoothing spline problem, followed by a conclusion in Section 4.

7.2 Problem Description

In this section, we describe the estimation problem discussed in this paper. In particular, we discuss the method of producing monotonically increasing curves that pass close to given waypoints while minimizing the cost of driving the curve between the points. Given the empirical probability distribution defined on $[0,T]$ subdivided into $N \leq 10$ intervals, we select a nodal value t_i in each subinterval and the relative frequency of the interval τ_i. For CDF estimation, we consider the data set

$$D = \{(t_i, \alpha_i) : i = 1, \cdots N\},$$

where $\alpha_i = \sum_{j=1}^{i} \tau_j \geq 0$, $0 < t_1 < \cdots < t_N \leq T$.

The paragraphs that follow discuss the control theoretic spline construction when no constraints are imposed on the control system and further when we have non-negative derivative constraints on the optimal curve.

7.2.1 Smoothing Splines

We assume a linear controllable and observable system of the form

$$\dot{x} = Ax + bu,\ y = cx, x(0) = x_0 \tag{7.1}$$

where $x \in \mathbf{R}^n$, A, b, and, c are constant matrices of compatible dimension, and u and y are scalar functions. The solution to (7.1) is

$$y(t) = ce^{At}x_0 + \int_0^t ce^{A(t-s)}bu(s)ds. \tag{7.2}$$

Our goal is to use control theoretic smoothing spline techniques, to determine $u^*(t)$ that minimizes the quadratic cost function

$$J(u;x_0) = \lambda \int_0^T u^2(t)dt + (\hat{y} - \hat{\alpha})'Q(\hat{y} - \hat{\alpha}) + x_0'Rx_0 \tag{7.3}$$

where Q=diag$\{\omega_i : i = 1,\ldots,N\}$ and R are positive definite matrices, the constant $\omega_i > 0$ is a weight that represents how important it is that $y(t_i)$ passes close to α_i and smoothing parameter $\lambda > 0$ controls the trade off between smoothness of the spline curve and closeness of this curve to the data. We consider the basis of linearly independent functions

$$l_i(s) = \begin{cases} ce^{A(t_i-s)}b : t_i \geq s \\ 0 \qquad\qquad : t_i < s \end{cases}$$

so that $y_i = \langle \beta_i, x_0 \rangle_R + \langle l_i, u \rangle_L$ where $\beta_i = R^{-1}e^{A't_i}c'$ and we define the inner products

$$\langle g, h \rangle_L = \int_0^T g(t)h(t)dt \text{ and } \langle z, w \rangle_Q = z'Qw.$$

The linear independence of the basis functions follows from the fact that the l_is vanish at different t_is.

When no constraints are imposed on the derivative of $y(t)$, the Hilbert space smoothing spline construction in [1] produces the unique, optimal control $u^* \in L^2[0,T]$ that minimizes the given cost function. This optimal control is given by

$$u^*(t) = -\frac{1}{\lambda} \sum_{i=1}^{N} \langle \hat{y} - \hat{\alpha}, e_i \rangle_Q l_i(t) \tag{7.4}$$

where the optimal smoothed data

$$\hat{y} = \left(I + FQ + \frac{1}{\lambda}GQ\right)^{-1} \left(FQ + \frac{1}{\lambda}GQ\right)\hat{\alpha},$$

$\hat{\alpha} = (\alpha_1, \cdots, \alpha_N)'$, and G and F are the Grammian matrices.

Throughout this paper, we shall concentrate on second order systems of the form

$$A = \begin{pmatrix} 0 & 1 \\ 0 & 0 \end{pmatrix}, b = \begin{pmatrix} 0 \\ 1 \end{pmatrix}, \text{ and } c = \begin{pmatrix} 1 & 0 \end{pmatrix},$$

which produces the classical cubic splines. For these splines, we have basis functions

$$l_i(s) = \begin{cases} t_i - s : \ t_i \geq s \\ 0 \qquad\ \ : \ t_i < s \end{cases}$$

for $i = 1, \cdots, N$. The Grammian matrix G has components

$$\begin{aligned} G_{ij} = G_{ji} &= \int_0^{min(t_i,t_j)} l_i(s)l_j(s)ds \\ &= \int_0^{min(t_i,t_j)} (t_i - s)(t_j - s)ds, \text{ for } i \neq j \\ G_{ii} &= \int_0^T l_i^2(s)ds = \int_0^{t_i} (t_i - s)^2 ds, \text{ for } i = j, \end{aligned}$$

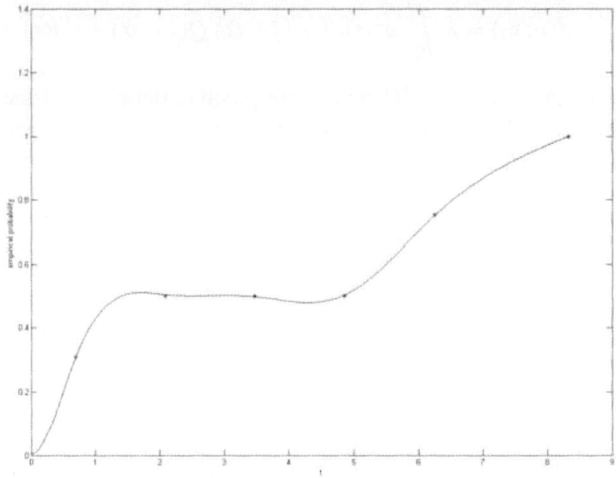

Fig. 7.1. The curve shown represents the optimal solution to the problem where monotonicity constraints have not been imposed. The asterisks represent the six way points (t_i, α_i). Here we take $\lambda = 0.001$

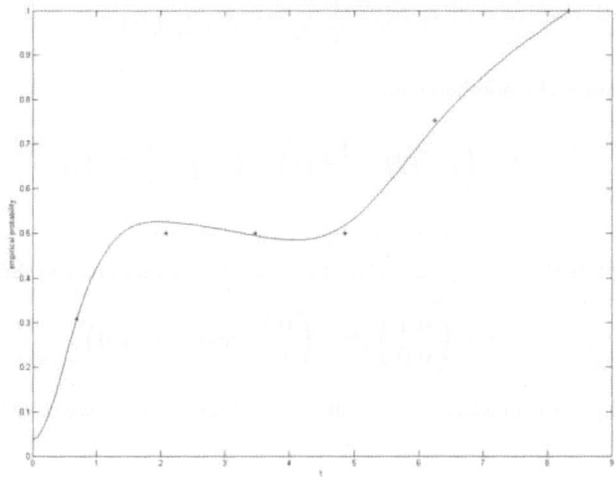

Fig. 7.2. This curve was obtained with the same construction as in 7.1 with $\lambda = 0.01$.

and the Grammian matrix $F = \beta\beta'$ where β is an $N \times n$ matrix with ith row given by $\beta_i = R^{-1}e^{A't_i}c'$ for $i = 1, \cdots, N$.

Using this type of optimization produces curves as shown in Figure 7.1 and 7.2.

The data was obtained from a cumulative distribution that is assumed to be constant on an interval. Here we observe that although the splines closely approximate the way points, there is no guarantee that the spline is monotone on the interval $[0,T]$ nor that the end conditions are satisfied. For CDF estimation we require that the spline is non-negative and monotonically increasing; thus, alternative construction is necessary.

7.2.2 Smoothing Splines with Derivative Constraints

We now consider formulation of the solution to the estimation problem while imposing monotonicity constraints on the optimal curve. This translates to finding a continuous curve, $y(t)$ that minimizes the cost in (7.3) that satisfies

$$y \in C^1[0,T]$$
$$\dot{y}(t) \geq 0$$

In this section, we describe the spline construction using the Hilbert space methods and via Lagrange multipliers. Here, we assume without loss of generality that $x_0 = 0$ in (7.1). Thus, our goal is to obtain a control function $u^* \in L^2[0,T]$ that minimizes the cost

$$J(u) = \lambda \int_0^T u^2(t)dt + \sum_1^N \omega_i \left(\int_0^T l_i(s)u(s)ds - \alpha_i \right)^2. \tag{7.5}$$

Hilbert Space Spline Construction

This spline construction seeks to obtain finite reparametrization of the minimization problem. We assume that the nodes $0 < t_1 < \cdots < t_N = T$ and begin this construction by considering the interval $[0,t_1)$. The problem thus reduces to fitting a curve $y(t)$ between the points (0,0) and (t_1, α_1) under the given constraints $y(0) = 0 = \delta_1$, $y(t_1) = \delta_2$, $\delta_1 \leq \delta_2$, and $\dot{y}(t) \geq 0$ for all $t \in [0,t_1)$. We define the constraint variety as

$$V_{\delta_2} = \left\{ u : \int_0^{t_1} (t_1 - s)u(s)ds = \delta_2 \right\}$$

and the orthogonal complement of V_0 is

$$V_0^{\perp} = \left\{ v : \int_0^{t_1} v(s)u(s)ds = 0 \forall u \in V_0 \right\}.$$

The optimal control that solves this problem is given by

$$u^*(s) = \begin{cases} k_1(t_1 - s) : s \in [0,t_1) \\ 0 \qquad\qquad : otherwise \end{cases}$$

where from the initial conditions we get

$$k_1 = \frac{\delta_2 - \delta_1}{\int_0^{t_1} (t_1 - s)^2 ds}.$$

Then, for the remaining intervals, that is $[t_j, t_{j+1})$ for $j = 1, \cdots, N-1$, we want to fit a curve $y(t)$ between the points (t_j, α_j) and (t_{j+1}, α_{j+1}) under the constraints $y(t_j) = \delta_{j+1}$, $y(t_{j+1}) = \delta_{j+2}$, $\delta_{j+2} \geq \delta_{j+1}$ and $\dot{y}(t) \geq 0$. Proceeding as before gives the control function

$$u^*_{j+1}(s) = \begin{cases} k_{j+1}(t_{j+1} - s) : s \in [t_j, t_{j+1}) \\ 0 \quad\quad\quad\quad\quad\quad : otherwise \end{cases}$$

where $k_{j+1} = \frac{\delta_{j+2} - \delta_{j+1}}{\int_{t_j}^{t_{j+1}} (t_{j+1} - s)^2 ds}$, for $j = 1, \cdots, N-1$. From this construction we get that the optimal spline that approximates the data is

$$y(t) = \begin{cases} k_1 \int_0^t (t-s)(t_1 - s)ds & : t \in [0, t_1) \\ k_{j+1} \int_{t_j}^t (t-s)(t_{j+1} - s)ds + \delta_{j+1} & : t \in [t_j, t_{j+1}) \\ \delta_{N+1} & : t = t_N = T. \end{cases}$$

The problem then becomes one of determining the $(N+1)$-vector δ that minimizes the cost

$$\begin{aligned} J(\delta) &= \lambda k_1 \int_0^{t_1} u(s)ds + \lambda \sum_1^{N-1} \int_{t_j}^{t_{j+1}} u_{j+1}(s)ds \\ &+ \sum_1^N \omega_j (\delta_{j+1} - \alpha_j)^2 \\ &= (\delta_2 - \delta_1)a_1 + \sum_1^{N-1} (\delta_{j+2} - \delta_{j+1})a_{j+1} \\ &+ \sum_1^N b_j (\delta_{j+1} - \alpha_j)^2 \end{aligned}$$

subject to $\delta_{j+2} - \delta_{j+1} \geq 0, \delta_1 = 0$, and $\delta_{N+1} = 1$. We may write this problem in equivalent matrix form

$$\min_\delta f'\delta + \frac{1}{2}\delta' H \delta,$$

where f is a $(N+1) \times 1$ vector with $f' = [-a_1, (a_1 - a_2 - 2\alpha_1 b_1), \cdots, (a_{N-1} - a_N - 2\alpha_N b_N), (a_N - 2\alpha_N b_N)]$, H is a $(N+1) \times (N+1)$ matrix with components $H = \text{diag}\{0, 2\omega_1, \cdots, 2\omega_N\}$ and constants

$$a_{j+1} = \begin{cases} \dfrac{\lambda \int_0^{t_1} (t_1 - s)ds}{\int_0^{t_1} (t_1 - s)^2} & : j = 0 \\ \dfrac{\lambda \int_{t_j}^{t_{j+1}} (t_{j+1} - s)ds}{\int_{t_j}^{t_{j+1}} (t_{j+1} - s)^2 ds} & : j = 1, \cdots, N-1 \end{cases}$$

Using this optimization routine produces curves as shown in Figure 7.3. Here we obtain an approximation which is monotonically increasing and satisfies the end conditions $y(0) = 0$ and $y(T) = 1$, however, the spline is not differentiable. This is an important property in CDF estimation since the derivative should produce an estimate for the continuous probability distribution function from which the data was sampled.

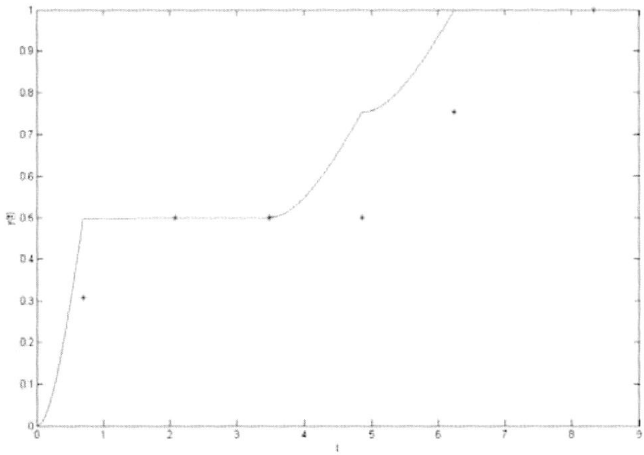

Fig. 7.3. Optimal spline with equal weight assigned to all way points and $\lambda = 0.001$.

Lagrangian Spline Construction

Based on the cost function defined in (7.5) we can form the associated Lagrangian

$$L(u,\nu) = \lambda \int_0^T u^2(t)dt - \int_0^T \dot{y}(t)d\nu(t) + \sum_1^N \omega_i \left(\int_0^T l_i(s)u(s)ds - \alpha_i \right)^2, \tag{7.6}$$

where $\dot{y}(t) = \int_0^t u(s)ds \geq 0$, $\nu \in BV[0,T]$, the space of functions of bounded variation on $[0,T]$, which is the dual space of $C[0,T]$ [5]. Integrating the Stieltjes integral by parts yields

$$L(u,\nu) = \lambda \int_0^T u^2(t)dt - \int_0^T (\nu(T) - \nu(t))u(t)dt + \sum_1^N \omega_i \left(\int_0^T l_i(s)u(s)ds - \alpha_i \right)^2, \tag{7.7}$$

Thus, the optimal curve [3] is determined by solving the problem

$$\max_{\nu \geq 0} \inf_{u} L(u, \nu). \tag{7.8}$$

It is shown in [3] and [4] that the set of control functions which solves this optimization problem exists and is unique. Moreover, due to the convexity of the problem, we can obtain the optimal control function by calculating the Frechet differential of L with respect to u as follows. Following [3], we let $L_\nu(u) = L(u, \nu)$, then for $h \in L^2[0,T]$

$$\partial L_\nu(u,h) = \frac{1}{\varepsilon} \lim_{\varepsilon \to 0} (L_\nu(u + \varepsilon h) - L_\nu(u)) =$$
$$\int_0^T (2\lambda u(t) - (\nu(T) - \nu(t)))h(t)dt +$$
$$2\sum_1^N \omega_i \int_0^T \left(\int_0^T (l_i(s)u(s)ds - \alpha_i) \right) l_i(t)h(t)dt$$

Hence, the differential is zero for all $h \in L^2[0,T]$ whenever

$$2\lambda u^*(t) + 2\sum_i^N \omega_i(y(t_i) - \alpha_i)l_i(t) - (\nu(T) - \nu(t)) = 0.$$

The above equation is true especially for the optimal $\nu = \nu^*$, which gives

$$2\lambda u^*(t) + 2\sum_i^N \omega_i(y(t_i) - \alpha_i)l_i(t) - C_t = 0,$$

where $C_t = \nu^*(T) - \nu^*(t) \geq 0$ from the positivity constraint on ν^* whenever $\dot{y}(t) > 0$. For the second order system, $l_i(t)$ is linear in t for $i = 1, \cdots, N$, and so the above equation gives that $u^*(t)$ has to piecewise linear. Based on the definition of $l_i(t)$, we have that the optimal control changes at the specified way points and whenever $\dot{y}(t) = 0$. Also, if $\dot{y}(t) = 0$ on an interval, then $u^*(t) = 0$. Therefore the optimal control is a piecewise linear function for all $t \in [0,T]$.

To determine the optimal control u^* that minimizes our cost using this Lagrangian method for spline construction, requires first determining the optimal function $\nu^* \in BV[0,T]$. This increases the difficulty of obtaining a solution to our problem and for this reason we have chosen to go no further with such construction.

7.3 Dynamic Programming

In this section, we illustrate the reformulation of the monotone problem in a finite setting that can be handled easily. Furthermore, since our main goal is to approximate the CDF, we will require

$$y(0) = 0 \text{ and } y(T) = 1.$$

Dividing the cost function (7.5) into an interpolation and smoothing part yields the optimal value function

$$
\begin{aligned}
\hat{S}_i(y_i, \dot{y}_i) &= \min_{y_{i+1} \geq y_i, y_{i+1} \geq 0} \{\lambda V_i(y_i, \dot{y}_i, y_{i+1}, \dot{y}_{i+1}) \\
&\quad + \hat{S}_{i+1}(y_{i+1}, \dot{y}_{i+1})\} \\
&\quad + \omega_i (y_i - \alpha_i)^2, i = 0, \cdots, N-1 \\
\hat{S}_N(y_N, \dot{y}_N) &= \omega_N (y_N - \alpha_N)^2
\end{aligned}
$$

subject to $\sum_0^N y_{i+1} - y_i = 1$, which is equivalent to $y(T) = 1$, where $V_i(y_i, \dot{y}_i, y_{i+1}, \dot{y}_{i+1})$ is the cost for driving the system between $(y_i, \dot{y}_i)$ and $(y_{i+1}, \dot{y}_{i+1})$ while keeping the derivative nonnegative.

The optimal solution is thus found by determining $\hat{S}(0,0)$ where we let $\omega_0 = 0$ and α_0 be an arbitrary number. Solving this dynamic programming problem reduces to determining the function $V_i(y_i, \dot{y}_i, y_{i+1}, \dot{y}_{i+1})$, which is equivalent to finding the $2 \times N$ variables $y_1, \cdots, y_N, \dot{y}_1, \cdots, \dot{y}_N$. This is the finite reparametrization of the infinite dimensional problem.

Under specified assumptions, in [3] and [4], the cost in the optimal value function reduces to

$$
\begin{aligned}
&V_i(y_i, \dot{y}_i, y_{i+1}, \dot{y}_{i+1}) = \\
&\qquad 4\frac{\dot{y}_i(t_{i+1} - t_i)^2 - 3(y_{i+1} - y_i)(t_{i+1} - t_i)(\dot{y}_i + \dot{y}_{i+1})}{(t_{i+1} - t_i)^3} \\
&\qquad + 4\frac{3(y_{i+1} - y_i)^2 + (t_{i+1} - t_i)^2 \dot{y}_{i+1}^2}{(t_{i+1} - t_i)^3}, \\
&\qquad \text{if } y_{i+1} - y_i \geq \chi(t_{i+1} - t_i, \dot{y}_i, \dot{y}_{i+1}), \\
&\qquad \frac{4(\dot{y}_{i+1}^{3/2} + \dot{y}_i^{3/2})^2}{9(y_{i+1} - y_i)}, \text{ if } y_{i+1} - y_i < \chi(t_{i+1} - t_i, \dot{y}_i, \dot{y}_{i+1})
\end{aligned}
$$

where $\chi(t_{i+1} - t_i, \dot{y}_i, \dot{y}_{i+1}) = \frac{t_{i+1} - t_i}{3}(\dot{y}_i + \dot{y}_{i+1} - \sqrt{\dot{y}_i \dot{y}_{i+1}})$ and $t_0 = y_0 = \dot{y}_0 = 0$. Using the dynamic programming algorithm for CDF approximation yields optimal curves shown in Figure 7.4.

7.4 Conclusion

In this paper, we have shown that the dynamic programming algorithm implemented for CDF estimation produces a spline $y(t)$ satisfies all required constraints of our problem, that is,

$$
y(0) = 0, y(T) = 1, \dot{y}(t) \geq 0, \text{ and } y \in C^1[0, T].
$$

The solution is produced with an easily implemented numerically sound algorithm. Later methods of monotone spline construction include work shown in [8] . For the second order system considered, the monotone cubic splines converge quadratically to the probability distribution function. We expect much faster convergence rates using monotone quintic splines, however, this construction is yet to be developed.

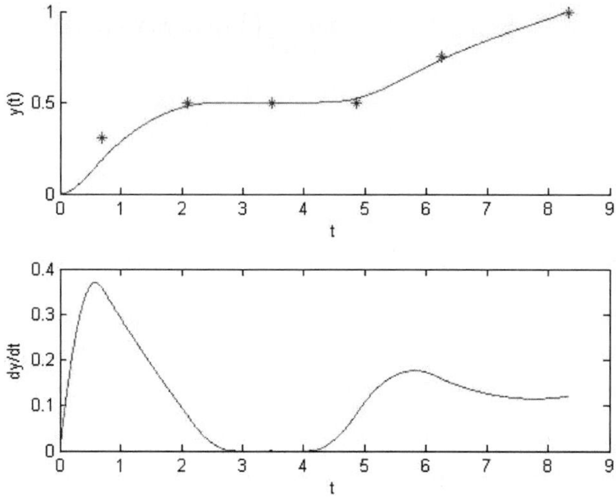

Fig. 7.4. The curve shown represents the optimal solution to the problem of estimating the cdf of data summarized with the empirical cdf. Here we take $\lambda = 0.01$

References

1. Charles, J.K.: Probability Distribution Estimation using Control Theoretic Smoothing Splines. Dissertation, Texas Tech University (2009)
2. Eubank, R.L.: Nonparametric Regression and Spline Smoothing. Statistics: Textbooks and Monographs, vol. 157. Marcel Dekker, Inc., New York (1999)
3. Egerstedt, M., Martin, C.F.: Monotone Smoothing Splines. In: Mathematical Theory of Networks and Systems, Perpignan, France (2000)
4. Egerstedt, M., Martin, C.F.: Control Theoretic Splines: Optimal Control, Statistics, and Path Planning. Princeton University Press (in press)
5. Luenbeger, D.G.: Optimization by Vector Space Methods. John Wiley & Sons, New York (1969)
6. Nagahara, M., Sato, K., Yamamoto, Y.: H^{∞} Optimal Nonparametric Density Estimation from Quantized Samples. Submitted to ISCIE SSS
7. Martin, C.F., Egerstedt, M.: Trajectory Planning for Linear Control Systems with Generalized Splines. In: Mathematical Theory of Networks and Systems, Padova, Italy (1998)
8. Meyer, M.C.: Inference Using Shape-Restricted Regression Splines. Annals of Applied Statistics 2(3), 1013–1033 (2008)
9. Silverman, B.W.: Density Esimation for Statistics and Data Analysis. Chapman and Hall, London (1986)
10. Silverman, B.W.: Spline Smoothing: The Equivalent Variable Kernel Method. Ann. Statist. 12, 898–916 (1984)
11. Silverman, B.W.: Some Aspects of the Spline Smoothing Approach Nonparametric Regression Curve Fitting. J. Royal Statist, Soc. B, 1–52 (1985)
12. Wahba, G.: Spline Models for Observational Data. In: CBMS-NSF Regional Conference Series in Applied Mathematics, 59, SIAM, SIAM, Philadelphia (1990)

8

Global Output Regulation with Uncertain Exosystems*

Zhiyong Chen[1] and Jie Huang[2]

[1] School of Electrical Engineering and Computer Science, The University of Newcastle, Callaghan, NSW 2308, Australia.

[2] Department of Mechanical and Automation Engineering, The Chinese University of Hong Kong, Shatin, New Territories, Hong Kong, China

Summary. The asymptotic tracking and disturbance rejection problem for a class of uncertain nonlinear lower triangular systems is studied when the reference trajectories and/or disturbances are finite combinations of sinusoids with arbitrary unknown amplitudes, phases, any frequencies. The explicit regulator design relies on the internal model principle and a newly developed robust adaptive technique.

8.1 Introduction

Consider a class of lower-triangular systems described as follows,

$$\begin{aligned} \dot{q}_o &= \kappa_o(\mathscr{Q}_1, v, w) \\ \dot{q}_i &= \kappa_i(\mathscr{Q}_i, v, w) + q_{i+1},\ i = 1, \cdots, r \\ e &= q_1 - q_d(v, w), \end{aligned} \tag{8.1}$$

where $\mathscr{Q}_i := \text{col}(q_o, q_1, \cdots, q_i)$, $q_o \in \mathbb{R}^{n_o}$ and $q_i \in \mathbb{R}$ are the states, $u(t) := q_{r+1} \in \mathbb{R}$ is the input, and $e(t) \in \mathbb{R}$ is the output representing the tracking error. All functions in the system (8.1) are polynomial. The disturbance and/or reference signal $v \in \mathbb{R}^q$ is produced by a linear exosystem described by

$$\dot{v} = A_1(\sigma)v,\ \ v(0) = v_o. \tag{8.2}$$

The unknown parameters $w \in \mathbb{R}^{p_1}$ and $\sigma \in \mathbb{R}^{p_2}$ are assumed to be in known compact sets $\mathbb{W}$ and $\mathbb{S}$, respectively. Also, we assume v_o is in a known compact set $\mathbb{V}_o$, and hence $v(t) \in \mathbb{V}, \forall t \geq 0$ for a compact set $\mathbb{V}$ due to the following assumption which means that the solution of the exosystem is a sum of finite many sinusoidal functions. Typically, σ represents the frequencies of these sinusoidal functions.

*The work of the first author was supported by the Australian Research Council under grant No. DP0878724. The work of the second author was supported by the Research Grants Council of the Hong Kong Special Administration Region under grant No. 412408.

X. Hu et al. (Eds.): Three Decades of Progress in Control Sciences, pp. 105–119, 2010.

Assumption 8.1.1. *For all $\sigma \in \mathbb{S}$, the exosystem is assumed to be neutrally stable in the sense that all the eigenvalues of $A_1(\sigma)$ are simple and have zero real part.* ∎

The output regulation problem is also called a servomechanism problem which aims to achieve asymptotically tracking a class of reference inputs and rejecting a class of disturbances. Here both the reference inputs and disturbances are generated by the exosystem (8.2). A necessary condition for output regulation problem is the existence of a sufficiently smooth function satisfying the so-called regulator equations. Due to the special lower triangular structure of the system (8.1), the solvability of the regulator equations reduces to the following assumption.

Assumption 8.1.2. *There exists $\mathbf{q}_o(v,w,\sigma)$, a polynomial function in v with coefficients depending on w and σ such that*

$$\frac{\partial \mathbf{q}_o(v,w,\sigma)}{\partial v} A_1(\sigma)v = \kappa_o(\mathbf{q}_o(v,w,\sigma), q_d(v,w), v, w)$$

for all $v \in \mathbb{V}$, $w \in \mathbb{W}$, and $\sigma \in \mathbb{S}$. ∎

Under Assumption 8.1.2, we can define, for all $v \in \mathbb{V}$, $w \in \mathbb{W}$, and $\sigma \in \mathbb{S}$, the functions $\mathbf{q}_1(v,w,\sigma) = q_d(v,w)$, and $\mathbf{q}_i(v,w,\sigma), i = 2,\cdots,r+1$, as follows:

$$\mathbf{q}_i(v,w,\sigma) = \frac{\partial \mathbf{q}_{i-1}(v,w,\sigma)}{\partial v} A_1(\sigma)v - \kappa_{i-1}(\mathbf{q}_o(v,w,\sigma),\cdots,\mathbf{q}_{i-1}(v,w,\sigma),v,w).$$

Let $\mathbf{q}(v,w,\sigma) = \mathrm{col}(\mathbf{q}_o(v,w,\sigma),\cdots,\mathbf{q}_r(v,w,\sigma))$ and $\mathbf{u}(v,w,\sigma) := \mathbf{q}_{r+1}(v,w,\sigma)$. Then the functions $\mathbf{q}(v,w,\sigma)$ and $\mathbf{u}(v,w,\sigma)$ constitute the solution of the regulator equations for the system composed of (8.1) and (8.2).

With $u = \mathbf{u}(v,w,\sigma)$, the solution $\mathbf{q}(v,w,\sigma)$ defines an invariant manifold

$$\{(q,v,w,\sigma)\,|\,q = \mathbf{q}(v,w,\sigma)\}$$

for the composite system (8.1), (8.2), $\dot{w} = 0$, and $\dot{\sigma} = 0$. The error output of the system is identically zero on this manifold, which is called the output zeroing invariant manifold. This manifold is a center manifold of the system when the exosystem satisfies Assumption 8.1.1, The objective of the output regulation problem is to further make this manifold globally attractive by feedback of available variables. To make the above statement more precise, we give the formulation of the global robust output regulation problem as follows.

Global Output Regulation Problem: For given $\mathbb{V}$, $\mathbb{W}$, and $\mathbb{S}$, which are compact subsets of $\mathbb{R}^q$, $\mathbb{R}^{p_1}$, and $\mathbb{R}^{p_2}$ containing the origins, respectively, find a state feedback controller such that, for all $v(t) \in \mathbb{V}$, $w \in \mathbb{W}$, and $\sigma \in \mathbb{S}$, the trajectories of the closed-loop system, starting from any initial states, exist and are bounded for all $t > 0$, and satisfy

$$\lim_{t\to\infty}\{q(t) - \mathbf{q}(v(t),w,\sigma)\} = 0 \text{ and } \lim_{t\to\infty}\{u(t) - \mathbf{u}(v(t),w,\sigma)\} = 0.$$

The above problem formulation implies the fulfilment of the asymptotic tracking $\lim_{t\to\infty} e(t) = 0$ of the closed-loop system by noting $e = q_1 - \mathbf{q}_1(v,w,\sigma)$. This definition is consistent with the case where the exosystem is known, e.g., in [1]. When the exosystem is exactly known, the robust output regulation problem has been extensively studied within the framework of internal model principle (see, e.g., [2, 3, 4, 5, 6]). Technically, the output regulation problem of a system can be converted into a stabilization problem of an augmented system composed of the given plant and a well defined dynamic compensator called internal model. Therefore, a key step in solving the output regulation problem is the stabilization of the augmented system. When the exosystem is known, the resulting stabilization problem of the augmented system can often be handled by various robust control techniques. However, when the exosystem is not exactly known, e.g., the matrix $A_1(\sigma)$ depends on an unknown constant vector σ, the resulting stabilization problem becomes more intriguing. Robust control techniques are not adequate to handle the uncertainties in the augmented system caused by the uncertain parameter σ. Various adaptive control techniques are needed to stabilize the augmented system. In this paper, we will apply a newly developed robust adaptive control design approach in [7] to stabilize the augmented system.

The output regulation problem for nonlinear systems with unknown exosystem has been handled elsewhere for some different scenarios. For example, the semi-global output regulation problem of lower-triangular systems by an output feedback control is given in [3], and the global output regulation problem of a class of interconnected output feedback systems by output feedback control is given in [8]. Also, a global asymptotic tracking problem of lower triangular systems is studied in [9]. The problem studied in [9] is a special case of the problem described above in that the exogenous signal v does not appear in the functions κ_i's in (8.1). The main difficulty encountered in our current problem is that we have to use the state feedback to handle the global output regulation of system (8.1) because the output feedback control cannot handle the global problem without some additional restrictive assumptions. However, the state feedback control entails the construction of a series of r internal models in contrast with the construction of a single internal model for the output feedback case. As a result, the resulting augmented system will also be more complex. In particular, the adaptive control of the augmented system has to be done by a recursive approach. Each recursion necessitates a dynamic coordinate transformation which leads to a newly augmented system with more complex parameter uncertainty. This phenomenon is called "propagation of uncertainties". The approach in [9] can only handle the special case where the functions κ_i's do not contain v. In this paper, by utilizing the newly developed robust adaptive control design approach in [7], we will extend the work in [9] to a full output regulation problem.

The rest of the paper is organized as follows. Section 8.2 provides a typical construction of the internal model to deal with the unknown parameter in the exosystem Based on the internal model and the robust adaptive control approach proposed in [7], the global robust output regulation problem formulated in this paper is solved in Section 8.3 followed by a numerical example. Finally, Section 8.4 closes this paper with some concluding remarks.

8.2 Problem Conversion

Let us first recall from the general framework established in [6] that the robust output regulation problem can be approached in two steps. In the first step, an appropriate dynamic compensator called internal model is designed. Attachment of the internal model to the given plant leads to an augmented system subject to an external disturbance. The internal model has the property that the solution of a well defined regulation problem of the augmented system will lead to the output regulation solution of the given plant and exosystem. Thus, once an appropriate internal model is available, it remains to tackle a global stabilization problem of the augmented system.

Let us first note that $\mathbf{q}_{i+1}(v,w,\sigma), i=1,\cdots,r$, are polynomial in v with coefficients depending on w and σ. Thus, there exists nonnegative integers r_i, functions

$$\vartheta_i(v,w,\sigma) := \operatorname{col}\left(\mathbf{q}_{i+1}(v,w,\sigma), \dot{\mathbf{q}}_{i+1}(v,w,\sigma), \cdots, \frac{d^{(r_i-1)}\mathbf{q}_{i+1}(v,w,\sigma)}{dt^{(r_i-1)}}\right),$$

and matrices $\Phi_i(\sigma) \in \mathbb{R}^{r_i \times r_i}$ such that

$$\dot{\vartheta}_i = \Phi_i(\sigma)\vartheta_i, \ \ \mathbf{q}_{i+1}(v,w,\sigma) = [1 \ 0 \ \cdots \ 0]\vartheta_i. \tag{8.3}$$

Moreover, all the eigenvalues $\Phi_i(\sigma)$ are simple with zero real parts.

Remark 8.2.1. For convenience, we allow some of r_i to be zero so that the above derivation also applied to the special case where $\mathbf{q}_{i+1}(v,w,\sigma)$ is identically zero. In this case, the dimension of ϑ_i is understood to be zero. ∎

Let

$$\begin{aligned}
\theta_i(v,w,\sigma) &:= T_i(\sigma)\vartheta_i(v,w,\sigma)\\
E_i(\sigma) &:= T_i(\sigma)\Phi_i(\sigma)T_i^{-1}(\sigma)\\
\Psi_i(\sigma) &:= [1 \ 0 \ \cdots \ 0]T_i^{-1}(\sigma).
\end{aligned}$$

Also, let

$$\begin{aligned}
\theta(v,w,\sigma) &= \operatorname{col}(\theta_1(v,w,\sigma),\cdots,\theta_r(v,w,\sigma))\\
\alpha(\sigma,\theta) &= \text{block diag}(E_1(\sigma)\theta_1,\cdots,E_r(\sigma)\theta_r)\\
\beta(\sigma,\theta) &= \operatorname{col}(\Psi_1(\sigma)\theta_1,\cdots,\Psi_r(\sigma)\theta_r).
\end{aligned} \tag{8.4}$$

Then, it can be verified that $\{\theta(v,w,\sigma), \alpha(\sigma,\theta), \beta(\sigma,\theta)\}$ satisfies

$$\dot{\theta}(v,w,\sigma) = \alpha(\sigma,\theta), \ \ \operatorname{col}(\mathbf{q}_2,\cdots,\mathbf{q}_r,\mathbf{u}) = \beta(\sigma,\theta).$$

The triplet $\{\theta(v,w,\sigma), \alpha(\sigma,\theta), \beta(\sigma,\theta)\}$ is called a steady-state generator of the system (8.1) and (8.2) with output $\operatorname{col}(q_2,\cdots,q_r,u)$ [6].

As, for each i, system (8.3) is linear and observable, it is possible to construct a so-called canonical internal model corresponding to each i suggested by [10]. For this purpose, pick any controllable pairs (M_i, N_i) with $M_i \in \mathbb{R}^{r_i \times r_i}$, $N_i \in \mathbb{R}^{r_i \times 1}$, and M_i Hurwitz, and solve $T_i(\sigma)$ from the Sylvester equation

$$T_i(\sigma)\Phi_i(\sigma) - M_i T_i(\sigma) = N_i[1\ 0\ \cdots\ 0].$$

Furthermore, let

$$\dot{\eta}_i = M_i \eta_i + N_i q_{i+1},\ \ i = 1, \cdots, r. \tag{8.5}$$

Then, (8.5) defines an internal model for the system (8.1) and (8.2) with output col $(q_2, \cdots, q_r, u)$. The composition of the system (8.1) and the internal model (8.5) is called an augmented system. Again, we note that the dimension of η_i is understood to be zero when $\mathbf{q}_{i+1}(v, w, \sigma)$ is identically zero.

If σ is known, performing on the augmented system the following coordinate and input transformation

$$\begin{aligned} \bar{q}_o &= q_o - \mathbf{q}_o(v, w, \sigma),\ \ \bar{q}_1 = e, \\ \bar{q}_{i+1} &= q_{i+1} - \Psi_i(\sigma)\eta_i,\ \bar{\eta}_i = \eta_i - \theta_i - N_i \bar{q}_i,\ \ i = 1, \cdots, r \end{aligned} \tag{8.6}$$

yields a system of the following form

$$\begin{aligned} \dot{\bar{q}}_o &= \bar{\kappa}_o(\bar{\mathcal{Q}}_1, d) \\ \dot{\bar{\eta}}_i &= M_i \bar{\eta}_i + \gamma_i(\bar{\zeta}_{i-1}, \mathcal{Q}_i, d) \\ \dot{\bar{q}}_i &= \bar{\kappa}_i(\bar{\mathcal{Q}}_i, d) + \bar{q}_{i+1},\ i = 1, \cdots, r \end{aligned} \tag{8.7}$$

where $d = \mathrm{col}(v, w)$, $\bar{\mathcal{Q}}_i := \mathrm{col}(\bar{q}_o, \bar{q}_1, \cdots, \bar{q}_i)$, $\bar{\zeta}_i := \mathrm{col}(\bar{\eta}_1, \cdots, \bar{\eta}_i)$, $\bar{\kappa}_i(0, d) = 0$, and $\gamma(0, 0, d) = 0$. If there is a controller of the form

$$\bar{u} = \bar{q}_{r+1} = g(\lambda, \bar{\mathcal{Q}}_r),\ \dot{\lambda} = \psi(\lambda, \bar{\mathcal{Q}}_r)$$

that solves the global stabilization problem of system (8.7), then the following controller

$$\begin{aligned} u &= g(\lambda, \bar{\mathcal{Q}}_r) + \Psi_r(\sigma)\eta_r \\ \dot{\lambda} &= \psi(\lambda, \bar{\mathcal{Q}}_r) \\ \dot{\eta}_i &= M_i \eta_i + N_i q_{i+1}, i = 1, \cdots, r \end{aligned} \tag{8.8}$$

solves the output regulation problem of the original system (8.1) [6]. Nevertheless, when σ is unknown, the controller (8.8) is not implementable. To overcome the difficulty caused by the uncertain exosystem, we consider the following coordinate transform:

$$\begin{aligned} &x_o = q_o - \mathbf{q}_o(v, w, \sigma),\ x_1 = e,\ x_i = q_i,\ i = 2, \cdots, r+1 \\ &z_i = \eta_i - \theta_i - N_i x_i,\ i = 1, \cdots, r \\ &d = \mathrm{col}(v, w, \sigma),\ \mu = \mathrm{col}(w, \sigma). \end{aligned}$$

Under the new coordinates, the augmented system (8.1) and (8.5) can be rewritten as follows:

$$\begin{aligned}\dot{x}_o &= f_o(x_o, x_1, d)\\ \dot{z}_i &= M_i z_i + \gamma_i(\zeta_{i-1}, \chi_i, d) + \delta g_i(\zeta_{i-1}, \chi_i, d)\\ \dot{x}_i &= f_i(\zeta_i, \chi_i, d) + \delta p_i(\zeta_i, \chi_i, d) + x_{i+1},\ i = 1, \cdots, r\end{aligned} \tag{8.9}$$

with $\chi_i := \mathrm{col}(x_o, \cdots, x_i)$ and $\zeta_i := \mathrm{col}(z_1, \cdots, z_i)$. The functions are defined as follows:

$$\begin{aligned}f_o(x_o, x_1, d) &= \kappa_o(q_o, q_1, v, w) - \kappa_o(\mathbf{q}_o, \mathbf{q}_1, v, w)\\ \gamma_1(\chi_1, d) &= M_1 N_1 x_1 - N_1 A_1,\ \ \delta g_1(\chi_1, d) = 0\\ f_1(\zeta_1, \chi_1, d) &= A_1 + \Psi_1(\sigma)\eta_1 - \Psi_1(\sigma)\theta_1\\ \delta p_1(\zeta_1, \chi_1, d) &= -\Psi_1(\sigma)\eta_1\end{aligned}$$

and for $i = 2, \ldots, r$,

$$\begin{aligned}\gamma_i(\zeta_{i-1}, \chi_i, d) &= M_i N_i x_i - N_i A_i + N_i \Psi_{i-1}(\sigma) E_{i-1}(\sigma)(N_{i-1} x_{i-1} + z_{i-1})\\ \delta g_i(\zeta_{i-1}, \chi_i, d) &= -N_i B_i - N_i \Psi_{i-1}(\sigma) E_{i-1}(\sigma)\eta_{i-1}\\ f_i(\zeta_i, \chi_i, d) &= A_i - \Psi_{i-1}(\sigma) E_{i-1}(\sigma)(N_{i-1} x_{i-1} + z_{i-1}) + \Psi_i(\sigma)(N_i x_i + z_i)\\ \delta p_i(\zeta_i, \chi_i, d) &= B_i + \Psi_{i-1}(\sigma) E_{i-1}(\sigma)\eta_{i-1} - \Psi_i(\sigma)\eta_i\end{aligned}$$

where

$$\begin{aligned}A_i &:= \kappa_i(q_o, q_1, \Psi_1(\sigma)\eta_1, \cdots, \Psi_{i-1}(\sigma)\eta_{i-1}, v, w) - \kappa_i(\mathbf{q}_o, \cdots, \mathbf{q}_i, v, w)\\ B_i &:= \kappa_i(\mathcal{Q}_i, v, w) - \kappa_i(q_o, q_1, \Psi_1(\sigma)\eta_1, \cdots, \Psi_{i-1}(\sigma)\eta_{i-1}, v, w).\end{aligned}$$

It can be verified that $f_i(0, 0, d) = 0, i = 0, 1, \cdots, r$, and $\gamma_i(0, 0, d) = 0, i = 1, \cdots, r$. What left is to find an adaptive controller $u := x_{i+1}$ for the system (8.9) such that the states of the closed-loop system from any initial condition are bounded, and $\lim_{t \to \infty} e(t) = 0$. Such a problem is called a global adaptive stabilization problem, whose solvability implies that for the original system (8.1). Various control problems of the class of systems (8.9) have been studied in several papers [11, 12, 13] under various assumptions. In particular, the problem studied recently in [7] is motivated by the adaptive stabilization problem of (8.9) and can be directly utilized in this paper. In other words, the global output regulation for the original system (8.1) is solved by combining the internal model introduced in this section and the adaptive regulator/stablizer proposed in [7].

8.3 Robust Adaptive Controller Design

As the system (8.9) involves both static and dynamic uncertainties and the dynamic uncertainty does not satisfy input-to-state stability assumption. These complexities entail an approach that integrates both robust and adaptive techniques. As always, the key for developing an adaptive control law is to find an appropriate Lyapunov function candidate for the system to be controlled. Such a Lyapunov function exists under the following assumption.

Assumption 8.3.1. *There exists a sufficiently smooth function $V(q_o)$ bounded by some class $\mathcal{K}_\infty$ polynomial functions, such that, along the trajectories of $\dot{q}_o = \kappa_o(q_o, q_1, v, w)$,*

$$\frac{dV(q_o)}{dt} \leq -\|q_o\|^2 + \pi(q_1) \tag{8.10}$$

for some polynomial positive definite function π. ∎

Let $\alpha_i(x_i)$, $i = 1, \cdots, r$, be some sufficiently smooth functions. Applying the following coordinate transformation

$$\tilde{x}_o = x_o,\ \tilde{x}_1 = x_1,\ \tilde{x}_{i+1} = x_{i+1} - \alpha_i(\tilde{x}_i),\ i = 1, \cdots, r \tag{8.11}$$

to the system (8.9) with $\delta g_i = 0$ and $\delta p_i = 0$ gives

$$\begin{aligned}
\dot{x}_0 &= f_0(x_o, \tilde{x}_1, d)\\
\dot{z}_i &= M_i z_i + \varphi_i(\zeta_{i-1}, \tilde{\chi}_i, d)\\
\dot{\tilde{x}}_i &= \phi_i(\zeta_i, \tilde{\chi}_i, d) + \alpha_i(\tilde{x}_i) + \tilde{x}_{i+1},\ i = 1, \cdots, r
\end{aligned} \tag{8.12}$$

where $\tilde{\chi}_i = \mathrm{col}(x_o, \tilde{x}_1, \cdots, \tilde{x}_i)$ and

$$\begin{aligned}
\varphi_i(\zeta_{i-1}, \tilde{\chi}_i, d) &= \gamma_i(\zeta_{i-1}, x_o, \tilde{x}_1, \tilde{x}_2 + \alpha_1(\tilde{x}_1), \tilde{x}_i + \alpha_{i-1}(\tilde{x}_{i-1}), d)\\
\phi_i(\zeta_i, \tilde{\chi}_i, d) &= f_i(\zeta_i, x_o, \tilde{x}_1, \tilde{x}_2 + \alpha_1(\tilde{x}_1), \cdots, \tilde{x}_i + \alpha_{i-1}(\tilde{x}_{i-1}), d)\\
&\quad -(\partial\alpha_{i-1}(\tilde{x}_{i-1})/\partial\tilde{x}_{i-1})(\phi_{i-1}(\tilde{\zeta}_{i-1}, \tilde{\chi}_{i-1}, d) + \alpha_{i-1}(\tilde{x}_{i-1}) + \tilde{x}_i)
\end{aligned}$$

with $\phi_1(\zeta_1, \tilde{\chi}_1, d) = f_1(\zeta_1, \chi_1, d)$.

The following proposition is from [13] with a slight modification for the need of this paper. A direct implication of this theorem is that the static controller $u = \alpha_r(\tilde{x}_r)$ globally stabilizes the system (8.9) with $\delta g_i = 0$ and $\delta p_i = 0$. Moreover, $V(\zeta_r) + W(\tilde{\chi}_r)$ is an Lyapunov function for the closed-loop system.

Proposition 8.3.1. *Under Assumption 8.3.1, there exist polynomial functions $\alpha_i(\cdot)$, $i = 1, \cdots, r$, and positive definite and radially unbounded functions $V(\zeta_r)$ and $W(\tilde{\chi}_r) = \Sigma_{i=0}^{r} W_i\ (\tilde{x}_i)$, such that, along the trajectories of the system (8.12) with $\tilde{x}_{r+1} = 0$,*

$$\frac{d(V(\zeta_r) + W(\tilde{\chi}_r))}{dt} \leq -k(\zeta_r, \tilde{\chi}_r) \tag{8.13}$$

for some positive definite function $k(\cdot, \cdot)$.

The major difficulty in solving our problem is to deal with the non-trivial terms δg_i and δp_i. This difficulty will be overcome by introducing an adaptive control technique. To this end, we require some uncertain functions to be linearly parameterized. Thus, we need the following assumptions:

Assumption 8.3.2. *There exist polynomial functions $m_i, \bar{m}_i, h_i, l_i$ such that, $y_i = m_i(\zeta_i, \chi_i, d)$ and $\bar{y}_i = \bar{m}_i(\zeta_{i-1}, \chi_i, d)$ are measurable, and*

$$\delta g_i(\zeta_{i-1}, \chi_i, d) = h_i(\bar{y}_i, \mu), \ \ \delta p_i(\zeta_i, \chi_i, d) = l_i(y_i, \mu) \tag{8.14}$$

for all ζ_i, χ_i. Moreover, for $i = 1, \cdots, r-1$, $\dot{y}_i = \varkappa_i(y_{i+1}, \mu)$ for some polynomial function $\varkappa_i$. ∎

Assumption 8.3.3. *There exists a polynomial function $\bar{f}_i$ such that*

$$\bar{f}_i(y_i, \zeta_i - \bar{\zeta}_i, \chi_i, \bar{\chi}_i, \mu) = f_i(\zeta_i, \chi_i, d) - f_i(\bar{\zeta}_i, \bar{\chi}_i, d),$$

for all $\zeta_i, \bar{\zeta}_i, \chi_i, \bar{\chi}_i$. ∎

Assumption 8.3.4. *There exists a polynomial function $\bar{\gamma}_i$ such that*

$$\bar{\gamma}_i(\bar{y}_i, \zeta_{i-1} - \bar{\zeta}_{i-1}, \chi_i, \bar{\chi}_i, \mu) = \gamma_i(\zeta_{i-1}, \chi_i, d) - \gamma_i(\bar{\zeta}_{i-1}, \bar{\chi}_i, d),$$

for all $\zeta_{i-1}, \bar{\zeta}_{i-1}, \chi_i, \bar{\chi}_i$. ∎

Under these assumptions, with the functions α_i, $i = 1, \ldots, r$, and the Lyapunov function $W(\cdot)$ obtained in Proposition 8.3.1, it is ready to recursively give the controller design algorithm following the steps proposed in [7]. During the recursive operations, we will encounter more difficulties caused by the uncertainties propagated from the previous steps. After the recursion, a closed-loop is obtained whose stability can be established by using the certainty equivalence principle. The procedure is detailed as follows.

Initial Step: By Assumption 8.3.2, h_1 is a polynomial function. Thus we can let

$$h_1(\bar{y}_1, \mu) = \rho_1(\bar{y}_1)\varpi_1(\mu) \tag{8.15}$$

for a sufficiently smooth function matrix ρ_1 and a column function vector ϖ_1. Let s_1 be a state matrix with the same dimension as that of ρ_1, which is governed by

$$\dot{s}_1 = M_1 s_1 + \rho_1(\bar{y}_1), \tag{8.16}$$

and define a coordinate transformation $\bar{z}_1 = z_1 - s_1\varpi_1(\mu)$. Next, under Assumption 8.3.3, we can define a polynomial function

$$\ell_1(y_1, s_1, \mu) = f_1(z_1, \chi_1, d) - f_1(\bar{z}_1, \chi_1, d) + l_1(y_1, \mu)$$

which is linearly parameterized in the sense that

$$\ell_1(y_1, s_1, \mu) = \rho_1(y_1, s_1)\omega_1(\mu)$$

for a sufficiently smooth row vector function ρ_1 and a column function vector ω_1. Then, a vector $\hat{\omega}_1$ is used to estimate $\omega_1(\mu)$, which is generated by an update law

$$\dot{\hat{\omega}}_1 = \psi_1(y_1, \xi_1) = k_1(dW_1(x_1)/dx_1)\rho_1^{\mathsf{T}}(y_1, s_1)$$

for any $k_1 > 0$. The estimation error is denoted as $\tilde{\omega}_1 = \hat{\omega}_1 - \omega_1$.

Recursive Step: The notations

$$(\bar{\chi}_i) - (\rho_i, \varpi_i) - (\bar{\zeta}_i, \xi_i) - (\rho_i, \omega_i) - (\psi_i) - (\lambda_i) \tag{8.17}$$

have been defined in the initial step for $i = 1$ with

$$\bar{\chi}_1 = \text{col}(x_o, \bar{x}_1), \ \bar{x}_1 = x_1, \ \bar{\zeta}_1 = \bar{z}_1, \ \xi_1 = s_1, \ \lambda_1 = \hat{\omega}_1.$$

For convenience, we let $\xi_o, \lambda_o \in \mathbb{R}^0$. For $i = 2, \cdots, r$, the notations (8.17) are defined recursively in the following order:

- $(\bar{\chi}_i)$: $\bar{\chi}_i := \text{col}(x_o, \bar{x}_1, \cdots, \bar{x}_i)$ where

$$\bar{x}_i = x_i + \rho_{i-1}(y_{i-1}, \xi_{i-1}, \lambda_{i-2})\hat{\omega}_{i-1} - \alpha_{i-1}(\bar{x}_{i-1}).$$

- (ρ_i, ϖ_i): under Assumptions 8.3.2 and 8.3.4, we can define a polynomial function

$$\begin{aligned}\bar{h}_i(\bar{y}_i, \xi_{i-1}, \lambda_{i-1}, \mu) &= h_i(\bar{y}_i, \mu) + \gamma_i(\zeta_{i-1}, \chi_i, d)\\ &\quad -\gamma_i(\bar{\zeta}_{i-1}, x_o, \bar{x}_1, \bar{x}_2 + \alpha_1(\bar{x}_1), \bar{x}_i + \alpha_{i-1}(\bar{x}_{i-1}), d).\end{aligned}$$

  Clearly, the function $\bar{h}_i(\bar{y}_i, \xi_{i-1}, \lambda_{i-1}, \mu)$ is linearly parameterized in the sense that

$$\bar{h}_i(\bar{y}_i, \xi_{i-1}, \lambda_{i-1}, \mu) = \rho_i(\bar{y}_i, \xi_{i-1}, \lambda_{i-1})\varpi_i(\mu) \tag{8.18}$$

  for a sufficiently smooth function matrix ρ_i and a column function vector ϖ_i.
- (ζ_i, ξ_i): $\zeta_i := \text{col}(\bar{z}_1, \cdots, \bar{z}_i)$ and $\xi_i := \text{col}(s_1, \cdots, s_i)$ where s_i is a square matrix governed by

$$\dot{s}_i = M_i s_i + \rho_i(\bar{y}_i, \xi_{i-1}, \lambda_{i-1})$$

  and $\bar{z}_i = z_i - s_i\varpi_i(\mu)$.
- (ρ_i, ω_i): By Assumption 8.3.2, we can denote the time derivative of $\rho_{i-1}(y_{i-1}, \xi_{i-1}, \lambda_{i-2})$ by $\bar{\rho}_{i-1}(y_i, \xi_{i-1}, \lambda_{i-2}, \mu)$ since

$$\begin{aligned}\frac{d\rho_{i-1}(y_{i-1}, \xi_{i-1}, \lambda_{i-2})}{dt} &= \frac{\partial \rho_{i-1}}{\partial y_{i-1}}\varkappa_{i-1}(y_i, \mu) + \sum_{j=1}^{i-1}\frac{\partial \rho_{i-1}}{\partial s_j}[M_j s_j + \rho_j(\bar{y}_j, \xi_{j-1}, \lambda_{j-1})]\\ &\quad + \sum_{j=1}^{i-2}\frac{\partial \rho_{i-1}}{\partial \hat{\omega}_j}\psi_j(y_j, \xi_j, \lambda_{j-1}).\end{aligned}$$

  Under Assumptions 8.3.2 and 8.3.3, we can define a polynomial function

$$\begin{aligned}\ell_i(y_i,\xi_i,\lambda_{i-1},\mu) = l_i(y_i,\mu) + f_i(\zeta_i,\chi_i,d) - f_i\left(\bar{\zeta}_i,x_o,\bar{x}_1,\cdots,\bar{x}_i+\alpha_{i-1}(\bar{x}_{i-1}),d\right)\\ +(\partial\alpha_{i-1}(\bar{x}_{i-1})/\partial\bar{x}_{i-1})\rho_{i-1}(y_{i-1},\xi_{i-1},\lambda_{i-2})\tilde{\omega}_{i-1}(\mu)\\ +\bar{\rho}_{i-1}(y_i,\xi_{i-1},\lambda_{i-2})\hat{\omega}_{i-1}\\ +\rho_{i-1}(y_{i-1},\xi_{i-1},\lambda_{i-2})\psi_{i-1}(y_{i-1},\xi_{i-1},\lambda_{i-2})\end{aligned}$$

where $\tilde{\omega}_{i-1} := \hat{\omega}_{i-1} - \omega_{i-1}$. Clearly, the function $\ell_i(y_i,\xi_i,\lambda_{i-1},\mu)$ is linearly parameterized in the sense that

$$\ell_i(y_i,\xi_i,\lambda_{i-1},\mu) = \rho_i(y_i,\xi_i,\lambda_{i-1})\omega_i(\mu)$$

for a sufficiently smooth row vector function ρ_i and a column function vector ω_i.

- (ψ_i): for any $k_i > 0$, let

$$\psi_i(y_i,\xi_i,\lambda_{i-1}) = k_i(dW_i(\bar{x}_i)/d\bar{x}_i)\rho_i^{\mathsf{T}}(y_i,\xi_i,\lambda_{i-1}).$$

- (λ_i): $\lambda_i := \mathrm{col}(\hat{\omega}_1,\cdots,\hat{\omega}_i)$, where $\hat{\omega}_i$ is vector variable governed by

$$\dot{\hat{\omega}}_i = \psi_i(y_i,\xi_i,\lambda_{i-1}).$$

With the notations defined above, the system (8.9) can be rewritten in the following form

$$\begin{aligned}\dot{x}_o &= f_o(x_o,x_1,d)\\ \dot{\bar{z}}_i &= M_i\bar{z}_i + \varphi_i(\bar{\zeta}_{i-1},\bar{\chi}_i,d)\\ \dot{\bar{x}}_i &= \phi_i(\bar{\zeta}_i,\bar{\chi}_i,d) + \alpha_i(\bar{x}_i) - \rho_i(y_i,\xi_i,\lambda_{i-1})\tilde{\omega}_i + \bar{x}_{i+1},\ i=1,\cdots,r.\end{aligned} \tag{8.19}$$

Observe that the closed-loop system (8.19) reduces to (8.12) when $\tilde{\omega}_i = 0$. The structure of the closed-loop system (8.19) makes it possible to apply the certainty equivalence principle to guarantee the stability property of the closed loop system with the control input u determined from $\bar{x}_{r+1} = 0$, i.e.,

$$u = -\rho_r(y_r,\xi_r,\lambda_{r-1})\hat{\omega}_r + \alpha_r(\bar{x}_r). \tag{8.20}$$

It further leads to the solvability of the original output regulation problem. The main result is summarized in the following theorem.

Theorem 8.3.1. *Under Assumptions 8.1.1 - 8.3.4, the global robust output regulation problem of (8.1) is solvable.*

Proof. It has been proved in [7] that, under the controller (8.20), all states of the closed-loop system are bounded and

$$\lim_{t\to 0}\mathrm{col}(\bar{\zeta}_i(t),\bar{\chi}_i(t)) = 0.$$

Denote

$$a_i(t) = q_i(t) - \mathbf{q}_i(v(t),w,\sigma),\ i=0,\cdots,r+1.$$

It remains to show $\lim_{t\to\infty} a_i(t) = 0$, which is obviously true for $i = 0, 1$. Now, we assume $\lim_{t\to\infty} a_j(t) = 0,\ i \geq j \geq 0$ is true for a given $0 < i < r+1$. Because all states in the closed-loop system are bounded, the signal $\ddot{a}_i(t)$ is bounded which implies $\dot{a}_i(t)$ is uniformly continuous in t. By using Barbalat's lemma, $\lim_{t\to\infty} \dot{a}_i(t) = 0$. Also, we note that

$$\begin{aligned}
\dot{a}_i &= \dot{q}_i - \frac{\partial \mathbf{q}_i(v,w,\sigma)}{\partial v} A_1(\sigma)v \\
&= \kappa_i(\mathcal{Q}_i, v, w) + q_{i+1} - \kappa_i(\mathbf{q}_o(v,w,\sigma), \cdots, \mathbf{q}_i(v,w,\sigma), v, w) - \mathbf{q}_{i+1}(v,w,\sigma) \\
&= \kappa_i(\mathcal{Q}_i, v, w) - \kappa_i(\mathbf{q}_o(v,w,\sigma), \cdots, \mathbf{q}_i(v,w,\sigma), v, w) + a_{i+1}.
\end{aligned}$$

By the assumption that $\lim_{t\to\infty} a_j(t) = 0,\ i \geq j \geq 0$, we have

$$\lim_{t\to\infty} \{\kappa_i(\mathcal{Q}_i(t), v, w) - \kappa_i(\mathbf{q}_o(v(t), w, \sigma), \cdots, \mathbf{q}_i(v(t), w, \sigma), v(t), w)\} = 0$$

which implies $\lim_{t\to\infty} a_{i+1}(t) = 0$. The proof is complete by using mathematical induction.

Moreover, denote $b_i(t) = \eta_i(t) - \theta_i(t)$. Then

$$\dot{b}_i = M_i b_i + N_i a_{i+1}$$

which implies

$$\lim_{t\to\infty} \{\eta_i(t) - \theta_i(t)\} = 0$$

as $\lim_{t\to\infty} a_{i+1}(t) = 0$ and M_i is Hurwitz. ∎

Example 8.3.1. We consider the global robust output regulation problem for the following lower triangular system with $r = 3$:

$$\begin{aligned}
\dot{q}_1 &= 0.2q_2 \\
\dot{q}_2 &= 0.5q_1 \sin q_2 + w_1 v_1 + w_2 v_2 + q_3 \\
\dot{q}_3 &= w_3(q_1^2 + q_2) + w_4 v_1 + u \\
e &= q_1
\end{aligned} \tag{8.21}$$

coupled with an exosystem

$$\dot{v}_1 = -\sigma v_2,\ \dot{v}_2 = \sigma v_1. \tag{8.22}$$

The objective is to design a state-feedback regulator to have the output q_1 of system (8.21) asymptotically converge to zero when the system is perturbed by a sinusoidal disturbance of unknown frequency with arbitrarily large fixed amplitude, produced by exosystem (8.22), in the presence of four uncertain parameters (w_1, w_2, w_3, w_4). We note that the regulator equations associated with (8.21) and (8.22) have a globally defined solution in polynomial form as follows

$$\begin{aligned}
&\mathbf{q}_1(v,w,\sigma) = 0,\ \mathbf{q}_2(v,w,\sigma) = 0,\ \mathbf{q}_3(v,w,\sigma) = -w_1 v_1 - w_2 v_2, \\
&\mathbf{u}(v,w,\sigma) = -w_2 \sigma v_1 + w_1 \sigma v_2 - w_4 v_1.
\end{aligned}$$

Therefore, the global robust output regulation can be formulated as in Section 8.1.

To explicitly give the controller, we will first construct the internal model for the purpose of problem conversion. We have the matrices

$$\Phi_2(\sigma) = \Phi_3(\sigma) = \begin{bmatrix} 0 & 1 \\ -\sigma^2 & 0 \end{bmatrix}, \; M_2 = M_3 = \begin{bmatrix} -1 & 0 \\ 0 & -2 \end{bmatrix} \; N_2 = N_3 = \begin{bmatrix} 0.2 \\ 0.5 \end{bmatrix}.$$

Next, we can solve $T_2(\sigma)$ and $T_3(\sigma)$ from the Sylvester equation as

$$T_2(\sigma) = T_3(\sigma) = \begin{bmatrix} \frac{0.2}{\sigma^2+1} & -\frac{0.2}{\sigma^2+1} \\ \frac{1}{\sigma^2+4} & -\frac{0.5}{\sigma^2+4} \end{bmatrix},$$

hence,

$$\Psi_2(\sigma) = \Psi_3(\sigma) = \begin{bmatrix} -5\sigma^2 - 5 & 2\sigma^2 + 8 \end{bmatrix}.$$

After the introduction of the internal model (8.5), we can convert the system into (8.9). Here we note $\eta_1 \in \mathbb{R}^o$ because $\mathbf{q}_2(v,w,\sigma) = 0$.

Next, we will give the detailed recursive calculation on the quantities used in the controller design for the system (8.9). In particular, we can choose

$$\alpha_1(\tilde{x}_1) = -\tilde{x}_1, \; \alpha_2(\tilde{x}_2) = -K_1\tilde{x}_2, \; \alpha_3(\tilde{x}_3) = -K_2\tilde{x}_3(1+\tilde{x}_3^2)$$

where K_1 and K_2 are determined by the bound of w_1, w_2, w_3, w_4 and σ.

Step 1: There is no internal model introduced in the first step.

Step 2: Let $\bar{y}_2 \in \mathbb{R}^o$, $y_2 = \eta_2$. Then, we have $\bar{h}_2(\bar{y}_2, \mu) = 0$. Since $\ell_2(y_2, \mu) = -\Psi_2(\sigma)\eta_2$, we have

$$\rho_2(y_2) = -\eta_2^{\mathrm{T}}\bar{\Psi}_2^{\mathrm{T}}, \; \bar{\Psi}_2 = \begin{bmatrix} -5 & 2 \\ -5 & 8 \end{bmatrix}, \; \omega_2(\mu) = \mathrm{col}(\sigma^2, 1).$$

The function $W_2(\tilde{x}_2) = (\tilde{x}_2^2 + \tilde{x}_2^4)/2$ is used to determine the function ψ_2.

Step 3: Let $\bar{y}_3 = \mathrm{col}(q_1, q_2, q_3, \eta_2)$, $y_3 = \mathrm{col}(q_1, q_2, q_3, \eta_2, \eta_3)$. Since

$$\Psi_2(\sigma)E_2(\sigma) = [-10\sigma^2 - 10, 2\sigma^2 + 8]$$

and

$$\bar{h}_3(\bar{y}_3, \hat{\omega}_2, \mu) = M_3N_3\eta_2^{\mathrm{T}}\bar{\Psi}_2^{\mathrm{T}}\hat{\omega}_2 - N_3\Psi_2(\sigma)E_2(\sigma)\eta_2,$$

we have

$$\rho_3(\bar{y}_3, \hat{\omega}_2) = \begin{bmatrix} 0 & M_3N_3\eta_2^{\mathrm{T}}\bar{\Psi}_2^{\mathrm{T}}\hat{\omega}_2 \end{bmatrix} - N_3\eta_2^{\mathrm{T}}\hat{\Psi}_2^{\mathrm{T}}, \; \hat{\Psi}_2 = \begin{bmatrix} -10 & 2 \\ -10 & 8 \end{bmatrix}, \; \varpi_3(\mu) = \mathrm{col}(\sigma^2, 1).$$

We note the derivative of $\rho_2(y_2)$ is $\bar{\rho}_2(y_3, \mu) = -\dot{\eta}_2^{\mathrm{T}}\bar{\Psi}_2^{\mathrm{T}} = -(M_2\eta_2 + N_2x_3)^{\mathrm{T}}\bar{\Psi}_2^{\mathrm{T}}$, which is well defined and is available to the control law. A calculation shows

$$\ell_3(y_3, \xi_3, \hat{\omega}_2, \mu) = [A_{11}\ A_{12} + A_{21}\ A_{22}]\mathrm{col}(\sigma^4, \sigma^2, 1) + B\mathrm{col}(\sigma^2, 1) + C$$

with

$$\begin{aligned} A &= \bar{\Psi}_2 s_3 \\ B &= -(N_3 \rho_2(y_2)\hat{\omega}_2)^{\mathsf T}\bar{\Psi}_2^{\mathsf T} + \eta_2^{\mathsf T}\hat{\Psi}_2^{\mathsf T} - \eta_3^{\mathsf T}\bar{\Psi}_2^{\mathsf T} - (\partial \alpha_2(\bar{x}_2)/\partial \bar{x}_2 \rho_2(y_2)) \\ C &= (\partial \alpha_2(\bar{x}_2)/\partial \bar{x}_2)\rho_2(y_2)\hat{\omega}_2 + \bar{\rho}_2(y_3, \mu)\hat{\omega}_2 + \rho_2(y_2)\psi_2(y_2). \end{aligned}$$

Therefore, we have

$$\rho_3(y_3, \xi_3, \hat{\omega}_2) = \left[0\ B\right] + \left[A_{11}\ A_{12} + A_{21}\ C + A_{22}\right],\ \omega_3 = \mathrm{col}(\sigma^4, \sigma^2, 1).$$

The function $W_3(\tilde{x}_3) = 3\tilde{x}_3^2/2$ is used to determine the function ψ_3.

In the numerical simulation, we can compare the non-adaptive controller and the adaptive one. The simulation is conducted with the parameters $w_1 = -0.4, w_2 = 0.8, w_3 = 0.3, w_4 = 1$, $v_1(0) = 10, v_2(0) = 0, q_1(0) = 5, q_2(0) = 8, q_3(0) = -1$, and the initial values of the remaining states being zero. The simulation condition is listed in Table 8.1.

Table 8.1. Simulation condition for the system (8.21)

Time (s)	0 – 100	100 – 200	200 – 300	300 – 400
Adaptive law	off	off	on	on
σ	2	1	1	2

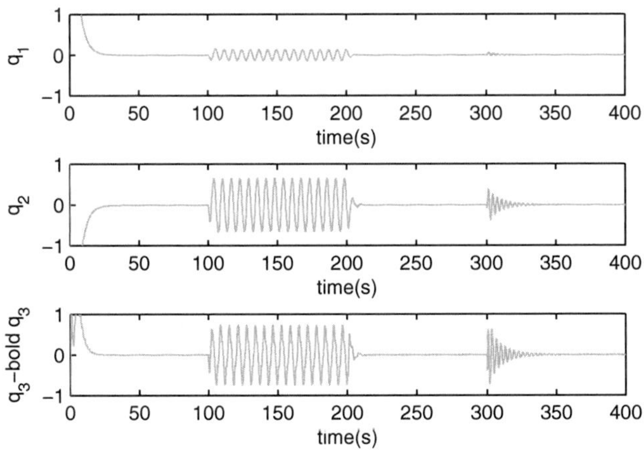

Fig. 8.1. Profile of the tracking errors for the plant states and input

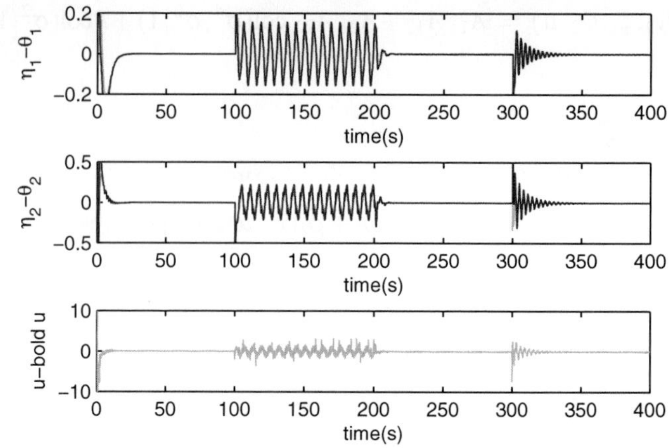

Fig. 8.2. Profile of the tracking errors for the internal model states

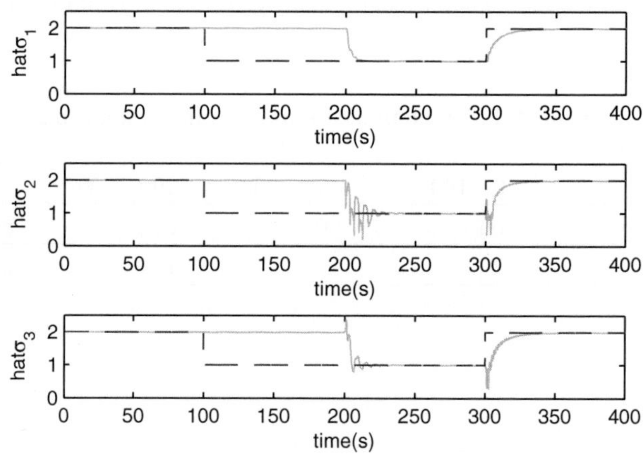

Fig. 8.3. Profile of the estimated frequencies

For the first 100 seconds, the value of σ is the same as that used for the controller design, the adaptive law is off. The desired tracking performance $\lim_{t\to\infty} e(t) = 0$ is shown in Figure 8.1. At $t = 100$, the parameter σ changes its value and the tracking performance degrades significantly. When the adaptive law is turned on at $t = 200$, the tracking error quickly converges to zero. Good tracking performance is maintained even after another step change of the parameter at $t = 300$. The tracking performance is shown in Figures. 8.1 and 8.2. We also observe the convergence of the parameter estimation in the simulation. Due to the over-parametrization, the unknown frequency σ is estimated three times in terms of $\hat{\sigma}_1 := \sqrt{\hat{\omega}_{21}}$, $\hat{\sigma}_2 := \sqrt[4]{\hat{\omega}_{31}}$, and $\hat{\sigma}_3 := \sqrt{\hat{\omega}_{32}}$. The convergence is plotted in Figure 8.3 when the adaptive law is on after 200s.

8.4 Conclusion

In this paper, we have presented a set of solvability conditions of the global robust output regulation for a class of lower triangular systems subject to uncertain exosystems. The construction of the controller relies on the recently developed approach integrating both the robust and adaptive techniques. The simulation results have illustrated the effectiveness of the controller. Also the convergence of the estimated parameters to the true values of the unknown parameters in the exosystem can be observed.

References

1. Chen, Z., Huang, J.: A general formulation and solvability of the global robust output regulation problem. IEEE Transactions on Automatic Control 50, 448–462 (2005)
2. Khalil, H.: Robust servomechanism output feedback controllers for feedback linearizable systems. Automatica 30, 1587–1589 (1994)
3. Serrani, A., Isidori, A., Marconi, L.: Semiglobal nonlinear output regulation with adaptive internal model. IEEE Transactions on Automatic Control 46, 1178–1194 (2001)
4. Byrnes, C.I., Isidori, A.: Limit sets, zero dynamics and internal models in the problem of nonlinear output regulation. IEEE Transactions on Automatic Control 48, 1712–1723 (2003)
5. Ding, Z.: Universal disturbance rejection for nonlinear systems in output feedback form. IEEE Transactions on Automatic Control 48, 1222–1226 (2003)
6. Huang, J., Chen, Z.: A general framework for tackling the output regulation problem. IEEE Transactions on Automatic Control 49, 2203–2218 (2004)
7. Chen, Z., Huang, J.: Robust adaptive regulation of polynomial systems with dynamic uncertainties. In: Proceedings of the 48st IEEE Conference on Decision and Control, pp. 5275–5280 (2009)
8. Ye, X.D., Huang, J.: Decentralized adaptive output regulation for a class of large-scale nonlinear systems. IEEE Transactions on Automatic Control 48, 276–281 (2003)
9. Chen, Z., Huang, J.: Global tracking of uncertain nonlinear cascaded systems with adaptive internal model. In: Proceedings of the 41st IEEE Conference on Decision and Control, pp. 3855–3862 (2002)
10. Nikiforov, V.O.: Adaptive non-linear tracking with complete compensation of unknown disturbances. European Journal of Control 4, 132–139 (1998)
11. Jiang, Z.P., Mareels, I.: A small-gain control method for nonlinear cascaded systems with dynamic uncertainties. IEEE Transactions on Automatic Control 42, 292–308 (1997)
12. Jiang, Z.P., Praly, L.: Design of robust adaptive controllers for nonlinear systems with dynamic uncertainties. Automatica 34, 825–840 (1998)
13. Chen, Z., Huang, J.: A Lyapunov's direct method for the global robust stabilization of nonlinear cascaded systems. Automatica 44, 745–752 (2008)

9

A Survey on Boolean Control Networks: A State Space Approach

Daizhan Cheng, Zhiqiang Li, and Hongsheng Qi

Key Laboratory of Systems and Control, AMSS, Chinese Academy of Sciences, Beijing 100190, P.R. China

Summary. Boolean network is a proper tool to describe the cellular network. The rising of systems biology stimulates the investigation of Boolean (control) networks. Since the bearing space of a Boolean network is not a vector space, to use state space analysis to the dynamics of Boolean (control) network a proper way to describe the state space and its subspaces becomes a challenging problem. This paper surveys a systematic description of the state space of Boolean (control) networks. Under this framework the state space is described as a set of logical functions. Its subspaces are subsets of this set of logical functions. Using semi-tensor product of matrices and the matrix expression of logic, state space and each subspaces are connected to their structure matrices, which are logical matrices. In the light of this expression, certain properties of state space and subspaces, which are closely related to control problems, are obtained. Particularly, the coordinate transformation of state space, the regular subspaces, which is generated by part of coordinate variables; the invariant subspaces etc. are proposed and the corresponding necessary and sufficient conditions are presented to verify them.

9.1 Introduction

Accompanying the flourishing of the systems biology, Boolean network received the most attention, not only from the biology community, but also physics, systems science, etc. Historically, in 1943, McCulloch and Pitts published a paper "A logical calculus of the ideas immanent in nervous activity", which claim that "the brain could be modeled as a network of logical operations such as and (conjunction), or (disjunction), not(negation) and so forth". Then "Jacob and Monod were publishing their first papers on genetic circuits in 1961 through 1963. It was the work for which they later won the Nobel Prize. ... any cell contains a number of 'regulatory' genes that act as switches and can turn on another on and off. ... if genes can turn one another on and off, then you can have genetic circuits." [21] Motivated by their works, Kauffman proposed firstly to use Boolean network to describe the cellular networks [15]. It has been then developed by [1, 2, 20, 14, 3, 19, 13] and many others, and becomes a powerful tool in describing, analyzing, and simulating the cellular networks. We refer to [16] for a tutorial introduction to Boolean network.

X. Hu et al. (Eds.): Three Decades of Progress in Control Sciences, pp. 121–139, 2010.

A Boolean network consists of n nodes, denoted by $\mathscr{V} = \{1,\cdots,n\}$, and a set of directed edges $\mathscr{E} \subset \mathscr{V} \times \mathscr{V}$. If $(i,j) \in \mathscr{E}$, then there is an information flow from node i to node j. Moreover, at each moment $t = 0,1,\cdots$, each node can take one of two values, $0 \sim F$(False) or $1 \sim T$(True). Denote $\mathscr{D} = \{0,1\}$, a node can be represented by a logical variable $x_i(t) \in \mathscr{D}$. A network graph can be used to describe the incidence relation. To describe the dynamics of a Boolean network, we need a set of logical dynamic equations as

$$\begin{cases} x_1(t+1) = f_1(x_1(t),\cdots,x_n(t)) \\ \vdots \\ x_n(t+1) = f_n(x_1(t),\cdots,x_n(t)), \quad x_i \in \mathscr{D}, \end{cases} \tag{9.1}$$

where f_i, $i = 1,\cdots,n$ are logical functions.

A logical function consists of logical variables and some logical operators. The followings are some commonly used logical operators [18]: $\neg$ (negation); $\wedge$ (conjunction); $\vee$ (disjunction); $\rightarrow$ (conditional); $\leftrightarrow$ (biconditional); $\leftrightarrow \neg$ (exclusive or).

We use an example to depict Boolean network.

Example 9.1.1. Consider a Boolean network depicted in Fig. 9.1. Its dynamics is described as

$$\begin{cases} x_1(t+1) = (x_1(t) \wedge x_2(t) \wedge \neg x_4(t)) \vee (\neg x_1(t) \wedge x_2(t)) \\ x_2(t+1) = x_2(t) \vee (x_3(t) \leftrightarrow x_4(t)) \\ x_3(t+1) = (x_1(t) \wedge \neg x_4(t)) \vee (\neg x_1(t) \wedge x_2(t)) \vee (\neg x_1(t) \wedge \neg x_2(t) \wedge x_4(t)) \\ x_4(t+1) = x_1(t) \vee \neg x_2(t) \vee x_4(t). \end{cases} \tag{9.2}$$

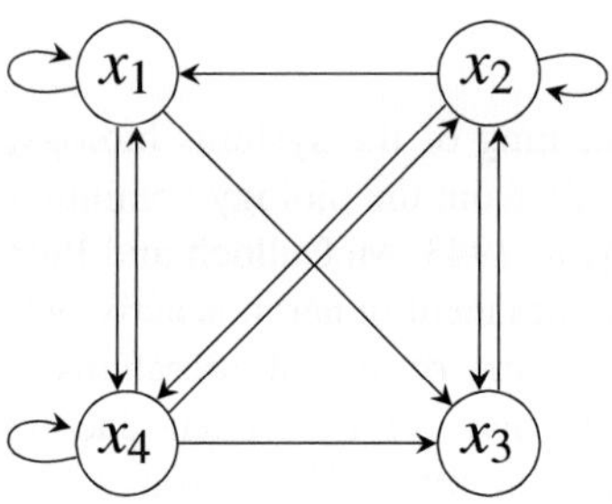

Fig. 9.1. Boolean network

□

For a Boolean network, if there are some additional inputs $u_i(t) \in \mathscr{D}$ and outputs $y_i(t) \in \mathscr{D}$, it becomes a Boolean control network. The dynamics of a Boolean control network can be described as

$$\begin{cases} x_1(t+1) = f_1(x_1(t),\cdots,x_n(t),u_1(t),\cdots,u_m(t)) \\ \vdots \\ x_n(t+1) = f_n(x_1(t),\cdots,x_n(t),u_1(t),\cdots,u_m(t)), \quad x_i \in \mathscr{D},\ u_i \in \mathscr{D} \\ y_j(t) = h_j(x_1(t),\cdots,x_n(t)), \quad j=1,\cdots,p,\ y_j \in \mathscr{D}, \end{cases} \tag{9.3}$$

where f_i, $i=1,\cdots,n$, h_j, $j=1,\cdots,p$ are logical functions.

We turn the Boolean network in Example 9.1.1 to a Boolean control network by adding inputs and outputs as in the following example.

Example 9.1.2. Consider a Boolean control network depicted in Fig. 9.2, which is obtained from Fig. 9.1 by adding two inputs, u_1,u_2, and one out output, y. Its dynamics is described as

$$\begin{cases} x_1(t+1) = (x_2(t) \leftrightarrow \neg u_1(t)) \leftrightarrow (x_2(t) \wedge x_4(t)) \\ x_2(t+1) = x_2(t) \vee (x_3(t) \leftrightarrow x_4(t)) \\ x_3(t+1) = ((x_1(t) \leftrightarrow \neg x_4(t)) \to u_2(t)) \leftrightarrow (x_2(t) \wedge x_4(t)) \\ x_4(t+1) = \neg(x_2(t) \wedge (x_4(t)) \\ y(t) = x_3(t) \leftrightarrow \neg x_4(t). \end{cases} \tag{9.4}$$

□

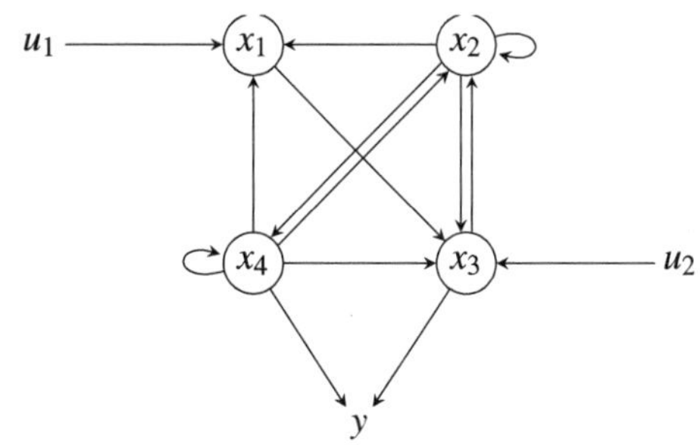

Fig. 9.2. Boolean control network

Recently, we proposed a new method for analyzing and synthesizing Boolean (control) networks. This new approach can be sketched as follows: A new matrix product, called the semi-tensor product of matrices, is proposed and via it a logical function can be expressed as an algebraic equation. Based on this expression, a Boolean (control) network can be converted into a discrete-time dynamic (control) system. State space and some meaningful subspaces are defined. Then the conventional state space analysis tools for control systems are applicable to them.

The purpose of this paper is to provide a survey on this new approach. The rest of this paper is organized as follows: Section 9.2 presents a method to describe the state space and its subspaces, which are not vector spaces. Section 9.3 introduces the algebraic expression of Boolean networks. Semi-tensor product and the matrix expression of logic are introduced first. Then they are used to produce the algebraic form of the dynamics of Boolean (control) networks. Under the state space framework, the coordinate transformation, regular subspace, invariant subspace etc. are investigated in Section 9.4. Easily verifiable formulas are obtained for testing them. In Section 9.5, the state space approach for Boolean (control) networks has been extended to multi-valued (control) networks. Section 9.6 is the conclusion.

9.2 State Space Structure

State space description of a control system, which is firstly proposed by Kalman, is one of the pillars of the modern control theory. Unfortunately, there is no vector space structure, such as subspaces of $\mathbb{R}^n$ for linear systems or tangent space of a manifold for nonlinear systems, for Boolean (control) networks. To use the state space approach, the state space and its subspaces have to be defined carefully. In the following definition, they are only defined as set and subsets. They can be considered as topological space and subspaces with discrete topology.

Let $x_1,\cdots,x_s \in \mathscr{D}$ be a set of logical variables. Denote by $F_\ell(x_1,\cdots,x_s)$ the set of logical functions of $x_1,\cdots,x_s$. It is obvious that F_ℓ is a finite set with cardinality 2^{2^s}.

Definition 9.2.1. *Consider Boolean network (9.1) (or Boolean control network (9.3)).*

(1) The state space of (9.1) or (9.3) is defined as

$$\mathscr{X} = F_\ell(x_1,\cdots,x_n). \tag{9.5}$$

(2) Let $y_1,\cdots,y_s \in \mathscr{X}$. Then

$$\mathscr{Y} = F_\ell(y_1,\cdots,y_s) \subset \mathscr{X} \tag{9.6}$$

is called a subspace of $\mathscr{X}$.

(3) Let $\{x_{i_1},\cdots,x_{i_s}\} \subset \{x_1,\cdots,x_n\}$. Then

$$\mathscr{Z} = F_\ell(x_{i_1},\cdots,x_{i_s}) \tag{9.7}$$

is called a s dimensional natural subspace of $\mathscr{X}$.

Remark 9.2.1. To understand this definition, we give the following explanation:

(1) Let $x_1,\cdots,x_n$ be a set coordinate variables of $\mathbb{R}^n$. Then, in dual sense, we can say that $\mathbb{R}^n$ is the set of all the linear functions of $x_1,\cdots,x_n$. We denote it as

$$L = \{r_1x_1 + r_2x_2 + \cdots + r_nx_n \,|\, r_1,\cdots,r_n \in \mathbb{R}\}.$$

Moreover, a subspace could be the set of all the linear functions of a subset $\{x_{i_1},\cdots,x_{i_s}\} \subset \{x_1,\cdots,x_n\}$, denoted by

$$L_0 = \{r_{i_1}x_{i_1} + \cdots + r_{i_s}x_{i_s} \,|\, r_{i_1},\cdots,r_{i_s} \in \mathbb{R}\}.$$

It is clear that L is an n-dimensional vector space and L_0 is its s-dimensional subspace.

Here we can identify a space (or subspace) with its domain.

(2) Similar to the argument in (1), we may identify the set of functions with their domain. Then from (9.5) we have $\mathscr{X} \sim \mathscr{D}^n$, and from (9.7) we have $\mathscr{Z} \sim \mathscr{D}^s$. As for (9.6), we do not have $\mathscr{Y} \sim \mathscr{D}^s$. To see this, say, $s = 2$ and $y_1 = x_1 \wedge x_2$ and $y_2 = x_1 \vee x_2$. Later on, one will see that the domain of $\mathscr{Y}$ is not $\mathscr{D}^2$.
(3) Under this understanding, we call $\{x_1, \cdots, x_n\}$ a basis of $\mathscr{X}$ or a coordinate frame of $\mathscr{D}^n$. Similarly, $\{x_{i_1}, \cdots, x_{i_s}\}$ is a basis of $\mathscr{Z}$ or a coordinate frame of $\mathscr{D}^s$. But we call $\{y_1, \cdots, y_s\}$ a generator of $\mathscr{Y}$.

Consider a logical mapping $G : \mathscr{D}^n \to \mathscr{D}^s$. It can be expressed as

$$z_i = g_1(x_1, \cdots, x_n), \quad i = 1, \cdots, s. \tag{9.8}$$

Definition 9.2.2. *Let $\mathscr{X} = F_\ell(x_1, \cdots, x_n)$ be the state space of (9.1) or (9.3). Assume there exist $z_1, \cdots, z_n \in \mathscr{X}$, such that*

$$\mathscr{X} = F_\ell(z_1, \cdots, z_n),$$

then the logical mapping $T : (x_1, \cdots, x_n) \mapsto (z_1, \cdots, z_n)$ is called a coordinate transformation of the state space.

The following proposition is obvious.

Proposition 9.2.1. *A mapping $T : \mathscr{D}^n \to \mathscr{D}^n$ is a coordinate transformation, iff T is one-to-one and onto (i.e., bijective).*

Definition 9.2.3. *Let $\mathscr{X} \subset \mathscr{Z}$ be as defined in (9.5) and (9.7) respectively.*

(1) A mapping $P : \mathscr{D}^n \to \mathscr{D}^s$, defined from (the domain of) $\mathscr{X}$ to (the domain of) $\mathscr{Z}$, as

$$P : (x_1, \cdots, x_n) \mapsto (x_{i_1}, \cdots, x_{i_s}),$$

is called a natural projection from $\mathscr{X}$ to $\mathscr{Z}$.
(2) Given $F : \mathscr{D}^n \to \mathscr{D}^n$, $\mathscr{Z}$ is called an invariant subspace (with respect to F), if there exists a mapping $\bar{F}$ such that graph in Fig. 9.3 is commutative.

□

Let $X := (x_1, \cdots, x_n)^T \in \mathscr{D}^n$, $U := (u_1, \cdots, u_m)^T \in \mathscr{D}^m$, and $Y = (y_1, \cdots, y_p)^T \in \mathscr{D}^p$. Then we can briefly denote system (9.1) as

$$X(t+1) = F(X(t)), \quad X \in \mathscr{D}^n. \tag{9.9}$$

Similarly, (9.3) can be expressed as

$$\begin{cases} X(t+1) = F(X(t), U(t)), & X \in \mathscr{D}^n,\ U \in \mathscr{D}^m \\ Y(t) = H(X(t)), & Y \in \mathscr{D}^p. \end{cases} \tag{9.10}$$

Definition 9.2.4.

(1) Consider system (9.1) (equivalently, (9.9)). $\mathscr{Z}$ is an invariant subspace, if it is invariant with respect to F.

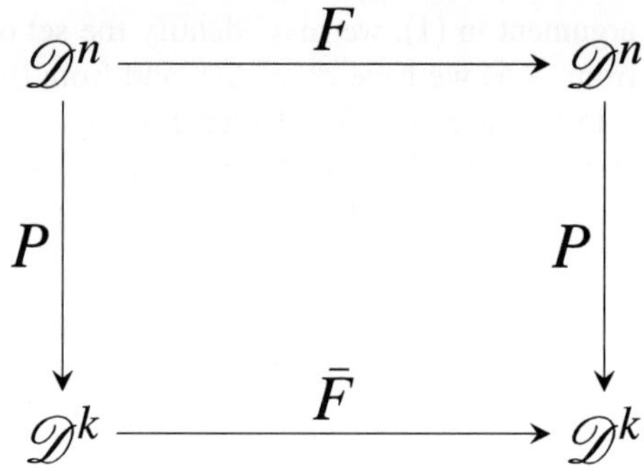

Fig. 9.3. Invariant subspace

(2) *Consider system (9.3) (equivalently, (9.10)). $\mathscr{X}$ is a control invariant subspace, if there exists a state feedback control $U(t) = G(X(t))$, such that for the closed-loop system*

$$X(t+1) = F(X(t), G(X(t))) := \tilde{F}(X(t)),$$

$\mathscr{X}$ is invariant with respect to $\tilde{F}$.

9.3 Algebraic Form of Boolean (Control) Networks

Converting the logic dynamics of a Boolean (control) network into a discrete time conventional dynamic (control) system via semi-tensor product of matrices was introduced firstly in [8] and [6]. We give a brief introduction to this.

First, we give some notations:

- δ_n^i: the i-th column of the identity matrix I_n;
- Δ_n: the set $\{\delta_n^i | i = 1, \cdots, n\}$ ($\Delta := \Delta_2$);
- $Col(A)$: set of columns of A;
- $Row(A)$: set of rows of A;
- $\mathscr{L}_{m\times n}$: $A \in M_{m\times n}$ is called a logical matrix, denoted by $A \in \mathscr{L}_{m\times n}$, if $Col(A) \subset \Delta_m$;
- if $A \in \mathscr{L}_{m\times n}$ is $A = [\delta_m^{i_1}, \cdots, \delta_m^{i_n}]$, it is briefly denoted as

$$A = \delta_m[i_1, \cdots, i_n].$$

9.3.1 Semi-tensor Product of Matrices

Definition 9.3.1.

(1) *Let X be a row vector of dimension np, and Y be a column vector with dimension p. Then we split X into p equal-size blocks as $X^1, \cdots, X^p$, which are $1 \times n$ rows. Define the semi-tensor product (STP), denoted by $\ltimes$, as*

$$\begin{cases} X \ltimes Y = \sum_{i=1}^{p} X^i y_i \in \mathbb{R}^n, \\ Y^T \ltimes X^T = \sum_{i=1}^{p} y_i (X^i)^T \in \mathbb{R}^n. \end{cases} \tag{9.11}$$

(2) *Let $A \in M_{m\times n}$ and $B \in M_{p\times q}$. If either n is a factor of p, say $nt = p$ and denote it as $A \prec_t B$, or p is a factor of n, say $n = pt$ and denote it as $A \succ_t B$, then we define the STP of A and B , denoted by $C = A \ltimes B$, as the following: C consists of $m \times q$ blocks as $C = (C^{ij})$ and each block is*

$$C^{ij} = A^i \ltimes B_j, \quad i = 1, \cdots, m, \quad j = 1, \cdots, q,$$

where A^i is i-th row of A and B_j is the j-th column of B.

We refer to [4, 5] for basic properties of $\ltimes$. Roughly speaking, it is a generalization of conventional matrix product, and all the major properties of conventional matrix product remain true. The following property is frequently used in the sequel.

Proposition 9.3.1. *Let $A \in M_{m\times n}$ and $Z \in \mathbb{R}^t$ be a column vector. Then*

$$Z \ltimes A = (I_t \ltimes A) \ltimes Z. \tag{9.12}$$

Definition 9.3.2. *An $mn \times mn$ matrix, denoted by $W_{[m,n]}$, is called a swap matrix, if it has the following structure: Label its columns by*

$$(11, 12, \cdots, 1n, \cdots, m1, m2, \cdots, mn),$$

and its rows by

$$(11, 21, \cdots, m1, \cdots, 1n, 2n, \cdots, mn).$$

Then its element in the position $((I,J),(i,j))$ is assigned as

$$w_{(IJ),(ij)} = \delta_{i,j}^{I,J} = \begin{cases} 1, & I = i \text{ and } J = j, \\ 0, & \text{otherwise.} \end{cases} \tag{9.13}$$

When $m = n$ we briefly denote $W_{[n]} := W_{[n,n]}$.

Example 9.3.1. Let $m = 2$ and $n = 3$, the swap matrix $W_{[2,3]}$ is

$$\delta_6[1, 3, 5, 2, 4, 6].$$

□

Proposition 9.3.2. *Let $X \in \mathbb{R}^m$ and $Y \in \mathbb{R}^n$ be two columns. Then*

$$W_{[m,n]} \ltimes X \ltimes Y = Y \ltimes X, \quad W_{[n,m]} \ltimes Y \ltimes X = X \ltimes Y. \tag{9.14}$$

9.3.2 Matrix Expression of Logic

To use the matrix expression of logic, we use vectors for logical values. Precisely,

$$T \sim 1 \sim \delta_2^1, \quad F \sim 0 \sim \delta_2^2; \quad \mathscr{D} \sim \Delta.$$

Let $f(x_1, \cdots, x_n)$ be a logical function. In vector form, f is a mapping $f : \Delta^n \to \Delta$.

Definition 9.3.3. *A 2×2^n matrix M_f is called the structure matrix of the logical function f, if*

$$f(x_1, \cdots, x_n) = M_f x_1 x_2 \cdots x_n, \quad x_i \in \Delta. \tag{9.15}$$

Theorem 9.3.1. *For any logical function $f(x_1, \cdots, x_n)$, there exists a unique structure matrix $M_f \in \mathscr{L}_{2\times 2^n}$, such that*

$$f(x_1, \cdots, x_n) = M_f x_1 x_2 \cdots x_n, \quad x_i \in \mathscr{D}. \tag{9.16}$$

The structure matrices of some basic logical operators are listed in the following table 9.1.

Table 9.1. Structure Matrix of Operators

LO	Structure Matrix	LO	Structure Matrix
$\neg$	$M_n = \delta_2[2\ 1]$	$\vee$	$M_d = \delta_2[1\ 1\ 1\ 2]$
$\rightarrow$	$M_i = \delta_2[1\ 2\ 1\ 1]$	$\leftrightarrow$	$M_e = \delta_2[1\ 2\ 2\ 1]$
$\wedge$	$M_c = \delta_2[1\ 2\ 2\ 2]$	$\leftrightarrow\neg$	$M_p = \delta_2[2\ 1\ 1\ 2]$

9.3.3 Algebraic Form of Boolean Networks

Let $G : \mathscr{D}^n \to \mathscr{D}^s$ be defined by

$$z_i = g_i(x_1, \cdots, x_n), \quad x_i \in \mathscr{D},\ i = 1, \cdots, s. \tag{9.17}$$

Then by identify $\mathscr{D} \sim \Delta$, we have their algebraic form as

$$z_i = M_i x_1 x_2 \cdots x_n, \quad x_i \in \Delta,\ i = 1, \cdots, s, \tag{9.18}$$

where M_i is the structure matrix of g_i. Denote by $z = \ltimes_{i=1}^s z_i$ and $x = \ltimes_{i=1}^n x_i$. Then we have

Theorem 9.3.2. *Given a logical mapping $G : \mathscr{D}^n \to \mathscr{D}^s$, described by (9.17) (equivalently, (9.18)). Then there is a unique matrix, $M_G \in \mathscr{L}_{2^s\times 2^n}$, called the structure matrix of G, such that*

$$z = M_G x. \tag{9.19}$$

Remark 9.3.1.

(1) (9.17), (9.18), and (9.19) are all equivalent. (9.17) is the logical form of functions, (9.18) is called the algebraic form of each functions, and (9.19) is the algebraic form of the mapping.
(2) We can get from any one form the other two forms. We refer to [8, 6] for the converting formulas.

Corollary 9.3.1.

(1) Consider Boolean network (9.1). There exists unique $L \in \mathcal{L}_{2^n \times 2^n}$ such that (9.1) can be expressed as

$$x(t+1) = Lx(t), \tag{9.20}$$

where $x(t) = \ltimes_{i=1}^n x_i(t)$. L is called the transition matrix of system (9.1).
(2) Consider Boolean control network (9.3). There exist unique $L \in \mathcal{L}_{2^n \times 2^{n+m}}$ and unique $H \in \mathcal{L}_{2^p \times 2^n}$ such that (9.3) can be expressed as

$$\begin{cases} x(t+1) = Lu(t)x(t), \\ y(t) = Hx(t), \end{cases} \tag{9.21}$$

where $x(t) = \ltimes_{i=1}^n x_i(t)$, $u(t) = \ltimes_{i=1}^m u_i(t)$, $y(t) = \ltimes_{i=1}^p y_i(t)$. L and H are called the transition matrix and output matrix of system (9.3) respectively.

Example 9.3.2.

(1) Consider Boolean network (9.2). It is easy to calculate its algebraic form as

$$x(t+1) = Lx(t), \tag{9.22}$$

where

$$L = \delta_{16}[11\ 1\ 11\ 1\ 11\ 13\ 15\ 9\ 1\ 2\ 1\ 2\ 9\ 15\ 13\ 11].$$

(2) Consider Boolean control network (9.4). Its algebraic form is

$$\begin{cases} x(t+1) = Lu(t)x(t) \\ y(t) = Hx(t), \end{cases} \tag{9.23}$$

where

$$\begin{aligned} L = \delta_{16}[&10\ \ 3\ 10\ \ 3\ 11\ 15\ 15\ 11\ 10\ \ 3\ 10\ \ 3\ 11\ 15\ 15\ 11 \\ &10\ \ 1\ 10\ \ 1\ 11\ 13\ 15\ \ 9\ 12\ \ 3\ 12\ \ 3\ \ 9\ 15\ 13\ 11 \\ &2\ 11\ \ 2\ 11\ \ 3\ \ 7\ \ 7\ \ 3\ \ 2\ 11\ \ 2\ 11\ \ 3\ \ 7\ \ 7\ 3 \\ &2\ \ 9\ \ 2\ \ 9\ \ 3\ \ 5\ \ 7\ \ 1\ \ 4\ 11\ \ 4\ 11\ \ 1\ \ 7\ \ 5\ 3]; \\ H = \ \ &\delta_2[2\ \ 1\ \ 1\ \ 2\ \ 2\ \ 1\ \ 1\ \ 2\ \ 2\ \ 1\ \ 1\ \ 2\ \ 2\ \ 1\ \ 1\ 2]. \end{aligned}$$

We refer to [8] for calculating the transition matrix etc.[1]

□

[1] A toolbox in Matlab is provided in `http://lsc.amss.ac.cn/~dcheng/stp/STP.zip` for the related computations.

9.4 State Space Analysis

Using state space description described in the previous section, easily verifiable formulas can be obtained to construct and/or test the properties of subspaces.

9.4.1 Testing Coordinate Transformation

Let $T : \mathscr{D}^n \to \mathscr{D}^n$ be a described as

$$z_i = t_i(x_1, \cdots, x_n), \quad x_i, z_i \in \mathscr{D}, \; i = 1, \cdots, n. \tag{9.24}$$

In vector form, we set $x = \ltimes_{i=1}^n x_i$ and $z = \ltimes_{i=1}^n z_i$, $\forall x_i, z_i \in \Delta$. Then we can have the algebraic form of this mapping as

$$z = M_T x, \tag{9.25}$$

where $M_T \in \mathscr{L}_{2^n \times 2^n}$ is the structure matrix of T.

It is easy to prove the following:

Theorem 9.4.1. *A mapping $T : \mathscr{D}^n \to \mathscr{D}^n$ is a coordinate transformation, iff its structure matrix $M_T \in \mathscr{L}_{2^n \times 2^n}$ is non-singular.*

It is easy to verify that if T is a coordinate transformation, then its structure matrix M_T is an orthogonal matrix. That is, the inverse mapping, $T^{-1} : (z_1, z_2, z_3, z_4) \mapsto (x_1, x_2, x_3, x_4)$, has its structure matrix $M_T^{-1} = (M_T)^T$.

Under a coordinate transformation, T, the algebraic form of the network (9.1) becomes

$$z(t+1) = M_T x(t+1) = M_T L x(t) = M_T L M_T^T z(t) := \tilde{L} z(t), \tag{9.26}$$

where

$$\tilde{L} = M_T L M_T^T.$$

Consider Boolean control network (9.3). We have

$$\begin{aligned} z(t+1) &= M_T x(t+1) \\ &= M_T L u(t) x(t) \\ &= M_T L u(t) M_T^T z(t) \\ &= M_T L (I_{2^m} \otimes M_T^T) u(t) z(t); \end{aligned}$$

and

$$y(t) = H(t) x(t) = H(t) M_T^T z(t).$$

We conclude that under coordinate frame $z = M_T x$ system (9.3) becomes

$$\begin{cases} z(t+1) = \tilde{L} u(t) z(t) \\ y(t) = \tilde{H} z(t), \end{cases} \tag{9.27}$$

where

$$\tilde{L} = M_T L (I_{2^m} \otimes M_T^T);$$

and

$$\tilde{H} = H(t) M_T^T.$$

Example 9.4.1. Consider the Boolean (control) network (9.2) (respectively, (9.4)) again.

(1) We may define a state space coordinate transformation, $T : (x_1, x_2, x_3, x_4) \mapsto (z_1, z_2, z_3, z_4)$, as

$$\begin{cases} z_1 = x_1 \leftrightarrow \neg x_4 \\ z_2 = \neg x_2 \\ z_3 = x_3 \leftrightarrow \neg x_4 \\ z_4 = x_4. \end{cases} \tag{9.28}$$

Denote the algebraic form of this mapping as

$$z = M_T x.$$

It is easy to calculate that

$$M_T = \delta_{16}[15\ 6\ 13\ 8\ 11\ 2\ 9\ 4\ 7\ 14\ 5\ 16\ 3\ 10\ 1\ 12],$$

which is non-singular. So, T is a coordinate transformation.

(2) Under coordinate frame z, the algebraic form of network (9.2) is

$$z(t+1) = \tilde{L}z(t), \tag{9.29}$$

where

$$\tilde{L} = \delta_8[3\ 3\ 7\ 7\ 15\ 15\ 15\ 15\ 1\ 1\ 5\ 5\ 5\ 6\ 5\ 6].$$

From (9.29), its logic form can be obtained as

$$\begin{cases} z_1(t+1) = z_1(t) \to z_2(t) \\ z_2(t+1) = z_2(t) \wedge z_3(t) \\ z_3(t+1) = \neg z_1(t) \\ z_4(t+1) = z_1(t) \vee z_2(t) \vee z_4(t). \end{cases} \tag{9.30}$$

(3) Under coordinate frame z, the algebraic form of network (9.4) is

$$\begin{cases} z(t+1) = \tilde{L}u(t)z(t) \\ y(t) = \tilde{H}z(t), \end{cases} \tag{9.31}$$

where

$$\begin{aligned} \tilde{L} = \delta_{16}[&1\ \ 1\ \ 5\ \ 5\ 14\ 13\ 14\ 13\ 1\ 1\ \ 5\ \ 5\ 14\ 13\ 14\ 13 \\ &3\ \ 3\ \ 7\ \ 7\ 16\ 15\ 16\ 15\ 1\ 1\ \ 5\ \ 5\ 14\ 13\ 14\ 13 \\ &9\ \ 9\ 13\ 13\ \ 6\ \ 5\ \ 6\ \ 5\ 9\ 9\ 13\ 13\ \ 6\ \ 5\ \ 6\ 5 \\ &11\ 11\ 15\ 15\ \ 8\ \ 7\ \ 8\ \ 7\ 9\ 9\ 13\ 13\ \ 6\ \ 5\ \ 6\ 5]; \\ \tilde{H} = \delta_{16}[&1\ \ 1\ \ 2\ \ 2\ \ 1\ \ 1\ \ 2\ \ 2\ 1\ 1\ \ 2\ \ 2\ \ 1\ \ 1\ \ 2\ 2]. \end{aligned}$$

From (9.31), its logic form can be obtained as

$$\begin{cases} z_1(t+1) = z_2(t) \leftrightarrow u_1(t) \\ z_2(t+1) = z_2(t) \wedge z_3(t) \\ z_3(t+1) = z_1(t) \rightarrow u_2(t) \\ z_4(t+1) = z_2(t) \vee (\neg z_4(t)) \\ \quad y(t) \quad = z_3(t). \end{cases} \tag{9.32}$$

□

9.4.2 Testing Regular Subspace

Let $\mathscr{Z}_0 = F_\ell(z_1,\cdots,z_s)$ be a subspace of the state space $\mathscr{X}$. Since $z_i \in \mathscr{X}$, $i = 1,\cdots,s$, they can be expressed as

$$z_i = g_i(x_1,\cdots,x_n), \quad i = 1,\cdots,s. \tag{9.33}$$

Equation (9.33) defined a mapping $G : \mathscr{D}^n \to \mathscr{D}^s$. Setting $z = \ltimes_{i=1}^s z_i$ and $x = \ltimes_{i=1}^n x_i$, the algebraic form of G is expressed as

$$z = M_G x := \begin{bmatrix} g_{11} & \cdots & g_{1,2^n} \\ \vdots & & \\ g_{2^s,1} & \cdots & g_{2^s,2^n} \end{bmatrix} x. \tag{9.34}$$

Then we have the following result.

Theorem 9.4.2. *Let $\mathscr{Z}_0 = F_\ell(z_1,\cdots,z_s)$, where z_i, $i = 1,\cdots,s$ are determined by (9.33)-(9.34). Then $\mathscr{Z}_0$ is a regular subspace, iff*

$$\sum_{j=1}^{2^n} g_{ij} = 2^{n-s}, \quad i = 1,\cdots,2^s. \tag{9.35}$$

Example 9.4.2. Consider $\mathscr{X} = F_\ell(x_1,x_2,x_3,x_4)$. Consider $\mathscr{Z}_0 = F_\ell(z_1,z_2)$.

(1) Assume

$$\begin{cases} z_1 = x_2 \wedge x_3 \\ z_2 = x_1 \vee x_4. \end{cases}$$

Let $y = z_1 z_2$ and $x = \ltimes_{i=1}^n x_i$. Then its algebraic form can be expressed as $y = Mx = \delta_4[1\ 1\ 3\ 3\ 3\ 3\ 3\ 3\ 1\ 2\ 3\ 4\ 3\ 4\ 3\ 4]$, equivalently

$$M = \begin{bmatrix} 1\,1\,0\,0\,0\,0\,0\,0\,1\,0\,0\,0\,0\,0\,0\,0 \\ 0\,0\,0\,0\,0\,0\,0\,0\,0\,1\,0\,0\,0\,0\,0\,0 \\ 0\,0\,1\,1\,1\,1\,1\,1\,0\,0\,1\,0\,1\,0\,1\,0 \\ 0\,0\,0\,0\,0\,0\,0\,0\,0\,0\,0\,1\,0\,1\,0\,1 \end{bmatrix}.$$

Since $\sum_{i=1}^{16} m_{1i} = 3$, $\sum_{i=1}^{16} m_{2i} = 1$, $\sum_{i=1}^{16} m_{3i} = 9$, $\sum_{i=1}^{16} m_{3i} = 4$, $\mathscr{Z}_0$ is not a regular subspace.

(2) Assume

$$\begin{cases} z_1 = x_2 \leftrightarrow x_3 \\ z_2 = \neg x_3. \end{cases}$$

Let $y = \ltimes_{i=1}^{2} z_i$ and $x = \ltimes_{i=1}^{4} x_i$. Then its algebraic form can be expressed as $y = Mx = \delta_4[2\ 2\ 3\ 3\ 4\ 4\ 1\ 1\ 2\ 2\ 3\ 3\ 4\ 4\ 1\ 1]$, equivalently

$$M = \begin{bmatrix} 0\,0\,0\,0\,0\,0\,1\,1\,0\,0\,0\,0\,0\,0\,1\,1 \\ 1\,1\,0\,0\,0\,0\,0\,0\,1\,1\,0\,0\,0\,0\,0\,0 \\ 0\,0\,1\,1\,0\,0\,0\,0\,0\,0\,1\,1\,0\,0\,0\,0 \\ 0\,0\,0\,0\,1\,1\,0\,0\,0\,0\,0\,0\,1\,1\,0\,0 \end{bmatrix}.$$

Since $\sum_{i=1}^{16} m_{ri} = 4, r = 1,2,3,4$, $\mathscr{Z}_0$ is a regular subspace.

□

9.4.3 Testing Invariant Subspace

Invariant subspace is particularly important in analyzing the cycles of a Boolean network [6]. It is also important in control design. We consider only the regular subspace, because so far, only such invariant subspaces are used.

Let $\mathscr{Z}_0 = F_\ell(z_1, \cdots, z_s)$ be a regular subspace of the state space $\mathscr{X}$, where z_i, $i = 1, \cdots, s$ are determined by (9.33)-(9.34). The algebra form of system (9.1) is

$$x(t+1) = Lx(t).$$

Using the above notations we have

Theorem 9.4.3. *$\mathscr{Z}_0$ is an invariant subspace with respect to system (9.1), iff one of the following two equivalent conditions is satisfied.*

(i).

$$Row(M_G L) \subset SpanRow(M_G). \tag{9.36}$$

(ii). There exists an $H \in \mathscr{L}_{2^s \times 2^s}$ such that

$$M_G L = H M_G. \tag{9.37}$$

Example 9.4.3.

(1) Consider system (9.2). Let $\mathscr{Z}_0 = F_\ell(z_1, z_2, z_3)$, where

$$\begin{cases} z_1 = x_1 \leftrightarrow \neg x_4 \\ z_2 = \neg x_2 \\ z_3 = x_3 \leftrightarrow \neg x_4. \end{cases} \tag{9.38}$$

Then $\mathscr{Z}_0$ is an invariant subspace of (9.2). To see this, set $x = \ltimes_{i=1}^{4} x_i$, $z = \ltimes_{i=1}^{3} z_i$. Then we have

$$z = M_G x,$$

where

$$M_G = \delta_8[8\ 3\ 7\ 4\ 6\ 1\ 5\ 2\ 4\ 7\ 3\ 8\ 2\ 5\ 1\ 6].$$

Then it is easy to see that

$$H = \delta_8[2\ 4\ 8\ 8\ 1\ 3\ 3\ 3]$$

verifies (9.37).

(2) $\mathscr{Z}_0$ is also a control invariant subspace of system (9.4). Because if we choose

$$u_1(t) = 1 \sim \begin{pmatrix}1\\0\end{pmatrix}, \quad u_2(t) = 1 \sim \begin{pmatrix}1\\0\end{pmatrix},$$

The dynamics of (9.4) becomes

$$\begin{cases} x_1(t+1) = (x_2(t) \leftrightarrow 0) \leftrightarrow (x_2(t) \wedge x_4(t)) \\ x_2(t+1) = x_2(t) \vee (x_3(t) \leftrightarrow x_4(t)) \\ x_3(t+1) = ((x_1(t) \leftrightarrow \neg x_4(t)) \rightarrow 1) \leftrightarrow (x_2(t) \wedge x_4(t)) \\ x_4(t+1) = \neg(x_2(t) \wedge (x_4(t)) \\ \\ y(t) = x_3(t) \leftrightarrow \neg x_4(t). \end{cases} \tag{9.39}$$

The algebraic form of network (9.39) is

$$x(t+1) = Lx(t), \tag{9.40}$$

where

$$L = \delta_{16}[10\ 3\ 10\ 3\ 11\ 15\ 15\ 11\ 10\ 3\ 10\ 3\ 11\ 15\ 15\ 11].$$

Setting $x = \ltimes_{i=1}^4 x_i$, $z = \ltimes_{i=1}^3 z_i$, there exists an $H \in \mathscr{L}_{2^3 \times 2^3}$ satisfies (9.37), where

$$H = \delta_8[1\ 3\ 7\ 7\ 1\ 3\ 7\ 7]$$

□

9.5 Multi-valued Network

Consider a network with n nodes: x_i, $i = 1, \cdots, n$. When $x_i(t) \in \mathscr{D}$, $\forall i, \forall t \geq 0$, the network is a Boolean network. Now we define

$$\mathscr{D}_k := \left\{ \frac{i}{k-1} \mid i = 0, 1, \cdots, k-1 \right\}, \quad k \geq 3.$$

If we allow $x_i(t) \in \mathscr{D}_k$, $\forall i, \forall t \geq 0$, the network becomes a k-valued network. The network graphs for Boolean network and for k-valued networks are the same. The general dynamic equation (9.1) ((9.3)) for Boolean network (Boolean control network) is still valid for k-valued network (k-valued control network), when $\mathscr{D}$ is replaced by $\mathscr{D}_k$. We refer to [17] for detailed discussion on multi-valued networks.

To give a brief survey, we first define some commonly used logical operators:

(i) Negation: $\neg : \mathscr{D}_k \to \mathscr{D}_k$, defined as

$$\neg p := 1 - p; \tag{9.41}$$

(ii) i-retriever: $\nabla_i : \mathscr{D}_k \to \mathscr{D}_k$, $i = 1, 2, \cdots, k$, defined as

$$\nabla_i(p) = \begin{cases} 1, & \text{when} \quad p = \frac{k-i}{k-1}, \\ 0, & \text{otherwise.} \end{cases} \tag{9.42}$$

(iii) Rotator: $\oslash : \mathscr{D}_k \to \mathscr{D}_k$, defined as

$$\oslash(p) := \begin{cases} p + \frac{1}{k-1}, & p \neq 1, \\ 0, & p = 1. \end{cases} \tag{9.43}$$

(iv) Conjunction: $\wedge : \mathscr{D}_k^2 \to \mathscr{D}_k$, defined as

$$p \wedge q := \min(p, q); \tag{9.44}$$

(v) Disjunction: $\vee : \mathscr{D}_k^2 \to \mathscr{D}_k$, defined as

$$p \vee q := \max(p, q); \tag{9.45}$$

(vi) Conditional: $\to : \mathscr{D}_k^2 \to \mathscr{D}_k$, defined as

$$p \to q := \neg p \vee q; \tag{9.46}$$

(vii) Biconditional: $\leftrightarrow : \mathscr{D}_k^2 \to \mathscr{D}_k$, defined as

$$p \leftrightarrow q := (p \to q) \wedge (q \to p). \tag{9.47}$$

To use matrix expression, we identify $\mathscr{D}_k$ with Δ_k. Precisely, we set an one-to-one correspondence between their entries as

$$\frac{i}{k-1} \sim \delta_k^{k-i}, \quad i = 0, 1, \cdots, k-1.$$

Then a k-valued logical variable $p \in \mathscr{D}$ has its vector form, still denoted by p, $p \in \Delta_k$.

Let $F : \mathscr{D}_k^n \to \mathscr{D}_k^m$ be described as

$$z_i = f_i(x_1, \cdots, x_n), \quad i = 1, \cdots, m. \tag{9.48}$$

In vector form, we have $x_i, z_j \in \Delta_k$. Setting $x = \ltimes_{i=1}^n x_i$, $z = \ltimes_{i=1}^m z_i$, we have the following result, which corresponds to Theorem 9.3.1.

Theorem 9.5.1. *Given a logical mapping $F : \mathscr{D}_k^n \to \mathscr{D}_k^m$, described by (9.48). Then there is a unique matrix, $M_F \in \mathscr{L}_{k^m \times k^n}$, called the structure matrix of F, such that*

$$z = M_F x. \tag{9.49}$$

Table 9.2. Structure Matrix of Operators (k=3)

Operator	Structure Matrix	Operator	Structure Matrix
$\neg$	$M_n = \delta_3[3\ 2\ 1]$	$\vee$	$M_d = \delta_3[1\ 1\ 1\ 1\ 2\ 2\ 1\ 2\ 3]$
$\oslash$	$M_o = \delta_3[3\ 1\ 2]$	$\wedge$	$M_c = \delta_3[1\ 2\ 3\ 2\ 2\ 3\ 3\ 3\ 3]$
∇_1	$M_{\nabla_1} = \delta_3[1\ 1\ 1]$	$\rightarrow$	$M_i = \delta_3[1\ 2\ 3\ 1\ 2\ 2\ 1\ 1\ 1]$
∇_2	$M_{\nabla_2} = \delta_3[2\ 2\ 2]$	$\leftrightarrow$	$M_e = \delta_3[1\ 2\ 3\ 2\ 2\ 2\ 3\ 2\ 1]$
∇_3	$M_{\nabla_3} = \delta_3[3\ 3\ 3]$		

Note that when $m = 1$ the mapping becomes a logical function and M_F is called the structure matrix of the function. Assume $k = 3$ the structure matrices of the previous fundamental operators are collected in the following table 9.2.

By replacing $\mathscr{D}$ with $\mathscr{D}_k$, (equivalently, replacing Δ with Δ_k), the mappings $T : \mathscr{D}_k^n \to \mathscr{D}_k^n$ and $G : \mathscr{D}_k^n \to \mathscr{D}_k^s$ defined in (9.24) and (9.33) become

$$z_i = t_i(x_1, \cdots, x_n), \quad x_i, z_i \in \mathscr{D}_k,\ i = 1, \cdots, n. \tag{9.50}$$

$$z_i = g_i(x_1, \cdots, x_n), \quad x_i, z_j \in \mathscr{D}_k,\ i = 1, \cdots, s,\ j = 1, \cdots, n. \tag{9.51}$$

As in Section 9.4.1, setting $z = \ltimes_{i=1}^n z_i$ and $x = \ltimes_{i=1}^n x_i$, the algebraic form of mapping T is

$$z = M_T x, \tag{9.52}$$

where $M_T \in \mathscr{L}_{k^n \times k^n}$ is the structure matrix of T.

As in Section 9.4.2, setting $z = \ltimes_{i=1}^s z_i$ and $x = \ltimes_{i=1}^n x_i$, the algebraic form of mapping G is

$$z = M_G x := \begin{bmatrix} g_{11} & \cdots & g_{1,k^n} \\ \vdots & & \\ g_{k^s,1} & \cdots & g_{k^s,k^n} \end{bmatrix} x. \tag{9.53}$$

It is easy to prove the following:

Theorem 9.5.2. *A mapping $T : \mathscr{D}_k^n \to \mathscr{D}_k^n$ is a coordinate transformation, iff its structure matrix $M_T \in \mathscr{L}_{k^n \times k^n}$ is non-singular.*

Theorem 9.5.3. *Let $\mathscr{Z}_0 = F_\ell(z_1, \cdots, z_s)$, where z_i, $i = 1, \cdots, s$ are determined by (9.51) (equivalently (9.53)). Then $\mathscr{Z}_0$ is a regular subspace, iff*

$$\sum_{j=1}^{k^n} g_{ij} = k^{n-s}, \quad i = 1, \cdots, k^s. \tag{9.54}$$

Let $\mathscr{Z}_0 = F_\ell(z_1, \cdots, z_s)$ be a regular subspace of the state space $\mathscr{X}$, where z_i, $i = 1, \cdots, s$ are determined by (9.51) (equivalently (9.53)). The algebra form of multi-valued system (9.1) is

$$x(t+1) = Lx(t),$$

where $L \in \mathscr{L}_{k^n \times k^n}, x = \ltimes_{i=1}^n x_i, x_i \in \mathscr{D}_k, \quad i = 1, \cdots, n.$

Using the above notations we have

Theorem 9.5.4. *$\mathscr{Z}_0$ is an invariant subspace with respect to multi-valued system (9.1), iff one of the following two equivalent conditions is satisfied.*

(i).

$$Row(M_G L) \subset SpanRow(M_G) \tag{9.55}$$

(ii). There exists an $H \in \mathscr{L}_{k^s \times k^s}$ such that

$$M_G L = H M_G. \tag{9.56}$$

We give an example to illustrate the above theorems for 3-valued network.

Example 9.5.1. Consider the network (9.2), if the logical variables $x_i \in \Delta_3, \quad i = 1, \cdots, 4$, and the logical operators are defined as (9.41)–(9.47). Define $x = \ltimes_{i=1}^{4} x_i$. Based on Theorem 9.5.1, there exists a unique matrix $L \in \mathscr{L}_{3^4 \times 3^4}$, such that

$$x(t+1) = Lx(t), \tag{9.57}$$

where

$$\begin{array}{rl} L = \delta_{81}[& 61\ 31\ \ 1\ 61\ 31\ \ 1\ 61\ 31\ \ 1\ 61\ 40\ 37\ 70\ 40\ 37\ 70\ 40\ 28 \\ & 61\ 67\ 73\ 70\ 67\ 64\ 79\ 67\ 55\ 31\ 32\ 32\ 31\ 32\ 32\ 31\ 32\ 32 \\ & 31\ 41\ 41\ 40\ 41\ 41\ 40\ 41\ 32\ 58\ 67\ 76\ 67\ 67\ 67\ 76\ 67\ 58 \\ & \ 1\ \ 2\ \ 3\ \ 1\ \ 2\ \ 3\ \ 1\ \ 2\ \ 3\ 31\ 41\ 41\ 40\ 41\ 41\ 40\ 41\ 32 \\ & 55\ 67\ 79\ 64\ 67\ 70\ 73\ 67\ 61\ \ ], \end{array}$$

For the mapping (9.28), its algebra form is

$$z = M_T x.$$

It is easy to calculate that

$$\begin{array}{rl} M_T = \delta_{81}[& 79\ 50\ 21\ 76\ 50\ 24\ 73\ 50\ 27\ 70\ 41\ 12\ 67\ 41\ 15\ 64\ 41\ 18 \\ & 61\ 32\ \ 3\ 58\ 32\ \ 6\ 55\ 32\ \ 9\ 52\ 50\ 48\ 49\ 50\ 51\ 46\ 50\ 54 \\ & 43\ 41\ 39\ 40\ 41\ 42\ 37\ 41\ 45\ 34\ 32\ 30\ 31\ 32\ 33\ 28\ 32\ 36 \\ & 25\ 50\ 75\ 22\ 50\ 78\ 19\ 50\ 81\ 16\ 41\ 66\ 13\ 41\ 69\ 10\ 41\ 72 \\ & \ 7\ 32\ 57\ \ 4\ 32\ 60\ \ 1\ 32\ 63\ \ ], \end{array}$$

which is singular. So, T is not a coordinate transformation in 3-valued network. □

9.6 Conclusion

This paper reviewed the state space description of Boolean (control) networks. Using semi-tensor product of matrices and the matrix expression of logic, the state space of a Boolean (control) network are defined as

$$\mathscr{X} = F_\ell(x_1, \cdots, x_n);$$

and a subspace $\mathscr{Z} = F_\ell(z_1, \cdots, z_k) \subset \mathscr{X}$ is expressed as

$$z = T_0 x, \quad \text{where } T_0 \in \mathscr{L}_{2^k \times 2^n}.$$

Through this way, a subspace is represented by a logical matrix.

Assume $k = n$ and T_0 is nonsingular, the mapping: $X = (x_1, \cdots, x_n) \mapsto Z = (z_1, \cdots, z_n)$ becomes a coordinate change.

Using this state space approach, the dynamics of a Boolean network can be converted into a discrete time (standard) dynamics [8]. Under this framework, several control problem have been investigated in detail. We list some recent works at the follows, which are based on this framework.

(i) the input-state structure analysis for cycles [6];
(ii) the controllability and observability of Boolean control networks [7];
(iii) the realization of Boolean control networks [9];
(iv) disturbance decoupling of Boolean control networks [10, 12];
(v) stability and stabilization of Boolean (control) networks, [11].

There are lots of control problems which have not been much investigated. For instance, system identification, optimal control, etc. The state space reviewed in this paper seems applicable to them.

References

1. Akutsu, T., Miyano, S., Kuhara, S.: Inferring qualitative relations in genetic networks and metabolic pathways. Bioinformatics 16, 727–773 (2000)
2. Albert, R., Barabasi, A.-L.: Dynamics of complex systems: scaling laws or the period of Boolean networks. Phys. Rev. Lett. 84, 5660–5663 (2000)
3. Aldana, M.: Boolean dynamics of networks with scale-free topology. Physica D 185, 45–66 (2003)
4. Cheng, D.: Semi-tensor product of matrices and its applications — A survey. In: ICCM 2007, vol. 3, pp. 641–668 (2007)
5. Cheng, D., Qi, H.: Semi-tensor Product of Matrices, Theorem and Applications (in Chinese). Science Press, Beijing (2007)
6. Cheng, D.: Input-state approach to Boolean networks. IEEE Trans. Neural Network 20(3), 512–521 (2009)
7. Cheng, D., Qi, H.: Controllability and observability of Boolean control networks. Automatica 45(7), 1659–1667 (2009)
8. Cheng, D., Qi, H.: A linear representation of dynamics of Boolean networks. IEEE Trans. Aut. Contr. (provitionally accepted)
9. Cheng, D., Li, Z., Qi, H.: Realization of Boolean control networks. Automatica (accepted)
10. Cheng, D.: Disturbance decoupling of Boolean control networks. IEEE Trans. Aut. Contr. (revised)
11. Cheng, D., Liu, J.: Stabilization of Boolean control networks. CDC-CCC'09 (to appear)
12. Cheng, D., Qi, H., Li, Z.: Canalyzing Boolean mapping and its application to disturbance decoupling of Boolean control networks, Proc. of ICCA09, Christchurch, New Zealand, 2009 (to appear)

13. Drossel, B., Mihaljev, T., Greil, F.: Number and length of attractors in a critical Kauffman model with connectivity one. Phys. Rev. Lett. 94, 088701 (2005)
14. Harris, S.E., Sawhill, B.K., Wuensche, A., Kauffman, S.: A model of transcriptional regulatory networks based on biases in the observed regulation rules. Complexity 7, 23–40 (2002)
15. Kauffman, S.A.: Metabolic stability and epigenesis in randomly constructed genetic nets. J. Theoretical Biology 22, 437–467 (1969)
16. Kauffman, S.A.: At Home in the Universe. Oxford Univ. Press, Oxford (1995)
17. Li, Z., Cheng, D.: Algebraic approach to dynamics of multi-valued networ. Int. J. Bif. Chaos 20(3) (to appear 2010)
18. Rade, L., Westergren, B.: Mathematics Handbook for Science and Engineering, 4th edn. Studentlitteratur, Lund (1998)
19. Samuelsson, B., Troein, C.: Superpolynomial growth in the number of attractots in Kauffman networks. Phys. Rev. Lett. 90, 90098701 (2003)
20. Shmulevich, I., Dougherty, R., Kim, S., Zhang, W.: Probabilistic Boolean neworks: A rule-based uncertainty model for gene regulatory networks. Bioinformatics 2(18), 261–274 (2002)
21. Waldrop, M.M.: Complexity. Touchstone, New York (1992)

10

Nonlinear Output Regulation: Exploring Non-minimum Phase Systems*

F. Delli Priscoli[1], A. Isidori[1,2], and L. Marconi[2]

[1] Dipartimento di Informatica e Sistemistica, Università di Roma "La Sapienza", Via Ariosto 25, 00185 Rome, Italy
[2] C.A.SY. – Dipartimento di Elettronica, Informatica e Sistemistica, University of Bologna, 40136 Bologna, Italy

Summary. The present paper presents a new contribution to the design of output regulators for a class of nonlinear systems characterized by a possibly unstable zero dynamics. It is shown how the problem in question is handled by addressing a stabilization problem for a suitably defined reduced auxiliary plant.

10.1 Introduction

The problem of tracking and asymptotic disturbance rejection (also known as the generalized *servomechanism problem* or as the *output regulation problem*) is to design a controller so as to obtain a closed-loop system in which all trajectories are bounded, and a regulated output asymptotically decays to zero as time tends to infinity. The peculiar aspect of this design problem is the characterization of the class of all possible exogenous inputs (disturbances, commands, uncertain constant parameters) as the set of all possible solutions of a fixed (finite-dimensional) differential equation. In this setting, *any source of uncertainty* (about actual disturbances affecting the system, about actual trajectories that are required to be tracked, about any uncertain constant parameters) is treated as *uncertainty in the initial condition* of a fixed autonomous finite dimensional dynamical system, known as the *exosystem*.

The body of theoretical results that was developed in this domain of research over about three decades has scored numerous important successes and has now reached a stage of full maturity. Remarkable, in this respect, are a series of contributions by C.I. Byrnes, together with the co-authors of this note, which can be considered as milestones in the design of regulators for nonlinear systems. They include the necessary conditions known as the nonlinear regulator equations (developed in [6]), the concept of immersion into a linear observable system for the design of internal models (developed in [3]), the concept of adaptive internal model (developed in [10], the concept of steady state behavior for nonlinear systems (developed in [1]).

*Dedicated to Chris Byrnes and Anders Linquist, outstanding scientists and most dear friends.

X. Hu et al. (Eds.): Three Decades of Progress in Control Sciences, pp. 141–152, 2010.

Most of the design methods proposed in the literature still address a restricted class of systems, namely systems in normal form with a (globally) stable zero dynamics. It was only recently that the issue of solving problems of output regulation for systems possessing an unstable zero dynamics has been addressed. In this respect, a promising approach is the one presented in [5], where – by enhancing the earlier approach discussed in [8] – the problem is handled by means of a design technique which has the advantage of keeping separate the influences of the (unstable) zero dynamics and of the parameters of the internal model. In the present paper, we show how the approach in question can be used to handle the case of a system whose zero dynamics as a feed-forward form.

10.2 The Setup

We begin with a summary of the setup and of the results of [5]. Consider a nonlinear system in normal form

$$\begin{aligned} \dot{z}_0 &= f_0(w, z_0, \xi_1, \ldots, \xi_r) \\ \dot{\xi}_1 &= \xi_2 \\ &\cdots \\ \dot{\xi}_{r-1} &= \xi_r \\ \dot{\xi}_r &= q_0(w, z_0, \xi_1, \ldots, \xi_r) + u \\ e &= \xi_1 \end{aligned} \tag{10.1}$$

with *control* input $u \in \mathbb{R}$, *regulated* output $e \in \mathbb{R}$, in which $w \in \mathbb{R}^s$ is a vector of *exogenous* inputs which cannot be controlled, solutions of a fixed ordinary differential equation of the form

$$\dot{w} = s(w). \tag{10.2}$$

In this setup, w can be viewed as a model of time-varying commands, external disturbances, and also uncertain constant plant parameters. The initial states of (10.1) and of (10.2) are assumed to range over a fixed compact sets X and W, with W invariant under the dynamics of (10.2). Motivated by well-known standard design procedures, we assume throughout that the *measured output* y coincides with the *partial state* $(\xi_1, \ldots, \xi_r)$. The states w and z_0 are, on the contrary, *not* available for measurement.

The problem of output regulation is to design a controller

$$\begin{aligned} \dot{\xi} &= \varphi(\xi, y) \\ u &= \gamma(\xi, y) \end{aligned}$$

with initial state in a compact set Ξ, yielding a closed-loop system in which

- the positive orbit of $W \times X \times \Xi$ is bounded,
- $\lim_{t \to \infty} e(t) = 0$, uniformly in the initial condition (on $W \times X \times \Xi$).

The standard point of departure in the analysis of the problem of output regulation is the identification of a (smooth) *controlled invariant* manifold entirely contained in the set of all states at which $e = 0$ (see [6]). In the present context, this can be specialized as follows. Let the aggregate of (10.1) and (10.2), be rewritten as

$$\begin{aligned} \dot{w} &= s(w) \\ \dot{z} &= f(w,z,\zeta) \\ \dot{\zeta} &= q(w,z,\zeta)+u \end{aligned} \tag{10.3}$$

in which $\zeta = \xi_r = e^{(r-1)}$ and $z = \mathrm{col}(z_0, \xi_1, \ldots, \xi_{r-1})$. Assume the existence of a smooth map $\pi_0 : W \to \mathbb{R}^{n-r}$ satisfying

$$\frac{\partial \pi_0}{\partial w} s(w) = f_0(w, \pi_0(w), 0, \ldots, 0) \qquad \forall w \in W.$$

and note that the map $\pi : W \to \mathbb{R}^{n-1}$ defined as $z = \mathrm{col}(\pi_0(w), 0, \ldots, 0)$ satisfies

$$\frac{\partial \pi}{\partial w} s(w) = f(w, \pi(w), 0) \qquad \forall w \in W. \tag{10.4}$$

Trivially, the smooth manifold

$$\{(w,z,\zeta) : w \in W, z = \pi(w), \zeta = 0\},$$

a subset of the set of all states at which $e = \xi_1 = 0$, can be rendered invariant by feedback, actually by the control

$$u = -q(w, \pi(w), 0). \tag{10.5}$$

The second step in the solution of the problem usually consists in making assumptions that make it possible to generate the control (10.5) by means of an *internal model*. In a series of recent papers, it was shown how these assumptions could be progressively weakened, moving from the so-called assumption of "immersion into a linear observable system", to "immersion into a nonlinear uniformly observable system (as in [2])" to the recent results of [7], in which it was shown that no assumption is in fact needed for the construction of an internal model if only continuous (thus possibly not locally Lipschitz) controllers are acceptable. Motivated by these recent advances, we assume the existence of a pair F_0, G_0, in which F_0 is a $d \times d$ Hurwitz matrix and G_0 is a $d \times 1$ column vector that makes the pair F_0, G_0 controllable, of a locally Lipschitz map $\gamma : \mathbb{R}^d \to \mathbb{R}$ and a continuously differentiable map $\tau : W \to \mathbb{R}^d$ satisfying

$$\begin{aligned} &\frac{\partial \tau}{\partial w} s(w) = F_0 \tau(w) + G_0 \gamma(\tau(w)) && \forall w \in W \\ &-q(w, \pi(w), 0) = \gamma(\tau(w)) && \forall w \in W. \end{aligned} \tag{10.6}$$

Properties (10.4) and (10.6) are instrumental in the design of a controller that solves the problem of output regulation.

10.3 The Design Method of [5]

10.3.1 The Controller and the Reduction Procedure

Consider, for the original plant, a controller of the form

$$\begin{aligned} u &= \dot{N}(\varphi) + \gamma(\eta) + v \\ v &= -k[\zeta - N(\varphi)] \\ \dot{\eta} &= F_0(\eta - G_0[\zeta - N(\varphi)]) + G_0[\gamma(\eta) + v] \\ \dot{\varphi} &= L(\varphi + M[\zeta - N(\varphi)]) - Mv \end{aligned} \tag{10.7}$$

which is a dynamic controller, with internal state (η, φ), "driven" only by the measured variable ζ. Change variables as

$$\begin{aligned} \theta &= \zeta - N(\varphi) \\ \chi &= \varphi + M\theta \\ x &= \eta - G_0\theta \end{aligned}$$

to obtain a system

$$\begin{aligned} \dot{w} &= s(w) \\ \dot{z} &= f(w, z, \theta + N(\chi - M\theta)) \\ \dot{\chi} &= L(\chi) + M[q(w, z, \theta + N(\chi - M\theta)) + \gamma(x + G_0\theta)] \\ \dot{x} &= F_0 x - G_0 q(w, z, \theta + N(\chi - M\theta)) \\ \dot{\theta} &= q(w, z, \theta + N(\chi - M\theta)) + \gamma(x + G_0\theta) - k\theta\,. \end{aligned} \tag{10.8}$$

This system can be seen as feedback interconnection of a system with input θ and state (w, z, χ, x) and of a system with input (w, z, χ, x) and state θ. The advantage of seeing system (10.8) in this form is that we can appeal to the following result (see e.g. [7]).

Proposition 10.3.1. *Consider a system of the form (10.8). Let P be an arbitrary fixed compact set of initial conditions for (w, z, χ, x). Suppose there exists a set $\mathscr{A}$ which is locally exponentially stable for*

$$\begin{aligned} \dot{w} &= s(w) \\ \dot{z} &= f(w, z, N(\chi)) \\ \dot{\chi} &= L(\chi) + M[q(w, z, N(\chi)) + \gamma(x)] \\ \dot{x} &= F_0 x - G_0 q(w, z, N(\chi))\,, \end{aligned} \tag{10.9}$$

with a domain of attraction that contains the set P. Suppose also that

$$q(w, z, N(\chi)) + \gamma(x) = 0, \qquad \forall (w, z, \chi, x) \in \mathscr{A}\,. \tag{10.10}$$

Then, for any choice of a compact set Θ, there is a number k^* such that, for all $k > k^*$, the set $\mathscr{A} \times \{0\}$ is locally exponentially stable, with a domain of attraction that contains $P \times \Theta$.

If the assumption of this Proposition are fulfilled and, in addition, the regulated variable $e = \xi_1$ vanishes $\mathscr{A}$, we conclude that the proposed controller is able to solve the problem of output regulation.

All of the above suggests the use of the degrees of freedom in the choice of the parameters of the controller in order to fulfill the hypotheses of Proposition 10.3.1. To this end, recall that, by assumption, there exists $\pi(w)$ and $\tau(w)$ satisfying (10.4) and (10.6). Hence, it is readily seen that if $L(0) = 0$ and $N(0) = 0$, the set

$$\mathscr{A} = \{(w, z, \chi, x) : w \in W, z = \pi(w), \chi = 0, x = \tau(w)\}$$

is a compact invariant set of (10.9). Moreover, by construction, the identity (10.10) holds. Trivially, also ξ_1 vanishes on this set. Thus, it is concluded that if the set $\mathscr{A}$ can be made local exponentially stable, with a domain of attraction that contains the compact set of all admissible initial conditions, the proposed controller, with large k solves the problem of output regulation.

System (10.9) is not terribly difficult to handle. As a matter of fact, it can be regarded as interconnection of three much simpler subsystems.

To see this, set

$$\begin{aligned} z_a &= z - \pi(w) \\ \tilde{x} &= x - \tau(w) \end{aligned}$$

and define

$$\begin{aligned} f_a(w, z_a, \zeta) &= f(w, z_a + \pi(w), \zeta) - f(w, \pi(w), 0, 0) \\ h_a(w, z_a, \zeta) &= q(w, z_a + \pi(w), \zeta) - q(w, \pi(w), 0, 0) . \end{aligned}$$

In the new coordinates thus introduced, the invariant manifold $\mathscr{A}$ is simply the set

$$\mathscr{A} = \{(w, z_a, \chi, \tilde{x}) : w \in W, (z_a, \chi, \tilde{x}) = (0, 0, 0)\} .$$

Bearing in mind (10.4) and (10.6), it is readily seen that

$$\dot{z}_a = f_a(w, z_a, N(\chi))$$

and

$$q(w, z, N(\chi)) = h_a(w, z_a, N(\chi)) - \gamma(\tau(w)) .$$

In view of this, using again (10.6), system (10.9) can be seen as a system with input v and output y_f defined as

$$\begin{aligned} \dot{w} &= s(w) \\ \dot{z}_a &= f_a(w, z_a, N(\chi)) \\ \dot{\chi} &= L(\chi) + M[h_a(w, z_a, N(\chi)) + v] \\ \dot{\tilde{x}} &= F_0 \tilde{x} - G_0 h_a(w, z_a, N(\chi)) \\ y_f &= \gamma(\tilde{x} + \tau(w)) - \gamma(\tau(w)) \end{aligned} \tag{10.11}$$

subject to unitary output feedback

$$v = y_{\mathrm{f}}.$$

System (10.11), in turn, can be seen as the cascade of an "inner loop" consisting of a subsystem, which we call the "auxiliary plant", modelled by equations of the form

$$\begin{aligned} \dot{w} &= s(w) \\ \dot{z}_{\mathrm{a}} &= f_{\mathrm{a}}(w, z_{\mathrm{a}}, u_{\mathrm{a}}) \\ y_{\mathrm{a}} &= h_{\mathrm{a}}(w, z_{\mathrm{a}}, u_{\mathrm{a}}), \end{aligned} \tag{10.12}$$

controlled by

$$\begin{aligned} \dot{\chi} &= L(\chi) + M[y_{\mathrm{a}} + v] \\ u_{\mathrm{a}} &= N(\chi), \end{aligned} \tag{10.13}$$

cascaded with a system, which we call a "weighting filter", modelled by equations of the form

$$\begin{aligned} \dot{\tilde{x}} &= F_0 \tilde{x} - G_0 y_{\mathrm{a}} \\ \overline{y} &= \gamma(\tilde{x} + \tau(w)) - \gamma(\tau(w)). \end{aligned} \tag{10.14}$$

All of this is depicted in Fig. 10.1.

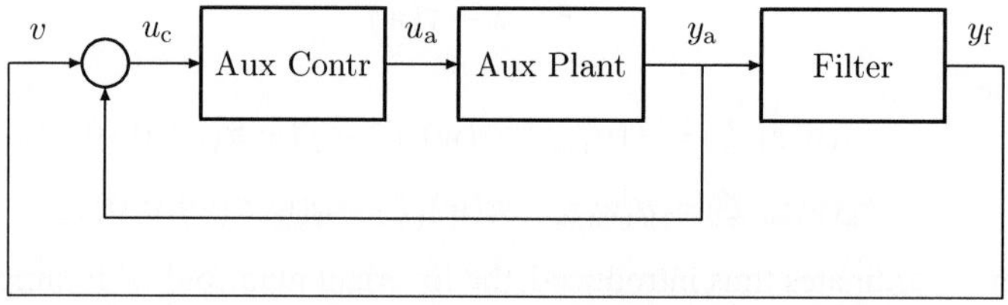

Fig. 10.1. The feedback structure of system (10.9)

We have in this way transformed the original design problem in the problem of rendering the equilibrium of the closed loop system (10.9) asymptotically stable, with a locally quadratic Lyapunov function, with a domain of attraction that contains the compact set of all admissible initial conditions.

10.3.2 The Case of Harmonic Exogenous Inputs

Consider now the case in which the internal model has a pair of purely imaginary eigenvalues at $\pm i\Omega$. This corresponds to a regulation problem in which the exogenous inputs (to be followed and/or rejected) are sinusoidal functions of time. Pick,

$$F_0 = \begin{pmatrix} 0 & 1 \\ -\Omega^2 & -2\Omega \end{pmatrix}, \quad G_0 = \begin{pmatrix} 0 \\ 1 \end{pmatrix}.$$

In this case, $\gamma(x) = \Psi x$ with Ψ, the unique vector which assigns to $F_0 + G_0 \Psi$ the characteristic polynomial $\lambda^2 + \Omega^2$ given by

$$\Psi = \begin{pmatrix} 0 & 2\Omega \end{pmatrix} .$$

With this choice of F_0, system (10.11) can be interpreted as the cascade connection of

$$\begin{aligned} \dot{w} &= s(w) \\ \dot{z}_a &= f_a(w, z_a, N(\chi)) \\ \dot{\chi} &= L(\chi) + M[h_a(w, z_a, N(\chi)) + v] \\ y &= \dot{h}_a(w, z_a, N(\chi), v) \end{aligned} \tag{10.15}$$

and

$$\begin{aligned} \dot{\tilde{x}}_1 &= -\Omega^2 z_2 - y \\ \dot{z}_2 &= \tilde{z}_1 - 2\Omega z_2 \\ y_f &= 2\Omega z_2 . \end{aligned} \tag{10.16}$$

in which

$$\dot{h}_a(w, z_a, N(\chi), v) :=$$

$$\frac{\partial h_a}{\partial w} s(w) + \frac{\partial h_a}{\partial z_a} f_a(w, z_a, N(\chi)) + \frac{\partial h_a}{\partial \zeta} \frac{\partial N}{\partial \chi} (L(\chi) + M[h_a(w, z_a, N(\chi)) + v]) .$$

System (10.16) a stable linear system, with a transfer function

$$\Phi'(s) = -\frac{2\Omega}{(s+\Omega)^2}$$

whose L_2-gain is equal to $2/\Omega$. Thus, by known facts, a sufficient condition for system (10.9) to be globally asymptotically stable (with a locally quadratic Lyapunov function), is that system (10.15) be globally asymptotically stable, with a locally quadratic Lyapunov function and an L_2-gain γ_0, between input v and output y, satisfying

$$2\gamma_0 < \Omega . \tag{10.17}$$

Therefore, as observed in [5], the following result holds.

Proposition 10.3.2. *Consider a problem of output regulation for a plant modelled by equations of the form (10.3), with an internal model with a pair of imaginary eigenvalues at* $\pm i\Omega$*. Let* $(L(\cdot), M, N(\cdot))$ *be such that the associated controller (10.13) renders system (10.15) globally asymptotically stable, a locally quadratic Lyapunov function and an* L_2 *gain g satisfying* $\gamma_0 < \Omega/2$*. Then, there exists a number* k^* *such that, for all* $k > k^*$*, the controller (10.7) solves the problem of (semiglobal) output regulation.*

10.4 Dealing with Systems Whose Zero Dynamics Are in Feed-Forward Form

As a continuation of the analysis initiated in [5], suppose that the auxiliary plant (10.12) is a system of the form

$$\begin{aligned}\dot{w} &= Sw \\ \dot{z}_1 &= a(z_1, z_2, w)z_2 \\ \dot{z}_2 &= b(z_2)u_\mathrm{a} \\ y_\mathrm{a} &= z_1\end{aligned} \tag{10.18}$$

and hence the output y of (10.15) is

$$y = \dot{y}_\mathrm{a} = a(z_1, z_2, w)z_2\,.$$

Assume the existence of two numbers $0 < \ell_1 < \ell_2$ such that

$$\ell_1 \le a(z_1, z_2, w) \le \ell_2, \qquad \ell_1 \le b(z_2) \le \ell_2$$

for all z_1, z_2, w.

We control this system by means of a linear controller having transfer function

$$-k\frac{s+\varepsilon}{1+s/g} = -kg\frac{s+\varepsilon}{s+g}$$

(note that this system is not strictly proper as in (10.13), but it can be rendered such by addition of a "far off" pole). A realization of this transfer function is

$$\begin{aligned}\dot{\xi} &= -g\xi + gu_\mathrm{c} \\ u_\mathrm{a} &= -k[(\varepsilon - g)\xi + gu_\mathrm{c}]\,.\end{aligned}$$

Bearing in mind that

$$u_\mathrm{c} = z_1 + v$$

we obtain for system (10.15) the form

$$\begin{aligned}\dot{\xi} &= -g\xi + g(z_1 + v) \\ \dot{z}_1 &= az_2 \\ \dot{z}_2 &= -bk[(\varepsilon - g)\xi + g(z_1 + v)] \\ y &= az_2\,.\end{aligned} \tag{10.19}$$

We perform first a linear change of variable, whose purpose is to see the system as a closed loop containing an integrator with a gain coefficient equal to ε. Set

$$z_3 = \xi - z_1$$

to obtain

$$\begin{aligned}\dot{z}_1 &= az_2 \\ \dot{z}_2 &= -bk[(\varepsilon - g)(z_3 + z_1) + g(z_1 + v)] = -bk[(\varepsilon - g)z_3 + \varepsilon z_1 + gv] \\ \dot{z}_3 &= -g(z_1 + z_3) + g(z_1 + v) - az_2 = -az_2 - gz_3 + gv \\ y &= az_2\,.\end{aligned}$$

Next, we proceed now with a nonlinear change of variable of the form

$$\tilde{z}_1 = z_1 + z_3 - \phi(z_2)$$

(which replaces the former z_1), with $\phi(z_2)$ satisfying

$$\frac{\partial\phi}{\partial z_2} b(z_2)k(\varepsilon - g) = g\,.$$

Such a $\phi(z_2)$ always exists (since $b(z_2)$ is bounded from below and from above), and can be found by direct integration.

This yields

$$\begin{aligned}
\dot{\tilde{z}}_1 &= az_2 + (-az_2 - gz_3 + gv) - \frac{\partial\phi}{\partial z_2}(-bk)[(\varepsilon - g)z_3 + \varepsilon z_1 + gv] \\
&= gv + \frac{g}{\varepsilon - g}[\varepsilon(\tilde{z}_1 - z_3 + \phi(z_2)) + gv] = \frac{g\varepsilon}{\varepsilon - g}[\tilde{z}_1 - z_3 + \phi(z_2)] + [g + \frac{g^2}{\varepsilon - g}]v \\
&= \frac{g\varepsilon}{g - \varepsilon}[-\tilde{z}_1 + z_3 - \phi(z_2) - v]
\end{aligned}$$

and

$$\begin{aligned}
\dot{z}_2 &= -bk[(\varepsilon - g)z_3 + \varepsilon(\tilde{z}_1 - z_3 + \phi(z_2) + gv] = -bk[-gz_3 + gv] - \varepsilon bk[\tilde{z}_1 + \phi(z_2)] \\
\dot{z}_3 &= -az_2 - gz_3 + gv \\
y &= az_2\,.
\end{aligned}$$

The upper subsystem is a system with state $\tilde{z}_1$ and inputs z_2, z_3, v whose gains, though, cannot be modified. The lower subsystem, can be seen as a system with state (z_2, z_3) and input v, modelled by

$$\begin{aligned}
\dot{z}_2 &= -bk[-gz_3 + gv] \\
\dot{z}_3 &= -az_2 - gz_3 + gv \\
y &= az_2
\end{aligned} \tag{10.20}$$

(in which $b = b(z_2)$ e $a = a(\tilde{z}_1 - z_3 + \phi(z_2), z_2, w)$) affected by a "perturbation" term of the form

$$-\varepsilon bk[\tilde{z}_1 + \phi(z_2)]\,.$$

Since $\phi(z_2)$ is by construction globally Lypschitz, the effect of this term (on the stability of the overall system (10.19) and on the gain between v and y) could be made negligible by lowering the parameter ε, provided that we are able to find a suitable Lyapunov function for the unperturbed system (10.20).

Rewrite the latter as

$$\begin{aligned}
\dot{z} &= F(z) + G(z) \\
y &= H(z)
\end{aligned}$$

and consider the positive definite function

$$V(z) = \frac{P_2}{2}(z_2 + b(z_2)kz_3)^2 + \frac{P_3}{2}z_3^2\,.$$

In view of the above, the inequality

$$L_F V + H^2 + \frac{1}{4\gamma_0^2}(L_G V)^2 < 0 \tag{10.21}$$

can be enforced, with γ_0 chosen so that $\gamma_0 < \Omega/2$, we can conclude that, if ε is sufficiently small, system (10.19) has the desired properties. As a consequence, the result of Proposition 10.3.2 applies and the controller (10.7) solves the problem of (semiglobal) output regulation.

In what follows, we show how to enforce (10.21) on an arbitrarily large compact set. Take the derivative of V along F, to obtain

$$\begin{aligned} L_F V &= P_2(z_2 + bkz_3)[-bk(-gz_3) + bk(-az_2 - gz_3) + b'kz_3(-bk(-gz_3))] \\ &\quad + P_3 z_3(-az_2 - gz_3) \end{aligned}$$

in which $b'(z_2)$ is the derivative of $b(z_2)$. Simplification yields

$$L_F V = P_2(z_2 + bkz_3)[-bkaz_2 + b'bk^2 gz_3^2] + P_3 z_3(-az_2 - gz_3).$$

With any arbitrarily large $R > 0$ we obtain, for $|z| < R$, the quadratic estimate

$$\begin{aligned} L_F V &\le -P_2(z_2 + bkz_3)bkaz_2 + P_3 z_3(-az_2 - gz_3) + P_2|z_2 + bkz_3| \cdot R|b'b|k^2 g|z_3| \\ &\le -P_2 abkz_2^2 - P_3 gz_3^2 + [P_2|ab^2|k^2 + P_3|a|]|z_2||z_3| \\ &\quad + P_2 R|b'b|k^2 g|z_2||z_3| + P_2|b|kR|b'b|k^2 gz_3^2 \end{aligned} .$$

Likewise, take the derivative of V along G, to obtain

$$L_G V = P_2(z_2 + bkz_3)[-bkg + bkg + b'kz_3(-bkg)] + P_3 z_3 g,$$

which, after suitable simplification, yields the estimate

$$|L_G V| \le P_2 R|b'b|k^2 g|z_2| + [P_2|b|kR|b'b|k^2 g + P_3 g]|z_3|$$

for $|z| < R$. Finally, observe that

$$H^2 = a^2 z_2^2.$$

Assuming

$$\begin{aligned} &P_2 abk - a^2 > 4/\varepsilon + 1 \\ &P_2|ab^2|k^2 + P_3|a| + P_2 R|b'b|k^2 g < 1 \\ &P_3 g - P_2|b|kR|b'b|k^2 g > 2\varepsilon \end{aligned} \tag{10.22}$$

we obtain

$$L_F V + H^2 \le -(4/\varepsilon + 1)z_2^2 - 2\varepsilon z_3^2 + |z_2||z_3|.$$

Assuming

$$\begin{aligned} &[P_2 R|b'b|k^2 g]^2 < 4\gamma^2 \\ &2[P_2 R|b'b|k^2 g][P_2|b|kR|b'b|k^2 g + P_3 g] < 4\gamma^2 \\ &[P_2|b|kR|b'b|k^2 g + P_3 g]^2 < \varepsilon 4\gamma^2 \end{aligned} \tag{10.23}$$

we obtain

$$\frac{1}{4\gamma_0^2} L_G V^2 \leq z_2^2 + |z_2||z_3| + \varepsilon z_3^2.$$

In summary, if (10.22) and (10.23) hold, we obtain, for $|z| < R$, the estimate

$$L_F V + H^2 + \frac{1}{4\gamma_0^2}(L_G V)^2 < -(4/\varepsilon) z_2^2 - \varepsilon z_3^2 + 2|z_2||z_3|$$

whose right-hand side is quadratic and negative definite. It remains to show that (10.22) and (10.23) can be fulfilled, by properly setting k, P_2 and P_3. To this end, replace the second of (10.22) by

$$P_3|a| < 0.5, \qquad P_2|ab^2|k^2 + P_2 R|b'b|k^2 g < 0.5$$

Then, the third of (10.22) and the third of (10.23), if P_2 is kept fixed and k is lowered, can be replaced by

$$P_3|a| < 0.5, \qquad P_3 g > 2\varepsilon, \qquad P_3^2 g^2 < \varepsilon 4\gamma_0^2$$

Thanks to the square of P_3 in the last inequality, it is seen that all of these can be enforced, for a given, γ_0 by proper choice of ε and P_3. Let these be fixed (note that they are independent of P_2) and consider again the first of (10.22) in which we choose $P_2 = P/k$, with P to make it fulfilled. It remains to settle the first and second of (10.23), which is indeed possible by lowering the k. In summary, on any arbitrarily large compact set and for any choice of γ_0, the left-hand side of (10.21) can be estimated by a quadratic negative definite function provided that P_2, P_3 are appropriately set and k is small enough.

10.5 Conclusions

Most of the design methods, proposed in recent years, for the design of controllers to the purpose of solving problems of asymptotic tracking and disturbance rejection, only address systems in normal form with a (globally) stable zero dynamics. In this paper, by pursuing the design strategy suggested in [8] and enhanced in [5], we shows how the problem in question can be handled in the case of systems possessing an unstable zero dynamics in feed-forward form.

References

1. Byrnes, C.I., Isidori, A.: Limit sets, zero dynamics and internal models in the problem of nonlinear output regulation. IEEE Trans. on Automatic Control 48, 1712–1723 (2003)
2. Byrnes, C.I., Isidori, A.: Nonlinear Internal Models for Output Regulation. IEEE Trans. Automatic Control 49, 2244–2247 (2004)
3. Byrnes, C.I., Delli Priscoli, F., Isidori, A.: Output regulation of uncertain nonlinear systems. Birkhäuser, Boston (1997)

4. Delli Priscoli, F., Marconi, L., Isidori, A.: A New Approach to Adaptive Nonlinear Regulation. SIAM J. Control and Optimization 45, 829–855 (2006)
5. Delli Priscoli, F., Marconi, L., Isidori, A.: A A dissipativity-based approach to output regulation of non-minimum phase systems. Systems and Control Letters 58, 584–591 (2009)
6. Isidori, A., Byrnes, C.I.: Output regulation of nonlinear systems. IEEE Trans. Automatic Control 25, 131–140 (1990)
7. Marconi, L., Praly, L., Isidori, A.: Output Stabilization via Nonlinear Luenberger Observers. SIAM J. Control and Optimization 45, 2277–2298 (2006)
8. Marconi, L., Isidori, A., Serrani, A.: Non-resonance conditions for uniform observability in the problem of nonlinear output regulation. Systems & Control Lett. 53, 281–298 (2004)
9. Pavlov, A., van de Wouw, N., Nijmeijer, H.: Uniform Output Regulation of Nonlinear Systems: a Convergent Dynamics Approach. Birkhauser, Boston (2006)
10. Serrani, A., Isidori, A., Marconi, L.: Semiglobal nonlinear output regulation with adaptive internal model. IEEE Trans. Automatic Control 46, 1178–1194 (2001)

11

Application of a Global Inverse Function Theorem of Byrnes and Lindquist to a Multivariable Moment Problem with Complexity Constraint*

Augusto Ferrante[1], Michele Pavon[2], and Mattia Zorzi[3]

[1] Dipartimento di Ingegneria dell'Informazione, Università di Padova, via Gradenigo 6/B, 35131 Padova, Italy
[2] Dipartimento di Matematica Pura ed Applicata, Università di Padova, via Trieste 63, 35131 Padova, Italy
[3] Dipartimento di Ingegneria dell'Informazione, Università di Padova, via Gradenigo 6/B, 35131 Padova, Italy

Summary. A generalized moment problem for multivariable spectra in the spirit of Byrnes, Georgiou and Lindquist is considered. A suitable parametric family of spectra is introduced. The map from the parameter to the moments is studied in the light of a global inverse function theorem of Byrnes and Lindquist. An efficient algorithm is proposed to find the parameter value such that the corresponding spectrum satisfies the moment constraint.

11.1 Introduction

This paper represents an attempt to pay a tribute to two great figures of Systems and Control Theory. It would be difficult to even mention the long string of benchmark contributions that we owe to Anders and Chris. It would entail listing results in linear and nonlinear control, deterministic and stochastic systems, finite and infinite dimensional problems, etc. This string, no matter how much compactification we drew from string theory, would simply be too long. So we leave this task to the many that are better qualified than us. We like to stress, instead, two other aspects of their long lasting influence in the systems and control community. One is that both have devoted a lot of time and energy to form young researchers. Their generous help and tutoring to students and junior scientists continues unabated to this day. A second peculiar aspect of Anders and Chris is that they embody at its best an American-European scientist, having strong cultural and scientific ties on both sides of the ocean. For instance, it is not by chance that both have contributed so much over the years to MTNS, one of the few conferences that belongs equally to the US and to Europe (and to the rest of the world).

*Work partially supported by the MIUR-PRIN Italian grant "New Techniques and Applications of Identification and Adaptive Control".

X. Hu et al. (Eds.): Three Decades of Progress in Control Sciences, pp. 153–167, 2010.

Over the past decade Anders, Chris and Tryphon Georgiou, together with a number of coworkers and students, have developed a whole new field that may be called *Moment Problems with Complexity Constraint*, see [4, 16] and references therein. Their generalized moment problems include as special cases some of the most central problems in our field such as the *covariance extension problem* (see the next section) and *Nevanlinna-Pick interpolation* of robust control. The mathematics, involving global inverse function theorems, differential geometry, analytic interpolation, convex optimization, homotopy methods, iterative numerical schemes, etc. is particularly rich and beautiful. Significant applications to spectral estimation have already been developed. One of the key to the success of this broad program has been the establishing by Anders and Chris of suitable global inverse functions theorems generalizing Hadamard's type theorems, see [3] and references therein. These can be applied in manifold ways. For the generalized moment problems with entropy-like criterions, they yield existence for the dual problem which is typically a convex optimization problem with *open, unbounded* domain.

In this paper, we try to exploit this result of Anders and Chris to circumvent one of the stumbling blocks in this field. We deal, namely, with the multivariable problem where the spectrum must satisfy a suitable generalized moment constraint and must be of limited complexity. We consider the situation where an "a priori" estimate Ψ of the spectrum is available. Motivated by the scalar case and multivariate, $\Psi = I$ case solutions, we introduce a suitable parametric family of spectra with bounded McMillan degree. We then establish properness of the map from the parameter to the moments. Injectivity, and hence surjectivity, of this map is then proven in a special case. A multivariate generalization of the efficient algorithm [21, 7, 9] is finally proposed.

We employ the following notation. For a complex matrix A, A^* denotes the transpose conjugate of A. We denote by $\mathfrak{H}_n$ the vector space of Hermitian matrices of dimension $n \times n$ endowed with the inner product $\langle P, Q\rangle := \mathrm{tr}(PQ)$, and by $\mathfrak{H}_{+,n}$ the subset of positive definite matrices. For a matrix-valued rational function $\chi(z) = H(zI-F)^{-1}G+J$, we define $\chi^*(z) = G^*(z^{-1}I-F^*)^{-1}H^*+J^*$. We denote by $\mathbb{T}$ the unit circle in the complex plane $\mathbb{C}$ and by $C(\mathbb{T})$ the family of complex-valued, continuous functions on $\mathbb{T}$. $C_+(\mathbb{T})$ denotes the subset of $C(\mathbb{T})$ whose elements are real-valued, positive functions. Finally, $C(\mathbb{T};\mathfrak{H}_m)$ stands for the space of $\mathfrak{H}_m$-valued continuous functions.

11.2 A Generalized Moment Problem

Consider the rational transfer function

$$G(z) = (zI-A)^{-1}B, \qquad A \in \mathbb{C}^{n\times n}, B \in \mathbb{C}^{n\times m}, \quad n \geq m \tag{11.1}$$

of the system

$$x(t+1) = Ax(t) + By(t),$$

where A is a stability matrix, i.e. has all its eigenvalues in the open unit disc, (A,B) is a *reachable pair*, and B is a full column rank matrix. The transfer function G

models a bank of filters fed by a stationary process $y(t)$ of unknown spectral density $\Phi(z)$. We assume that we know (or that we can reliably estimate) the steady-state covariance Σ of the state x of the filter. We have

$$\Sigma = \int G\Phi G^*,$$

where, here and in the sequel, integration occurs on the unit circle with respect to the normalized Lebesgue measure. Let $\mathscr{S}_m = \mathscr{S}_+^{m\times m}(\mathbb{T})$ be the family of $\mathfrak{H}_{+,m}$-valued functions defined on the unit circle which are bounded and coercive. We consider the following *generalized moment problem*:

Problem 11.2.1. Let $\Sigma \in \mathfrak{H}_{+,n}$ and $G(z)(zI-A)^{-1}B$ of dimension $n \times m$ with the same properties as in (11.1). Find Φ in $\mathscr{S}_m$ that satisfies

$$\int G\Phi G^* = \Sigma. \tag{11.2}$$

The question of *existence* of $\Phi \in \mathscr{S}_m$ satisfying (11.2) and, when existence is granted, the *parametrization* of all solutions to (11.2), may be viewed as a generalized moment problem. For instance, let $C_k := E\{y(n)y^*(n+k)\}$, and take

$$A = \begin{bmatrix} 0 & I_m & 0 & \dots & 0 \\ 0 & 0 & I_m & \dots & 0 \\ \vdots & \vdots & & \ddots & \vdots \\ 0 & 0 & 0 & \dots & I_m \\ 0 & 0 & 0 & \dots & 0 \end{bmatrix}, \quad B = \begin{bmatrix} 0 \\ 0 \\ \vdots \\ 0 \\ I_m \end{bmatrix}, \quad \Sigma = \begin{bmatrix} C_0 & C_1 & C_2 & \dots & C_{n-1} \\ C_1^* & C_0 & C_1 & \ddots & C_{n-2} \\ C_2^* & \ddots & \ddots & \ddots & \ddots \\ \vdots & \ddots & \ddots & \ddots & \ddots \\ C_{n-1}^* & C_{n-2}^* & \ddots & \ddots & C_0 \end{bmatrix},$$

so that $G(z)$ is a block-column with k-th component being $G_k(z) = z^{k-n-1}I$. This is the classical *covariance extension problem*, where the information available is the finite sequence of covariance lags $C_0, C_1, \dots, C_{n-1}$ of the process y. It is known that the set of densities consistent with the data is nonempty if $\Sigma \geq 0$ and contains infinitely many elements if $\Sigma > 0$ [17], see also [10, 1, 2, 11]. Other important problems of Systems and Control Theory, such as the Nevanlinna-Pick interpolation problem, may be cast in the frame of Problem 11.2.1, see [15].

It may be worthwhile to recall that moment problems form a special class of inverse problems that are typically *not* well-posed in the sense of Hadamard[1]. When Problem 11.2.1 is feasible, a unique solution may be obtained by minimizing a suitable criterion: We mention the Kullback-Leibler type criterion employed in [15] and a suitable multivariable Hellinger-type distance introduced in [8, 23]. The reader is referred to these papers for full motivation, and to [22] for results on the well-posedness of these optimization problems. In [5, 15, 14, 3], a different, interesting viewpoint if taken. It is namely there shown that all solutions to Problem 11.2.1 may

[1] A problem is said to be well-posed, in the sense of Hadamard, if it admits a solution, such a solution is unique, and the solution depends continuously on the data.

be obtained as minimizers of a suitable entropy-like (pseudo-)distance from an "a priori" spectrum Ψ as the latter varies in $\mathscr{S}_m$. There, Ψ is thought of as a parameter. This viewpoint leads to the more challenging moment problem with degree constraint. The latter consists in finding solutions to Problem 11.2.1 whose McMillan degree is "a priori" bounded.

Existence of $\Phi \in \mathscr{S}_m$ satisfying (11.2) in the general case is a nontrivial matter. It has been shown that the following conditions are equivalent [13]:

1. The family of $\Phi \in \mathscr{S}_m$ satisfying constraint (11.2) is nonempty;
2. there exists $H \in \mathbb{C}^{m\times n}$ such that

$$\Sigma - A\Sigma A^* = BH + H^*B^*; \tag{11.3}$$

3. the following rank condition holds

$$\operatorname{rank}\begin{pmatrix} \Sigma - A\Sigma A^* & B \\ B^* & 0 \end{pmatrix} = \operatorname{rank}\begin{pmatrix} 0 & B \\ B^* & 0 \end{pmatrix}. \tag{11.4}$$

A fourth equivalent condition is based on the linear operator $\Gamma : C(\mathbb{T};\mathfrak{H}_m) \to \mathfrak{H}_n$ that will play a crucial role in the rest of the paper:

$$\Gamma : \Phi \mapsto \int G\Phi G^*. \tag{11.5}$$

Existence of $\Phi \in C(\mathbb{T} : \mathfrak{H}_m)$ satisfying $\int G\Phi G^* = \Sigma$ can be expressed as

$$\Sigma \in \operatorname{Range}\Gamma. \tag{11.6}$$

It is has been shown in [12] that when there is a spectrum Φ in $\mathscr{S}_m$ satisfying (11.2), then there exists also $\Phi^o \in C(\mathbb{T};\mathfrak{H}_m)$ (the *maximum entropy* spectrum (11.15) below) satisfying (11.2). Thus, condition (11.6) will be a standing assumption in this paper. For $X \in \mathfrak{H}_n$ and $\Phi \in C(\mathbb{T};\mathfrak{H}_m)$, we have

$$\left\langle X, \int G\Phi G^* \right\rangle = \operatorname{tr}\left(X \int G\Phi G^*\right) = \operatorname{tr}\left(\int (G^*XG)\Phi\right).$$

We conclude that $\Gamma^* : \mathfrak{H}_n \to C(\mathbb{T};\mathfrak{H}_m)$, the *adjoint map* of Γ, is given by

$$\Gamma^* : X \mapsto G^*XG, \tag{11.7}$$

and

$$(\operatorname{Range}\Gamma)^\perp = \left\{X \in \mathfrak{H}_n | G^*(e^{j\vartheta})XG(e^{j\vartheta}) = 0,\ \forall e^{j\vartheta} \in \mathbb{T}\right\}. \tag{11.8}$$

11.3 Kullback-Leibler Approximation of Spectral Densities

In this section, we recall some important result obtained in the *scalar case*, i.e. the case when $m = 1$. In [15], a Kullback-Leibler type of distance for spectra in $\mathscr{S}_1 := \mathscr{S}_+^{1\times 1}(\mathbb{T})$ was introduced:

$$\mathrm{d}(\Psi\|\Phi) = \int \Psi \log\left(\frac{\Psi}{\Phi}\right).$$

As is well known, this pseudo-distance originates in hypothesis testing, where it represents the mean information for observation for discrimination of an underlying probability density from another [19, p.6]. It also plays a central role in several other fields of science such as information theory, identification, stochastic processes, statistical mechanics, etc., where it goes under different names such as *divergence*, *relative entropy*, *information distance*, etc. If $\int \Phi = \int \Psi$, we have $\mathrm{d}(\Psi\|\Phi) \geq 0$. The choice of $\mathrm{d}(\Psi\|\Phi)$ as a distance measure, even for spectra that have different zeroth moment, is discussed in [15, Section III]. Minimizing $\Phi \to \mathrm{d}(\Psi\|\Phi)$ rather than $\Phi \to \mathrm{d}(\Phi\|\Psi)$ is unusual with respect to the statistics-probability-information theory world. Minimizing with respect to the first argument, however, leads to a non-rational solution even when Ψ is rational (see below). Moreover, this atypical minimization includes as special case ($\Psi \equiv 1$) maximization of entropy. In [15], the following problem is considered:

Problem 11.3.1. Given $\Psi \in \mathscr{S}_1$ and $\Sigma \in \mathfrak{H}_{+,n}$,

$$\begin{aligned} &\text{minimize} \quad \mathrm{d}(\Psi\|\Phi) \\ &\text{over} \quad \left\{\Phi \in \mathscr{S}_1 \mid \int G\Phi G^* = \Sigma\right\}. \end{aligned}$$

Let

$$\mathscr{L}_+ := \{\Lambda \in \mathfrak{H}_n : \; G^*\Lambda G > 0, \; \forall \mathrm{e}^{\mathrm{j}\vartheta} \in \mathbb{T}\}. \tag{11.9}$$

For $\Lambda \in \mathscr{L}_+$, consider the *unconstrained* minimization of the *Lagrangian function*

$$\begin{aligned} L(\Phi,\Lambda) &= \mathrm{d}(\Psi\|\Phi) + \mathrm{tr}\left(\Lambda\left(\int G\Phi G^* - \Sigma\right)\right) \\ &= \mathrm{d}(\Psi\|\Phi) + \int G^*\Lambda G\Phi - \mathrm{tr}(\Lambda\Sigma),. \end{aligned} \tag{11.10}$$

This is a convex optimization problem. The variational analysis in [15] shows that the unique minimizer is given by

$$\hat{\Phi}_{KL} = \frac{\Psi}{G^*\Lambda G}. \tag{11.11}$$

Thus, the original Problem 11.3.1 is now reduced to finding $\hat{\Lambda} \in \mathscr{L}_+$ satisfying

$$\int G \frac{\Psi}{G^*\hat{\Lambda}G} G^* = \Sigma. \tag{11.12}$$

This is accomplished via duality theory. The dual problem turns out to be equivalent to minimizing a strictly convex function on the open and unbounded set $\mathscr{L}_+^\Gamma = \mathscr{L}_+ \cap \mathrm{Range}(\Gamma)$. A global inverse function theorem of Byrnes and Lindquist is then used to establish existence and uniqueness for the dual problem under the assumption of feasibility of the primal problem, see [3], references therein and [7]. Notice that, when Ψ is rational, (11.11) shows that the degree of the solution is "a priori" bounded by $2n$ plus the degree of Ψ.

In practical applications, the solution of the dual problem is a numerical challenge. In fact, the dual variable is an Hermitian matrix and, as discussed in [15], the reparametrization in vector form may lead to a loss of convexity. Moreover, the dual functional and its gradient tend to infinity at the boundary. To efficiently deal with the dual problem, the following algorithm has been proposed in [21] and further discussed in [7]:

$$\Lambda_{k+1} = \Theta(\Lambda_k) := \int \Lambda_k^{1/2} G \left[\frac{\Psi}{G^* \Lambda_k G} \right] G^* \Lambda_k^{1/2}, \qquad \Lambda_0 = \frac{1}{n} I. \tag{11.13}$$

It has been shown in [21] that Θ maps *density matrices* to density matrices, i.e. if Λ is a positive semi-definite Hermitian matrix with trace equal to 1, then $\Theta(\Lambda)$ has the same properties. Moreover, Θ maintains positive definiteness, i.e., if $\Lambda > 0$, then $\Theta(\Lambda) > 0$. If the sequence $\{\Lambda_k\}$ converges to a limit point $\hat{\Lambda} > 0$ then such a $\hat{\Lambda}$ is a fixed point for the map Θ and hence satisfies (11.12). It has been recently shown in [9] that $\{\Lambda_k\}$ is locally asymptotically convergent to a limit point $\hat{\Lambda}$ that satisfies (11.12).

11.4 The Multivariable Case

Let us go back to the multivariable setting of Problem 11.2.1. Inspired by the Umegaki relative entropy of statistical quantum mechanics [20], we define $\mathrm{d}(\Psi\|\Phi)$ for Φ and Ψ in $\mathscr{S}_m$

$$\mathrm{d}(\Psi\|\Phi) = \int \mathrm{tr}\,(\Psi(\log\Psi - \log\Phi)). \tag{11.14}$$

Consider first the case where $\Psi = I$ the identity matrix. Then Problem 11.3.1 turns into the *maximum entropy problem*:

Problem 11.4.1. Given $\Sigma \in \mathfrak{H}_{+,n}$,

$$\text{maximize} \quad \int \mathrm{tr}\log\Phi = \int \log\det\Phi = -\mathrm{d}(I\|\Phi)$$

$$\text{over} \quad \left\{ \Phi \in \mathscr{S}_m \mid \int G\Phi G^* = \Sigma \right\}.$$

In [12], the following result was established that (considerably) generalizes Burg's maximum entropy spectrum [6]: Assume feasibility of Problem 11.2.1. Then, the unique solution of Problem 11.4.1 is given by

$$\hat{\Phi} = \left[G^* \Sigma^{-1} B \left(B^* \Sigma^{-1} B \right)^{-1} B^* \Sigma^{-1} G \right]^{-1}. \tag{11.15}$$

Unfortunately, it appears quite problematic generalizing this result to the case of a general $\Psi \in \mathscr{S}_m$. Indeed, as pointed out in [14], the variational analysis cannot be carried through. To overcome this difficulty, in [8] a new metric was introduced that is induced by a sensible generalization to the multivariable case of the *Hellinger distance*. In [8, 23], the problem of computing the spectral density Φ minimizing this generalized Hellinger distance from a prior Ψ, under constraints (11.2), has been analyzed and it has been shown that the solution is still a rational function with an *a priori* bound on its McMillan degree. A different strategy is connected to *homotopy methods* that are described in [14] to find a spectrum that satisfies the constraint when such a family in nonempty.

In this paper, in the spirit of [14, Section IV], and motivated by the scalar case and $\Psi = I$ results, we start by introducing explicitly a parametric family of spectra $\Phi_\Lambda, \Lambda \in \mathscr{L}_+$ in which to look for a solution of Problem 11.2.1. In order to do that, we need first the following result:

Lemma 11.4.1. *Let $G(z) = (zI - A)^{-1}B$ with $A \in \mathbb{C}^{n\times n}$, $B \in \mathbb{C}^{n\times m}$, and let (A,B) be a reachable pair. Let $\Lambda \in \mathscr{L}_+$. Then, the algebraic Riccati equation*

$$\Pi = A^*\Pi A - A^*\Pi B(B^*\Pi B)^{-1}B^*\Pi A + \Lambda, \tag{11.16}$$

admits a unique stabilizing solution $P \in \mathfrak{H}_n$. The corresponding matrix B^*PB is positive definite and the spectrum of closed loop matrix

$$Z := A - B(B^*PB)^{-1}B^*PA \tag{11.17}$$

lays in the open unit disk. Let L be the unique (lower triangular) right Choleski factor of B^*PB (so that $B^*PB = L^*L$).

The following factorization holds:

$$G^*\Lambda G = W_\Lambda^* W_\Lambda, \tag{11.18}$$

where

$$W_\Lambda(z) := L^{-*}B^*PA(zI - A)^{-1}B + L. \tag{11.19}$$

The rational function $W_\Lambda(z)$ is the unique stable and minimum phase right spectral factor of $G^*\Lambda G$, such that $W_\Lambda(\infty)$ is lower triangular and with positive entries in the main diagonal.

We are now ready to introduce our class of multivariate spectral density functions:

$$\Phi_\Lambda := W_\Lambda^{-1}\Psi W_\Lambda^{-*}, \quad \Lambda \in \mathscr{L}_+. \tag{11.20}$$

Notice that the optimal Kullback-Leibler approximant in the scalar case (11.11) and in the multivariate, $\Psi = I$ case (11.15) do belong to this class. This class, however, is different from the one proposed in [14, Section IV]. Although the latter is fully justified by general geometric considerations (Krein-Nudelmann theory [18]), our class is more suitable for implementation of the following matricial version of the efficient algorithm (11.13):

$$\Lambda_{k+1} = \Theta(\Lambda_k) := \int \Lambda_k^{1/2} G \left[W_{\Lambda_k}^{-1} \Psi W_{\Lambda_k}^{-*} \right] G^* \Lambda_k^{1/2}, \qquad \Lambda_0 = \frac{1}{n} I. \tag{11.21}$$

It is easy to see that this map preserves trace and positivity as in the scalar case. We have performed a limited number of simulations in this general setting. In all these simulations, the sequence Λ_k converges very fast to a matrix $\hat{\Lambda}$, for which the corresponding spectral density (given by (11.20)) solves Problem 11.2.1.

Before addressing the computational aspects of the problem, we need first to investigate the following question:

Problem 11.4.2. Let $\Sigma \in \mathrm{Range}_+\Gamma := \mathrm{Range}\,\Gamma \cap \mathfrak{H}_{+,n}$. Let $G(z) = (zI - A)^{-1}B$ with the same properties as in Problem 11.2.1, and let $\Psi \in \mathscr{S}_m$. Find $\Lambda \in \mathscr{L}_+$ such that Φ_Λ given by (11.20) satisfies

$$\int G \Phi_\Lambda G^* = \Sigma. \tag{11.22}$$

Most of this paper is devoted to this question. In particular we show that in the case when $\Psi(z) = \psi(z) Q$ with $\psi(z) \in C_+(\mathbb{T})$ and Q is a constant positive definite matrix, Problem 11.4.2 is feasible. To this aim we need some preliminary results. Consider the map $\omega : \mathscr{L}_+^\Gamma \longrightarrow \mathrm{Range}_+\Gamma$ given by

$$\omega : \Lambda \mapsto \int G \Phi_\Lambda G^*. \tag{11.23}$$

Notice that ω is a continuos map between open subsets of the linear space $\mathrm{Range}\,\Gamma$. It is clear that Problem 11.4.2 is feasible if and only if the map ω is surjective. We are now precisely in the setting of Theorem 2.6 in [3]. It states that if ω is *proper* and injective than it is surjective. We first show that ω is *proper*, i.e. the preimage of every compact set in $\mathrm{Range}_+\Gamma$ is compact in $\mathscr{L}_+^\Gamma$. For this purpose, we need the following lemma.

Lemma 11.4.2. *If $G^* \Lambda G > 0$, $\forall e^{j\vartheta} \in \mathbb{T}$, then there exists $\Lambda_+ \in \mathfrak{H}_{+,n}$ such that $G^* \Lambda G = G^* \Lambda_+ G$.*

Proof. As shown in Lemma 11.4.1, we can perform the factorization

$$G^* \Lambda G = W_\Lambda^* W_\Lambda, \tag{11.24}$$

where the (right) spectral factor $W_\Lambda(z)$ is given by (11.19). The spectral factor $W_\Lambda(z)$ may be easily rewritten as

$$W_\Lambda = L^{-*} \left[B^* P A (zI - A)^{-1} B + B^* P B \right] = L^{-*} B^* P \left[A (zI - A)^{-1} + I \right] B. \tag{11.25}$$

It is immediate to check that $A(zI - A)^{-1} + I = z(zI - A)^{-1}$ so that

$$W_\Lambda = z L^{-*} B^* P (zI - A)^{-1} B. \tag{11.26}$$

and thus

$$G^* \Lambda G = W_\Lambda^* W_\Lambda = W_{\Lambda_1}^* W_{\Lambda_1}, \tag{11.27}$$

with

$$W_{\Lambda_1} := z^{-1} W_\Lambda = L^{-*} B^* P (zI - A)^{-1} B. \tag{11.28}$$

So there exists a matrix $C_\circ = (L^{-*}B^*P)^* \in \mathbb{C}^{n\times m}$ such that

$$G^* \Lambda G = G^* C_\circ C_\circ^* G. \tag{11.29}$$

We observe that on the unit circle $\mathbb{T}$, $G^*\Lambda G$ is continuous and positive definite so that there exists a positive constant μ such that

$$G(z)^* \Lambda G(z) > \mu I, \quad \forall z \in \mathbb{T}.$$

Similarly, on the unit circle $\mathbb{T}$, G^*G is continuous and hence there exists a positive constant ν such that

$$G(z)^* G(z) < \nu I, \quad \forall z \in \mathbb{T}.$$

Let $\varepsilon := \frac{\mu}{4\nu}$. Now let $\Lambda_1 := \frac{1}{2}\Lambda - \varepsilon I$. Clearly, $\forall\, z \in \mathbb{T}$, we have

$$G(z)^* \Lambda_1 G(z) = \frac{1}{2} G(z)^* \Lambda G(z) - \varepsilon G(z)^* G(z) \geq \left(\frac{\mu}{2} - \frac{\mu}{4\nu}\nu\right) I = \frac{\mu}{4} I > 0. \tag{11.30}$$

Hence, by resorting to the same argument that led to (11.29), we conclude that there exists $C_1 \in \mathbb{C}^{n\times m}$ such that

$$G^* (\frac{1}{2}\Lambda - \varepsilon I) G = G^* C_1 C_1^* G.$$

Therefore we have

$$\begin{aligned}
G^* \Lambda G &= \frac{1}{2} G^* C_\circ C_\circ^* G + \frac{1}{2} G^* C_\circ C_\circ^* G + \varepsilon G^* G - \varepsilon G^* G \\
&= G^* (\frac{1}{2} C_\circ C_\circ^* + \varepsilon I) G + \frac{1}{2} G^* C_\circ C_\circ^* G - \varepsilon G^* G \\
&= G^* (\frac{1}{2} C_\circ C_\circ^* + \varepsilon I) G + \frac{1}{2} G^* \Lambda G - \varepsilon G^* G \\
&= G^* (\frac{1}{2} C_\circ C_\circ^* + \varepsilon I) G + G^* (\frac{1}{2}\Lambda - \varepsilon I) G \\
&= G^* (\frac{1}{2} C_\circ C_\circ^* + \varepsilon I) G + G^* C_1 C_1^* G \\
&= G^* (\frac{1}{2} C_\circ C_\circ^* + \varepsilon I + C_1 C_1^*) G = G^* \Lambda_+ G,
\end{aligned}$$

where $\Lambda_+ := \frac{1}{2} C_\circ C_\circ^* + \varepsilon I + C_1 C_1^*$ is clearly positive definite. □

Theorem 11.4.1. *The map ω is proper.*

Proof. We observe that $\mathscr{L}_+^\Gamma$ and $\mathrm{Range}_+\Gamma$ are subsets of a finite dimensional linear space so that compact sets in $\mathscr{L}_+^\Gamma$ and $\mathrm{Range}_+\Gamma$ are characterized by being closed and bounded. Accordingly, to prove the statement is sufficient to show that $\omega^{-1}(K)$ is closed and bounded for any compact set K. To see that $\omega^{-1}(K)$ is bounded we choose an arbitrary sequence $\{\Lambda_n\}$ such that $\Lambda_n \in \mathscr{L}_+^\Gamma$, $\|\Lambda_n\| \to \infty$ and we show that the minimum eigenvalue of $\omega(\Lambda_n)$ approaches zero as n tends to infinity. This means that, as n tends to infinity, $\omega(\Lambda_n)$ approaches the boundary of the co-domain $\mathrm{Range}_+\Gamma$ which is a subset of the positive definite matrices. Therefore, there does not exist a compact set K in $\mathrm{Range}_+\Gamma$ such that $\omega^{-1}(K)$ contains the sequence Λ_n. Similarly, to see that $\omega^{-1}(K)$ is closed we choose an arbitrary sequence $\Lambda_n \in \mathscr{L}_+^\Gamma$ approaching the boundary of $\mathscr{L}_+$, and we show that there does not exist a compact set K in $\mathrm{Range}_+\Gamma$ such that $\omega^{-1}(K)$ contains the sequence Λ_n. The proof, which is detailed only for the case $\|\Lambda_n\| \to \infty$, will be divided in four steps.

Step 1: Observing that $\Psi(z)$ is bounded (i.e. $\exists\, m:\ \Psi \le mI$), we have

$$\begin{aligned} 0 \le \omega(\Lambda) &= \int GW_\Lambda^{-1}\Psi W_\Lambda^{-*}G^* \\ &\le m\int GW_\Lambda^{-1}W_\Lambda^{-*}G^* = m\int G(G^*\Lambda G)^{-1}G^*. \end{aligned} \tag{11.31}$$

It is therefore sufficient to consider the map

$$\begin{aligned} \tilde{\omega} : \mathscr{L}_+^\Gamma &\longrightarrow \quad \mathrm{Range}_+\Gamma \\ \Lambda &\mapsto \int G(G^*\Lambda G)^{-1}G^* \end{aligned} \tag{11.32}$$

and to show that the minimum eigenvalue of $\tilde{\omega}(\Lambda_n)$ approaches zero.

Step 2: By (11.8), $(\mathrm{Range}\,\Gamma)^\perp = \ker \Gamma^*$. Hence, the minimum singular value ρ of the map Γ^* restricted to $\mathrm{Range}\,\Gamma$ is strictly positive. Accordingly, since $\mathrm{Range}_+\Gamma \subset \mathrm{Range}\Gamma$, we have

$$\|G^*\Lambda_n G\| \ge \rho\|\Lambda_n\| \to \infty. \tag{11.33}$$

Step 3: By Lemma 11.4.2, we know that there exists $\Lambda_{n+} = \Lambda_{n+}^* > 0$ such that

$$G^*\Lambda_{n+}G = G^*\Lambda_n G, \quad \forall\, n. \tag{11.34}$$

We have $\|\Lambda_{n+}\| \to \infty$. In fact, let μ_n be the maximum eigenvalue of Λ_{n+}, so that $\Lambda_{n+} < \mu_n I$. It follows that

$$\mu_n\|G^*G\| \ge \|G^*\Lambda_{n+}G\| = \|G^*\Lambda_n G\| \longrightarrow +\infty. \tag{11.35}$$

Since $\|G^*G\| > 0$, the latter implies $\mu_n \to +\infty$ and hence $\|\Lambda_{n+}\| \to \infty$.

Step 4: By Lemma 11.4.2 and recalling that $\Pi \le I$ for any orthogonal projection matrix Π, we have

$$\begin{aligned}\tilde{\omega}(\Lambda_n) &= \int G(G^*\Lambda_n G)^{-1}G^* \\ &= \int G(G^*\Lambda_{n+}G)^{-1}G^* \\ &= \Lambda_{n+}^{-1/2}\left[\int \Lambda_{n+}^{1/2}G(G^*\Lambda_{n+}G)^{-1}G^*\Lambda_{n+}^{1/2}\right]\Lambda_{n+}^{-1/2} \\ &= \Lambda_{n+}^{-1/2}\left[\int \Pi_{\Lambda_{n+}^{1/2}G}\right]\Lambda_{n+}^{-1/2} \\ &\le \Lambda_{n+}^{-1}, \end{aligned} \tag{11.36}$$

where we denote by $\Pi_{\Lambda_{n+}^{1/2}G}$ the orthogonal projection on $\Lambda_{n+}^{1/2}G$. Finally, as shown in Step 3, $\|\Lambda_{n+}\| \to \infty$ so that the minimum eigenvalue of Λ_{n+}^{-1} and, *a fortiori*, the minimum eigenvalue of $\tilde{\omega}(\Lambda_n)$, approaches zero. □

As already mentioned, if the map ω were also injective, then we could conclude that ω is surjective and hence Problem (11.4.2) is feasible. As a preliminary result, we show injectivity in the case when $\Psi(z)$ is a scalar spectral density (i.e. $\Psi(z) = \psi(z)I_m$ with $\psi(z) \in C_+(\mathbb{T})$).

Theorem 11.4.2. *Let $\Psi(z)$ be a scalar spectral density. Then the map ω is injective and hence surjective.*

Proof. Let

$$\Lambda_1, \Lambda_2 \in \mathcal{L}_+^{\Gamma} \subset \operatorname{Range}\Gamma, \tag{11.37}$$

and assume that

$$\omega(\Lambda_1) - \omega(\Lambda_2) = 0. \tag{11.38}$$

Define

$$\Phi_1 := \psi W_{\Lambda_1}^{-1} W_{\Lambda_1}^{-*} = \psi(G^*\Lambda_1 G)^{-1}, \tag{11.39}$$

and

$$\Phi_2 := \psi W_{\Lambda_2}^{-1} W_{\Lambda_2}^{-*} = \psi(G^*\Lambda_2 G)^{-1}. \tag{11.40}$$

Thus,

$$0 = \omega(\Lambda_1) - \omega(\Lambda_2) = \Gamma(\Phi_1) - \Gamma(\Phi_2) = \Gamma(\Phi_1 - \Phi_2) \tag{11.41}$$

so that $(\Phi_1 - \Phi_2) \in \ker\Gamma$.

The adjoint transform of Γ is easily seen to be given by

$$\begin{aligned} \Gamma^* : \mathfrak{H}_n &\longrightarrow C(\mathbb{T}, \mathfrak{H}_m) \\ M &\mapsto G^*MG. \end{aligned} \tag{11.42}$$

Thus, condition $(\Phi_1 - \Phi_2) \in \ker\Gamma = (\operatorname{Range}\Gamma^*)^{\perp}$ reads

$$\langle G^*MG, \Phi_1 - \Phi_2 \rangle = \operatorname{tr}\int G^*MG(\Phi_1 - \Phi_2) = 0, \ \forall M \in \mathfrak{H}_n. \tag{11.43}$$

In particular, by choosing $M = \Lambda_2 - \Lambda_1$, we get

$$\begin{aligned}
0 &= \operatorname{tr}\int [G^*(\Lambda_2-\Lambda_1)G]\,(\Phi_1-\Phi_2)\\
&= \operatorname{tr}\int [G^*(\Lambda_2-\Lambda_1)G]\,\psi\,[(G^*\Lambda_1 G)^{-1}-(G^*\Lambda_2 G)^{-1}]\\
&= \operatorname{tr}\int \psi\,[G^*(\Lambda_2-\Lambda_1)G]\,(G^*\Lambda_1 G)^{-1}\,[G^*\Lambda_2 G-G^*\Lambda_1 G]\,(G^*\Lambda_2 G)^{-1}\\
&= \operatorname{tr}\int \psi\,[G^*(\Lambda_2-\Lambda_1)G]\,(G^*\Lambda_1 G)^{-1}\,[G^*(\Lambda_2-\Lambda_1)G]\,W_{\Lambda_2}^{-1}W_{\Lambda_2}^{-*}\\
&= \operatorname{tr}\int \psi W_{\Lambda_2}^{-*}\,[G^*(\Lambda_2-\Lambda_1)G]\,(G^*\Lambda_1 G)^{-1}\,[G^*(\Lambda_2-\Lambda_1)G]\,W_{\Lambda_2}^{-1}. \qquad (11.44)
\end{aligned}$$

Since $\psi \in C_+(\mathbb{T})$, and $(G^*\Lambda_1 G)^{-1}$ is positive definite on $\mathbb{T}$, the integrand function is positive semi-definite. Therefore, (11.44) implies

$$[G^*(\Lambda_2-\Lambda_1)G]\,(G^*\Lambda_1 G)^{-1}\,[G^*(\Lambda_2-\Lambda_1)G] \equiv 0, \qquad (11.45)$$

that, in turn, yields

$$G^*(\Lambda_2-\Lambda_1)G \equiv 0. \qquad (11.46)$$

By (11.8), $\Lambda_2-\Lambda_1 \in (\operatorname{Range}\Gamma)^\perp$. The latter, together with (11.37), yields

$$\Lambda_2-\Lambda_1 \in \operatorname{Range}\Gamma \cap (\operatorname{Range}\Gamma)^\perp = \{0\}, \qquad (11.47)$$

so that $\Lambda_1=\Lambda_2$. □.

We are now ready to prove our main result

Theorem 11.4.3. *Let $\Psi(z)=\psi(z)Q$ with $\psi(z)\in C_+(\mathbb{T})$ and $Q\in\mathfrak{H}_{+,n}$. Then the map ω is surjective.*

Proof. We first observe that, since B is assumed to be full column rank, we may perform a change of basis and assume, without loss of generality, that $B=\begin{bmatrix} I\\ 0\end{bmatrix}$. Secondly, notice that, it is sufficient to extend the domain of ω to the whole set $\mathscr{L}_+$ and prove the result for the map with extended domain. In fact, if $\omega(\Lambda)=\Sigma$ for a certain $\Lambda\in\mathscr{L}_+$, and $\Lambda_\Gamma\in\mathscr{L}_+^\Gamma$ is the orthogonal projection of Λ in $\operatorname{Range}\Gamma$, then also $\omega(\Lambda_\Gamma)=\Sigma$. Next, we need to compute GW_Λ^{-1}. We observe that:

$$W_\Lambda^{-1}=L^{-1}-(B^*PB)^{-1}B^*PA(zI-Z)^{-1}BL^{-1}, \qquad (11.48)$$

where Z, defined in (11.17), is a stability matrix. Hence,

$$GW_\Lambda^{-1}=-(zI-A)^{-1}B(B^*PB)^{-1}B^*PA(zI-Z)^{-1}BL^{-1}+(zI-A)^{-1}BL^{-1} \qquad (11.49)$$

Notice that

$$B(B^*PB)^{-1}B^*PA=A-Z=(zI-Z)-(zI-A).$$

Plugging this expression into (11.49) we get

$$GW_\Lambda^{-1}=(zI-Z)^{-1}BL^{-1}=(zI-Z)^{-1}\begin{bmatrix} L^{-1}\\ 0\end{bmatrix}, \qquad (11.50)$$

where we have used the fact that $B = \begin{bmatrix} I \\ 0 \end{bmatrix}$. We now partition P conformably with B as

$$P = \begin{bmatrix} P_1 & P_{12} \\ P_{12}^* & P_2 \end{bmatrix}. \tag{11.51}$$

Then we immediately see that $B^*PB = P_1$ so that $L^{-1} = L_{P_1^{-1}}$ is the Choleski factor[2] of P_1^{-1}. Moreover, the matrix Z has the following expression

$$Z = \begin{bmatrix} 0 & -P_1^{-1}P_{12} \\ 0 & I \end{bmatrix} A. \tag{11.52}$$

Consider now an arbitrary $\bar{\Sigma} \in \mathrm{Range}_+ \Gamma$ and let $\bar{\Psi}(z) = \psi(z)I$. In wiew of Theorem 11.4.2, the map $\bar{\omega} : \Lambda \mapsto \int GW_\Lambda^{-1}\bar{\Psi}W_\Lambda^{-*}G^*$ is surjective. Hence there exists $\bar{\Lambda} \in \mathscr{L}_+$ such that $\bar{\omega}(\bar{\Lambda}) = \bar{\Sigma}$. Let $\bar{P}$ be the corresponding stabilizing solution of the ARE (11.16) and $\bar{Z}$ be the associated closed-loop matrix whose spectrum is contained in the open unit disk.

We are now ready to address the case when $\Psi(z) = \psi(z)Q$: Define

$$\tilde{P}_1 := (L_{\bar{P}_1^{-1}} L_Q^{-1} L_Q^{-*} L_{\bar{P}_1^{-1}}^*)^{-1}, \quad \tilde{P}_{12} := \tilde{P}_1 \bar{P}_1^{-1} \bar{P}_{12}, \quad \tilde{P}_2 := \bar{P}_2, \tag{11.53}$$

and let $\tilde{P}$ be the corresponding 2×2 block matrix. Moreover, let

$$\tilde{\Lambda} := \tilde{P} - (A^*\tilde{P}A - A^*\tilde{P}B(B^*\tilde{P}B)^{-1}B^*\tilde{P}A) \tag{11.54}$$

We have the following facts:

1. If $\Lambda = \tilde{\Lambda}$ then $\tilde{P}$ is, by construction, solution of the ARE (11.16).
2. The corresponding closed-loop matrix $\tilde{Z}$ is immediately seen to be equal to $\bar{Z}$ whose spectrum is contained in the open unit disk.
3. Since $\tilde{P}_1 = B^*\tilde{P}B$ is, by construction positive definite and $\tilde{Z}$ is a stability matrix, we can associate to $\tilde{P}$ a spectral factorization of $G^*\tilde{\Lambda}G$ of the form (11.18) so that $G^*\tilde{\Lambda}G$ is positive definite on $\mathbb{T}$ or, equivalently, $\tilde{\Lambda} \in \mathscr{L}_+$.
4. Since a product of Choleski factors and the inverse of a Choleski factor are Choleski factors, and taking into account that the Choleski factor is unique, from the definition of $\tilde{P}_1$, we get $L_{\tilde{P}_1^{-1}} = L_{\bar{P}_1^{-1}} L_Q^{-1}$.
5. As a consequence of the previous observation we get

$$GW_{\tilde{\Lambda}}^{-1} L_Q = GW_{\bar{\Lambda}}^{-1} \tag{11.55}$$

In conclusion, $\tilde{\Lambda} \in \mathscr{L}_+$ and, as it follows immediately from (11.55),

$$\omega(\tilde{\Lambda}) = \bar{\omega}(\bar{\Lambda}) = \bar{\Sigma} \tag{11.56}$$

which concludes the proof. □

[2] We denote by L_Ξ the lower triangular left Choleski factor of a positive definite matrix Ξ, i.e. the unique lower triangular matrix having positive entries in the main diagonal and such that $\Xi = LL^*$.

References

1. Byrnes, C.I., Gusev, S., Lindquist, A.: A convex optimization approach to the rational covariance extension problem. SIAM J. Control and Opimization 37, 211–229 (1999)
2. Byrnes, C.I., Gusev, S., Lindquist, A.: From finite covariance windows to modeling filters: A convex optimization approach. SIAM Review 43, 645–675 (2001)
3. Byrnes, C.I., Linquist, A.: Interior point solutions of variational problems and global inverse function theorems. International Journal of Robust and Nonlinear Control 17, 463–481 (2007)
4. Byrnes, C.I., Linquist, A.: Important moments in systems and control. SIAM J. Control and Optimization 47(5), 2458–2469 (2008)
5. Byrnes, C.I., Linquist, A.: A convex optimization approach to generalized moment problems. In: Control and Modeling of Complex Systems: Cybernetics in the 21st Century, pp. 3–21. Birkhäuser, Boston (2003)
6. Cover, T.M., Thomas, J.A.: Information Theory. Wiley, New York (1991)
7. Ferrante, A., Pavon, M., Ramponi, F.: Further results on the Byrnes-Georgiou-Lindquist generalized moment problem. In: Chiuso, A., Ferrante, A., Pinzoni, S. (eds.) Modeling, Estimation and Control: Festschrift in honor of Giorgio Picci on the occasion of his sixty-fifth birthday, pp. 73–83. Springer, Heidelberg (2007)
8. Ferrante, A., Pavon, M., Ramponi, F.: Hellinger vs. Kullback-Leibler multivariable spectrum approximation. IEEE Trans. Aut. Control 53, 954–967 (2008)
9. Ferrante, A., Ramponi, F., Ticozzi, F.: On the convergence of an efficient algorithm for Kullback-Leibler approximation of spectral densities. IEEE Trans. Aut. Control, Submitted for publication (2009)
10. Georgiou, T.: Realization of power spectra from partial covariance sequences. IEEE Trans. on Acoustics, Speech, and Signal Processing 35, 438–449 (1987)
11. Georgiou, T.: The interpolation problem with a degree constraint. IEEE Trans. Aut. Control 44, 631–635 (1999)
12. Georgiou, T.: Spectral analysis based on the state covariance: the maximum entropy spectrum and linear fractional parameterization. IEEE Trans. Aut. Control 47, 1811–1823 (2002)
13. Georgiou, T.: The structure of state covariances and its relation to the power spectrum of the input. IEEE Trans. Aut. Control 47, 1056–1066 (2002)
14. Georgiou, T.: Relative entropy and the multivariable multidimensional moment problem. IEEE Trans. Inform. Theory 52, 1052–1066 (2006)
15. Georgiou, T., Lindquist, A.: Kullback-Leibler approximation of spectral density functions. IEEE Trans. Inform. Theory 49, 2910–2917 (2003)
16. Georgiou, T., Lindquist, A.: A convex optimization approach to ARMA modeling. IEEE Trans. Aut. Control 53, 1108–1119 (2008)
17. Grenander, U., Szegö, G.: Toeplitz Forms and Their Applications. University of California Press, Berkeley (1958)
18. M. G. Kreĭn and A. A. Nudel'man. *The Markov Moment Problem and Extremal Problems*. Amer. Math. Soc., Providence, RI, 1977.
19. Kullback, S.: Information Theory and Statistics, 2nd edn. Dover, Mineola (1968)
20. Nielsen, M.A., Chuang, I.L.: Quantum Computation and Quantum Information. Cambridge Univ. Press, Cambridge (2000)
21. Pavon, M., Ferrante, A.: On the Georgiou-Lindquist approach to constrained Kullback-Leibler approximation of spectral densities. IEEE Trans. Aut. Control 51, 639–644 (2006)

22. Ramponi, F., Ferrante, A., Pavon, M.: On the well-posedness of multivariate spectrum approximation and convergence of high- resolution spectral estimators. Systems and Control Letters, to appear (March 2009)
23. Ramponi, F., Ferrante, A., Pavon, M.: A globally convergent matricial algorithm for multivariate spectral estimation. IEEE Trans. Aut. Control 54, 2376–2388 (2009)

12

Unimodular Equivalence of Polynomial Matrices

P.A. Fuhrmann[1,*] and U. Helmke[2,†]

[1] Department of Mathematics, Ben-Gurion University of the Negev, Beer Sheva, Israel
[2] Universität Würzburg, Institut für Mathematik, Würzburg, Germany

Summary. In Gauger and Byrnes [10], a characterization of the similarity of two $n \times n$ matrices in terms of rank conditions was given. This avoids the use of companion or Jordan canonical forms and yields effective decidability criteria for similarity. In this paper, we generalize this result to an explicit characterization when two polynomial models are isomorphic. As a corollary, we derive necessary and sufficient rank conditions for strict equivalence of arbitrary matrix pencils. We also briefly discuss the related equivalence problem for group representations. The techniques we use are based on the tensor products of polynomial models and related characterizations of intertwining maps.

12.1 Introduction

The task of classifying square matrices up to similarity is one of the core problems in linear algebra. Standard approaches for deciding similarity depend upon the Jordan canonical form, the invariant factor algorithm and the Smith form, or the closely related rational canonical form. In numerical linear algebra, this leads to deep algorithmic problems, unsolved even to this date, that are caused by numerical instabilities in solving non-symmetric eigenvalue problems or by the inability to effectively compute sizes of the Jordan blocks or degrees of invariant factors, if the matrix entries are not known precisely.

In a pioneering paper Gauger and Byrnes [10], Chris Byrnes and Michael Gauger derived a new type of rank conditions for algebraically deciding similarity of arbitrary pairs of matrices over a field $\mathbb{F}$. Their main result is that two matrices $A, B \in \mathbb{F}^{n\times n}$ are similar if and only if the following two conditions hold:

1. The characteristic polynomials coincide, i.e. $\det(zI - A) = \det(zI - B)$.
2.

$$\operatorname{rank}(I \otimes A - A \otimes I) = \operatorname{rank}(I \otimes B - B \otimes I) = \operatorname{rank}(B \otimes I - I \otimes A). \qquad (12.1)$$

*Partially supported by the ISF under Grant No. 1282/05.
†Partially supported by the DFG SPP 1305 under Grant HE 1858/12-1.

X. Hu et al. (Eds.): Three Decades of Progress in Control Sciences, pp. 169–185, 2010.

Here $A \otimes B$ denotes the usual Kronecker product of two matrices. In subsequent work by Dixon [1] it was shown that the first condition on the characteristic polynomials is superfluous, so that similarity can be solely decided based on rank computations. Moreover, Dixon improved the result by Byrnes and Gauger in two different directions. First, he replaced the above rank condition by the seemingly more complicated quadratic rank constraint

$$r_{A,B}^2 = r_{A,A} r_{B,B}.$$

Here $r_{A,B} = \text{rank}(A \otimes I - I \otimes B)$ and similarly for $r_{A,A}, r_{B,B}$. This equality rank constraint has the appealing form of a Cauchy-Schwarz condition, as Dixon proved that the inequality $r_{A,B}^2 \leq r_{A,A} r_{B,B}$ holds for arbitrary matrices. Moreover, Dixon extended the result to an isomorphism criterion for finite-length modules over a principal ideal domain.

Over an algebraically closed field, Friedland [7] showed the closely related linear dimension inequality

$$2\dim\text{Ker}\,(B \otimes I - I \otimes A) \leq \dim\text{Ker}\,(I \otimes A - A \otimes I) + \dim\text{Ker}\,(I \otimes B - B \otimes I), \quad (12.2)$$

and proved that equality holds if and only if A, B are similar. This inequality can also be deduced from Dixon's quadratic rank constraint. A generalization and proof of the linear dimension inequality for matrices A, B of sizes $n \times n$ and $m \times m$, respectively, appears in the unpublished book manuscript Friedland [8]

One of the main important aspects of the Byrnes-Gauger work is that their criterion allows to decide similarity by completely algebraic means, i.e. by computing minors of differences of associated Kronecker product matrices. In contrast, the invariants occuring in the classical Jordan canonical forms or rational canonical forms cannot be computed solely in terms of algebraic functions of the entries of A, B. We would also like to stress that – since ranks of real matrices can be effectively determined via the singular value decomposition– this may open new perspectives to robustly decide upon the approximate similarity of two matrices. Such an approach may thus bypass intrinsic numerical difficulties with deciding similarity via the Jordan canonical form, see Edelman and Kagström [2].

In this paper, we generalize the Byrnes-Gauger result to one that characterizes strict equivalence of regular matrix pencils. Explicitly, we prove that two pencils $zE - F, z\overline{E} - \overline{F} \in \mathbb{F}[z]^{n \times n}$ are strict equivalent if and only if

$$\text{rank}(E \otimes F - F \otimes E) = \text{rank}(\overline{E} \otimes \overline{F} - \overline{F} \otimes \overline{E}) = \text{rank}(F \otimes \overline{E} - E \otimes \overline{F}) \quad (12.3)$$

This contains the Byrnes-Gauger result as a special case, but may look only as a minor extension. We prove this result by actually proving a more general characterization of unimodular equivalence of nonsingular polynomial matrices $D(z)$ and $\overline{D}(z)$ in terms of the equality of dimensions of spaces of intertwining maps of polynomial models. Such polynomial model spaces are defined for any nonsingular polynomial matrix and the theory of such models has been extensively developed by the first author, see Fuhrmann [4]. We prove that two polynomial models $X_D, X_{\overline{D}}$ are isomorphic as $\mathbb{F}[z]$-modules if and only if

$$\dim \mathrm{Hom}_{\mathbb{F}[z]}(X_D, X_D) = \dim \mathrm{Hom}_{\mathbb{F}[z]}(X_{\bar{D}}, X_{\bar{D}}) = \dim \mathrm{Hom}_{\mathbb{F}[z]}(X_D, X_{\bar{D}}). \qquad (12.4)$$

A similar condition is derived for characterizing equivalence of finite-dimensional complex Lie group representations. In particularly, this leads to an effective decidability condition when two complex representations of $SL_2(\mathbb{C})$ are equivalent. Our main tools in deriving such results are explicit formula for the tensor product of two polynomial models, a theory that has been recently developed in our joint paper Furmann and Helmke [6].

This paper is dedicated to Chris Byrnes and Anders Lindquist on occassion of their recent birthdays. Our research in this paper has been initiated and stimulated by discussions with Chris Byrnes during the Symposium in honor of G. Picci in Venice and the last Oberwolfach Control Theory meeting at the Mathematical Research Centre. It is a pleasure to thank him for sharing his ideas and interest with us, in the past as well as in the present, and for most enjoyable collaborations that the second author enjoyed with him during the past decades. Happy birthday, Chris and Anders!

12.2 Polynomial Models and Intertwining Maps

We begin with a brief summary on polynomial models and its connection to the matrix similarity problem; for the theory of functional models and its applications to linear algebra and systems theory see Fuhrmann [4, 5]. A detailed exposition for tensor products in connection with intertwining maps see [6], while a standard reference to tensor products is Hungerford [11].

Given a linear transformation $A : X \longrightarrow X$ on an n–dimensional vector space, the vector space can be endowed with an $\mathbb{F}[z]$-module structure by defining, for $p(z) \in \mathbb{F}[z]$ and $x \in X$, $p \cdot x = p(A)x$. Of course, this construction is very well known and goes back at least to the early work by Krull [13]. It leads to the standard approach for classifying linear operators. We denote by X_A the n-dimensional vector space with the induced module structure. One can generalize this construction in a rather straightforward way for arbitrary nonsingular polynomial matrices. Thus, given a nonsingular polynomial matrix $D(z) \in \mathbb{F}[z]^{p\times p}$, the corresponding polynomial model is defined as

$$X_D = \{f \in \mathbb{F}[z]^p | D^{-1}f \text{ strictly proper}\}.$$

These functional models are suitable for realization theory; see Fuhrmann [4] for details. The action of z, defined by

$$z \cdot f = D(z)\pi_-(D(z)^{-1}zf(z))$$

on polynomial vectors $f(z) \in X_D$, then yields a canonical $\mathbb{F}[z]$-module structure on X_D. Associated with this action of the polynomial z there is a canonically defined linear operator $S_D : X_D \longrightarrow X_D$ as

$$(S_D f)(z) = zf(z) - D(z)\xi_f, \qquad f \in X_D, \qquad (12.5)$$

where $\xi_f = (D^{-1}f)_{-1}$ is the residue of $D^{-1}f$. We refer to S_D as the shift operator on X_D. We have the module isomorphism

$$X_D \simeq \mathbb{F}[z]^p / D(z)\mathbb{F}[z]^p \tag{12.6}$$

and thus can interpret X_D as a concretization of the above quotient module. The link between this circle of ideas and linear algebra is made by associating with a linear operator A the uniquely defined matrix pencil $D(z) := zI - A$. Then X_A can be identified with X_{zI-A}. It is important to note that we have the similarity of linear operators

$$A \simeq S_{zI-A}, \tag{12.7}$$

which links the similarity problem for matrices A to the classification problem of polynomial models X_{zI-A} up to module isomorphisms. A result that is closely related and which will appear later on, is that two linear maps $A : \mathbb{F}^n \longrightarrow \mathbb{F}^n$ and $B : \mathbb{F}^n \longrightarrow \mathbb{F}^n$ are similar if and only if the pencils $zI - A$ and $zI - B$ are unimodular equivalent. The latter condition is in turn equivalent to the polynomial models X_{zI-A} and X_{zI-B} are isomorphic.

The isomorphism of polynomial models $X_A \simeq X_{zI-A}$ and the related similarity of shift operators (12.7) leads immediately to the cyclic decomposition of linear transformations. In fact, given a nonsingular polynomial matrix $D(z) \in \mathbb{F}[z]^{p\times p}$, there exist unimodular polynomial matrices $U(z), V(z)$, such that $U(z)D(z) = \Delta(z)V(z)$, where $\Delta(z) = \mathrm{diag}(d_1, \ldots, d_p)$ is the Smith form of $D(z)$, i.e. $d_1, \ldots, d_p$ are the invariant factors of $D(z)$. This implies the following isomorphism result, see Fuhrmann [4],

$$X_D \simeq X_\Delta \simeq \oplus_{i=1}^p X_{d_i} \tag{12.8}$$

and hence

$$\dim X_D = \dim X_\Delta = \sum_{i=1}^p \dim X_{d_i} = \sum_{i=1}^p \deg d_i = \deg(\det D). \tag{12.9}$$

Given nonsingular polynomial matrices $D_i(z) \in \mathbb{F}[z]^{p_i \times p_i}$, $i = 1,2$, the two corresponding polynomial models are isomorphic if and only if there exist polynomial matrices $N_1(z), N_2(z) \in \mathbb{F}[z]^{p_2 \times p_1}$ satisfying the equality

$$N_2(z)D_1(z) = D_2(z)N_1(z) \tag{12.10}$$

which is embeddable in the doubly coprime factorization

$$\begin{pmatrix} Y_2(z) & -X_2(z) \\ -N_2(z) & D_2(z) \end{pmatrix} \begin{pmatrix} D_1(z) & X_1(z) \\ N_1(z) & Y_1(z) \end{pmatrix} = \begin{pmatrix} I & 0 \\ 0 & I \end{pmatrix}. \tag{12.11}$$

In this case, the isomorphism $Z : X_{D_1} \longrightarrow X_{D_2}$ is given by

$$Zf = \pi_{D_2} N_2 f, \qquad f \in X_{D_1}. \tag{12.12}$$

This characterization of intertwining maps can be presented in a more abstract way using tensor products of polynomial models over the ring $\mathbb{F}[z]$. This yields the isomorphism

$$Y_B \otimes_{\mathbb{F}[z]} X^*_{A^\top} \simeq \mathrm{Hom}_{\mathbb{F}[z]}(X_A, Y_B). \tag{12.13}$$

We note that $Z \in \mathrm{Hom}_{\mathbb{F}[z]}(X_A, Y_B)$ if and only if $ZA = BZ$.

One other thing to note is that if $A_2 = PA_1P^{-1}$ and $B_2 = QB_1Q^{-1}$, then we have the isomorphism

$$\mathrm{Hom}_{\mathbb{F}[z]}(X_{A_1}, Y_{B_1}) \simeq \mathrm{Hom}_{\mathbb{F}[z]}(X_{A_2}, Y_{B_2}), \tag{12.14}$$

given, for $Z \in \mathrm{Hom}_{\mathbb{F}[z]}(X_{A_1}, Y_{B_1})$, by $Z \mapsto QZP^{-1}$.

As in the case of tensor product over the field $\mathbb{F}$, we can get a concrete representation of the tensor product of two polynomial models over the polynomial ring $\mathbb{F}[z]$, see Fuhrmann and Helmke [6]. However, it is not needed for our present purpose. Instead, we use the fact that, given a commutative ring with identity R and R-modules M,N having the direct sum representations $M = \oplus_{i=1}^k M_i$ and $N = \oplus_{j=1}^l N_j$, the tensor product has the following distributivity property

$$(\oplus_{i=1}^k M_i) \otimes_R (\oplus_{j=1}^l N_j) \simeq \oplus_{i=1}^k \oplus_{j=1}^l (M_i \otimes_R N_j) \tag{12.15}$$

We proceed to apply (12.15) to the tensor product of polynomial models over the polynomial ring $\mathbb{F}[z]$. Given nonsingular polynomial matrices $D_i(z) \in \mathbb{F}[z]^{p_i \times p_i}$, $i = 1,2$, there exist unimodular polynomial matrices $U_i(z), V_i(z)$, $i = 1,2$, such that $U_i(z)D_i(z) = \Delta_i(z)V_i(z)$, where $\Delta_i(z) = \mathrm{diag}(d_1^{(i)}, \dots, d_{p_i}^{(i)})$ is the Smith form of $D_i(z)$, i.e. $d_1^{(i)}, \dots, d_{p_i}^{(i)}$ are the invariant factors of $D_i(z)$. Since the polynomial model X_{D_i} is isomorphic to X_{Δ_i} as $\mathbb{F}[z]$-modules, we have, by (12.14), the isomorphism

$$\mathrm{Hom}_{\mathbb{F}[z]}(X_{D_1}, X_{D_2}) \simeq \mathrm{Hom}_{\mathbb{F}[z]}(X_{\Delta_1}, X_{\Delta_2}). \tag{12.16}$$

This isomorphism is useful in the computation of dimension formulas.

$$\begin{aligned} \dim \mathrm{Hom}_{\mathbb{F}[z]}(X_{D_1}, X_{D_2}) &= \dim \mathrm{Hom}_{\mathbb{F}[z]}(X_{\Delta_1}, X_{\Delta_2}) \\ &= \dim X_{\Delta_2} \otimes_{\mathbb{F}[z]} X_{\Delta_1} \\ &= \textstyle\sum_{i=1}^{p_2} \sum_{j=1}^{p_1} \dim X_{d_i^{(2)}} \otimes_{\mathbb{F}[z]} X_{d_j^{(1)}}. \end{aligned} \tag{12.17}$$

To apply (12.17), given scalar polynomials d, e, we need to compute the tensor product $X_d \otimes_{\mathbb{F}[z]} X_e$. For this we use general results concerning the tensor product of quotient modules. Let M_1, M_2 be R modules, with R a commutative ring. Let $N_i \subset M_i$ be submodules. The quotient spaces M_i/N_i have a natural R module structure. Let N be the submodule generated in $M_1 \otimes_R M_2$ by $N_1 \otimes_R M_2$ and $M_1 \otimes_R N_2$. Then we have the isomorphism

$$M_1/N_1 \otimes_R M_2/N_2 \simeq (M_1 \otimes_R M_2)/N \tag{12.18}$$

We apply this to our situation, using the isomorphism (12.6) and noting that $d\mathbb{F}[z] + e\mathbb{F}[z] = (d \wedge e)\mathbb{F}[z]$, with $d \wedge e$ the g.c.d. of d and e. This implies

$$\begin{aligned} X_d \otimes_{\mathbb{F}[z]} X_e &\simeq \mathbb{F}[z]/d(z)\mathbb{F}[z] \otimes_{\mathbb{F}[z]} \mathbb{F}[z]/e(z)\mathbb{F}[z] \\ &\simeq \mathbb{F}[z]/(d(z)\mathbb{F}[z] + e(z)\mathbb{F}[z]) = \mathbb{F}[z]/(d \wedge e)(z)\mathbb{F}[z] \simeq X_{d\wedge e}. \end{aligned} \tag{12.19}$$

As a result, combining (12.17) and (12.19), we obtain the dimension formula

$$\dim \mathrm{Hom}_{\mathbb{F}[z]}(X_{D_1}, X_{D_2}) = \dim(X_{D_2} \otimes_{\mathbb{F}[z]} X_{D_1^\top}) = \sum_{i=1}^{p_2} \sum_{j=1}^{p_1} \deg(d_i^{(2)} \wedge d_j^{(1)}). \quad (12.20)$$

The dimension formula (12.20) is quite old, see Frobenius [3].

Next, we specialize the dimension formula (12.20) to the case that $D_2(z) = D_1(z)$. Given a nonsingular $D(z) \in \mathbb{F}[z]^{p \times p}$, let $d_1, \ldots, d_p$ be the invariant factors of $D(z)$ ordered so that $d_i | d_{i-1}$. Let $e_{ij} = d_i \wedge d_j = g.c.d.(d_i, d_j) = d_{\max\{i,j\}}$. Let $\delta_i = \deg d_i$ and let $n = \sum_{i=1}^p \delta_i = \deg(\det D(z))$. Then we have

$$\dim \mathrm{Hom}_{\mathbb{F}[z]}(X_D, X_D) = \dim(X_D \otimes_{\mathbb{F}[z]} X_{D^\top}) = \delta_1 + 3\delta_2 + \cdots + (2p-1)\delta_p. \quad (12.21)$$

This formula is due to Shoda [14] and appears in Gantmacher [9].

12.3 The Similarity Characterization

We quote, with a trivial modification, the following elementary combinatorial lemma from Gauger and Byrnes [10].

Lemma 12.3.1. *Let $n_1 \geq \cdots \geq n_p \geq 0$ and $m_1 \geq \cdots \geq m_m \geq 0$ nonincreasing sequences of integers. Then*

$$\sum_{i=1}^{p} \sum_{j=1}^{p} \min(n_i, n_j) + \sum_{i=1}^{m} \sum_{j=1}^{m} \min(m_i, m_j) \geq 2 \sum_{i=1}^{p} \sum_{j=1}^{m} \min(n_i, m_j). \quad (12.22)$$

Equality occurs if and only if the number of positive integers in both sequences is the same and $m_i = n_i$ for all such i.

Theorem 12.3.1. *Given $D_1(z) \in \mathbb{F}[z]^{m \times m}$ and $D_2(z) \in \mathbb{F}[z]^{p \times p}$ be nonsingular. Assume $d_1^{(1)}, \ldots, d_m^{(1)}$ and $d_1^{(2)}, \ldots, d_p^{(2)}$ to be the invariant factors of $D_1(z)$ and $D_2(z)$ respectively, taken to be monic and ordered so that $d_i^{(\nu)} | d_{i-1}^{(\nu)}$, $\nu = 1, 2$. Let also $m_i = \deg d_i^{(1)}$ and $n_i = \deg d_i^{(2)}$. Then we have*

$$\dim \mathit{Hom}_{\mathbb{F}[z]}(X_{D_1}, X_{D_1}) + \dim \mathit{Hom}_{\mathbb{F}[z]}(X_{D_2}, X_{D_2}) \geq 2 \dim \mathit{Hom}_{\mathbb{F}[z]}(X_{D_1}, X_{D_2}). \quad (12.23)$$

The following statements are equivalent.

1. *There exists an $\mathbb{F}[z]$-isomorphism*

$$X_{D_1} \simeq X_{D_2} \quad (12.24)$$

2. *The polynomial matrices $D_1(z), D_2(z)$ are equivalent, i.e. their nontrivial invariant factors are equal.*
3. *$\mathit{Hom}_{\mathbb{F}[z]}(X_{D_1}, X_{D_1})$, $\mathit{Hom}_{\mathbb{F}[z]}(X_{D_2}, X_{D_2})$ and $\mathit{Hom}_{\mathbb{F}[z]}(X_{D_1}, X_{D_2})$ are isomorphic as $\mathbb{F}[z]$-modules.*

4.

$$\dim Hom_{\mathbb{F}[z]}(X_{D_1}, X_{D_1}) = \dim Hom_{\mathbb{F}[z]}(X_{D_2}, X_{D_2}) = \dim Hom_{\mathbb{F}[z]}(X_{D_1}, X_{D_2}). \tag{12.25}$$

5.

$$\dim Hom_{\mathbb{F}[z]}(X_{D_1}, X_{D_1}) + \dim Hom_{\mathbb{F}[z]}(X_{D_2}, X_{D_2}) = 2\dim Hom_{\mathbb{F}[z]}(X_{D_1}, X_{D_2}). \tag{12.26}$$

Proof. Inequality (12.23) follows from (12.20) and (12.22).

Next we prove the equivalence of the above statements. From (12.20) and (12.21), we conclude

$$\begin{aligned} \dim \mathrm{Hom}_{\mathbb{F}[z]}(X_{D_1}, X_{D_1}) &= \textstyle\sum_{i=1}^{p} \sum_{j=1}^{p} \min(n_i, n_j), \\ \dim \mathrm{Hom}_{\mathbb{F}[z]}(X_{D_2}, X_{D_2}) &= \textstyle\sum_{i=1}^{m} \sum_{j=1}^{m} \min(m_i, m_j) \\ \dim \mathrm{Hom}_{\mathbb{F}[z]}(X_{D_1}, X_{D_2}) &= \textstyle\sum_{i=1}^{p} \sum_{j=1}^{m} \min(n_i, m_j). \end{aligned}$$

From Lemma 12.3.1, (12.23) follows and equality holds if and only if $p = m$ and $n_i = m_i, i = 1, \ldots, p$. This shows the equivalence of (2) and (5). Obviously, (4) implies (5), but in turn also (5) implies (2) and hence also (4). Obviously, (2) implies (1). On the other hand, clearly (1) implies (4) and therefore implies (2). That (1) implies (3) is trivial. By the second part of Lemma 12.3.1, we conclude that $m_i = n_i$ for all i, i.e. that $\deg d_i^{(1)} = \deg d_i^{(2)}$. From (12.25) it follows that

$$\sum_{i=1}^{p} \sum_{j=1}^{m} \deg(d_i^{(1)} \wedge d_j^{(1)}) = \sum_{i=1}^{p} \sum_{j=1}^{m} \deg(d_i^{(2)} \wedge d_j^{(2)}) = \sum_{i=1}^{p} \sum_{j=1}^{m} \deg(d_i^{(2)} \wedge d_j^{(1)}).$$

Hence, necessarily, $d_j^{(2)} = d_j^{(1)}$ for all j, for otherwise we have $\deg(d_i^{(2)} \wedge d_j^{(1)}) < \deg(d_i^{(2)} \wedge d_j^{(2)})$ for some i, j which contradicts (12.25). This implies the similarity of S_{D_1} and S_{D_2}, i.e. (3) implies (1). The result follows. □

The following is well known, see Gantmacher [9] or Fuhrmann [5].

Proposition 12.3.1. *Let $A, B \in \mathbb{F}^{n\times n}$. Then $zI - A$ and $zI - B$ are unimodularly equivalent if and only if A and B are similar.*

Corollary 12.3.1. *Let $D_1(z) \in \mathbb{F}[z]^{m\times m}$ and $D_2(z) \in \mathbb{F}[z]^{p\times p}$ be nonsingular with $\deg\det D_1(z) = \deg\det D_2(z) = n$. Let $A_1, A_2 \in \mathbb{F}^{n\times n}$ be matrix representations for the shift operators $S_{D_1} : X_{D_1} \longrightarrow X_{D_1}$, $S_{D_2} : X_{D_2} \longrightarrow X_{D_2}$ for any choice of bases in X_{D_1}, X_{D_2} respectively. Then the polynomial models X_{D_1}, X_{D_2} are isomorphic $\mathbb{F}[z]$-modules if and only if the following rank condition holds.*

$$rank(I_n \otimes A_1 - A_1 \otimes I_n) = rank(I_n \otimes A_2 - A_2 \otimes I_n) = rank(I_n \otimes A_1 - A_2 \otimes I_n). \tag{12.27}$$

Proof. By definition of the module structure on X_{D_i}, an $\mathbb{F}$-linear map $Z : X_{D_i} \longrightarrow X_{D_i}$ is $\mathbb{F}[z]$-linear if and only if $ZS_{D_i} = S_{D_i}Z$, $i = 1, 2$. Similarly, for maps $Z : X_{D_1} \longrightarrow X_{D_2}$. Thus, for $i = 1, 2$

$$\dim_{\mathbb{F}} \mathrm{Hom}_{\mathbb{F}[z]}(X_{D_i}, X_{D_i}) = \dim_{\mathbb{F}}\{X \in \mathbb{F}^{n\times n} | A_i X = X A_i\}$$

and also

$$\dim_{\mathbb{F}} \mathrm{Hom}_{\mathbb{F}[z]}(X_{D_1}, X_{D_2}) = \dim_{\mathbb{F}}\{X \in \mathbb{F}^{n\times n} | A_2 X = X A_1\}.$$

Since any matrix $A \in \mathbb{F}^{n\times n}$ is similar to its transpose $A^\top \in \mathbb{F}^{n\times n}$ (as oth have the same invariant factors), we conclude that

$$\begin{aligned}\dim\{X \in \mathbb{F}^{n\times n} | A_i X = X A_i\} &= \dim\{X \in \mathbb{F}^{n\times n} | A_i X = X A_i^T\} \\ &= \dim\{X \in \mathbb{F}^{n\times n} | A_1 X = X A_2^T\}.\end{aligned}$$

Note that $I \otimes A_i - A_i \otimes I$ acts on $\mathbb{F}^{n\times n}$ as $X \mapsto X A_i^T - A_i X$. Therefore

$$\begin{aligned}\mathrm{rank}(I \otimes A_i - A_i \otimes I) &= n^2 - \dim_{\mathbb{F}} \mathrm{Hom}_{\mathbb{F}[z]}(X_{D_i}, X_{D_i}) \\ \mathrm{rank}(I \otimes A_2 - A_1 \otimes I) &= n^2 - \dim_{\mathbb{F}} \mathrm{Hom}_{\mathbb{F}[z]}(X_{D_1}, X_{D_2}).\end{aligned}$$

The result now follows from Theorem 12.3.1. □

From this result we immediately conclude the following decidability criterion of Gauger and Byrnes [10] for similarity of two matrices $A_1, A_2 \in \mathbb{F}^{n\times n}$. Note that Gauger and Byrnes assumed additionally that the characteristic polynomials $\det(zI - A_1), \det(zI - A_2)$ coincide. However, this assumption is in fact not needed.

Corollary 12.3.2. *Let $A, B \in \mathbb{F}^{n\times n}$. Then A and B are similar if and only if the following equalities hold.*

$$\dim\{X | AX = XA\} = \dim\{X | BX = XB\} = \dim\{X | BX = XA\}. \tag{12.28}$$

Equivalently, if and only if the rank conditions are satisfied:

$$rank(I_n \otimes A - A \otimes I_n) = rank(I_n \otimes B - B \otimes I_n) = rank(I_n \otimes A - B \otimes I_n). \tag{12.29}$$

Proof. We use the similarity (12.7) and apply Corollary 12.3.1 to the pencils $D_1(z) = zI - A, D_2(z) = zI - B$. □

Subsequent to the work of Gauger and Byrnes, Dixon [1] derived another equivalent characterization of similarity by proving that two matrices $A_1, A_2 \in \mathbb{F}^{n\times n}$ are similar if and only if

$$r^2_{A_1,A_2} = r_{A_1,A_1} r_{A_2,A_2}, \tag{12.30}$$

where for any two matrices A, B, we have

$$r_{A,B} = \mathrm{rank}(I \otimes A - B \otimes I). \tag{12.31}$$

This leads to the following combinatorial identity as an amusing consequence.

Proposition 12.3.2. *Let $n_1 \geq \cdots \geq n_p \geq 0$ and $m_1 \geq \cdots \geq m_m \geq 0$ nonincreasing sequences of integers with $\sum_i n_i = \sum_j m_j$. Then*

$$\left(\sum_{i=1}^{p}\sum_{j=1}^{m} \min(n_i, m_j)\right)^2 \leq \left(\sum_{i=1}^{p}\sum_{j=1}^{p} \min(n_i, n_j)\right)\left(\sum_{i=1}^{m}\sum_{j=1}^{m} \min(m_i, m_j)\right) \tag{12.32}$$

and equality holds if and only if $m = p$ and $m_i = n_i$ for $i = 1, \ldots, p$.

Proof. For any pair of matrices $A,B \in \mathbb{F}^{n\times n}$, Dixon proved the inequalities

$$\begin{aligned} r_{A,B}^2 &\le r_{A,A} r_{B,B} \\ s_{A,B}^2 &\le s_{A,A} s_{B,B}, \end{aligned} \tag{12.33}$$

and showed that equality holds if and only if A and B are similar. Here

$$\begin{aligned} r_{A,B} &= \operatorname{rank}(I\otimes B - A\otimes I) \\ s_{A,B} &= \dim\operatorname{Ker}(I\otimes B - A\otimes I). \end{aligned} \tag{12.34}$$

Let $n_1 \ge \cdots \ge n_p$, $m_1 \ge \cdots \ge m_m$ denote the degrees of the invariant factors of $zI - A$ and $zI - B$ respectively. Then

$$\begin{aligned} s_{A,A} &= \textstyle\sum_{i=1}^{p}\sum_{j=1}^{p} \min(n_i,n_j) \\ s_{B,B} &= \textstyle\sum_{i=1}^{m}\sum_{j=1}^{m} \min(m_i,m_j) \\ s_{A,B} &= \textstyle\sum_{i=1}^{p}\sum_{j=1}^{m} \min(n_i,m_j). \end{aligned} \tag{12.35}$$

The result follows from the equality on $s_{A,B}^2$, as any integer sequences $n_1 \ge \cdots \ge n_p$, $m_1 \ge \cdots \ge m_m$ can be realized as the degrees of the invariant factors of $A,B \in \mathbb{F}^{n\times n}$, provided $\sum_i n_i = \sum_j m_j = n$. □

Remark: It remains a challenge to find a direct proof of Proposition 12.3.2, as this would immediately imply a proof of Dixon's result. Dixon's result in turn implies (12.23), i.e. Friedland's result. On the other hand, squaring both sides in the inequality (12.22) for $p = m$ leads to the inequality

$$(\sum_{i=1}^{p}\sum_{j=1}^{m} \min(n_i,m_j))^2 \le (\sum_{i=1}^{p}\sum_{j=1}^{p} \min(n_i,n_j))(\sum_{k=1}^{p}\sum_{l=1}^{p} \min(m_k,m_l)) + (\sum_{i<j}(n_j - m_j))^2, \tag{12.36}$$

which is weaker than Dixon.

We now turn to generalizing the above similarity conditions to strict equivalence of regular matrix pencils. Given matrices $E,F \in \mathbb{F}^{n\times n}$, we recall that the pencil $zE - F$ is called regular if $\det(zE - F)$ is not the zero polynomial. Two such pencils $zE - F$ and $z\overline{E} - \overline{F}$ are called **strict equivalent** if there exist constant invertible matrices $L,R \in GL_n(\mathbb{F})$ such that $zE - F = L(z\overline{E} - \overline{F})R^{-1}$.

In the sequel, we will assume that our ground field $\mathbb{F}$ has infinitely many elements (as is the case for every algebraically closed field), although the weaker assumption that the field has $|\mathbb{F}| > 2n$ elements would suffice. This assumption assures that, for every two regular pencils $zE - F, z\overline{E} - \overline{F}$, there exists $z_0 \in \mathbb{F} - \{0\}$ such that $\det(z_0E - F) \ne 0$ and $\det(z_0\overline{E} - \overline{F}) \ne 0$.

Theorem 12.3.2. *Let $\mathbb{F}$ be any field with $|\mathbb{F}| > 2n$. For regular pencils $zE - F, z\overline{E} - \overline{F}$, $E,F,\overline{E},\overline{F} \in \mathbb{F}^{n\times n}$ the following conditions are equivalent.*

1. *The pencils $zE - F, z\overline{E} - \overline{F}$ are strict equivalent.*
2. *We have the following equality*

$$\operatorname{rank}(F\otimes E - E\otimes F) = \operatorname{rank}(\overline{F}\otimes\overline{E} - \overline{E}\otimes\overline{F}) = \operatorname{rank}(F\otimes\overline{E} - E\otimes\overline{F}). \tag{12.37}$$

Proof. The implication (1) $\Rightarrow$ (2) is trivial. For (2) $\Rightarrow$ (1), choose any alement $c \in \mathbb{F} - \{0\}$ such that $\det(cE - F) \neq 0$ and $\det(c\bar{E} - \bar{F}) \neq 0$. Let $A = (cE - F)^{-1}F$ and $B = (c\bar{E} - \bar{F})^{-1}\bar{F}$. Then, by multiplying with $(cE - F) \otimes (cE - F)$, we have

$$\operatorname{rank}(A \otimes I - I \otimes A) = \operatorname{rank}(F \otimes (cE - F) - (cE - F) \otimes F) = \operatorname{rank}(F \otimes E - E \otimes F).$$

and similarly

$$\begin{aligned} \operatorname{rank}(B \otimes I - I \otimes B) &= \operatorname{rank}(\bar{F} \otimes \bar{E} - \bar{E} \otimes \bar{F}) \\ \operatorname{rank}(B \otimes I - I \otimes A) &= \operatorname{rank}(\bar{F} \otimes E - \bar{E} \otimes F). \end{aligned}$$

Applying Corollary 12.3.2 shows that A is similar to B. Let $R \in GL_n(\mathbb{F})$ with $B = RAR^{-1}$. For $S = cE - F, \bar{S} = c\bar{E} - \bar{F} \in GL_n(\mathbb{F})$, we obtain

$$\bar{F} = \bar{S}B = \bar{S}RAR^{-1} = \bar{S}RS^{-1}FR^{-1} = LFR^{-1},$$

with $L = \bar{S}RS^{-1}$. Similarly, for E we have

$$L(cE - F)R^{-1} = LSR^{-1} = \bar{S} = c\bar{E} - \bar{F} = c(LER^{-1} - \bar{E})$$

and, as $LFR^{-1} = \bar{F}$, we have $c(LER^{-1} - \bar{E}) = 0$. Since $c \neq 0$, we conclude $\bar{E} = LER^{-1}$, i.e. $zE - F$ is strict equivalent to $z\bar{E} - \bar{F}$. □

The above rank condition can be reformulated in a somewhat neater form using Bezout type polynomials. Let $D(z) = zE - F$, $\bar{D}(z) = z\bar{E} - \bar{F}$. Then, for any z, w

$$B(D,D)(z,w) = \frac{D(z) \otimes D(w) - D(w) \otimes D(z)}{z - w} = F \otimes E - E \otimes F \tag{12.38}$$

$$B(\bar{D},\bar{D})(z,w) = \frac{\bar{D}(z) \otimes \bar{D}(w) - \bar{D}(w) \otimes \bar{D}(z)}{z - w} = \bar{F} \otimes \bar{E} - \bar{E} \otimes \bar{F}$$

$$B(D,\bar{D})(z,w) = \frac{D(z) \otimes \bar{D}(w) - D(w) \otimes \bar{D}(z)}{z - w} = F \otimes \bar{E} - E \otimes \bar{F}$$

Thus $zE - F$ and $z\bar{E} - \bar{F}$ are strict equivalent if and only if, for all z, w

$$\operatorname{rank} B(D,D)(z,w) = \operatorname{rank} B(\bar{D},\bar{D})(z,w) = \operatorname{rank} B(D,\bar{D})(z,w). \tag{12.39}$$

More generally, if $D(z) \in \mathbb{F}[z]^{p \times p}$, $\bar{D}(z) \in \mathbb{F}[z]^{m \times m}$ are arbitrary nonsingular polynomial matrices, then (12.39) is a necessary condition for strict equivalence, for any $z, w \in \bar{\mathbb{F}}$. Here $\bar{\mathbb{F}}$ is the algebraic closure of $\mathbb{F}$. It would be interesting to see if this condition is also sufficient for strict equivalence. Just, for curiosity, we compute the Bezoutian polynomials for quadratic matrix pencils, $D(z) = Az^2 + Bz + C$ and $\bar{D}(z) = \bar{A}z^2 + \bar{B}z + \bar{C}$ as

$$\begin{aligned} B(D,D)(z,w) &= (A \otimes B - B \otimes A)zw + (A \otimes C - C \otimes A)(z + w) + (B \otimes C - C \otimes B) \\ B(\bar{D},\bar{D})(z,w) &= (\bar{A} \otimes \bar{B} - \bar{B} \otimes \bar{A})zw + (\bar{A} \otimes \bar{C} - \bar{C} \otimes \bar{A})(z + w) + (\bar{B} \otimes \bar{C} - \bar{C} \otimes \bar{B}) \\ B(D,\bar{D})(z,w) &= (A \otimes \bar{B} - B \otimes \bar{A})zw + (A \otimes \bar{C} - C \otimes \bar{A})(z + w) + (B \otimes \bar{C} - C \otimes \bar{B}). \end{aligned}$$

It is interesting to relate the above circle of ideas to the Weierstrass decomposition. Thus, assuming that $D(z) = zE - F$ and $\overline{D}(z) = z\overline{E} - \overline{F}$ are regular pencils, we associate with them the reflected pencils given by $D^{\sharp}(z) = E - zF$ and $\overline{D}^{\sharp}(z) = \overline{E} - z\overline{F}$. We first state and prove the existence and uniqueness result for the Weierstrass form over an arbitrary field; see Gantmacher [9] (vol II, Ch. XII, Theorem 3) for a proof. The shorter proof given here is based on systems realization theory and appears to be new. To begin with, we recall that an arbitrary $p \times m$ matrix G(z) of rational functions has a coprime factorization $G(z) = N(z)D(z)^{-1}$, unique up to a common unimodular right factor of the polynomial matrices N, D, such that the leading coefficient matrix of

$$\hat{G}(z) := \begin{pmatrix} N(z) \\ D(z) \end{pmatrix}$$

has full column rank. The McMillan degree $\delta(G)$ then is defined as the maximal degree of the $m \times m$ minors of $\hat{G}(z)$. From this definition it is obvious that the McMillan degree of $G(z)$ is invariant under arbitrary Möbius transformations.

Theorem 12.3.3 (Weierstrass). *Assume $zE - F \in \mathbb{F}[z]^{n\times n}$ is a regular pencil, i.e. $\det(zE - F)$ is a nonzero polynomial. There exist, up to similarity transformations, unique matrices $A \in \mathbb{F}^{r\times r}$ and a nilpotent $N \in \mathbb{F}^{(n-r)\times(n-r)}$ such that $zE - F$ is strict equivalent to* $\begin{pmatrix} zI_r - A & 0 \\ 0 & I_{n-r} - zN \end{pmatrix}$.

Proof. Since $zE - F$ is a regular pencil,

$$G(z) = (zE - F)^{-1} = \frac{\operatorname{adj}(zE - F)}{\det(zE - F)} \tag{12.40}$$

is a nontrivial rational matrix function. We decompose it as

$$G(z) = H(z) + P(z), \tag{12.41}$$

with $H(z)$ strictly proper and $P(z)$ polynomial. We choose a minimal realization

$$H(z) = C(zI - A)^{-1}B \tag{12.42}$$

where $A \in \mathbb{F}^{r\times r}$, $r = \deg\det(zE - F) = \delta(H)$ and $\delta(H)$ is the McMillan degree of $H(z)$. Next, let $P(z) = P_0 + \cdots + P_s z^s$ which implies $z^{-1}P(z^{-1}) = P_0 z^{-1} + \cdots + P_s z^{-s-1}$. Let

$$z^{-1}P(z^{-1}) = C_\infty(zI - N)^{-1}B_\infty \tag{12.43}$$

be a minimal realization. Clearly $N \in \mathbb{F}^{k\times k}$, where $k = \delta(z^{-1}P(z^{-1}))$. Combining (12.42) and (12.43), we get from (12.41)

$$G(z) = (zE - F)^{-1} = \begin{pmatrix} C & C_\infty \end{pmatrix} \begin{pmatrix} zI - A & 0 \\ 0 & I - zN \end{pmatrix}^{-1} \begin{pmatrix} B \\ B_\infty \end{pmatrix} \tag{12.44}$$

Here, by construction, $\begin{pmatrix} C & C_\infty \end{pmatrix} \in \mathbb{F}[z]^{n\times(r+k)}$ and $\begin{pmatrix} B \\ B_\infty \end{pmatrix} \in \mathbb{F}[z]^{(r+k)\times n}$. Since the pencil $zE - F$ has full rank n, so has $G(z)$. Thus we conclude $r + k \geq n$.

We proceed to prove the converse inequality. In case E is nonsingular, this is trivial as $\delta(zE-F)^{-1}=n$ which forces $r=n$ and $k=0$. In the case that E is singular, our assumption on regularity of $zE-F$ guarantees the existence of a nonzero $z_0\in\overline{\mathbb{F}}$ in the algebraic closure $\overline{\mathbb{F}}$ for which z_0E-F is nonsingular. We make now a change of variable via the fractional linear transformation $w=(z-z_0)^{-1}+z_0$ and its inverse $z=z_0+(w-z_0)^{-1}$. This moves the pole at infinity to z_0, while leaving all other poles finite. Noting that $\left[(z_0+(w-z_0)^{-1})E-F\right]^{-1}$ is finite at infinity, we compute

$$(zE-F)^{-1}=\left[(z_0+(w-z_0)^{-1})E-F\right]^{-1}=(w-z_0)\left[(w-z_0)(z_0E-F)+E\right]^{-1}$$

which is a proper rational function in w of McMillan degree $\leq n$. This shows that $r+k\leq n$ and, in turn, this implies $r+k=n$. Since $\begin{pmatrix} C & C_\infty\end{pmatrix}$ and $\begin{pmatrix} B \\ B_\infty\end{pmatrix}$ have full row and column rank respectively, they are both invertible. Thus, from (12.44), we conclude that

$$(zE-F)=\begin{pmatrix} B \\ B_\infty\end{pmatrix}^{-1}\begin{pmatrix} zI-A & 0 \\ 0 & I-zN\end{pmatrix}\begin{pmatrix} C & C_\infty\end{pmatrix}^{-1} \tag{12.45}$$

which shows the claimed strict equivalence. □

Theorem 12.3.4. *Let $\mathbb{F}$ be any field with $|\mathbb{F}|>2n$. Let $D(z):=zE-F, \overline{D}(z):=z\overline{E}-\overline{F}, D^\sharp(z):=E-zF, \overline{D}^\sharp(z):=\overline{E}-z\overline{F}$ be regular pencils, with $E,F,\overline{E},\overline{F}\in\mathbb{F}^{n\times n}$. The following conditions are equivalent.*

1. *The pencils $zE-F$ and $z\overline{E}-\overline{F}$ are strict equivalent.*
2. *The finite and infinite invariant factors of $zE-F$ and $z\overline{E}-\overline{F}$ coincide.*
3. *The polynomial models $X_D, X_{\overline{D}}$ and $X_{D^\sharp}, X_{\overline{D}^\sharp}$ are $\mathbb{F}[z]$-isomorphic, respectively.*
4. *For any $z_0\in\mathbb{F}-\{0\}$ such that $\det(z_0E-F)\neq0$ and $\det(z_0\overline{E}-\overline{F})\neq0$, the matrices $A=(z_0E-F)^{-1}F$ and $\overline{A}=(z_0\overline{E}-\overline{F})^{-1}\overline{F}$ are similar.*
5. *There exists a $z_0\in\mathbb{F}-\{0\}$ such that $\det(z_0E-F)\neq0$ and $\det(z_0\overline{E}-\overline{F})\neq0$, and such that the matrices $A=(z_0E-F)^{-1}F$ and $\overline{A}=(z_0\overline{E}-\overline{F})^{-1}\overline{F}$ are similar.*

Proof. For the equivalence of (1) and (2) we refer to Gantmacher [9] (vol II, Ch. XII, Theorem 2, p. 27). The equivalence of (1) with (4), (5) is shown in Theorem 12.3.2.

(1) $\Longrightarrow$ (3) If $D(z)$ and $\overline{D}(z)$ are strict equivalent, so are $D^\sharp(z)$ and $\overline{D}^\sharp(z)$. Thus $X_D\simeq X_{\overline{D}}$ and also $X_{D^\sharp}\simeq X_{\overline{D}^\sharp}$.

(3) $\Longrightarrow$ (1) Conversely, if $X_D\simeq X_{\overline{D}}$ and $X_{D^\sharp}\simeq X_{\overline{D}^\sharp}$ then $D(z),\overline{D}(z)$ have the same (finite) invariant factors and the same holds for $D^\sharp(z),\overline{D}^\sharp(z)$. To relate the infinite invariant polynomials with the invariant polynomials of $D^\sharp(z)$ and $\overline{D}^\sharp(z)$, we use the Weierstrass form. Thus, for any regular pencil $zE-F$, there exist $L,R\in GL_n(\mathbb{F})$, so that we have the following strict equivalence

$$L(zE-F)R^{-1}=\begin{pmatrix} zI_r-A & 0 \\ 0 & zN-I_{n-r}\end{pmatrix}, \tag{12.46}$$

where $A\in\mathbb{F}^{r\times r}$ and $N\in\mathbb{F}^{(n-r)\times(n-r)}$ nilpotent, $r=\deg\det(zE-F)$. Assume that $d_1,\ldots,d_r$ are the invariant factors, of degree δ_i of zI_r-A, i.e. of A, and thus

$d_1,\dots,d_r,1,\dots,1$ are those of $zE-F$. Then the invariant factors of $E-zF$ are $d_1^\sharp,\dots,d_r^\sharp,z^{\nu_1},\dots,z^{\nu_s}$, where $\nu_1,\dots,\nu_s$ are the nilpotency indices of N. Then $\sum_{i=1}^r \delta_i + \sum_{j=1}^s \nu_j = n$ and $z^{\nu_1},\dots,z^{\nu_s}$ are the infinite invariant factors of $D(z)$. Therefore, equality of the invariant factors of $D^\sharp(z)$ and $\bar{D}^\sharp(z)$ implies equality of the invariant factors of $D(z)$ and $\bar{D}(z)$ and we are done. □

The next result, an immediate consequence of Theorem 12.3.1, relates strict equivalence and unimodular equivalence of pencils. By comparison of part 3 in Theorem 12.3.4 with Theorem 12.3.5, the difference between unimodular equivalence and strict equivalence of matrix pencils becomes evident.

Theorem 12.3.5. *Two regular pencils $zE-F$ and $z\bar{E}-\bar{F}$ are unimodular equivalent equivalent, i.e. there exist unimodular matrices $U(z),V(z)\in GL_n(\mathbb{F}[z])$ for which*

$$U(z)(zE-F) = (z\bar{E}-\bar{F})V(z),$$

if and only if the polynomial models X_{zE-F} and $X_{z\bar{E}-\bar{F}}$ are $\mathbb{F}[z]$-isomorphic. Equivalently, for any matrix representations $A,\bar{A}\in\mathbb{F}^{r\times r}$ of the shift operators S_D and $S_{\bar{D}}$ on X_{zE-F} and $X_{z\bar{E}-\bar{F}}$ respectively, we have

$$rank(A\otimes I - I\otimes A) = rank(\bar{A}\otimes I - I\otimes \bar{A}) = rank(A\otimes I - I\otimes \bar{A}). \tag{12.47}$$

As we have shown in this paper, the tasks of classifying regular matrix pencils up to strict or unimodular equivalence are closely related to studying modules of intertwining maps of polynomial models. On the other hand, the space of module homomorphisms $\mathrm{Hom}_{\mathbb{F}[z]}(X_D,X_D)$ has a richer structure than just being a module over $\mathbb{F}[z]$ – it is also an $\mathbb{F}$-algebra. Shoda [14] has shown that $\mathrm{Hom}_{\mathbb{F}[z]}(X_{D_1},X_{D_1})$ and $\mathrm{Hom}_{\mathbb{F}[z]}(X_{D_2},X_{D_2})$ are isomorphic as $\mathbb{F}$–algebras if and only if the shift operators S_{D_1} and S_{D_2} are polynomially equivalent, i.e. S_{D_1} is similar to a polynomial $p(S_{D_2})$ in S_{D_2} and vice versa. A proof along the lines of this paper would be desirable. Of course, in another direction, it would be also very desirable to have a rank test available for unimodular equivalence that works directly on the pencil matrices E,F and $\bar{E},\bar{F}$, similarly to the one for strict equivalence. We leave these tasks for future research.

12.4 Classification of Group Representations

In this last section we briefly discuss the closely related and in fact more general problem of classifying finite–dimensional group representations. We begin by recalling some basic terminology; see e.g. Vinberg [15] for a textbook reference. Let G denote a group and V,W be finite dimensional complex vector spaces. A representation of G on V is a group homomorphism

$$\rho: G\to \mathrm{GL(V)}. \tag{12.48}$$

Every group action of G on a complex vector space V thus defines a representation, and conversely. V is then also called a G-module and we write $g\cdot v$ for $\rho(g)v$. If

$\rho : G \to \mathrm{GL(V)}$ and $\tilde{\rho} : G \to \mathrm{GL(W)}$ are two representations, a G-homomorphism is a complex linear map $T : V \to W$ such that for all $g \in G, v \in V$

$$T(g \cdot v) = g \cdot T(v), \tag{12.49}$$

i.e., equivalently if $T(\rho(g)v) = \tilde{\rho}(g)T(v)$ holds for all g, v. We refer to T then also as an intertwining map of the representations. Two representations $\rho : G \to \mathrm{GL(V)}$ and $\tilde{\rho} : G \to \mathrm{GL(W)}$ are called equivalent, if there exists an bijective intertwining map $T : V \to W$ (the inverse is then automatically intertwining). A G-invariant subspace of a representation (12.48) is a complex subspace $W \subset V$ with $\rho(g)W \subset W$ for all $g \in G$. A representation (12.48) is called **completely reducible**, if every G-invariant subspace W has a G-invariant complement W' with $W \oplus W' = V$, or, equivalently, if and only if V decomposes as the finite direct sum of G-invariant subspaces. A G-invariant subspace $W \subset V$ of a representation (12.48) is called **irreducible**, if $\{0\}$ and W are the only G-invariant subspace contained in W. Any completely reducible representation thus admits direct sum decomposition

$$V = n_1V_1 \oplus \cdots \oplus n_rV_r \tag{12.50}$$

by irreducible, pairwise non-equivalent subspaces V_i; here $mV := V \oplus \cdots \oplus V$ denotes the m-fold direct sum. It is well-known that compact Lie groups and complex semisimple Lie groups are completely reducible.

Classifying finite-dimensional group representations is closely related to the matrix similarity problem. In fact, it is a natural generalization of the latter. To see this, let A denote any complex $n \times n$-matrix. Let $\exp(\mathrm{A})$ denote the matrix exponential series of A. We associate with A the representation of the additive abelian group $\mathbb{C}^n$ as

$$\rho_A : \mathbb{C}^n \to \mathrm{GL}(\mathbb{C}^n), \quad (\rho_{\mathrm{A}})(\mathrm{t}_1, \cdots, \mathrm{t}_n) : \mathrm{v} \mapsto \exp(\sum_{\mathrm{i}=1}^{\mathrm{n}} \mathrm{t}_\mathrm{i}\mathrm{A}^{\mathrm{i}-1})\mathrm{v}. \tag{12.51}$$

It easily seen by inspection, that two matrices A, B are similar if and only if the representations ρ_A, ρ_B on $\mathbb{C}^n$ are equivalent. In fact, an intertwining map from ρ_A to ρ_B is just a matrix $T \in \mathbb{C}^{n \times n}$ satisfying $TA = BT$. Thus we can re-interpret the decidability problem for matrix similarity as the task of deciding equivalence for group representations ρ_A.

We now present a result that solves this problem for arbitrary compact Lie groups.

Let

$$\mathrm{Hom_G(V, W)}$$

denote the finite-dimensional $\mathbb{C}$-vector space of all intertwining maps from V to W.

Theorem 12.4.1. *Let V, W be completely reducible, finite-dimensional, complex representations of a group G. Then*

$$2\dim Hom_{\mathrm{G}}(\mathrm{V}, \mathrm{W}) \leq \dim Hom_{\mathrm{G}}(\mathrm{V}, \mathrm{V}) + \dim Hom_{\mathrm{G}}(\mathrm{W}, \mathrm{W}). \tag{12.52}$$

Equality holds if and only if V is equivalent to W.

Before proving this, a comment is in order. A representation ρ_A of a complex matrix A is completely reducible if and only if A is diagonalizable. As we have shown above, the above result holds without any such diagonalizability assumption. This raises the question in how far the complete reducibility assumption is actually necessary.

Proof. The necessity part is trivial, so let us assume that (12.52) holds. Decompose V,W into direct sums of irreducible G-invariant subspaces V_i, W_j as

$$V = n_1V_1 \oplus \cdots \oplus n_rV_r \quad , \quad W = m_1W_1 \oplus \cdots \oplus m_sW_s \tag{12.53}$$

By the Schur Lemma, see Vinberg [15], p. 46, Theorem 4.2.3, we have for any irreducible representations X,Y

$$\mathrm{dimHom_G(X,Y)} = 0 \tag{12.54}$$

if X,Y are not equivalent and

$$\mathrm{dimHom_G(X,Y)} = 1, \tag{12.55}$$

if X,Y are equivalent. Without loss of generality we can assume for some $d \leq \mathrm{min(r,s)}$ that $V_i \simeq W_i$ are equivalent for $i = 1,\cdots d$, while for $k,j > d$ none of the V_j is equivalent to any of the W_k. Therefore the Schur Lemma implies

$$\mathrm{dimHom_G(V,V)} = \sum_{i=1}^{r} n_i^2 \tag{12.56}$$

$$\mathrm{dimHom_G(W,W)} = \sum_{j=1}^{s} m_j^2 \tag{12.57}$$

$$\mathrm{dimHom_G(V,W)} = \sum_{i=1}^{d} n_i m_i. \tag{12.58}$$

Note that $2ab \leq a^2 + b^2$ with equality if and only if $a = b$. Applying this to each summand implies

$$2\sum_{i=1}^{d} n_i m_i \leq \sum_{i=1}^{d} (n_i^2 + m_i^2) \leq \sum_{i=1}^{r} n_i^2 + \sum_{j=1}^{s} m_j^2 \tag{12.59}$$

and equality holds if and only if $n_i = m_i$ for $i = 1,\cdots,d$, $n_j, m_k = 0$ for $k,j > d$. The result follows. □

The condition in the theorem is always satisfied for a compact Lie group G, and more generally for a complex semisimple Lie group. In particular, the result holds for any representations of $SL_n(\mathbb{C})$. Note, that an abelian group is–by definition–never semisimple. This implies the following result that generalizes Theorem 12.3.1.

Corollary 12.4.1. *Let G be either a compact Lie group or a complex semisimple Lie group. For any finite-dimensional representations V,W of G we have*

$$2\mathrm{dim}Hom_{\mathrm{G}}\mathrm{(V,W)} \leq \mathrm{dim}Hom_{\mathrm{G}}\mathrm{(V,V)} + \mathrm{dim}Hom_{\mathrm{G}}\mathrm{(W,W)} \tag{12.60}$$

and equality holds if and only if V is equivalent to W.

As a final result we show how to apply the above theory to the classification of $SL(2)$–representations. Note, that a complex $n \times n$-matrix A is called unipotent, if $I - A$ is nilpotent. Equivalently, A is unipotent if and only if $A = \exp(\mathrm{N})$ for a nilpotent matrix. Let $SL_2(\mathbb{C})$ denote the semisimple Lie group of complex 2×2-matrices of determinant one.

Corollary 12.4.2. *Let $\rho, \tilde{\rho} : SL_2(\mathbb{C}) \to SL_n(\mathbb{C})$ be representations and $A := \rho(E)$, $B := \tilde{\rho}(E)$, with*

$$E := \begin{pmatrix} 1 & 1 \\ 0 & 1 \end{pmatrix}. \tag{12.61}$$

Then $\rho, \tilde{\rho}$ are equivalent if and only if

$$rank(I_n \otimes A - A \otimes I_n) = rank(I_n \otimes B - B \otimes I_n) = rank(I_n \otimes A - B \otimes I_n). \tag{12.62}$$

Proof. The necessity of (12.62) is obvious. Thus, assume the condition holds. By the Byrnes-Gauger result we then know that the unipotent matrices A, B are similar. Let $d\rho, d\tilde{\rho} : \mathfrak{sl}_2(\mathbb{C}) \to \mathfrak{sl}_n(\mathbb{C})$ be the associated Lie algebra representations, obtained by differentiating $\rho, \tilde{\rho}$. Let $J := E - I$, so that $\exp(\mathrm{J}) = \mathrm{E}$. The nilpotent matrices $N_1 = d\rho(J), N_2 = d\tilde{\rho}(J)$ then satisfy $\exp(\mathrm{N}_1) = \mathrm{A}, \exp(\mathrm{N}_2) = \mathrm{B}$. By similarity of A, B we conclude that N_1, N_2 are similar. In fact, $N_1 = \sum_{k=1}^{n-1} \frac{(-1)^{k+1}}{k}(A - I)^k$, and similarly for N_2, B. Kostant has shown that then the Lie algebra representations $d\rho, d\tilde{\rho}$ are equivalent. This implies equivalence of ρ, $\tilde{\rho}$ and the result follows. □

One wonders, whether it is possible, to deduce the Byrnes-Gauger result from purely representation theoretic reasoning. In fact, this seems quite reasonable, in view of the Jacobson-Morozov lemma [12] for Lie algebra representations of $\mathfrak{sl}_2(\mathbb{C})$. Also, extensions to infinite dimensional unitary representations and operators on Hardy spaces are a challenge, that one can perhaps approach via the Peter-Weyl theorem and functional models. We leave these problems for future work.

References

1. Dixon, J.D.: An isomorphism criterion for modules over a principal ideal domain. Lin. and Multilin. Alg., 8, 69–72
2. Edelman, A., Elmroth, E., Kagström, B.: A Geometric Approach to Perturbation Theory of Matrices and Matrix Pencils. Part I: Versal Deformations. SIAM J. Matrix Anal. Appl., 18(3), 653–692
3. Frobenius, G.: Über die mit einer Matrix vertauschbaren Matrizen. Sitzungsberichte der Akad. der Wiss. zu Berlin
4. Fuhrmann, P.A.: Algebraic system theory: An analyst's point of view. J. Franklin Inst. 301, 521–540
5. Fuhrmann, P.A.: A Polynomial Approach to Linear Algebra. Springer, New York (1996)
6. Fuhrmann, P.A., Helmke, U.: Tensored polynomial models. Lin. Alg. Appl., 432, 678–721

7. Friedland, S.: Analytic similarity of matrices. In: Byrnes, C.I., Martin, C.F. (eds.) Algebraic and Geometric Methods in Linear Systems Theory. Lectures in Applied Math. Amer. Math. Soc, vol. 18, pp. 43–85 (1980)
8. Friedland, S.: Matrices. A book draft in preparation, `http://www2.math.uic.edu/~friedlan/bookm.pdf`
9. Gantmacher, F.R.: The Theory of Matrices, vols. I/II, Chelsea Publishing Company, New York
10. Gauger, M.A., Byrnes, C.I.: Characteristic free, improved decidability criteria for the similarity problem. Linear and Multilinear Algebra, 5, 153–158
11. Hungerford, T.W.: Algebra. Springer, New York (2003)
12. Jacobson, N.: Lie Algebras. Wiley, New York
13. Krull, W.: Theorie und Anwendung der verallgemeinerten Abelschen Gruppen. Sitzungsberichte der Heidelberger Akad. Wiss. Math.-Naturw. Kl.1 Abh., 1–32
14. Shoda, K.: Über mit einer Matrix vertauschbaren Matrizen. Math. Z. 29, 696–712
15. Vinberg, E.B.: Linear Representations of Groups. Birkhäuser Verlag, Basel

13

Sparse Blind Source Deconvolution with Application to High Resolution Frequency Analysis*

Tryphon T. Georgiou[1] and Allen Tannenbaum[2,3]

[1] University of Minnesota, Minneapolis, MN, USA
[2] Georgia Institute of Technology, Atlanta, GA, USA
[3] Technion, Israel Institute of Technology, Haifa, Israel

Summary. The title of the paper refers to an extension of the classical blind source separation where the mixing of unknown sources is assumed in the form of convolution with impulse response of unknown linear dynamics. A further key assumption of our approach is that source signals are considered to be sparse with respect to a known dictionary, which suggests a mixed L_1/L_2-optimization as a possible formalism for solving the un-mixing problem. We demonstrate the effectiveness of the framework numerically.

13.1 Introduction

One of the most powerful tools in signal analysis which surfaced in recent years is a collection of theories and techniques that allows sparse representations of signals. Fundamental contributions from a number of researchers [3, 4, 5, 6, 7, 8, 9, 22] sparked this rapidly developing field driven by a wide spectrum of applications from robust statistics to data compression, compressed sensing, image processing, estimation, and high resolution signal analysis. The present work builds on the well-paved paradigm of sparse representations by focusing on a problem of system/source identification known as blind source separation.

Blind source separation refers to the problem of separating sources from linear mixtures of these with unknown coefficients. For the special case where sources represent speech signals, the separation of voices corresponding to individual speakers is often referred to as the "cocktail party problem". Early work was based on the assumption that such signals are often statistically independent, and explored properties of second and higher order statistics. Typically, the required "un-mixing matrix" was sought as a solution to a suitable optimization problem which either

*This work was supported in part by grants from NSF, AFOSR, ARO, as well as by NIH (NAC P41 RR 13218) through Brigham and Women's Hospital. This work is part of the National Alliance for Medical Image Computing (NAMIC), funded by the National Institutes of Health through the NIH Roadmap for Medical Research, Grant U54 EB005149. Information on the National Centers for Biomedical Computing can be obtained from `http://nihroadmap.nih.gov/bioinformatics`.

X. Hu et al. (Eds.): Three Decades of Progress in Control Sciences, pp. 187–202, 2010.

maximizes the distance from "Gaussianity" of the individual components or their statistical independence. More recently, in light of advances in the aforementioned sparse representation theory, the idea of using prior information about the sources in the form of membership in a dictionary became an attractive alternative and was proposed already by Zibulevsky and Pearlmutter [24] in 2001, and more recently by Li *et al.* [17] and others, using various combinations of well-studied tools (K-means, Bayesian formalism, etc.) in combination with ℓ_1-optimization.

Our approach is rather direct as it assumes that the observed signals are outputs of an unknown dynamical system driven by unknown inputs, albeit these being sparse mixtures from a known dictionary. Thus, a salient feature of our formulation is that the "mixing" of signals has a structure inherited by linear dynamics and, the "mixing matrix" has entries that are Toeplitz matrices by themselves. Clearly, this is an under determined nonlinear problem with many possible solutions. Therefore, it is both natural and meaningful to seek, beside sparse representations of signals, small complexity of the intervening dynamics. Our formalism addresses this last dictum by imposing a penalty on the time-constant of the sought dynamical interactions. A suitable optimization problem is proposed and examples highlight the type of applications where this may be appropriate.

13.2 Sparse Representations and ℓ_1-Optimization

In order to motivate our methodology, we will give here some of the relevant background on sparse representations using the ℓ_1 norm. Full details may be found in [2, 3, 4, 5, 6, 7, 8, 9, 22], and the references therein.

Consider an under determined problem

$$\mathbf{H}x = y \tag{13.1a}$$

where the vector x represents the model, $\mathbf{H}$ is a linear ill-posed operator and the vector y contains data obtained from measurement. One possible way to regularize the problem is by using a Tikhonov-like regularization scheme [14]. However, it is often natural to assume that the model x is a linear combination of a small number of possible vectors that are collected into a $n \times N$ matrix B, where typically $N >> n$ (referred to as an "over-complete" basis or dictionary), and thus

$$x = Bv.$$

Therefore, the model-complexity is quantified by the number of nonzero entries of v called the sparsity $\|v\|_0$ of v. Seeking a solution with minimal number of nonzero entries can also be thought of as a form of regularization. Despite what the notation may suggest, $\|v\|_0$ is not a norm. In fact, the problem of minimizing $\|v\|_0$ subject to (13.1a) is combinatorial in nature and practically infeasible. However, it has been recognized for some time and, in recent years has formed the basis of the powerful theory of compressed sensing, that the ℓ_1-norm $\|\cdot\|_{\ell_1}$ can be thought of as a relaxation of $\|\cdot\|_0$ and that minimizing $\|\cdot\|_{\ell_1}$ in practice as well as in theory, for many interesting cases, leads with overwhelming probability to sparse solutions.

In practice, a more natural problem includes measurement noise ε and that equality in (13.1a) is not exact, that is

$$\mathbf{H}x + \varepsilon = y. \tag{13.1b}$$

Recently, it has been suggested that such problems can be effectively treated by one of the following formulations:

(i) $v = \text{argmin}\{\|v\|_1 \mid \|\mathbf{H}Bv - y\|^2 \leq \tau\}$, known as ***Basis Pursuit Denoising***,
(ii) $v = \text{argmin}\{\|\mathbf{H}Bv - y\|^2 \mid \|v\|_1 \leq \sigma\}$, known as ***Least Absolute Shrinkage and Selection Operator (LASSO)***, and
(iii) $v = \text{argmin}\{\mu\|v\|_1 + \|\mathbf{H}Bv - y\|^2\}$, known as ***Relaxed Basis Pursuit***.

The parameters τ, σ, μ are chosen such that the solution does not over-fit the data. In practice, the optimal choice for σ, τ and μ can be complicated, especially when the noise level is not well known. The interesting feature of these solutions is that they yield a sparse v, that is v with very few non-zero entries, and this has been explained and justified in a series of papers, see e.g., [6, 9, 13, 19] and the references therein.

On the numerical side of things, although problems of this nature had already come up in the 70's, recent work has shown dramatic improvements. Currently, there are three approaches that seem most efficient:

(a) ***Methods based on shrinkage***. Such methods (softly) threshold the solution at each iteration [11]. This may be considered to be an expectation maximization (EM) type algorithm that is very effective especially for image deconvolution problems.
(b) ***Interior point methods***. This is a class of algorithms based on work of Karmarkar [15] for linear programming. The technique uses a self-concordant barrier function in order encode the convex set; see [1] and the references therein.
(c) ***Methods based on reformulation and projection***. In these methods one sets $v = p - q$ where $p, q \geq 0$ and solves the corresponding optimization problem by a projection method [10]. We will give details about this method below since this we have found to be most effective for the type of problems considered in this note.

In the compressed sensing literature the matrix $\mathbf{H}$ and the dictionary B are typically known. However, formulating blind source separation in a similar setting, one needs to estimate $\mathbf{H}$ as well as recover x [17, 24]. The well-posedness of such a problem, draws on additional sets of natural assumptions. As we explain below, we are interested in the convolutive blind source problem where it is natural to assume linear time invariance, and therefore a Toeplitz or circulant structure for $\mathbf{H}$. This will lead to an iterative method of alternating least squares (for the $\mathbf{H}$) and ℓ_1 optimization (for the sparsity of the coefficient vector v) to be described below.

Typically an additional problem is the choice of the basis dictionary B. Obviously, the problem of MRI denoising requires a different basis than denoising an image of stars. The appropriate choice for B should be dictated by the problem and is a recent topic of extensive research [12, 18]. For the purposes of this note and because we focus on system identification, we consider sinusoids to be sufficient and tailor our example around these.

Our approach to blind source separation is cast below in the spirit of relaxed basis pursuit. Thus, we briefly summarize here a numerical method that can be used for solving such problems. This is a simple strategy that was recently investigated in [10] (see also, [16] and the references therein). The idea in [10] is that the non-smooth ℓ_1-norm is replaced by a smooth optimization problem with inequality constraints. The basic step is to reformulate the problem as a quadratic program problem by setting $v = p - q$ with both $p, q \geq 0$ (v is decomposed into its positive and negative parts). Then is easy to show that the relaxed basis pursuit problem is equivalent to the following optimization problem

$$\begin{aligned} \min_{p,q} \ & \frac{1}{2}\|\mathbf{H}B(p-q) - y\|^2 + \mu e^\top (p+q) \\ \text{s.t} \ & p, q \geq 0, \end{aligned} \tag{13.2}$$

where $e = [1, \ldots, 1]^\top$. In [10], a natural solution for this ***bound-constrained quadratic program*** is found using a natural gradient projection method [1, 20]. We have found that for the convolutive blind source separation problem of interest to us which has high level of sparsity, convergence is typically very fast.

13.3 Toeplitz Algebra and Some Notation

All indices are nonnegative integers; typically i, j, p, m, n are positive integers whereas t, τ represent time and take values in $\{0, 1, 2, \ldots\}$. Small (possibly indexed) letters without an argument $x, y, x_j, \ldots$ represent column vectors while $x(i), y(i), x_j(i), \ldots$ represent their entries, capitals Y, H represent block vectors (i.e., "bigger" vectors, composed of stacked-up "smaller" vectors y_i, h_{ij}, respectively) defined explicitly below. Boldface lower case letters $\mathbf{x}$ are reserved for square matrices with a lower triangular Toeplitz structure and boldface capital $\mathbf{X}, \mathbf{H}$ are used for structured block matrices with lower triangular Toeplitz entries. Thus, typically,

$$x = \begin{bmatrix} x(0) \\ x(1) \\ \vdots \\ x(n-1) \end{bmatrix} \in \mathbb{R}^{n \times 1}$$

while

$$\mathbf{x} = \begin{bmatrix} x(0) & 0 & \ldots & 0 \\ x(1) & x(0) & \ldots & 0 \\ \vdots & \vdots & & \vdots \\ x(n-1) & x(n-2) & \ldots & x(0) \end{bmatrix}$$

and this correspondence will be implicit throughout. The indexing of vector elements $x(0),\ x(1),\ \ldots$ reflects the nature of these entries as elements of a time-function $x(t)$ where $t = 0, 1, \ldots$, as they will typically represent either a time-signal or an impulse response of a linear time-invariant (LTI) dynamical system.

Let now $h(0), h(1), \ldots$ represent the impulse response of such a scalar LTI-system, $x(0), x(1), \ldots$ the input to the system when this is at rest, and $y(0), y(1), \ldots$ the resulting output. Restricting our attention to the finite time-window from $t = 0$ to $t-1$, the convolution relation

$$y(t) = h(t) * x(t) := \sum_{\tau=0}^{t} h(\tau)x(t-\tau),$$

for $t \in \{0, n-1\}$ can be expressed either as

$$\begin{bmatrix} y(0) \\ y(1) \\ \vdots \\ y(n-1) \end{bmatrix} = \begin{bmatrix} h(0) & 0 & \ldots & 0 \\ h(1) & h(0) & \ldots & 0 \\ \vdots & \vdots & & \vdots \\ h(n-1) & h(n-2) & \ldots & h(0) \end{bmatrix} \begin{bmatrix} x(0) \\ x(1) \\ \vdots \\ x(n-1) \end{bmatrix},$$

or, equivalently, following our earlier notation

$$y = \mathbf{h}x, \tag{13.3a}$$

as well as, in a purposefully redundant expression, by

$$\mathbf{y} = \mathbf{h} \cdot \mathbf{x} \tag{13.3b}$$

where the added columns when compared to (13.3a) do not impose any additional constraint. It is important to note that the latter expression inherits the commutativity of the convolution, i.e., that

$$\mathbf{h} \cdot \mathbf{x} = \mathbf{x} \cdot \mathbf{h},$$

and that the ring of $n \times n$ lower triangular Toeplitz matrices is in fact a commutative algebra.

13.4 Multivariable-System Identification

In this section we provide alternative ways to write down the dependence between inputs, outputs, and impulse responses of multivariable linear systems. Our interest in these stems from the need to express the various relationships in a conveniently unconstrained way for ease of carrying out the optimization of certain functionals in subsequent sections.

Consider a m-input, p-output (MIMO) LTI system and let x_j ($j \in \{1, \ldots m\}$) and y_i ($i \in \{1, \ldots p\}$) denote the respective inputs and outputs. Our goal is to write the input-output relationship in a form which is convenient for computing the elements of the $p \times m$ impulse responses $h_{ij}(t)$ ($i \in \{1, \ldots p\}, j \in \{1, \ldots m\}, t \in \{0, \ldots n-1\}$).

Making use of our earlier notation and restricting our attention to the time interval $\{0,\dots n-1\}$ we may use the representation

$$\begin{bmatrix} \mathbf{y}_1 \\ \mathbf{y}_2 \\ \vdots \\ \mathbf{y}_p \end{bmatrix} = \begin{bmatrix} \mathbf{h}_{11} & \mathbf{h}_{12} & \dots & \mathbf{h}_{1m} \\ \mathbf{h}_{21} & \mathbf{h}_{22} & \dots & \mathbf{h}_{2m} \\ \vdots & \vdots & & \vdots \\ \mathbf{h}_{p1} & \mathbf{h}_{p2} & \dots & \mathbf{h}_{pm} \end{bmatrix} \begin{bmatrix} \mathbf{x}_1 \\ \mathbf{x}_2 \\ \vdots \\ \mathbf{x}_m \end{bmatrix}, \tag{13.4a}$$

to express the dynamical dependence between variables. For ease of referencing, denote the elements of the above equation by bold capitals and write

$$\mathbf{Y} = \mathbf{H}\mathbf{X}. \tag{13.4b}$$

These are all structured matrices. Making use of the commutativity between their lower triangular Toeplitz entries, equation (13.4b) may be re-written as

$$\begin{bmatrix} \mathbf{y}_1 \\ \mathbf{y}_2 \\ \vdots \\ \mathbf{y}_p \end{bmatrix} = \left(I_p \otimes \left[\mathbf{x}_1,\, \mathbf{x}_2,\, \dots\, \mathbf{x}_m\right]\right) \begin{bmatrix} \mathbf{h}_{11} \\ \vdots \\ \mathbf{h}_{1m} \\ \mathbf{h}_{21} \\ \vdots \\ \mathbf{h}_{2m} \\ \vdots \\ \vdots \\ \mathbf{h}_{p1} \\ \vdots \\ \mathbf{h}_{pm} \end{bmatrix}, \tag{13.5a}$$

where I_p denotes the $p \times p$ identity matrix. At this point, we may also suppress the redundant set of equations that correspond to the columns other than the first on both sides of the above, and deduce the equivalent expression

$$\begin{bmatrix} y_1 \\ y_2 \\ \vdots \\ y_p \end{bmatrix} = \left(I_k \otimes \left[\mathbf{x}_1,\, \mathbf{x}_2,\, \dots\, \mathbf{x}_k\right]\right) \begin{bmatrix} h_{11} \\ \vdots \\ h_{1m} \\ h_{21} \\ \vdots \\ h_{2m} \\ \vdots \\ \vdots \\ h_{p1} \\ \vdots \\ h_{pm} \end{bmatrix} \tag{13.5b}$$

which no longer requires a specific structure for the vector containing all $p \times m$ impulse responses $h_{ij}(t)$ ($t \in \{0, \ldots n-1\}$). For convenience, define

$$\mathbb{X} := \left(I_p \otimes \left[\mathbf{x}_1,\, \mathbf{x}_2,\, \ldots\, \mathbf{x}_m\right]\right)$$

formed out of the Toeplitz matrices for the input time-history. This is a structured matrix which is formed out of input-values—the symbol $\otimes$ denotes the Kronecker product and the entries $\mathbf{x}_i$ have a Toeplitz structure. For simplicity we use analogous symbols to denote vectors formed out of stacking up vectors of outputs and impulse responses, respectively,

$$Y := \left[y_1^T\; y_2^T\; \ldots\; y_p^T\right]^T, \text{ and}$$
$$H := \left[h_{11}^T\; h_{12}^T\; \ldots\; h_{1m}^T\; h_{21}^T\; \ldots h_{pm}^T\right]^T,$$

and observe that (13.4b) is equivalent to the following

$$Y = \mathbb{X}H. \tag{13.6}$$

The advantage is that H is now a vector in $\mathbb{R}^N$, with $N = p \cdot m \cdot n$, which is unconstrained except for satisfying the linear equation (13.6). Therefore, if the input and output values in X, Y are known, one can easily obtain the "least squares" solution $(\mathbb{X}^T\mathbb{X})^{-1}\mathbb{X}Y$ for the impulse responses that minimize the quadratic error between observed outputs and estimated ones.

13.5 The Problem of Blind Source Deconvolution

In many physical situations measurements are due to excitations of known "signature," e.g., impulses, sinusoids, damped sinusoids, etc., that propagate and scatter before they are recorded at sensor locations. The effect of propagation through the medium can often be modeled using linear dynamics. When this is the case, the problem we are facing is that of identifying both the impulse response of linear-convolution maps as well as the de-convolved excitation input. The salient feature of our problem is that inputs are sparse linear combinations from a known directory (i.e., linear combinations that are made up of very few samples from the directory). When the directory is selected properly and if it satisfies the so-called restricted isometry property (see e.g., [3]) sparsity is inherited by ℓ_1-optimization, and the problem of simultaneous system identification and deconvolution can be cast as follows.

Problem 13.5.1 (blind source deconvolution). Consider p time-series $y_1(t)$, $y_2(t)$, $\ldots, y_p(t)$, $t \in \{0, \ldots n-1\}$, representing sensor readings e.g., at p different locations. Assume that the exact number m of expected input sources is known, as is an estimate λ, with $0 < \lambda < 1$, of the time-constant for the dynamics from source-locations to sensor-locations. Assume that a "dictionary" $n \times N$ matrix B is also known, and that all anticipated input signals are sparse linear combination of the columns of B. Determine values for the impulse responses $h_{ij}(t)$ ($i \in \{1, \ldots p\}$, $j \in \{1, \ldots m\}$) from the m

input-locations to the p sensor-locations, as well as values for $x_1(t), x_2(t), \ldots, x_m(t)$ over the same time-support $t \in \{0, \ldots n-1\}$, that minimize the functional

$$J(v_i, h_{ij}) := \sum_{i=1}^{m} \|v_i\|_{\ell_1} + \sum_{i,j,t} |h_{ij}(t)|^2 \lambda^{-2t} + \|\mathbb{X}H - Y\|_{\ell_2}^2$$
$$\text{where } \mathbb{X}, H, Y \text{ are as defined in Section 13.4,}$$
$$\text{and } v_i \in \mathbb{R}^n \text{ is subject to}$$
$$x_i = Bv_i, \text{ for } i \in \{1, \ldots, m\}. \tag{13.7}$$

The first term leads to sparsity for the solution vectors v_i as discussed in Section 13.2. The second term of the functional penalizes the values of the entries of $h_{ij}(t)$ in a way that is inversely proportional to powers of λ. This weight dictates a time-constant for the entries of h_{ij}'s which is approximately bounded by λ. Finally, the third term penalizes the mismatch between observed outputs and the values corresponding to the particular choice of inputs x_i and impulse responses h_{ij}.

13.6 Alternating Projections

The functional $J(v,h)$ in Problem 13.5.1 is clearly convex in each argument (though not jointly convex). Thus, we proceed to seek minima (which may only be local) by alternating between the optimizing with respect to one variable at a time.

The expression $\|\mathbb{X}H - Y\|_{\ell_2}$ involves H as an unconstrained vector. Therefore, when the v_i's are fixed, and hence so are the entries of $\mathbb{X}$, then J is quadratic in the h_{ij}'s. That is, except for a constant,

$$J \sim \|H\|_{\ell_2,\Lambda}^2 + \|\mathbb{X}H - Y\|_{\ell_2}^2,$$

since

$$\sum_{i,j,t} |h_{ij}(t)|^2 \lambda^{-2t} = \|H\|_{\ell_2,\Lambda}^2 := H^T \Lambda H$$

with a "weight matrix"

$$\Lambda = I_k \otimes \operatorname{diag}(1, \lambda^{-2}, \lambda^{-4}, \ldots \lambda^{-2m}).$$

Therefore, the optimal value for H when we minimize J with respect to the h_{ij}'s while keeping the v_i's fixed is

$$H_{\text{opt}} = (\Lambda + \mathbb{X}^T \mathbb{X})^{-1} \mathbb{X} Y. \tag{13.8}$$

We now consider fixing the h_{ij}'s and optimize with respect to the v_i's. This is a convex problem and can be solved as indicated earlier in Section 13.2. Indeed, we first define

$$V = \begin{bmatrix} v_1 \; v_2 \; \dots \; v_m \end{bmatrix} = \begin{bmatrix} v_1(0) & v_2(0) & \dots & v_m(0) \\ \vdots & & & \vdots \\ v_1(n-1) & v_2(n-1) & \dots & v_m(n-1) \end{bmatrix}$$

and

$$v := \mathrm{vec}(V) = \begin{bmatrix} v_1^T \; \dots \; v_m^T \end{bmatrix}^T,$$

and similarly,

$$x := \begin{bmatrix} x_1^T \; \dots \; x_m^T \end{bmatrix}^T.$$

Then, $\mathbb{X}H = \mathbf{H}x$ while $x = (I_m \otimes B) \cdot v$. Therefore, assuming H fixed, the functional J is the sum of $\|v\|_{\ell_1}$ and $\| (\mathbf{H} \cdot (I_\ell \otimes B)) \cdot v - Y \|_{\ell_2}$, and we can compute

$$v = \mathrm{argmin} \|v\|_{\ell_1} + \| (\mathbf{H} \cdot (I_\ell \otimes B)) \cdot v - Y \|_{\ell_2} \tag{13.9}$$

using either interior point or gradient projection methods discussed earlier.

13.7 Application

The following simplified example highlights and abstracts a number of case studies that we have conducted following the above program. Two noisy sinusoidal source signals are responsible for the readings at three sensor-outputs. These are

$$\begin{aligned} x_1(t) &= 2\sin(2\pi t/20) + w_1(t) \\ x_2(t) &= \sin(2\pi(t-3)/7) + w_2(t) \end{aligned}$$

with $w_i(t)$ ($i \in \{1,2\}$) independent Gaussian white noise with variance equal to 0.1 and $t \in \{1,2,\dots 100\}$. These are mixed through dynamics with impulse responses h_{ij} from jth source to ith sensor given in the table of Figure 13.1. The precise numbers approximate the values in a physical experiment we have analyzed at an earlier time.

We use as a dictionary, a matrix B which contains 140 sinusoidal components (70 cosines and 70 sines), spaced at $\frac{\Delta f}{2} + k\Delta f$ for $k = 0, 1, \dots$, with $\Delta f = 2\pi/70$ [radians/sample]. We observe that, despite the limited time-window over which measurement signals are available (Figure 13.6), two source signals are reliably estimated (shown in Figure 13.2) that are very close to the original ones (Figure 13.1)–it should be emphasized that the resolution is not limited by the $2\pi/n$ uncertainty of Fourier theory. Also, closer examination reveals a similarity between the impulse response maps (i.e., the one used to generate the data and the one estimated via blind deconvolution). In comparing signals and maps, it should kept in mind that there is an inherent indeterminacy of a scaling factor between the two–hence, the discrepancy in amplitude between signals in Figures 13.1 and 13.2. A more telling comparison can be based on Figures 13.6 and 13.7, which show the available measurements and the reconstructed $\mathbf{H}x$ values, respectively.

Table 13.1. h_{ij} from jth source to ith sensor

t	0	1	2	3	4	5	6	7
h_{11}	1	-.7	0	0	0	0	0	0 ...
h_{12}	1	-.7	-.5	.3	0	0	0	0 ...
h_{21}	0	0	1	.7	-.5	.3	0	0 ...
h_{22}	1	0	-1	0	.5	0	-2	0 ...
h_{31}	1	-.7	0	.3	-.2	0	0	0 ...
h_{32}	-2	1.5	0	-.5	.3	0	0	0 ...

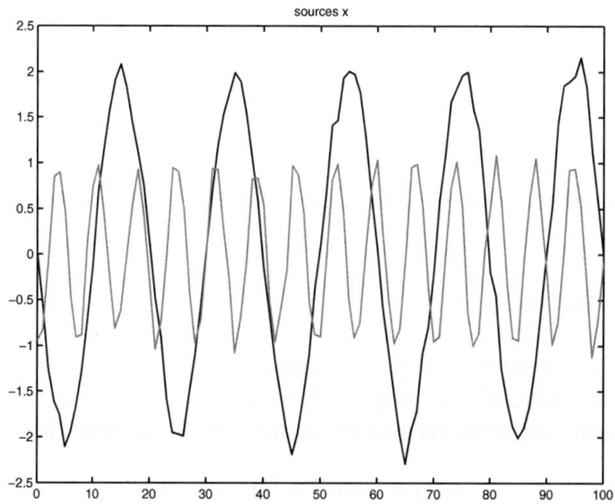

Fig. 13.1. Original source signals

13.8 Remarks

Invariably, the performance of such mathematical tools, where the solution of an optimization problem provides a possible explanation of the data, depends on the choice and relative importance placed on various terms. For instance, an added weight that accentuates the contribution of $\|v_i\|_{\ell_1}$ in the functional J has as effect to improve the sparsity of the relevant source signal. Also, it is possible to modify the functional so that the various source signals are "encouraged" or "discouraged" from sharing the same elements from the dictionary B. This circle of possibilities will be explored further in future work. The main conclusion of the present note is that the tools of the ℓ_1/compressed sensing theory can be directly applied to the blind source deconvolution problem and that with relatively straightforward optimization one can obtain reasonably consistent results.

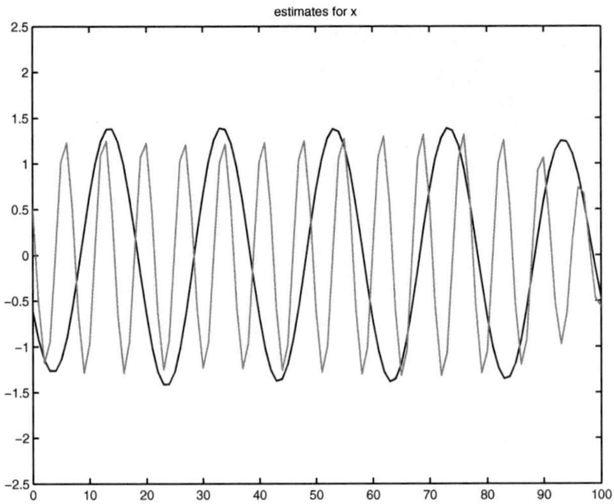

Fig. 13.2. Estimated source signals $x_i(t)$ for $i \in \{1,2\}$ (determined up to scaling)

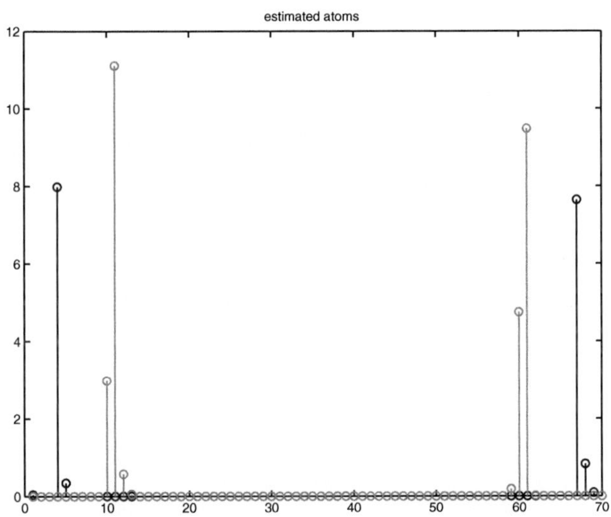

Fig. 13.3. Estimated atoms v_1 and v_2

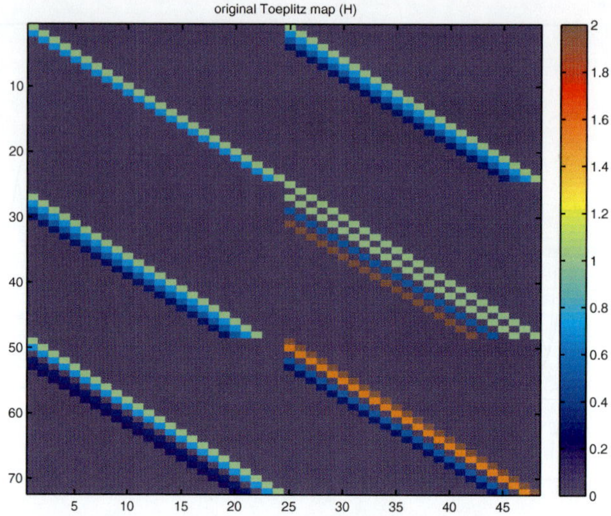

Fig. 13.4. Absolute values of **H**-entries

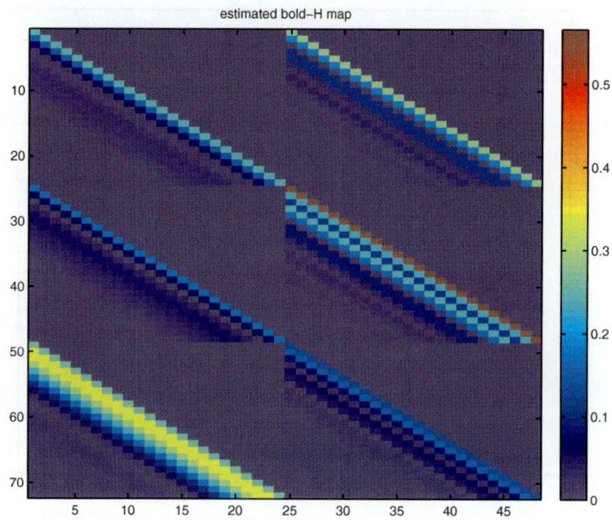

Fig. 13.5. Absolute values of estimated **H**-entries

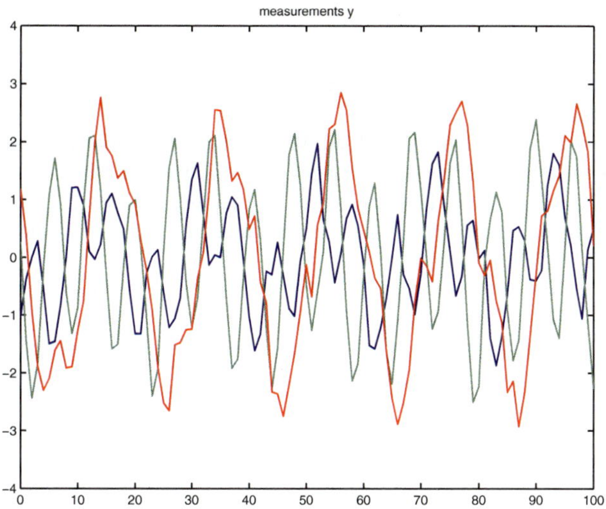

Fig. 13.6. Output measurements $y_i(t)$ for $i = \{1,2,3\}$

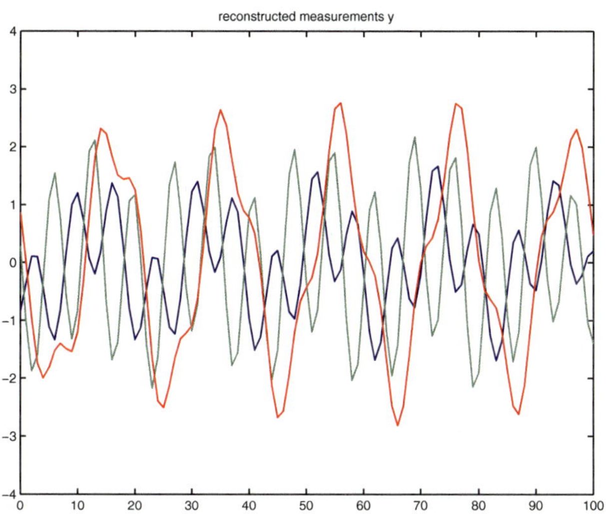

Fig. 13.7. Reconstructed values for output measurements $\mathbf{H}x$

References

1. Boyd, S., Vandenberghe, L.: Convex Optimization. Cambridge University Press, Cambridge (2004)
2. Bruckstein, A.M., Donoho, D.L., Elad, M.: From sparse solutions of systems of equations to sparse modeling of signals and images. SIAM Review (2008)
3. Candes, E., Romberg, J., Tao, T.: Robust uncertainty principles: Exact signal reconstruction from highly incomplete frequency information. IEEE Transactions on Information Theory 52, 489–509 (2006)
4. Candes, E., Tao, T.: Decoding by linear programming. IEEE Transactions on Information Theory 51, 4203–4215 (2005)
5. Candes, E., Tao, T.: Near optimal signal recovery from random projections: universal encoding strategies. IEEE Transactions on Information Theory 52, 5406–5425 (2006)
6. Candes, E., Braun, N., Wakin, M.: Sparse signal and image recovery from compressive samples. In: 4th IEEE International Symposium on Biomedical Imaging: From Nano to Macro, pp. 976–979 (2007)
7. Chen, S.S., Donoho, D.L., Saunders, M.A.: Atomic decomposition by basis pursuit. SIAM Journal on Scientific Computing 20(1), 33–61 (1998)
8. Donoho, D.L.: For most large underdetermined systems of linear equations the minimal ‘1-norm solution is also the sparsest solution. Communications on Pure and Applied Mathematics 59(6), 797–829 (2006)
9. Donoho, D.L., Elad, M.: Optimally sparse representation in general (nonorthogonal) dictionaries via ℓ_1 minimization. PNAS 100(5), 2197–2202 (2003)
10. Figueiredo, M.A.T., Nowak, R.D., Wright, S.J.: Gradient projection for sparse reconstruction: Application to compressed sensing and other inverse problems. IEEE Journal of Selected Topics in Signal Processing 1, 586–597 (2007)
11. Figueiredo, M., Nowak, R.: An EM algorithm for wavelet-based image restoration. IEEE Trans. Image Processing 12, 906–916 (2003)
12. Haber, E., Ascher, U.M., Oldenburg, D.: On optimization techniques for solving nonlinear inverse problems. Inverse Problems 16, 1263–1280 (2000)
13. Hyvarinen, A., Hoyer, P., Oja, E.: Sparse code shrinkage: denoising of nongaussian data by maximum likelihood estimation. Neural Computation 11(7), 1739–1768 (1999)
14. Johansen, T.A.: On Tikhonov regularization, bias and variance in nonlinear system identification. Automatica 33(3), 441–446 (1997)
15. Karmarkar, N.: A new polynomial-time algorithm for linear programming. Combinatorica 4, 373–395 (1984)
16. Koh, K., Kim, S.J., Boyd, S.: An interior-point method for large-scale l1-regularized logistic regression. J. Mach. Learn. Res. 8, 1519–1555 (2007)
17. Li, Y., Cichocki, P., Amari, S.: Sparse component analysis and blind source separation of underdetermined mixtures. Neural Computation 16, 1193–1234 (2004)
18. Malioutov, D.M., Cetin, M., Willsky, A.S.: Optimal sparse representations in general overcomplete base. In: IEEE International Conference on Acoustics, Speech, and Signal Processing (2004)
19. Mallat, S., Zhang, Z.: Matching pursuits with time-frequency dictionaries. IEEE Transactions on Signal Processing 41(12), 3397–3415 (1993)
20. Nocedal, J., Wright, S.: Numerical Optimization. Springer, New York (1999)
21. Romberg, J.K.: Sparse signal recovery via ℓ_1 minimization. In: Proceedings of the 40th Annual Conference on Information Sciences and Systems, March 2006, pp. 213–215 (2006)

22. Tsaig, Y., Donoho, D.L.: Breakdown of equivalence between the minimal l1-norm solution and the sparsest solution. Signal Processing 86(3), 533–548 (2006)
23. Wang, J., Sacchi, M.D.: High-resolution wave-equation amplitude-variation-withray- parameter imaging with sparseness constraints. Geophysics 72(1), S11–S18 (2007)
24. Zibulevsky, M., Pearlmutter, B.A.: Blind source separation by sparse decomposition in a signal dictionary. Neural Computation 13, 863–882 (2001)

14

Sequential Bayesian Filtering via Minimum Distortion Quantization*

Graham C. Goodwin[1], Arie Feuer[2], and Claus Müller[3]

[1] School of Electrical Engineering and Computer Science, The University of Newcastle, Australia
[2] Electrical Engineering Department, Technion-Israel Institute of Technology, Haifa, Israel
[3] School of Electrical Engineering and Computer Science, The University of Newcastle, Australia

Summary. Bayes Rule provides a conceptually simple, closed form, solution to the sequential Bayesian nonlinear filtering problem. The solution, in general, depends upon the evaluation of high dimensional multivariable integrals and is thus computationally intractable save in a small number of special cases. Hence some form of approximation is inevitably required. An approximation in common use is based upon the use of Monte Carlo sampling techniques. This general class of methods is referred to as Particle Filtering. In this paper we advocate an alternative deterministic approach based on the use of minimum distortion quantization. Accordingly we use the term Minimum Distortion Nonlinear Filtering (MDNF) for this alternative class of algorithms. Here we review the theoretical support for MDNF and illustrate its performance via simulation studies.

14.1 Introduction

Nonlinear Filtering appears in a wide variety of applied problems including radar tracking, GPS positioning, economics, telecommunications and biotechnology [19]. Due to the pivotal role of nonlinear filtering in these diverse problems, substantial research effort has been dedicated to finding computationally tractable solutions.

A typical state space formulation for the filtering problem is based on the following nonlinear Markovian model:

$$x_k = f(x_{k-1}, \omega_{k-1}) \tag{14.1}$$

$$y_k = h(x_k, v_k) \tag{14.2}$$

where $x_k \in \mathbb{R}^{n_x}$, $\omega_k \in \mathbb{R}^{n_\omega}$, $y_k \in \mathbb{R}^{n_y}$, $v_k \in \mathbb{R}^{n_v}$, are the state, (i.i.d.) process noise, measured output and (i.i.d.) measurement noise respectively. It is assumed that the probability distributions for ω_k and v_k are known. Hence (14.1), (14.2) are surrogates for the conditional distribution of x_k given x_{k-1} and the conditional distribution of y_k

*This chapter is dedicated to Chris Byrnes and Anders Lindquist.

X. Hu et al. (Eds.): Three Decades of Progress in Control Sciences, pp. 203–213, 2010.

given x_k respectively. We also assume knowledge of the prior distribution $p(x_0)$ of the initial state.

In the sequel, and when the meaning is clear from the context, we will use a short hand notation $p(\alpha|\beta)$ to describe the probability density of a random variable A taking the value α, given that another random variable B takes the value β.

We seek to evaluate $p(x_k|\mathscr{Y}_k)$ where $\mathscr{Y}_k = \{y_0, \dots, y_k\}$.

The solution to the above problem can be described using the Chapman Komolgorov equations. These equations are a statement of Bayes rule [11], [23], and take the following form:

State Update

$$p(x_{k+1}|\mathscr{Y}_k) = \int p(x_{k+1}|x_k)\, p(x_k|\mathscr{Y}_k)\, dx_k \tag{14.3}$$

Observation Update

$$p(x_{k+1}|\mathscr{Y}_{k+1}) = \frac{p(x_{k+1}|\mathscr{Y}_k)\, p(y_{k+1}|x_{k+1})}{\int p(x_{k+1}|\mathscr{Y}_k)\, p(y_{k+1}|x_{k+1})\, dx_{k+1}} \tag{14.4}$$

Whilst (14.3), (14.4) provide a complete conceptual solution to the problem of sequential nonlinear filtering, these equations are not useful in practical problems. The reason is that (14.3), (14.4) depend upon high dimensional multivariable integrals which are, in general, impossible to evaluate save in very special cases e.g., linear Gaussian problems where the Kalman filter [5] is applicable. However, in general, some form of approximation is necessary.

Three classes of approximation are in common use, [2], namely:

(i) Extended Kalman Filtering based upon local linearization and Gaussian distributions.
(ii) Monte Carlo - Particle Filtering based on discretization via random sampling.
(iii) Deterministic Methods - based on deterministic discretization of posterior density functions.

Here we will focus on classes (ii) and (iii) with emphasis on class (iii).

Methods in class (ii) and (iii) utilize the common strategy of approximating the posterior probability density function, $p(x)$, by a discrete density function, $\hat{p}(x)$, where

$$\hat{p}(x) = \sum_{i=1}^{N} p_i \delta(x - x_i) \tag{14.5}$$

The methods differ in the way the grid $\{x_i,\ i = 1, \dots, N\}$ is determined and the way the associated probabilities $\{p_i,\ i = 1, \dots, N\}$ are calculated.

The layout of the remainder of the chapter is as follows:

In Section 14.2 we briefly review Monte Carlo - Particle Filtering Methods. In Section 14.3 we outline Deterministic Methods. In Section 14.4 we review methods for Vector Quantization using Distortion Minimization strategies. In Section 14.5 we describe Minimum Distortion Nonlinear Filters. In Section 14.6, we describe an on-line griding technique for use in Minimum Distortion Nonlinear Filtering. In Section 14.7 we present an example. Finally, in Section 14.8 we draw Conclusions.

14.2 Monte Carlo Particle Filtering Methods

The basic idea of the Monte Carlo approach to nonlinear filtering is to draw a random sample of size N from the posterior distribution. An early paper describing this approach is [10]. The associated weights are typically then chosen as $\frac{1}{N}$. However a key problem is that the posterior distribution is not, in general, available. Thus, the strategy usually employed is to draw from an alternative distribution, $q(x)$. This alternative distribution is known by several names including *importance density*, *proposal distribution* or *importance function* [2], [1]. The samples are drawn from $q(x)$ and p_i, $i = 1, \ldots, N$ in (14.5) are chosen as

$$p_i = \frac{p(x_i)}{q(x_i)} \quad ; \quad i = 1, \ldots, N \tag{14.6}$$

When used in recursive filtering, the method is commonly referred to as Sequential Importance Sampling (SIS).

A common problem with SIS particle filtering methods is the "degeneracy phenomenon", whereby, after a few iterations, all but a few weights have negligible values. There are a number of remedies proposed in the literature for this problem. One strategy is to make a more elaborate choice of the importance density. Another strategy is to use resampling, in which case, the method is called Sequential Importance Resampling (SIR). The basic idea of SIR is that, at each time k, based on an approximate posterior distribution, N new samples are drawn using this distribution [7].

The topic of particle filtering is a very broad one and there are many embellishments of the above basic ideas.

An important property of Monte Carlo methods in that they converge, in a well defined sense, to the true distribution as the number of samples grows [9]. However, a difficulty is that if N is constrained, say by computational requirements, then it is doubtful that Monte Carlo type methods provide the best approximation. This motivates deterministic approaches which choose the grid and associated weights in a more deliberate fashion.

14.3 Deterministic Methods

A widely used deterministic method for nonlinear filtering is:

- Extended Kalman Filter [19]. This method uses a local linearization to propagate the mean and covariance via the standard Kalman Filter.

Other deterministic methods are based on the use of some form of griding. Common to all grid based methods is the notion that the grids are chosen deterministically. Among the griding methods we have:

- Iterative Quadrature (see e.g. [22], [2]). This method is widely used in computer graphics and physics. It is based on approximating a finite integral by a weighted sum of samples. The integrand is then based on a quadrature formula

$$\int f(x)p(x)dx \approx \sum_{i=1}^{m} c_i f(x_i) \tag{14.7}$$

  where $p(x)$ is treated as the weighting function and x_i as the quadrature point.
- Multigrid (see e.g. [1]). The general idea is to choose a finite grid to represent the continuous state space and this is then used to approximate the filtering problem as a Hidden Markov model (see e.g. [12]).
- Gaussian sum approximation (see e.g. [[20]]). The idea is to approximate the posterior probability density function by a weighted sum of Gaussians namely:

$$p(x) = \sum_{j=1}^{m} c_j N(\hat{x}_j, \Sigma_j) \tag{14.8}$$

  where the weights $c_j > 0$ and $\sum_{j+1}^{m} c_j = 1$. This is motivated by the observation that any non-Gaussian pdf can be approximated to any desired degree of accuracy by a sum of Gaussians.
- Unscented transformation (see e.g. [21]). This is similar, in spirit, to the Extended Kalman Filter in the sense that it propagates an approximation of the mean and covariance of the probability density function. However, rather then using a local linear approximation, the unscented filter uses samples which are passed through the appropriate nonlinear function. The sample points are deterministically chosen as the, so called, sigma points of a Gaussian distribution.
- Quantization based methods. This class of method uses vector quantization ideas to approximate a given probability density function. These methods will be the principal focus of the remainder of this contribution.

14.4 Vector Quantization via Distortion Minimization

To form a basis for the subsequent development, we outline below the essential ideas involved in vector quantization. The methods are widely used in signal processing, control and telecommunication, see [3]. For example, the current authors have utilized the idea of vector quantization in the context of stochastic optimization and control, see [6], [17].

The basic problem is as follows:

Given a random vector $X \in \mathbb{R}^{n_x}$ with probability density function $p(x)$ we seek a finite set of values $\mathscr{W}_x = \{x_1, \ldots, x_N\}$ and an associated collection of sets $\mathscr{S}_x = \{S_1, \ldots, S_N\}$ such that $\bigcup_1^N S_i = \mathbb{R}^{n_x}$ and $S_i \bigcap S_j = \emptyset; i \neq j$. (i.e., a tessellation of $\mathbb{R}^{n_x}$.)

The choice of $\mathscr{W}_x$ and $\mathscr{S}_x$ is based on minimization of an expected distortion measure[1], J, where

$$J(\mathscr{W}_x, \mathscr{S}_x) = \sum_{i=1}^{n} E\left\{\| x - x_i \|^2 | x \in \mathscr{S}_i\right\} \tag{14.9}$$

It is easily seen [3], [8] that given $\mathscr{W}_x$, the optimal sets S_i are the so called, Voronoi cells, defined as follows:

$$S_i = \left\{x \mid \| x - x_i \|^2 \leq \| x - x_j \|^2\right\} \tag{14.10}$$

(with some appropriate rule for assigning boundary points.)

Utilizing (14.10), the distortion measure becomes a function of $\mathscr{W}_x$ only. We thus write $J(\mathscr{W}_x)$.

Alternatively, if we are given $\mathscr{S}_x$, then the points x_i minimizing (14.9) satisfy the, so called centroid condition, i.e.,

$$x_i = E\{x \mid x \in S_i\} \tag{14.11}$$

In this case, the distortion measure becomes a function of $\mathscr{S}_x$ only. We thus write $J(\mathscr{S}_x)$.

Necessary and sufficient conditions for optimizing J are not available in the literature. Accordingly, a significant amount of research has been directed at numerical methods for obtaining stationary solutions in which (14.10) and (14.11) are simultaneously satisfied. A commonly deployed class of algorithms iterates between (14.10) and (14.11). These methods are generally called Lloyd algorithms [3], [8], [13].

Lloyd algorithms are based on the search for a fixed point. However, until recently, no convergence theory has been available. Recent work has established expressions for both the gradient, [15] and second derivative [14] of the distortion measure. The gradient can be used to develop gradient search algorithms with guaranteed convergence to a local minimum. Also, the availability of second derivative [14] opens the door to second order algorithms with accelerated convergence properties. Moreover, in very recent work, the current authors have shown [4] that Lloyd algorithms can be viewed as generalized gradient algorithms with a pre-specified step size. This insight reveals the mechanism underlying the known convergence phenomenon exhibited by Lloyd algorithms and motivates new algorithms.

14.5 Minimum Distortion Nonlinear Filtering

We next show how the ideas of Section 14.4 can be applied in the context of recursive nonlinear filtering. We call the resultant algorithms Minimum Distortion Nonlinear

---

[1] Here we use a quadratic norm but any reasonable distance measure can be utilized.

Filters (MDNF). The core principle of these filters is to utilize vector quantization as a (deterministic) mechanism for generating an optimal approximation to the posterior density function in nonlinear filtering.

In existing literature [16], [18], two strategies have been described. Common to both strategies is the fact that the (key) quantization step is carried out off-line. Brief details of the two methods are as follows:

(a) Marginal Quantization. Using (14.1) and given $p(x_o)$ and $p(x_k|x_{k-1})$, one can propagate the Markovian chain for the probability density functions for x_k via

$$p(x_k) = \int p(x_k|x_{k-1})p(x_{k-1})dx_{k-1} \tag{14.12}$$

To each of these prior distributions we apply optimal vector quantization. We then calculate the transition probability matrix $\hat{P}_k = [\hat{P}_k^{j,i}]$ of $\hat{x}_k$ given $\hat{x}_{k-1}$, i.e.,

$$\begin{aligned} \hat{P}_k^{j,i} &= P_r\left[x_k = x_k^j | x_{k-1} = x_{k-1}^i\right] \\ &= \int_{S_j^k} p\left(x_k|x_{k-1}^i\right) dx_k \qquad i = 1,\ldots,N_{k-1},\ j = 1,\ldots,N_k \end{aligned} \tag{14.13}$$

Here, S_j^k is the jth Voronoi cell corresponding to the grid $\mathscr{W}_k = \{x_{1,k},\ldots,x_{N,k}\}$. So, if we write

$$p\left(x_{k-1}|\mathscr{Y}_{k-1}\right) \simeq \sum_{i=1}^{N} p_{i,k-1}\delta\left(x_{k-1} - x_{i,k-1}\right) \tag{14.14}$$

then we obtain

$$p\left(x_k|\mathscr{Y}_k\right) \simeq \sum_{j=1}^{N} p_{j,k}\delta\left(x_k - x_{j,k}\right) \tag{14.15}$$

where

$$p_{j,k} = cp\left(y_k|x_k^j\right)\sum_{i=1}^{N} P_k^{j,i} p_{i,k-1} \tag{14.16}$$

where c is a normalization chosen to ensure that the sum of the probabilities is one.

Note that the Markovian property is not preserved by the above approximation.

(b) Markovian Quantization. This alternative algorithm sequentially builds the grid of points off-line. Thus, beginning with an approximation to $p(x_{k-1})$ of the form:

$$p\left(x'_{k-1}\right) = \sum_{i=1}^{N} p_{i,k-1}\delta\left(x'_{k-1} - x_{i,k-1}\right) \tag{14.17}$$

one calculates

$$p(x_k) \simeq \sum_{i=1}^{N} \int_{\Omega_i} p(\omega_k)p_{i,k-1} \tag{14.18}$$

where $\Omega_i = \{\omega_k | x_k = f(x_{i,k-1}, \omega_k)\}$. One then uses optimal quantization to generate the equivalent of (14.17) at time k. Finally, given the grids at each time, one follows the on-line steps given earlier in (14.14) to (14.16).

There are two disadvantages to methods (i) and (ii) above. Firstly, the quantization is based on prior data only and secondly one needs to store the grids for all times at which the filter is to be used.

We therefore purpose an alternative in which only $p(\omega_k)$ and $p(x_0)$ are quantized off-line. Then other distributions are quantized on-line. We term this approach online griding. This is described in the next Section.

14.6 On-Line Griding (OLG) MDNF

We assume that the distribution of the process noise is stationary. Hence, we grid $p(\omega.)$ to N' points. This is done off-line and stored. Then, beginning with a quantization of $p(x_{k-1})$ on N points we simply use equation (14.3) and (14.4) in discrete form to generate an approximation to $p(x_k)$ having $N * N'$ grid points. One can then continue the algorithm by optimal quantizing back to N grid points with their associated probabilities. The advantages of this approach are twofold, namely; (i) griding of $p(x_k, \mathscr{Y}_k)$ is done on-line and is adapted to the measured data and (ii) one need only store the prior calculated grid for $p(\omega_k)$. We will show via simulations (see Section 14.7) that this on-line griding algorithm has superior properties relative to earlier schemes described in Section 14.5. Theoretical support for the algorithm is currently being investigated.

14.7 Example

To demonstrate some of the features of the algorithms described above we consider the following simple nonlinear state estimation problem:

$$x_{k+1} = ax_k + \omega_k \tag{14.19}$$

$$y_k = x_k^2 + v_k \tag{14.20}$$

where $\{\omega_k\}$, $\{v_k\}$ are i.i.d. sequence distributed $N(0,1)$. The initial condition is also assured $N(0,1)$. We take $a = 0.9$.

Our goal is not to carry out a comprehensive comparison of all methods for nonlinear filtering. Instead, we will concentrate on the use of quantization. We will compare algorithms (a) and (b) from Section 14.5 with the algorithm of Section 14.6. The example is also simple enough to allow one to compute, to reasonable accuracy, the "true" posterior distribution.

A feature of the problem (14.19) and (14.20) is that the measurement does not give information about the sign of the state. Hence, the posterior distribution is generally bimodal. This can be seen in the "true" posterior distribution shown in Figure 14.1 for $k = 2$.

Also note that in Figure 14.1 we show both the density function and distribution function. The bimodal property is reflected in the distribution by a region of small change between regions of large change. In subsequent plots we will only

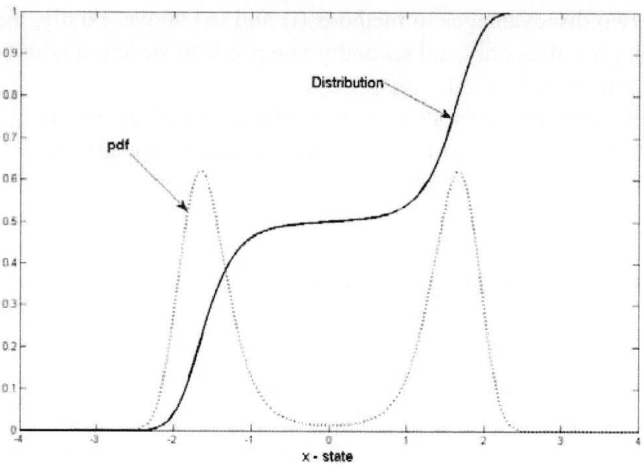

Fig. 14.1. Posterior pdf and distribution at time $k = 2$.

show distribution functions since these can be used for both discrete and continuous probability distributions.

We will compare the following four filters:

(i) The "true" filter based on direct use of (14.3), (14.4) with a fine uniform grid covering the range (-10, 10) with 1,000 points.

(ii) Algorithm (a) (Marginal Quantization). Here the quantization is applied off-line on the prior distribution generated by (14.19). Eleven quantization points were chosen.

(iii) Algorithm (b) (Markovian Quantization). Here the quantization is applied off-line on the prior distribution generated by (14.19). Again, eleven quantization points were chosen.

(iv) OLG - MDNF. Here the distribution of $p(\omega_k)$ is quantized, off-line, to 25 points and $p(x_k|\mathscr{Y}_k)$ is quantized, on-line, to 11 points.

The number of quantization points has been deliberately chosen as a relatively small number for the above example. This has been done so as (i) to highlight the relative merits of different algorithms and (ii) to open the door to higher order examples where the quantization density is necessarily restricted by computational issues.

The results are shown in Figures 14.2 to 14.5. These figures correspond to $k = 3,\ 25,\ 45,\ 80$ respectively. The figures show the posterior distribution for the four algorithms listed above, i.e. true (solid line), marginal quantization (dashed line), Markovian quantization (dotted line) and the OLG - MDNF algorithm (dash-dot line).

From all figures we clearly see that the OLG - MDNF algorithm correctly captures the bimodal nature of the posterior distribution. Indeed, in almost all cases, the OLG - MDNF algorithm tracks the "true" nature of the distribution to a high degree of accuracy. This is remarkable given the low quantization density used. On the other

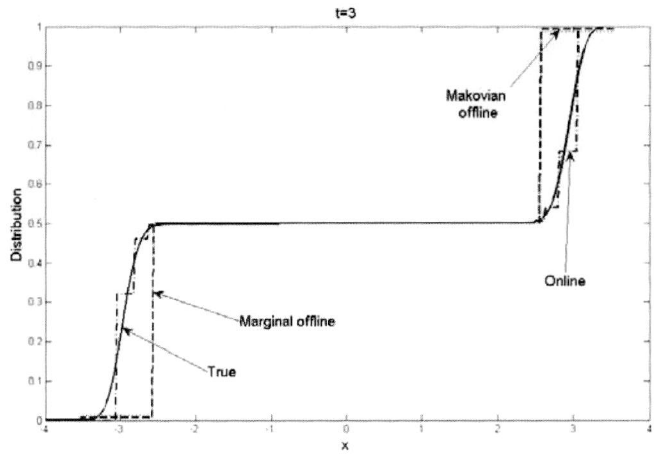

Fig. 14.2. Posterior distribution at time $k = 3$.

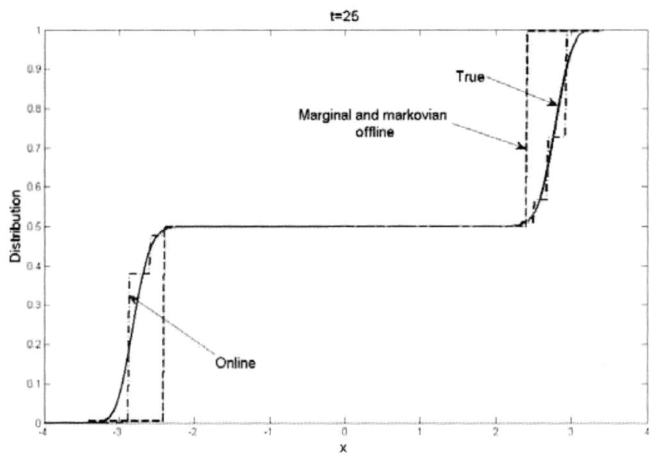

Fig. 14.3. Posterior distribution at time $k = 25$.

hand, the Marginal Quantization algorithm and the Markovian Quantization algorithm give a coarse approximation to the posterior distribution. This is not surprising given that the grid points are chosen off-line based on the open-loop model only.

14.8 Conclusions

This paper has briefly reviewed methods for sequential Bayesian nonlinear filtering with emphasis on methods which use approximations based on minimum distortion quantization. Several new results have been described and a simulation study presented.

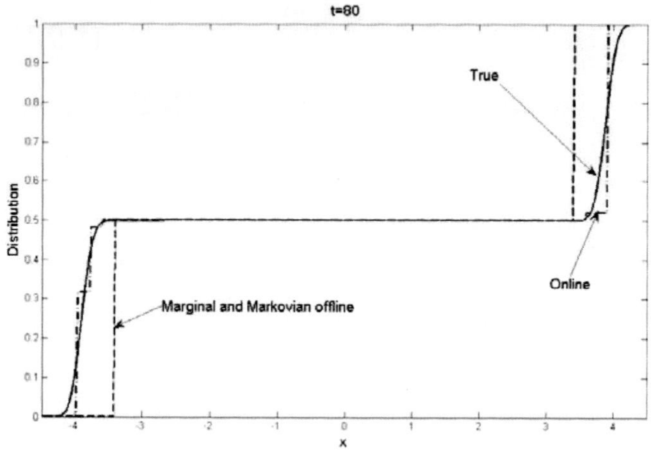

Fig. 14.4. Posterior distribution at time $k = 45$.

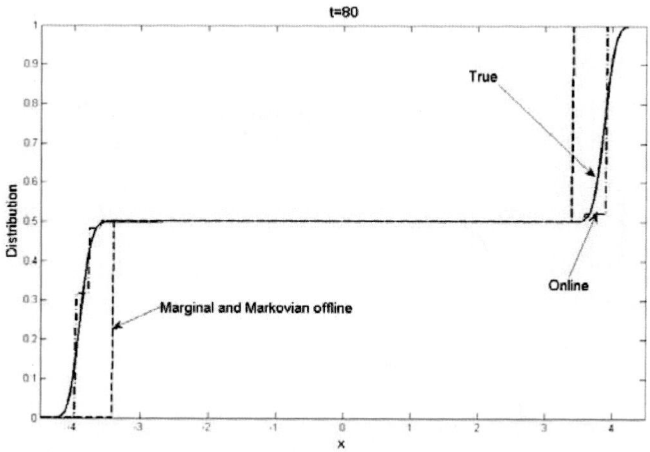

Fig. 14.5. Posterior distribution at time $k = 80$.

References

1. Arulampalam, S., Maskell, S., Gordon, N., Clapp, T.: A Tutorial on Particle Filters for On-line Non-linear/Non-Gaussian Bayesian Tracking. IEEE Trans. On Signal Proces. 50(2), 174–188 (2002)
2. Chen, Z.: Bayesian Filtering: From Kalman Filters to Particle Filters, and Beyond. Available at: `http://users.isr.ist.utl.pt/~jpg/tfc0607/chen_bayesian.pdf`
3. Gersho, A., Gray, R.M.: Vector Quantization and Signal Compression. Kluwer Academic Publishers, London (1992)
4. Goodwin, G.C., Feuer, A., Müller, C.: Gradient interpretation of the Lloyd algorithm in vector quantization. Available at: `http://livesite.newcastle.edu.au/sites/cdsc/profiles/goodwin_feuer_mueller.pdf`

5. Goodwin, G.C., Sin, K.S.: Adaptive Filtering Prediction and Control. Prentice-Hall, Englewood Cliffs (1984)
6. Goodwin, G.C., Østergaard, J., Quevedo, D.E., Feuer, A.: A vector quantization approach to scenario generation for stochastic NMPC. In: Int. Workshop on Assessment and Future Directions of NMPC, Pavia, Italy (September 2008)
7. Gordon, N., Salmond, D., Smith, A.F.M.: Novel Approach to Non-linear and Non-Gaussian Baysian State Estimation. IEE Proceedings-F 140, 107–113 (1993)
8. Graf, S., Luschgy, H.: Foundation of Quantization for Probability Distribution. Lecture Notes in Mathematics, vol. 1730. Springer, Berlin (2000)
9. Hammersley, J.M., Hanscomb, D.C.: Monte Carlo Methods. Chapman&Hall, London (1964)
10. Handschin, J.E., Mayne, D.Q.: Monte Carlo techniques to estimate the conditional expectation in multistage nonlinear filtering. International Journal of Control 9(5), 547–559 (1969)
11. Ho, Y.C., Lee, R.C.K.: A Baysian approach to problems in stochastic estimation and control. IEEE Trans. Automatic Control 9, 333–339 (1964)
12. Kulhavy, R.: Quo vadis, Baysian identification? Int. J. Adaptive Control and Signal Processing 13, 469–485 (1999)
13. Lloyd, S.P.: Least squares quantization in PCM. IEEE Trans. Inform. Theory 28, 127–135 (1982)
14. Müller, C., Goodwin, G.C., Feuer, A.: The gradient and Hessian of distortion measures in vector quantization. Available at: `http://livesite.newcastle.edu.au/sites/cdsc/profiles/mueller_goodwin_feuer.pdf`
15. Pages, G.: A space vector quantization method for numerical integration. Journal of Computational and Applied Mathematics, 89, 19997, 1–38
16. Pages, G., Pham, H.: Optimal quantization methods for nonlinear filtering with discrete-time observations. Bernoulli 5, 893–932 (2005)
17. Rojas, C.R., Goodwin, G.C., Seron, M.M.: Open-cut mine planning via closed loop receding horizon optimal control. In: Sánchez-Peño, R., Quevedo, J., Puig Cayuela, V. (eds.) Identification and Control: The gap between theroy and practice, Springer, Heidelberg (2007)
18. Sellami, A.: Comparative survey on nonlinear filtering methods: the quantization and the particle filtering approaches. J. Statist. Comp. and Simul. 78(2), 93–113 (2008)
19. Sorenson, H.W.: On the development of practical nonlinear filters. Inform. Sci. 7, 253–270 (1974)
20. Sorenson, H.W., Alspach, D.L.: Recursive Bayesian estimation using Gaussian sums. Automatica 7, 465–479 (1971)
21. Wan, E., van der Merwe, R.: The unscented Kalman filter. In: Haykin, S. (ed.) Kalman Filtering and Neural Networks, Wiley, New York (2001)
22. Wang, A.H., Klein, R.L.: Optimal quadrature formula nonlinear estimators. Inform. Sci. 16, 169–184 (1978)
23. West, M., Harrison, J.: Bayesian Forecasting and Dynamic Models, 2nd edn. Springer, New York (1997)

15

Pole Placement with Fields of Positive Characteristic*

Elisa Gorla[1] and Joachim Rosenthal[2]

[1] Department of Mathematics, University of Basel, Basel, Switzerland
[2] Institute of Mathematics, University of Zürich, Zürich, Switzerland

Summary. The pole placement problem belongs to the classical problems of linear systems theory. It is often assumed that the ground field is the real numbers $\mathbb{R}$ or the complex numbers $\mathbb{C}$. The major result over the complex numbers derived in 1981 by Brockett and Byrnes states that arbitrary static pole placement is possible for a generic set of m-inputs, p-outputs and McMillan degree n system as soon as $mp \geq n$. Moreover the number of solutions in the situation $mp = n$ is an intersection number first computed by Hermann Schubert in the 19th century. In this paper we show that the same result with slightly different proofs holds over any algebraically closed field.

15.1 Introduction

Let $\mathbb{F}$ be an arbitrary field and let A,B,C be matrices of size $n \times n$, $n \times m$ and $p \times n$, with entries in $\mathbb{F}$. These matrices define a discrete time dynamical system through the equations:

$$\begin{aligned} x(t+1) &= Ax(t) + Bu(t) \\ y(t) &= Cx(t). \end{aligned} \tag{15.1}$$

An $m \times p$ matrix K with entries in $\mathbb{F}$ defines the feedback law:

$$u(t) = Ky(t). \tag{15.2}$$

Applying (15.2) to the system (15.1), one gets the closed loop system:

$$x(t+1) = (A + BKC)x(t). \tag{15.3}$$

The static output pole placement problem asks for conditions on the matrices A,B,C which guarantee that the characteristic polynomial of the closed loop system, i.e., the characteristic polynomial of the matrix $(A+BKC)$ can be made arbitrary.

We can explain this problem also in terms of the so-called pole placement map. For this, identify the set of monic polynomials of degree n of the form:

*The work of both authors is supported by the Swiss National Science Foundation through grants #123393, #113251, and #107887.

X. Hu et al. (Eds.): Three Decades of Progress in Control Sciences, pp. 215–231, 2010.

$$s^n + a_{n-1}s^{n-1} + \cdots + a_1 s + a_0 \in \mathbb{F}[s]$$

with the vector space $\mathbb{F}^n$. Then we are seeking conditions which guarantee that the pole placement map:

$$\chi_{(A,B,C)} : \mathbb{F}^{m\times p} \longrightarrow \mathbb{F}^n, \quad K \longmapsto \det(sI - A - BKC) \tag{15.4}$$

is surjective, or at least that the image contains a non-empty Zariski-open set.

Many facets of this problem have been studied in the literature and the reader is referred to [2, 10, 14, 15] where also more references to the literature can be found.

If the base field is the complex numbers, then the major result is due to Brockett and Byrnes [1]:

Theorem 15.1.1. *If the base field $\mathbb{F}$ equals $\mathbb{C}$, the complex numbers, then χ is surjective for generic matrices A,B,C if and only if $mp \geq n$. Moreover if $mp = n$ and χ is surjective, then the general fiber $\chi^{-1}(\phi)$ has cardinality*

$$d(m,p) = \frac{1!2!\cdots(p-1)!(mp)!}{m!(m+1)!\cdots(m+p-1)!}. \tag{15.5}$$

In the next section we will go over the proof of Theorem 15.1.1 in the situation when the base field $\mathbb{F}$ is algebraically closed and has characteristic zero. In Section 15.3 we will address the difficulties which occur in positive characteristic. The main result of the paper is a proof that Theorem 15.1.1 holds over any algebraically closed field in the case $n = mp$.

15.2 Connection to Geometry and a Proof of Theorem 15.1.1 in Characteristic Zero

Consider the transfer function $G(s) := C(sI - A)^{-1}B$ and a left coprime factorization:

$$G(s) = D^{-1}(s)N(s) = C(sI - A)^{-1}B.$$

Over any field $\mathbb{F}$ we have the property that the $p \times (m+p)$ matrix $[N(s)\ D(s)]$ has rank p when evaluated at an arbitrary element of the algebraic closure $\overline{\mathbb{F}}$ of $\mathbb{F}$. In other words if $\lambda \in \overline{\mathbb{F}}$ then

$$\operatorname{rank}\,[N(\lambda)\ D(\lambda)] = p.$$

It was the insight of Hermann and Martin [5] to realize that every linear system $G(s)$ naturally defines a rational map into the Grassmann variety $\mathrm{Grass}(p, \mathbb{F}^{m+p})$:

$$h : \mathbb{P}^1 \longrightarrow \mathrm{Grass}(p, \mathbb{F}^{m+p}), \ s \longmapsto \mathrm{rowsp}\,[N(s)\ D(s)].$$

The map h does not depend on the coprime factorization, and two different linear systems $G_1(s)$ and $G_2(s)$ have different associated rational maps. By the previous remark, the map is well defined for every element $\lambda \in \overline{\mathbb{F}}$. For this reason one usually refers to h as the Hermann-Martin map associated to the linear system $G(s)$.

In order to arrive at an algebraic geometric formulation of the pole placement problem, consider a left coprime factorization $G(s) = D^{-1}(s)N(s)$ with the property that $\det(sI - A) = \det D(s)$. Then it is well known that the closed loop characteristic polynomial can also be written as:

$$\det(sI - A - BKC) = \det \begin{bmatrix} I & K \\ N(s) & D(s) \end{bmatrix}. \tag{15.6}$$

Assume now that a desired closed loop characteristic polynomial $\phi(s)$ factors over the algebraic closure as:

$$\phi(s) = \prod_{i=1}^{n}(s - s_i), \; s_i \in \overline{\mathbb{F}}, \; i = 1, \ldots, n.$$

The condition $\det(sI - A - BKC) = \phi(s)$ then translates into the geometric condition:

$$\operatorname{rowsp}[I\ K] \bigcap \operatorname{rowsp}[N(s_i)\ D(s_i)] \neq \{0\}, \; i = 1, \ldots, n.$$

This formulation is closely connected to a theorem due to Hermann Schubert:

Theorem 15.2.1. *Given n p-dimensional subspaces $U_i \subset \mathbb{C}^{m+p}$. If $n \leq mp$, then there is an m-dimensional subspace $V \subset \mathbb{C}^{m+p}$ such that*

$$V \bigcap U_i \neq \{0\}, \; i = 1, \ldots, n. \tag{15.7}$$

Moreover if $n = mp$ and the subspaces U_i are in "general position", then there are exactly $d(m,p)$ (see Equation (15.5)*) different solutions $V \subset \mathbb{C}^{m+p}$ satisfying Condition* (15.7).

Theorem 15.2.1 was derived by Hermann Schubert towards the end of the 19th century [11, 12]. The mathematicians at the time were not convinced with the proofs Schubert was providing. The verification of the statements constituted Hilbert's 15th problem, which he presented at the International Congress of Mathematics in 1900 in Paris. Theorem 15.2.1 has been later verified rigorously and we refer to Kleiman's survey article [4].

It is not completely obvious how the geometric result of Schubert implies Theorem 15.1.1 of Brockett and Byrnes. The following questions have to be addressed:

1. Given an m-dimensional subspace $\operatorname{rowsp}[K_1\ K_2] \subset \mathbb{C}^{m+p}$, where K_1 is an $m \times m$ matrix and K_2 is an $m \times p$ matrix. Assume $\operatorname{rowsp}[K_1\ K_2] \subset \mathbb{C}^{m+p}$ is a geometric solution, i.e.,
$$\det \begin{bmatrix} K_1 & K_2 \\ N(s_i) & D(s_i) \end{bmatrix} = 0, \; i = 1, \ldots, n. \tag{15.8}$$
Does it follow that $[K_1\ K_2]$ is row equivalent to $[I\ K]$ and K represents a feedback law? For this to happen it is necessary and sufficient that K_1 is invertible.

2. Assume $\text{rowsp}[K_1\, K_2] \subset \mathbb{C}^{m+p}$ is a geometric solution in the sense of (15.8). Does it follow that $\det \begin{bmatrix} I & K \\ N(s) & D(s) \end{bmatrix}$ is NOT the zero polynomial?
3. How is it possible to deal with multiple roots?

These questions were all addressed in [1]. A key ingredient is the notion of non-degenerate system.

Definition 15.2.1. *An m-input, p-output linear system $G(s) = D^{-1}(s)N(s)$ is called degenerate, if there exist an $m \times m$ matrix K_1 and an $m \times p$ matrix K_2 such that $[K_1\, K_2]$ has full rank m and*

$$\det \begin{bmatrix} K_1 & K_2 \\ N(s) & D(s) \end{bmatrix} = 0. \tag{15.9}$$

A system $G(s)$ which is not degenerate will be called non-degenerate.

In more geometric terms, the Hermann Martin curve associated to a non-degenerate system does not lie in any Schubert hyper-surface.

If $[N(s)\, D(s)]$ represents a non-degenerate system of McMillan degree n, then

$$\det \begin{bmatrix} K_1 & K_2 \\ N(s) & D(s) \end{bmatrix} \neq 0$$

for any $[K_1\, K_2]$ of full rank. If in addition $[K_1\, K_2]$ is a geometric solution, then Condition (15.8) is satisfied and $\det \begin{bmatrix} K_1 & K_2 \\ N(s) & D(s) \end{bmatrix}$ is a polynomial of degree at least n. All the full size minors of $[N(s)\, D(s)]$ have degree less than $n-1$, with the exception of the determinant of $D(s)$, which has degree n. So the polynomial $\det \begin{bmatrix} K_1 & K_2 \\ N(s) & D(s) \end{bmatrix}$ cannot have degree n unless K_1 is invertible. Hence it follows that a geometric solution for a non-degenerate system results in a feedback solution $u = Ky$ on the systems theory side.

Non-degenerate systems are therefore very desirable. The following theorem was formulated in [1] in the case when the base field is the complex numbers.

Theorem 15.2.2. *Let $\mathbb{F}$ be an arbitrary field. If $n < mp$ then every system (A,B,C) defined over $\mathbb{F}$ with m-inputs, p-outputs and McMillan degree n is degenerate. If $\mathbb{F}$ is an algebraically closed field and $n \geq mp$, then a generic system $(A,B,C) \in \mathbb{F}^{n^2+n(m+p)}$ is non-degenerate.*

The proof of the first part of the statement follows from basic properties of coprime factorizations of transfer functions. Indeed let the $p \times (m+p)$ polynomial matrix $M(s) = [N(s)\, D(s)]$ represent an m-inputs, p-outputs system of McMillan degree $n < mp$. Then, possibly after some row reductions, we find a row of $M(s)$ whose degree is at most $m-1$. Using this row one readily constructs a full rank $m \times (m+p)$ matrix such that (15.9) holds. This shows that $G(s) = D^{-1}(s)N(s)$ is degenerate.

The second part of the statement, namely that a generic system defined over $\mathbb{F}$ is non-degenerate, will be established through a series of lemmas. Here $\mathbb{F}$ is an algebraically closed field of characteristic zero.

Notice that it is enough to show that the set of degenerate systems is contained in a proper algebraic set of $\mathbb{F}^{n^2+n(m+p)}$. In order to prove this, we establish an algebraic relation between the polynomial matrix $[N(s)\ D(s)]$ and the matrices (A,B,C). The following lemma is an ingredient of classical realization theory. The proof and the concept of basis matrix is found in [9].

Lemma 15.2.1. *Assume $G(s)=D^{-1}(s)N(s)$ is a left coprime factorization of a $p\times m$ transfer function of McMillan degree n. Then for every $p\times n$ basis matrix $X(s)$ there are matrices $A\in\mathbb{F}^{n\times n}, B\in\mathbb{F}^{n\times m}$ and $C\in\mathbb{F}^{p\times n}$ such that:*

$$\ker_{\mathbb{F}(s)}\left[X(s)\mid N(s)\mid D(s)\right] = \operatorname{im}_{\mathbb{F}(s)}\begin{bmatrix}(sI_n-A) & B\\ 0 & I_m\\ C & 0\end{bmatrix}. \tag{15.10}$$

Furthermore (A,B,C) is a minimal realization of $G(s)$, i.e.,

$$G(s)=C(sI-A)^{-1}B,$$

and for every minimal realization (A,B,C) of $G(s)$ there exists a basis matrix $X(s)$ such that (15.10) *is satisfied.*

As pointed out in [9], for certain basis matrices $X(s)$ it is possible to compute (A,B,C) just "by inspection".

Using the previous lemma, one readily establishes the following:

Lemma 15.2.2. *Assume that (A,B,C) is a minimal realization of $G(s)=D^{-1}(s)N(s)$ and* $\det(sI-A)=\det D(s)$. *Then*

$$\det\begin{bmatrix}K_1 & K_2\\ N(s) & D(s)\end{bmatrix}=\det\begin{bmatrix}sI-A & B\\ K_2C & K_1\end{bmatrix}. \tag{15.11}$$

As before, identify an m-inputs, p-outputs system (A,B,C) of McMillan degree n with a point of $\mathbb{F}^{n^2+n(m+p)}$. Let S be the set:

$$\left\{((K_1,K_2),(A,B,C))\in \operatorname{Grass}(m,\mathbb{F}^{m+p})\times\mathbb{F}^{n^2+n(m+p)} : \det\begin{bmatrix}sI-A & B\\ K_2C & K_1\end{bmatrix}=0\right\}. \tag{15.12}$$

Since $\operatorname{Grass}(m,\mathbb{F}^{m+p})$ is a projective variety, the projection of S onto $\mathbb{F}^{n^2+n(m+p)}$ is an algebraic set. This follows from the main theorem of elimination theory (see, e.g., [6]). We have therefore established that the set of degenerate systems inside $\mathbb{F}^{n^2+n(m+p)}$ is an algebraic set. We establish the genericity result as soon as we can show the existence of one non-degenerate system, under the assumption that $n\geq mp$.

Remark 15.2.1. In the case of proper transfer functions, the dimension of the coincidence set S was computed in [8, Theorem 5.5]. With this result it was then shown in [8] that the set of non-degenerate systems inside the quasi-projective variety of proper transfer functions contains a dense Zariski-open set as soon as $n\geq mp$.

Definition 15.2.2. *Let* $\mathbb{F}$ *be an algebraically closed field of characteristic* 0. *The* osculating normal curve $C_{p,m}$ *is the closure of the image of the morphism*

$$\begin{aligned} \mathbb{F} &\longrightarrow \mathrm{Grass}(p,\mathbb{F}^{m+p}) \\ s &\longmapsto \mathrm{rowsp}\left[\tfrac{d}{d^i}s^j\right]_{i=0,\dots,p-1;\, j=0,\dots,m+p-1.} \end{aligned} \tag{15.13}$$

We denote by d/d^i the i-th derivative with respect to s, i.e.,

$$\frac{d}{d^i}s^j = \begin{cases} \prod_{k=0}^{i-1}(j-k)s^{j-i} & \text{if } j \geq i \\ 0 & \text{if } j < i. \end{cases}$$

The osculating normal curve is an example of a non-degenerate curve in the Grassmannian $\mathrm{Grass}(p,\mathbb{F}^{m+p})$. An elementary matrix proof of this fact was first given in [7]. We will say more about it in the next section. If $n > mp$ one constructs a non-degenerate system by simply multiplying the last column of the matrix representing the osculating normal curve by s^{n-mp}.

In the case $p = 1$, this is the rational normal curve of degree m in $\mathbb{P}^m \cong \mathrm{Grass}(1,\mathbb{F}^{m+1})$. In the case $m = 1$, the osculating normal curve is isomorphic to the rational normal curve of degree p in $\mathbb{P}^p \cong \mathrm{Grass}(p,\mathbb{F}^{p+1})$.

So far we have shown that if $mp \geq n$, then a generic system is non-degenerate. Moreover, if $n = mp$, the system is non-degenerate and the desired closed loop polynomial has distinct roots, then pole placement is possible with $d(m,p)$ different feedback compensators.

It remains to be addressed the question of multiple roots in the closed loop polynomial. This has been done in the literature by lifting the pole placement map (15.4) from $\mathbb{F}^{m\times p}$ to the Grassmann variety $\mathrm{Grass}(m,\mathbb{F}^{m+p})$. We follow the arguments in [10].

We can expand the closed loop characteristic polynomial as:

$$\det\begin{bmatrix} K_1 & K_2 \\ N(s) & D(s) \end{bmatrix} = \sum_\alpha k_\alpha g_\alpha(s), \tag{15.14}$$

where k_α are the Plücker coordinates of $\mathrm{rowsp}[K_1\ K_2] \in \mathrm{Grass}(m,\mathbb{F}^{m+p})$ and where the polynomials $g_\alpha(s)$ are (up to sign) the corresponding Plücker coordinates of $[N(s)\ D(s)]$. Let $\mathbb{P}^N$ be the projective space $\mathbb{P}(\wedge^m\mathbb{F}^{m+p})$ and let

$$E_{(A,B,C)} := \left\{k \in \mathbb{P}^N \mid \sum_\alpha g_\alpha(s)k_\alpha = 0\right\}.$$

As shown in [15], one has an extended pole placement map with the structure of a central projection:

$$L_{(A,B,C)} : \mathbb{P}^N - E_{(A,B,C)} \longrightarrow \mathbb{P}^n, \qquad k \longmapsto \sum_\alpha k_\alpha g_\alpha(s). \tag{15.15}$$

A system $[N(s)\ D(s)]$ is non-degenerate if and only if:

$$E_{(A,B,C)} \cap \mathrm{Grass}(m, \mathbb{F}^{m+p}) = \{\}.$$

For a non-degenerate system, the extended pole placement map $L_{(A,B,C)}$ induces a finite morphism:

$$\hat{\chi}_{(A,B,C)} : \mathrm{Grass}(m, \mathbb{F}^{m+p}) \longrightarrow \mathbb{P}^n, \quad \mathrm{rowsp}[K_1\ K_2] \longmapsto \det \begin{bmatrix} K_1 & K_2 \\ N(s) & D(s) \end{bmatrix}. \qquad (15.16)$$

The inverse image of a closed loop polynomial $\phi(s) \in \mathbb{P}^n$ under the map $L_{(A,B,C)}$ is a linear space which intersects the Grassmann variety $\mathrm{Grass}(m, \mathbb{F}^{m+p})$ in as many points (counted with multiplicity) as the degree of the Grassmann variety. This is equal to Schubert's number $d(m,p)$.

This completes the proof of Theorem 15.1.1 of Brockett and Byrnes in the case $n = mp$ not only for the field of complex numbers, but also in the case when the base field is algebraically closed and has characteristic zero. In Remark 15.3.7 we will discuss how to extend the proof to the case when $n < mp$.

In the next section we will discuss how to extend Theorem 15.2.2 to the case of an algebraically closed field of positive characteristic. We will show that it is much more tricky to establish the existence of non-degenerate systems in the case when the base field has positive characteristic.

15.3 A Proof of Theorem 15.1.1 in Positive Characteristic

Let $\mathbb{F}$ be an algebraically closed field of characteristic $q > 0$. Lemma 15.2.1 and Lemma 15.2.2 as formulated in the last section only depend on techniques from linear algebra and are true over an arbitrary field, so in particular over an algebraically closed field.

If X is a projective variety, Y is a quasi-projective variety, and $S \subset X \times Y$ is an algebraic subset, then the projection of S onto Y is a Zariski-closed subset of Y (see, e.g., [13, Chapter I, Section 5.2]). This shows that Theorem 15.2.2 also holds over an algebraically closed field. In order to establish Theorem 15.1.1, we have to show that there exists at least one non-degenerate system for any choice of the parameters $p, m, n \geq mp$. We will also show that a generic fiber contains $d(m,p)$ elements when $n = mp$.

The last statement is true as soon as the extended pole placement map $\hat{\chi}_{(A,B,C)}$ is separable [13, Chapter II, Section 6.3]. This is indeed the case: $\hat{\chi}_{(A,B,C)}$ can be seen as the composition of the Plücker embedding (which just involves the computation of minors) and the linear map $L_{(A,B,C)}$. Both maps are separable, and we conclude therefore that the composition map is separable.

So there remains the problem of establishing the existence of non-degenerate systems in the case $n \geq mp$. As we will show next, the osculating normal curve may be degenerate in characteristic $q > 0$. In Section 15.3.2 we will provide alternative examples of non-degenerate systems in positive characteristic, while in Section 15.3.3 we will discuss the case of finite fields.

15.3.1 The Osculating Normal Curve

Although the osculating normal curve is defined over a field of characteristic zero, its reduction modulo q defines a curve in $\mathrm{Grass}(p, \mathbb{F}^{m+p})$, which can again be regarded as the closure of the image of morphism (15.13). If $p = 1$, the curve is the rational normal curve of degree m in $\mathrm{Grass}(1, \mathbb{F}^{m+1}) \cong \mathbb{P}^m$. In particular it is non-degenerate. Notice however that the reduction of the osculating normal curve is degenerate whenever $q \leq p + m$, provided that $p \geq 2$. This is easily checked if $q < p$, since in this case the $(q+1)$-st row of the matrix defining the curve is identically zero. If $p \leq q \leq p+m$, consider the minor of the sub-matrix consisting of columns $1, \ldots, p-1, q$. This sub-matrix has the form:

$$\begin{bmatrix} 1 & s & \ldots & s^{p-2} & s^q \\ 0 & 1 & & (p-2)s^{p-3} & 0 \\ \vdots & & \ddots & \vdots & \vdots \\ \vdots & & & (p-2)! & \vdots \\ 0 & \ldots & & 0 & 0 \end{bmatrix}$$

It follows that the corresponding minor is zero. By choosing a compensator $[K_1 \; K_2]$ whose sub-matrix consisting of the "complementary columns" $p, p+1, \ldots, q-1, q+1, \ldots, p+m$ is the identity matrix and where all other elements are zero, one verifies that the osculating normal curve is also degenerate in this situation.

Remark 15.3.1. If at least one minor of the matrix $[N(s) \; D(s)]$ is 0, then the system $G(s) = D^{-1}(s)N(s)$ is degenerate.

Notice that if $q \gg 0$, then the reduction modulo q of the osculating normal curve is non-degenerate. This reflects the usual fact that "fields with large enough characteristic behave like fields of characteristic zero".

The appearance of many zero entries in the matrix over a field $\mathbb{F}$ of "small" positive characteristic q is due to the fact that many derivatives vanish. More precisely, let $h \in \{0, \ldots, q-1\}$ s.t. $j = h \bmod q$. Then

$$\frac{d}{d^i} s^j = \begin{cases} \prod_{k=0}^{i-1}(j-k)s^{j-i} & \text{if } h \geq i \\ 0 & \text{if } h < i \end{cases}$$

This was one of the reasons that motivated Hasse to introduce the following concept.

Definition 15.3.1. *The i-th* Hasse derivative *of a polynomial $u(s) = \sum_{j=0}^{d} u_j s^j$ is defined as:*

$$\frac{\partial}{\partial^i} u(s) = \sum_{j=i}^{d} \binom{j}{i} u_i s^{j-i}.$$

Observe that in characteristic 0 one has

$$\frac{\partial}{\partial^i} = \frac{1}{i!} \frac{d}{d^i}.$$

Moreover, none of the Hasse derivatives vanishes identically for all polynomials, regardless of the characteristic of the base field, whereas in characteristic $q > 0$, the i-th derivative of any polynomial is identically zero for all $i \geq q$.

It is therefore natural that we define the osculating normal curve in positive characteristic using the Hasse derivative instead of the normal derivative.

Definition 15.3.2. *Let* $\mathbb{F}$ *be an algebraically closed field of characteristic* $q > 0$. *The* osculating normal curve $C_{p,m}$ *is the closure of the image of the morphism*

$$\begin{aligned} \mathbb{F} &\longrightarrow \text{Grass}(p, \mathbb{F}^{m+p}) \\ s &\longmapsto \text{rowsp}\left[\frac{\partial}{\partial^i} s^j\right]_{i=0,\ldots,p-1;\, j=0,\ldots,m+p-1.} \end{aligned} \tag{15.17}$$

where ∂ *denotes the Hasse derivative.*

For $p \leq 2$ the definition agrees with the one given at the beginning of this section. In particular, for $p = 1$ we have a non-degenerate rational normal curve of degree m in $\text{Grass}(1, \mathbb{F}^{m+1}) \cong \mathbb{P}^m$. Notice also that the curve is well defined even if $p > q$, as we do not generate a zero row in the defining matrix.

Unfortunately, even with this adapted definition the osculating normal curve is degenerate for many choices of the parameters, as the following result points out:

Proposition 15.3.1. *Let* $\mathbb{F}$ *be an algebraically closed field of characteristic* $q > 0$. *Assume that* $q \leq m$. *Then the osculating normal curve* $C_{p,m}$ *is degenerate.*

Proof. By Remark 15.3.1, it suffices to show that one of the minors of the matrix:

$$\left[\binom{j}{i} s^{j-i}\right]_{i=0,\ldots,p-1;\, j=0,\ldots,m+p-1}$$

is zero. Consider the sub-matrix consisting of columns $0, \ldots, p-2, c$, where c is a multiple of q, $c \in \{p+1, \ldots, p+m\}$. The corresponding minor is:

$$\det\left[\binom{j}{i}\right]_{i=0,\ldots,p-1;\, j=0,\ldots,p-2,c} s = \binom{c}{p-1} s = 0.$$

The first equality follows from the observation that the matrix is upper triangular with ones on the diagonal, except for the entry in the lower right corner which equals $\binom{c}{p-1}$. □

Remark 15.3.2. If $q \mid p$ and $m \geq p$, the minor of the sub-matrix consisting of columns $p-1, p+1, \ldots, 2p-1$ equals

$$\det\left[\binom{j}{i}\right]_{i=0,\ldots,p-1;\, j=p-1,p+1,\ldots,2p-1} s^{p^2-1} = ps^{p^2-1} = 0.$$

The first equality follows from Lemma 9 in [3].

Remark 15.3.3. Degeneracy of the osculating normal curve over the field $\mathbb{F}_q$ with $q \leq \max\{p, m\}$ also follows from Theorem 15.3.1.

In Proposition 15.3.1 we saw that the osculating normal curve may be degenerate over a field $\mathbb{F}$ of positive characteristic q. Notice however that the curve may be non-degenerate for certain choices of the parameters p,m. The following example shows, e.g., that if the field $\mathbb{F}$ has characteristic 2, $m = 1$ and p is odd, then $C_{p,1}$ is non-degenerate.

Example 15.3.1. The curve $C_{p,1}$ is the closure of the image of the morphism

$$\begin{aligned} \mathbb{F} &\longrightarrow \mathrm{Grass}(p, \mathbb{F}^{p+1}) \\ s &\longmapsto \mathrm{rowsp}\left[\tbinom{j}{i} s^{j-i}\right]_{i=0,\dots,p-1;\, j=0,\dots,p.} \end{aligned}$$

The minors of the matrix that defines the morphism are

$$\binom{p}{i} s^i \quad \text{for } i = 0, \dots, p.$$

Hence the curve is non-degenerate if and only if all the minors are non-zero, if and only if

$$q \nmid \binom{p}{i} \quad \text{for any } i = 0, \dots, p.$$

Over a field of even characteristic, this is in fact the only case when the osculating normal curve is non-degenerate.

Corollary 15.3.1. *Let $\mathbb{F}$ be an algebraically closed field of characteristic 2. Then the osculating normal curve $C_{p,m}$ is degenerate, unless $m = 1$ and p is odd. In the latter case, $C_{p,1}$ is isomorphic to the rational normal curve of degree p in $\mathbb{P}^p$.*

15.3.2 Monomial Systems and MDS Matrices

Definition 15.3.3. *A matrix $M(s) = [N(s)\ D(s)]$ is* monomial *if the minors of all sizes of $M(s)$ are monomials. A system $G(s)$ associated to a monomial matrix $M(s)$ is called a* monomial system.

A monomial matrix $M(s) = [\alpha_{i,j} s^{d_{i,j}}]$ is determined by:

- the *coefficient matrix* $M = [\alpha_{i,j}]$,
- the *degree matrix* $[d_{i,j}]$.

The degree matrix has the property that $d_{i,j} + d_{k,l} = d_{i,l} + d_{k,j}$ for all i, j, k, l.

Example 15.3.2. The osculating normal curve defines a monomial system.

Example 15.3.3. Let $\mathbb{F}$ be a field which contains at least three distinct elements $0, 1, \alpha$. The matrix

$$M(s) = \begin{bmatrix} 1 & 0 & s^2 & \alpha s^3 \\ 0 & 1 & s & s^2 \end{bmatrix}$$

has minors

$$1, s, s^2, -s^2, \alpha s^3, (1-\alpha)s^4.$$

It therefore follows that $M(s)$ is a monomial matrix. A direct calculation shows that this system is non-degenerate.

Definition 15.3.4. *A matrix M with entries in $\mathbb{F}$ is* Maximum Distance Separable (MDS) *if all its maximal minors are non-zero.*

Remark 15.3.4. In coding theory a linear code $C \subset \mathbb{F}^n$ is called an MDS code if all the maximal minors of the generator matrix of C are non-zero. This explains the choice of the name for these matrices.

Remark 15.3.5. Let $M(s)$ be a monomial matrix. If the system associated to $M(s)$ is non-degenerate, then M is an MDS matrix. This follows from Remark 15.3.1. It is not always the case that a monomial matrix $M(s)$ with MDS coefficient matrix M is non-degenerate.

An example of degenerate $M(s)$ with MDS coefficient matrix is given in the following example.

Example 15.3.4. Let $\mathbb{F} = \mathbb{F}_5$, the finite field of 5 elements. The following monomial system defined by the matrix

$$M(s) = \begin{bmatrix} 1 & s & s & s^2 \\ 0 & 1 & 2 & 3s \end{bmatrix}$$

is left prime and has an MDS matrix as coefficient matrix. Nonetheless the system is degenerate as, e.g.,

$$[K_1\ K_2] := \begin{bmatrix} 0 & 1 & 2 & 0 \\ 0 & 0 & 0 & 1 \end{bmatrix}$$

results in the zero characteristic polynomial.

In the next theorem we show that an MDS matrix of given size defined over a field $\mathbb{F}$ exists only if the ground field $\mathbb{F}$ has enough elements.

Theorem 15.3.1. *Let $p,m \geq 2$ and let $M(s)$ be a monomial matrix of size $p \times (m+p)$ defined over a field $\mathbb{F}$ with q elements. If $q \leq \max\{p,m\}$, then $M(s)$ is degenerate.*

Proof. If $M(s)$ is non-degenerate, then its coefficient matrix M is MDS. Let $M^\perp$ be an $m \times (p+m)$-matrix defined over $\mathbb{F}$, such that $\text{rowsp}(M) = \ker(M^\perp)$. Let $C^\perp$ be the dual code of C. The generator matrix of $C^\perp$ is then $M^\perp$. It is well known that C is MDS if and only if $C^\perp$ is MDS. Therefore, $M^\perp$ is an MDS matrix.

We want to show that if M (resp. $M^\perp$) is MDS of size $p \times (m+p)$ (resp. $m \times (m+p)$) defined over $\mathbb{F}_q$, then $q \geq \max\{p+1, m+1\}$. The statement is symmetric in p,m, hence we can assume without loss of generality that $2 \leq p \leq m$. It suffices to prove that $q \geq m+1$.

We first consider the case $p = 2$. Since M is MDS, every pair of columns must be linearly independent. Over a field of q elements, there are $q^2 - 1$ choices for the first column, $q^2 - q$ for the second, $q^2 - 2(q-1) - 1$ choices for the third, and so forth. Since there are $q^2 - (m+1)(q-1) - 1$ choices for the $m+2$-nd column, it must be $q^2 - (m+1)(q-1) - 1 = (q-1)(q-m) \geq 1$, hence $q \geq m+1$.

For an arbitrary p, we can assume that the matrix M is of the form $[I_p\ A]$, where I_p is the $p \times p$ identity matrix and A is a matrix of size $p \times m$. The MDS property of M translates into the property that all the minors of all sizes of A are non-zero. Consider the submatrix N obtained from M by deleting the last $p-2$ rows and the columns $3,\dots,p$, $N=[I_2\ B]$ where B consists of the first two rows of A. N is a $2\times(m+2)$ MDS matrix, since all the minors of all sizes of B are non-zero. It follows that $q\geq m+1$ for every $p\geq 1$. $\square$

From the proposition it follows, e.g., that every monomial $M(s)$ of size $2\times(m+2)$ defined over $\mathbb{F}_2$ is degenerate, unless $m=1$. Clearly, there may be non-degenerate matrices which are not monomial. E.g., the following is an example of a 2×4 system defined over $\mathbb{F}_2$ which is non-degenerate:

Example 15.3.5. Consider the system defined over $\mathbb{F}_2$ by the matrix

$$M(s)=\begin{bmatrix}0 & s & s+1 & s^2\\ 1 & s^2+1 & 1 & s\end{bmatrix}.$$

The minors, listed in lexicographic order, are

$$s, s+1, s^2, (s^3+s+1), s^4, s.$$

A direct computation shows that the system is non-degenerate.

We conclude the paper with the main result.

Theorem 15.3.2. *Let $M(s)=[N(s)\ D(s)]$ be a monomial system having an MDS coefficient matrix M of the form $M=[I_p\ R]$. Let the degrees of the coefficient matrix be $d_{i,j}=j-i$ if $j\geq i$ and zero else. Then $M(s)$ is non-degenerate of degree mp.*

Proof. Denote by α a multi-index $\alpha=(\alpha_1,\dots,\alpha_p)$ with the property that

$$1\leq\alpha_1<\cdots\cdots<\alpha_p\leq m+p.$$

Denote by $m_1(s),\dots,m_p(s)$ the p row vectors of $M(s)$ and denote by $e_1,\dots,e_{m+p}$ the canonical basis of $\mathbb{F}^{m+p}$. One readily verifies that the Plücker expansion of $M(s)$ has the form:

$$m_1(s)\wedge\ldots\wedge m_p(s)=\sum_{\alpha\in\left\{{n \atop p}\right\}} m_\alpha e_{\alpha_1}\wedge\ldots\wedge e_{\alpha_p}s^{|\alpha|}$$

where $|\alpha|:=\sum_{i=1}^p(\alpha_i-i)$ and m_α is the minor of M corresponding to the columns $\alpha_1,\dots,\alpha_p$.

The multi-indices α have a natural partial order, coming from componentwise comparison of their entries. If $\beta=(\beta_1,\dots,\beta_p)$ is a multi-index, then one defines:

$$\alpha\leq\beta\ :\Longleftrightarrow\ \alpha_i\leq\beta_i \text{ for } i=1,\dots,p.$$

By contradiction assume now that $M(s)$ is degenerate. Let $[K_1\ K_2]$ be a compensator which leads to the closed loop characteristic polynomial zero:

$$\det\begin{bmatrix}K_1 & K_2\\ N(s) & D(s)\end{bmatrix}=\sum_\alpha k_\alpha g_\alpha(s)=0. \tag{15.18}$$

In the last expansion k_α denotes up to sign the $m \times m$ minor of $[K_1\ K_2]$ corresponding to the columns $1 \leq \hat{\alpha}_1 < \ldots < \hat{\alpha}_m \leq (m+p)$, $\hat{\alpha}_i \notin \{\alpha_1, \ldots, \alpha_p\}$.

$[K_1\ K_2]$ has a well defined row reduced echelon form with Pivot indices $\hat{\beta} = (\hat{\beta}_1, \ldots, \hat{\beta}_m)$. It follows that $k_\alpha = 0$ for $\alpha \not\leq \beta$. But this means that the term $m_\beta s^{|\beta|}$ cannot cancel in the expansion (15.18) and this is a contradiction. $M(s)$ is therefore non-degenerate. □

Remark 15.3.6. If $n > mp$ choose $d_{i,m+p} = n - mp + m + p - i$ in order to obtain once more a non-degenerate system of degree n.

By establishing the existence of a non-degenerate system, we have shown that Theorem 15.1.1 holds true for any algebraically closed field for $n = mp$.

Remark 15.3.7. In order to prove Theorem 15.1.1 in the situation when $n < mp$, one can show that for a generic system (A, B, C) the set of dependent compensators, i.e., the set of compensators which results in a zero closed loop characteristic polynomial, has minimum possible dimension, namely $mp - n - 1$. This is clearly sufficient to establish the result. In order to prove this statement, one can proceed in two ways. Either one shows that the condition is algebraic and constructs an example of a system of degree n satisfying the condition. Alternativeley one shows that the coincidence set S introduced in (15.12) has dimension $n^2 + n(m+p) + mp - n - 1$. The generic fiber of the projection onto the second factor has then dimension $mp - n - 1$. This last argument was developed for the dynamic pole placement problem in [8].

15.3.3 Non-degenerate Systems over Finite Fields

In this last subsection, we show that in general non-degeneracy does not guarantee that the pole placement map is surjective over a finite field.

Theorem 15.3.3. *Let $\mathbb{F}_2$ be the binary field. Then no non-degenerate system defined over $\mathbb{F}_2$ induces an onto pole placement map:*

$$\mathrm{Grass}(2, \mathbb{F}_2^4) \longrightarrow \mathbb{P}^4(\mathbb{F}_2).$$

Proof. Let $M(s)$ be a non-degenerate matrix with entries in $\mathbb{F}_2[s]$. Let $\mathbb{F}$ denote the algebraic closure of $\mathbb{F}_2$ and let

$$\chi : \mathrm{Grass}(2, \mathbb{F}^4) \longrightarrow \mathbb{P}^4(\mathbb{F})$$

be the pole placement map associated to $M(s)$ over $\mathbb{F}$. χ is a morphism, since $M(s)$ is non-degenerate. We will now show that the restriction of χ to $\mathbb{F}_2$-rational points

$$\mathrm{Grass}(2, \mathbb{F}_2^4) \longrightarrow \mathbb{P}^4(\mathbb{F}_2)$$

is never surjective.

Let $\mathrm{rowsp}(A) \in \mathrm{Grass}(2, \mathbb{F}_2^4)$. Denote by $A_{i,j}$ the determinant of the sub-matrix of A consisting of columns i and j. Then:

$$\chi(A) = \left[\sum_{i<j} \chi_{ijk} A_{i,j} \right]_{k=0,\ldots,4}$$

where

$$\det \begin{bmatrix} M(s) \\ A \end{bmatrix} = \sum_{k=0}^{4} \sum_{i<j} \chi_{ijk} A_{i,j} s^k.$$

Since the system is non-degenerate, the 5×6 matrix:

$$C = \begin{bmatrix} \chi_{120} & \cdots & \chi_{340} \\ \vdots & & \vdots \\ \chi_{124} & \cdots & \chi_{344} \end{bmatrix}$$

has full rank, hence its kernel is 1-dimensional and generated by a unique element of $\mathbb{F}_2^6$. By non-degeneracy, the generator of the kernel corresponds to a point in $\mathbb{P}^6(\mathbb{F}_2)$ which does not belong to $\mathrm{Grass}(2,\mathbb{F}^4)$. Hence we have the following possibilities for the generator of $\ker C$:

$$(1,0,0,0,0,1),(0,1,0,0,1,0),(0,0,1,1,0,0),(1,1,1,1,1,1),$$

$$(1,1,0,0,0,1),(1,0,1,0,0,1),(1,0,0,1,0,1),(1,0,0,0,1,1),$$

$$(1,1,0,0,1,0),(0,1,1,0,1,0),(0,1,0,1,1,0),(0,1,0,0,1,1),$$

$$(1,0,1,1,0,0),(0,1,1,1,0,0),(0,0,1,1,1,0),(0,0,1,1,0,1),$$

$$(0,0,1,1,1,1),(0,1,0,1,1,1),(0,1,1,0,1,1),(0,1,1,1,0,1),$$

$$(1,0,0,1,1,1),(1,0,1,0,1,1),(1,0,1,1,1,0),(1,1,0,1,0,1),$$

$$(1,1,0,1,1,0),(1,1,1,0,0,1),(1,1,1,0,1,0),(1,1,1,1,0,0).$$

Observe that the problem is symmetric with respect to the following changes of basis of $\mathbb{F}_2^6 = \langle e_{12},\ldots,e_{34}\rangle$ (which correspond to automorphisms of $\mathrm{Grass}(2,\mathbb{F}^4)$) and composition thereof:

- exchange e_{12} and e_{34} and leave the rest unaltered,
- exchange e_{13} and e_{24} and leave the rest unaltered,
- exchange e_{14} and e_{23} and leave the rest unaltered,
- exchange e_{12} and e_{13}, exchange e_{34} and e_{24},
- exchange e_{12} and e_{14}, exchange e_{34} and e_{23},
- exchange e_{13} and e_{14}, exchange e_{24} and e_{23}.

Hence, reducing to the analysis of the following possibilities is non-restrictive:

$$(1,0,0,0,0,1),(1,1,0,0,0,1),(0,0,1,1,1,1),(1,1,1,1,1,1).$$

Up to a change of coordinates in $\mathbb{P}^4$, we may assume that the corresponding matrix C is respectively:

$$\begin{bmatrix} 1&0&0&0&0&1 \\ 0&1&0&0&0&0 \\ 0&0&1&0&0&0 \\ 0&0&0&1&0&0 \\ 0&0&0&0&1&0 \end{bmatrix}, \begin{bmatrix} 1&0&0&0&0&1 \\ 0&1&0&0&0&1 \\ 0&0&1&0&0&0 \\ 0&0&0&1&0&0 \\ 0&0&0&0&1&0 \end{bmatrix}, \begin{bmatrix} 1&0&0&0&0&0 \\ 0&1&0&0&0&0 \\ 0&0&1&0&0&1 \\ 0&0&0&1&0&1 \\ 0&0&0&0&1&1 \end{bmatrix}, \begin{bmatrix} 1&0&0&0&0&1 \\ 0&1&0&0&0&1 \\ 0&0&1&0&0&1 \\ 0&0&0&1&0&1 \\ 0&0&0&0&1&1 \end{bmatrix}.$$

Analyzing each case, it is now easy to prove that the corresponding χ is not onto. E.g., in the first case we have:

$$\chi(A) = [A_{1,2} + A_{3,4}, A_{1,3}, A_{1,4}, A_{2,3}, A_{2,4}]$$

which is surjective if and only if the equations

$$A_{1,2} + A_{3,4} = \alpha_0, A_{1,3} = \alpha_1, A_{1,4} = \alpha_2, A_{2,3} = \alpha_3, A_{2,4} = \alpha_4,$$

$$A_{1,2}A_{3,4} + A_{1,3}A_{2,4} + A_{1,4}A_{2,3} = 0$$

have a solution in $\mathbb{F}_2^6$ for any choice of $[\alpha_0 : \ldots : \alpha_4] \in \mathbb{P}^4(\mathbb{F}_2)$. Letting $x = A_{1,2}$, the equations reduce to:

$$x^2 + \alpha_0 x + \alpha_1\alpha_4 + \alpha_2\alpha_3 = 0,$$

which has no solution over $\mathbb{F}_2$ for $\alpha_0 = \alpha_1 = \alpha_2 = \alpha_4 = 1, \alpha_3 = 0$.

In the second case we have:

$$\chi(A) = [A_{1,2} + A_{3,4}, A_{1,3} + A_{3,4}, A_{1,4}, A_{2,3}, A_{2,4}]$$

which is surjective if and only if the equations

$$A_{1,2} + A_{3,4} = \alpha_0, A_{1,3} + A_{3,4} = \alpha_1, A_{1,4} = \alpha_2, A_{2,3} = \alpha_3, A_{2,4} = \alpha_4,$$

$$A_{1,2}A_{3,4} + A_{1,3}A_{2,4} + A_{1,4}A_{2,3} = 0$$

have a solution in $\mathbb{F}_2^6$ for any choice of $[\alpha_0 : \ldots : \alpha_4] \in \mathbb{P}^4(\mathbb{F}_2)$. Letting $x = A_{3,4}$, the equations reduce to:

$$x^2 + (\alpha_0 + \alpha_4)x + \alpha_1\alpha_4 + \alpha_2\alpha_3 = 0,$$

which has no solution over $\mathbb{F}_2$ for $\alpha_0 = \alpha_1 = \alpha_2 = \alpha_3 = 1, \alpha_4 = 0$.

In the third case we have:

$$\chi(A) = [A_{1,2}, A_{1,3}, A_{1,4} + A_{3,4}, A_{2,3} + A_{3,4}, A_{2,4} + A_{3,4}]$$

which is surjective if and only if the equations

$$A_{1,2} = \alpha_0, A_{1,3} = \alpha_1, A_{1,4} + A_{3,4} = \alpha_2, A_{2,3} + A_{3,4} = \alpha_3, A_{2,4} + A_{3,4} = \alpha_4,$$

$$A_{1,2}A_{3,4} + A_{1,3}A_{2,4} + A_{1,4}A_{2,3} = 0$$

have a solution in $\mathbb{F}_2^6$ for any choice of $[\alpha_0 : \ldots : \alpha_4] \in \mathbb{P}^4(\mathbb{F}_2)$. Letting $x = A_{3,4}$, the equations reduce to:

$$x^2 + (\alpha_0 + \alpha_1 + \alpha_2 + \alpha_3)x + \alpha_1\alpha_4 + \alpha_2\alpha_3 = 0,$$

which has no solution over $\mathbb{F}_2$ for $\alpha_0 = \alpha_2 = \alpha_3 = 0, \alpha_1 = \alpha_4 = 1$.

In the last case we have:

$$\chi(A) = [A_{1,2} + A_{3,4}, A_{1,3} + A_{3,4}, A_{1,4} + A_{3,4}, A_{2,3} + A_{3,4}, A_{2,4} + A_{3,4}]$$

which is surjective if and only if the equations

$$A_{1,2} + A_{3,4} = \alpha_0, A_{1,3} + A_{3,4} = \alpha_1, A_{1,4} + A_{3,4} = \alpha_2, A_{2,3} + A_{3,4} = \alpha_3, A_{2,4} + A_{3,4} = \alpha_4,$$

$$A_{1,2}A_{3,4} + A_{1,3}A_{2,4} + A_{1,4}A_{2,3} = 0$$

have a solution in $\mathbb{F}_2^6$ for any choice of $[\alpha_0 : \ldots : \alpha_4] \in \mathbb{P}^4(\mathbb{F}_2)$. Letting $x = A_{3,4}$, the equations reduce to:

$$x^2 + (\alpha_0 + \alpha_1 + \alpha_2 + \alpha_3 + \alpha_4)x + \alpha_1\alpha_4 + \alpha_2\alpha_3 = 0,$$

which has no solution over $\mathbb{F}_2$ for $\alpha_0 = \alpha_1 = \alpha_4 = 1, \alpha_2 = \alpha_3 = 0$. □

References

1. Brockett, R.W., Byrnes, C.I.: Multivariable Nyquist criteria, root loci and pole placement: A geometric viewpoint. IEEE Trans. Automat. Control 26, 271–284 (1981)
2. Byrnes, C.I.: Pole assignment by output feedback. In: Nijmeijer, H., Schumacher, J.M. (eds.) Three Decades of Mathematical System Theory. Lecture Notes in Control and Information Sciences, vol. 135, pp. 31–78. Springer, Heidelberg (1989)
3. Gessel, I., Viennot, G.: Binomial determinants, paths, and hook length formulae. Adv. in Math. 58(3), 300–321 (1985)
4. Kleiman, S.L.: Problem 15: Rigorous foundations of Schubert's enumerative calculus. Proceedings of Symposia in Pure Mathematics 28, 445–482 (1976)
5. Martin, C.F., Hermann, R.: Applications of algebraic geometry to system theory: The McMillan degree and Kronecker indices as topological and holomorphic invariants. SIAM J. Control Optim. 16, 743–755 (1978)
6. Mumford, D.: Algebraic Geometry I: Complex Projective Varieties. Springer, New York (1976)
7. Rosenthal, J.: Geometric methods for feedback stabilization of multivariable linear systems. Ph.D. thesis, Arizona State University (1990)
8. Rosenthal, J.: On dynamic feedback compensation and compactification of systems. SIAM J. Control Optim. 32(1), 279–296 (1994)
9. Rosenthal, J., Schumacher, J.M.: Realization by inspection. IEEE Trans. Automat. Contr. 42(9), 1257–1263 (1997)
10. Rosenthal, J., Sottile, F.: Some remarks on real and complex output feedback. Systems & Control Letters 33(2), 73–80 (1998)
11. Schubert, H.: Anzahlbestimmung für lineare Räume beliebiger Dimension. Acta Math. 8, 97–118 (1886)
12. Schubert, H.: Beziehungen zwischen den linearen Räumen auferlegbaren charakteristischen Bedingungen. Math. Ann. 38, 598–602 (1891)

13. Shafarevich, I.R.: Basic algebraic geometry, 1, 2nd edn. Springer, Berlin (1994), Varieties in projective space, Translated from the 1988 Russian edition and with notes by Miles Reid
14. Wang, X.: Pole placement by static output feedback. Journal of Math. Systems, Estimation, and Control 2(2), 205–218 (1992)
15. Wang, X.: Grassmannian, central projection and output feedback pole assignment of linear systems. IEEE Trans. Automat. Contr. 41(6), 786–794 (1996)

16

High-Speed Model Predictive Control: An Approximate Explicit Approach

Colin N. Jones and Manfred Morari

Automatic Control Lab, ETH Zurich, CH-8092, Zurich, Switzerland

Summary. A linear quadratic model predictive controller (MPC) can be written as a parametric quadratic optimization problem whose solution is a piecewise affine (PWA) map from the state to the optimal input. While this 'explicit solution' can offer several orders of magnitude reduction in online evaluation time in some cases, the primary limitation is that the complexity can grow quickly with problem size. In this paper we introduce a new method based on bilevel optimization that allows the direct approximation of the non-convex receding horizon control law. The ability to approximate the control law directly, rather than first approximating a convex cost function leads to simple control laws and tighter approximation errors than previous approaches. Furthermore, stability conditions also based on bilevel optimization are given that are substantially less conservative than existing statements.

16.1 Introduction

This paper considers the implementation of an MPC controller for a linearly constrained linear system with a quadratic performance index. Standard practice is to compute the optimal control action in this case by solving a quadratic program at each time instant for the current value of the state. It was shown in [20, 10, 4] that this quadratic program can be posed as a parametric problem (pQP), where the parameter is the state of the system and that this pQP results in a piecewise affine function that maps the state to the optimal input; the so-called 'explicit solution'.

The motivation for computing the explicit solution is that the resulting piecewise affine function can be much faster and simpler to evaluate online than solving a quadratic program, which can in some cases lead to several orders of magnitude reduction in computation time making MPC applicable to very high-speed applications. The primary limitation of this approach is that the complexity, or number of affine pieces, of this explicit solution can grow very quickly with problem size.

In this paper, we propose an approximation approach that generates a low-complexity piecewise affine function directly from the optimal MPC formulation (i.e. without computing the optimal solution first). The approach is simple in that

X. Hu et al. (Eds.): Three Decades of Progress in Control Sciences, pp. 233–248, 2010.

it proceeds by choosing a small number of states and then interpolates the optimal control action at these points. The key questions to be answered are then: Which points to select, how to certify that the sub-optimal controller is stabilizing and how to define and compute the resulting level of sub-optimality.

Several authors have proposed similar interpolated approximations that operate by exploiting the convexity of the optimal cost function. Given a finite set of sampled points X, these methods differ primarily in the manner in which they choose which points to interpolate to define the approximate control law, or equivalently how they partition the feasible set. In [3], the authors propose a recursive simplical partition, in [14, 12] the partition is formed as a result of incremental convex hull algorithms and in [11, 9] a box decomposition scheme is used. The common feature amongst these proposals is the method of choosing the set X, which is done in an incremental fashion by inserting at each step the state for which the error between the approximate cost function and the optimal one is the greatest. The main motivation for this is that these points can be found by solving convex problems. However, these approaches require that the entire optimal input sequence over the prediction horizon is approximated, rather than just the first step, which defines the control law in a receding-horizon controller. This requirement of approximating the entire optimal sequence leads to a very conservative test for the approximation error.

This paper introduces a new method that approximates only the first step of the optimal control sequence and leaves the remainder defined implicity as the result of a secondary parametric optimization problem. The result is a decrease in the conservatism of the approximation error, which in general results in a significant decrease in approximation complexity. The cost of this improvement is that the optimization problems to be solved are no longer convex, but are indefinite *bilevel* quadratic optimization problems. Bilevel problems are those in which some of the optimization variables are constrained to be optimizers of a secondary optimization and are, even in the simplest case, NP-hard to solve. We show that the indefinite bilevel QPs required for this approach can be re-written as mixed-integer linear programs (MILPs) and hence solved using very efficient methods. The paper also introduces an improved test for stability for the resulting interpolated control laws over those given previously, which is also based on bilevel optimization.

Proposals have also been made in the literature to derive simpler approximate explicit control laws using methods other than interpolation. The reader is referred to the recent survey [1] for a complete review.

The key challenges for the proposed class of control schemes are: 1. How to guarantee that the approximate controller satisfies constraints? 2. How to select the best states to interpolate? and 3. How to verify that the resulting approximate controller is stabilizing? We cover each of these questions in Sections 16.3, 16.4 and 16.5 respectively.

16.2 Background

The goal is to control the linear system

$$x^+ = Ax + Bu \ , \tag{16.1}$$

where the state x and input u are constrained to lie in the polytopic sets $\mathbb{X} = \{x|Fx \leq f\} \subset \mathbb{R}^n$ and $\mathbb{U} = \{u|Gu \leq g\} \subset \mathbb{R}^m$ respectively.

Consider the following finite horizon optimal control problem:

$$\begin{aligned} J^\star(x) := \min J(\mathbf{x},\mathbf{u}) & & \\ \text{s.t. } x_{i+1} = Ax_i + Bu_i \ , & \quad \forall i = 0,\dots,N-1 \\ x_i \in \mathbb{X} \ , \ u_i \in \mathbb{U} \ , & \quad \forall i = 0,\dots,N-1 \\ x_N \in \mathbb{X}_N \ , \ x_0 = x \ , & \end{aligned} \tag{16.2}$$

where $\mathbb{X}_N = \{x|Hx \leq h\} \subset \mathbb{X}$ is a polytopic invariant set for the system $x^+ = Ax + B\nu(x)$ for some given linear control law $\nu : \mathbb{R}^n \mapsto \mathbb{R}^m$. For the remainder of the paper, we will use a bold symbol $\mathbf{x}$ to describe an ordered sequence $\mathbf{x} := (x_0,\dots,x_N)$, where x_i is the ith element of $\mathbf{x}$. We define $\mathscr{X} \subset \mathbb{R}^n$ to be the set of states x for which there exists a feasible solution to (16.2). The quadratic cost function J is defined as

$$J(\mathbf{x},\mathbf{u}) := \frac{1}{2} x_N^T Q_N x_N + \frac{1}{2}\sum_{i=0}^{N-1} u_i^T R u_i + x_i^T Q x_i \tag{16.3}$$

where the matrices $Q_N \in \mathbb{R}^{n\times n} \geq 0$ and $Q \subset \mathbb{R}^{n\times n} \geq 0$ and $R \in \mathbb{R}^{m\times m} > 0$ define the cost function.

If $\mathbf{u}^\star(x)$ is the optimal input sequence of (16.2) for the state x, and $u_0^\star(x)$ is the resulting receding horizon control law, then $J^\star$ is a Lyapunov function for the system $x^+ = Ax + Bu_0^\star(x)$ under the assumption that $V_N(x) = x^T Q_N x$ is a Lyapunov function for the system $x^+ = Ax + B\nu(x)$ and that the decay rate of V_N is greater than the stage cost $l(x,u) = u^T Ru + x^T Qx$ within the set $\mathbb{X}_N$ [18].

The goal in this paper is to compute a PWA function of low complexity that approximates the control law $u_0^\star(x)$ as closely as possible while still guaranteeing stability and satisfaction of all constraints.

16.3 Interpolated Control

We define an interpolated control law by choosing a finite set of distinct feasible states and then interpolating amongst the optimal control action at each of these states. In order to formalize this idea, we first define the regions over which this interpolation will occur.

Definition 16.3.1 (Triangulation). *A triangulation of a finite set of points $V \subset \mathbb{R}^n$ is a finite collection $T_V := \{S_0, \ldots, S_L\}$ such that*

- $S_i = \text{conv}(V_i)$ *is an n–dimensional simplex for some* $V_i \subset V$
- $\text{conv}(V) = \cup S_i$ *and* $\text{int}\, S_i \cap \text{int}\, S_j = \emptyset$ *for all* $i \neq j$
- *If $i \neq j$ then there is a common (possibly empty) face F of the boundaries of S_i and S_j such that $S_i \cap S_j = F$.*

There are various triangulations possible, most of which are compatible with the proposed approach. For example, the recursive triangulation developed in [6, 3] has the strong property of generating a simple hierarchy that can significantly speed online evaluation of the resulting control law. The Delaunay triangulation [7], which has the nice property of minimizing the number of 'skinny' triangles, or those with small angles is a common choice for which incremental update algorithms are well-studied and readily available (i.e. computation of $T_{V \cup \{v\}}$ given T_V). A particularly suitable weighted Delaunay triangulation can be defined by using the optimal cost function as a weighting term, which causes the resulting triangulation to closely match the optimal partition [14].

Given a discrete set of states V, we can now define an interpolated control law $\mu_{T_V} : \mathbb{R}^n \mapsto \mathbb{R}^m$.

Definition 16.3.2. *If $V \subset \mathscr{X} \subset \mathbb{R}^n$ is a finite set and $T = \{\text{conv}(V_1), \ldots, \text{conv}(V_L)\}$ is a triangulation of V, then the* interpolated control law $\mu_T : \text{conv}(V) \mapsto \mathbb{R}^m$ *is*

$$\mu_T(x) := \sum_{v \in V_j} u_0^\star(v)\lambda_v \ , \ \textit{if}\ x \in \text{conv}(V_j) \tag{16.4}$$

where $\lambda_v \geq 0$, $\sum \lambda_v = 1$ and $x = \sum_{v \in V_j} v\lambda_v$.

The following lemma describes the relevant properties of interpolated control laws.

Lemma 16.3.1. *If $V \subset \mathscr{X} \subset \mathbb{R}^n$ is a finite set and T_V is a triangulation of V, then the interpolated control law μ_{T_V} is:*

1. *Feasible for all $x \in \text{conv}(V)$:*
   $\exists\, \mathbf{x}, \mathbf{u}$ *feasible for* (16.2) *such that $x_0 = x$ and $u_0 = \mu_{T_V}(x)$*
2. *An affine function in each simplex* $\text{conv}(V_i)$
3. *Continuous*

Our goal is to compute an interpolated control law μ_{T_V} for the system (16.1) that is as close to the optimal $u_0^\star$ as possible by sampling a set of points V such that T_V is of a pre-specified complexity. As proposed in various papers [3, 14, 12, 11, 9], we do this in an incremental fashion beginning from any inner approximation $\text{conv}(V)$ of the feasible set $\mathscr{X}$ of (16.2) and the resulting initial triangulation $T_V = \{V_0, \ldots, V_l\}$.

At each iteration of the algorithm, we maintain a point set V and the resulting triangulation T_V. This is then used to compute a state x_{T_V} that is a maximizer for some function $\gamma : \mathbb{R}^m \times \mathbb{R}^m \mapsto \mathbb{R}$ that measures the error between the optimal control law $u_0^\star$ and the interpolated one μ_{T_V}.

$$x_{T_V} := \arg\max_{x \in \text{conv}(V)} \gamma(\mu_{T_V}(x), u_0^\star(x)) \ . \tag{16.5}$$

The point set $V' := V \cup \{x_{T_V}\}$ and triangulation $T_{V'}$ are then updated to include this worst-fit point. This simple procedure repeats until some specified approximation error ε is achieved, or the complexity of the triangulation has exceeded some bound. The general method is given as Algorithm 16.3.1 below.

Algorithm 16.3.1 Construction of an interpolated control law

Require: A finite subset V of the feasible set of (16.2), an error function $\gamma : \mathbb{R}^m \times \mathbb{R}^m \mapsto \mathbb{R}$ and an approximation error ε or a complexity specification *COMP*.

Ensure: A point set V such that $|T_V| \leq COMP$ or $\max_x \gamma(\mu_{T_V}(x), u_0^\star(x)) \leq \varepsilon$.

  Compute T_V

  repeat

    Compute a point $x_{T_V} \in \arg\max_x \gamma(\mu_{T_V}(x), u_0^\star(x))$

    Set $err \leftarrow \gamma(\mu_{T_V}(x_{T_V}), u_0^\star(x_{T_V}))$

    Update $V \leftarrow V \cup \{x_{T_V}\}$

    Compute T_V

  until $|T_V| \geq COMP$ or $err \leq \varepsilon$

  Return V and T_V

Remark 16.3.1. An initial inner approximation of the feasible set $\mathscr{X}$ can be computed by, for example, using the projection approximation approach proposed in [14].

Remark 16.3.2. Several methods are available to compute $T_{V \cup \{x_{T_V}\}}$ given the previous triangulation T_V in an incremental fashion [7].

The key questions we seek to answer here are how best to define the function γ and how to compute the maximization (16.5).

16.4 Error Computation

In this section we compute the error between the optimal and sub-optimal control laws by defining the error function as

$$\gamma(x) := \|u_0^\star(x) - \mu_{T_V}(x)\|_\infty \ , \tag{16.6}$$

which is natural for measuring the worst-case fit between two functions. In Section 16.5 we will examine a second possible function that provides a method to certify the stability of the sub-optimal control law.

The key requirement of using a function such as (16.6) is that we be able to solve the optimization problem (16.5). One immediate method of doing this is to first compute the optimal solution $u_0^\star(x)$ to the pQP (16.2) using one of a number of standard methods available [1]. This would provide an explicit PWA representation of $u_0^\star(x)$ and would then make the computation of (16.5) straightforward. However, it is likely that this optimal control law is too complex to be computed directly, and so here we aim to find an *implicit* representation of $u_0^\star(x)$. This section will outline how this can be done by writing (16.5) as a bilevel optimization problem, which can in turn be solved using a mixed-integer linear solver.

Similar or related methods have been proposed in the literature [3, 14, 12, 11, 9] that propose to approximate the control law indirectly, by first approximating the optimal cost function $J^\star$ and then using this result to define an interpolated control law. As discussed in the introduction, the primary limitation of such approaches is that one must approximate the entire optimal sequence $\mathbf{u}^\star$, rather than just the first step $u_0^\star$, which defines the control law. The sequence $\mathbf{u}^\star$ is significantly more complex than the control law $u_0^\star$, and hence these approaches will in general result in more complex interpolated control laws. The benefit, however, of approximating the entire sequence is that it is possible to write a version of (16.5) as a convex optimization problem, and hence more easily solve it.

Bilevel Optimization

Bilevel optimization problems have been extensively studied in the literature, and the reader is referred to the recent survey [5] for background details. Bilevel problems are hierachical in that the optimization variables are split into upper y and lower z parts, with the lower level variables constrained to be an optimal solution to a secondary optimization problem.

$$\begin{aligned} \min_{y} \ & V_U(y,z) \\ \text{s.t.} \ & G_U(y,z) \leq 0 \\ & z = \arg\min_{z} \ \{V_L(y,\hat{z}) \mid G_L(y,\hat{z}) \leq 0\} \end{aligned} \tag{16.7}$$

Remark 16.4.1. The bilevel formulation given here makes the implicit assumption that the optimizer of the lower-level problem is unique, which is valid since the optimizer will be unique as long as the matrix R in (16.3) is positive definite.

Solution Methods

Bilevel optimization problems are in general very difficult to solve. Even the simplest case where all functions are linear is NP-hard [8]. Several computational methods have been proposed for various types of bilevel optimization problems (see [5] for a survey), but for the purposes of this paper, the most relevant is that originally given in [2] for quadratic bilevel problems. The key observation is that if the lower level problem is convex and regular, then it can be replaced by its necessary and sufficient Karush-Kuhn-Tucker (KKT) conditions, yielding a standard single-level optimization problem.

$$\min_{y,z,\lambda} \quad V_U(y,z) \tag{16.8a}$$

$$\begin{aligned} \text{s.t.} \quad G_U(y,z) &\leq 0 \\ G_L(y,z) &\leq 0 \\ \lambda &\geq 0 \\ \lambda^T G_L(y,z) &= 0 \qquad\qquad (16.8b) \\ \nabla_z \mathscr{L}(y,z,\lambda) &= 0 \end{aligned}$$

where $\mathscr{L}(y,z,\lambda) := V_L(y,z) + \lambda^T G_L(y,z)$ is the Lagrangian function associated with the lower-level problem. For the special case of linear constraints and a quadratic cost, all constraints of (16.8) are linear and the complementarity condition (16.8b) is a set of disjunctive linear constraints, which can be described using binary variables, and thus leads to mixed-integer linear constraints.

Error Computation via Bilevel Optimization

To compute the maximum of the function γ given in (16.6) while maintaining an implicit representation of the optimal control law $u_0^\star$, we set the upper level cost V_U to γ and the lower level V_L to J. If the triangulation defining the interpolated control law is $T_V := \{S^0, \ldots, S^L\}$ and the interpolated control law in simplex S^i is $\mu_{S^i} = T^i x + t^i$, then for each simplical region S^i, we can compute the maximum of the error function γ with the following bilevel optimization:

$$\begin{aligned} \gamma_i := \max \quad & \|\mu - u_0\|_\infty \\ \text{s.t.} \quad & x \in S^i \\ & \mu = T^i x + t^i \\ & u_0 = \arg\min \quad \frac{1}{2} x_N^T Q_N x_N + \frac{1}{2} \sum_{i=0}^{N-1} u_i^T R u_i + x_i^T Q x_i \\ & \qquad\quad \text{s.t.} \quad x_{i+1} = A x_i + B u_i \\ & \qquad\qquad\quad F x_i \leq f \;, \quad G u_i \leq g \;, \quad H x_N \leq h \\ & \qquad\qquad\quad x_0 = x \end{aligned}$$

The lower-level optimization problem is clearly strictly convex, and can therefore be solved by replacing it with its KKT conditions, which results in the disjunctive optimization problem (16.9).

$$
\begin{aligned}
\gamma_i := \max\ & \|\mu - u_0\|_\infty \\
\text{s.t.}\ & \text{Upper-level constraints} && (16.9a)\\
& \left\lfloor \begin{array}{l} x \in S^i \\ \mu = T^i x + t^i \end{array} \right. \\
& \text{Primal constraints} && (16.9b)\\
& \left\lfloor \begin{array}{l} x_{i+1} = Ax_i + Bu_i \\ \quad x_0 = x \\ Fx_i \le f\ ,\quad Gu_i \le g\ ,\quad Hx_N \le h \end{array} \right. \\
& \text{Dual constraints} && (16.9c)\\
& \left\lfloor \begin{array}{l} \lambda_i^x \ge 0\ ,\quad \lambda_i^u \ge 0\ ,\quad \lambda_N^x \ge 0 \\ \nu_i \text{ free} \end{array} \right. \\
& \text{First order optimality } \nabla\mathscr{L} = 0 && (16.9d)\\
& \left\lfloor \begin{array}{l} 0 = Qx_i + A^T\nu_i - \nu_{i-1} + F^T\lambda_i^x \\ 0 = Ru_i + B^T\nu_i + G^T\lambda_i^u \\ 0 = Q_N x_N + H^T\lambda_N^x - \nu_{N-1} \end{array} \right. \\
& \text{Complementarity conditions} && (16.9e)\\
& \left\lfloor \begin{array}{l} \lambda_j^{x_i} = 0 \quad \text{or} \quad F_j x_i = f_j \\ \lambda_j^{u_i} = 0 \quad \text{or} \quad G_j u_i = g_j \\ \lambda_j^{x_N} = 0 \quad \text{or} \quad H_j x_N = h_j \end{array} \right.
\end{aligned}
$$

Remark 16.4.2. As written, (16.9) is a mixed integer linear programming problem (MILP) with logic constraints. Standard techniques exist to convert such logical constraints to linear ones with binary variables, and the reader is referred to e.g. [17] for details. While the resulting MILP is NP-hard to solve, there are both free and commercial solvers available that can tackle very large problems.

16.5 Stability

If the matrices of the cost function (16.3) and the set $\mathbb{X}_N$ are defined appropriately, then the optimal control law $u_0^\star$ is stabilizing for (16.1) and the optimal cost function $J^\star$ is a Lyapunov function for the resulting closed-loop system [18]. In this section, we seek verifiable conditions under which $J^\star$ is also a Lyapunov function for the approximate closed-loop system $x^+ = Ax + B\mu_V(x)$. We begin by giving a minor modification of the standard condition for an approximate control law to be stabilizing [19], which requires only that the approximate control law be specified, rather than an entire approximate input sequence.

Theorem 16.5.1 ([19, 15]). *Let $J^\star$ be the optimal solution of* (16.2) *and a Lyapunov function for $x^+ = Ax + Bu_0^\star(x)$. If $\mu(x) : \mathbb{R}^n \mapsto \mathbb{R}^m$ is a control law defined over the* *set S, then $J^\star$ is also a Lyapunov function for $x^+ = Ax + B\mu(x)$ if for all $x_0 \in S$, there exists a feasible state/input sequence $(\mathbf{x}, \mathbf{u})$ to* (16.2) *such that $u_0 = \mu(x_0)$ and*

$$J(\mathbf{x},\mathbf{u}) - J^\star(x_0) \le \frac{1}{2}x_0^T Q x_0 + \frac{1}{2}u_0^T R u_0 \tag{16.10}$$

Remark 16.5.1. The condition given in Theorem 16.5.1 is essentially the same as that used in several other papers on approximate MPC [19]. The standard approach is to define a function $\hat{J}$, which is the interpolation of the optimal cost at the vertices of the triangulation and then test this cost under condition (16.10). This test is efficient because the function $\hat{J}$ is piecewise affine and hence (16.10) can be evaluated as a series of convex optimizations. Rather than taking a linear interpolation, we here assume that the candidate Lyapunov function $\tilde{J}$ is given implicity by the optimization $\tilde{J}(x) := \min\{J(\mathbf{x},\mathbf{u}) | (16.2),\ x_0 = x,\ u_0 = \mu(x)\}$, which makes $\tilde{J}$ a convex piecewise quadratic function. Clearly, the condition $\tilde{J} \le \hat{J}$ holds, which makes the condition given in Theorem 16.5.1 less conservative than previous proposals and often significantly so. The cost is that condition (16.10) can no longer be verified by solving convex problems.

Theorem 16.5.1 gives a condition under which the optimal cost function $J^\star$ is a Lyapunov function for the closed-loop system under the interpolated control law $\mu_{T_V}(x)$. This condition is not trivial to test, but can be confirmed by solving a series of bilevel programs, which we demonstrate below.

Let $V \subset \mathscr{X}$ be a finite set that defines the interpolated control law μ_{T_V} over the triangulation $T_V := \{S_1, \ldots, S_L\}$. Define ξ_i to be the optimal cost of the bilevel optimization problem (16.11) for each $i = 1, \ldots, L$, where $\mu := T^i x + t^t$ is the affine control law in the simplical region S_i.

$$\xi_i := \min \ \frac{1}{2}\tilde{x}_0^T Q \tilde{x}_0 + \frac{1}{2}\tilde{u}_0^T R \tilde{u}_0 + J(\mathbf{x},\mathbf{u}) - J(\tilde{\mathbf{x}},\tilde{\mathbf{u}}) \tag{16.11a}$$

$$\begin{aligned}
\text{s.t. } & x_0 \in S_i \\
& \text{Constraints (16.2) on } \mathbf{x},\mathbf{u} \\
& (\tilde{\mathbf{x}},\tilde{\mathbf{u}}) = \arg\min \ J(\tilde{\mathbf{x}},\tilde{\mathbf{u}}) \qquad\qquad (16.11b)\\
& \qquad\qquad \text{s.t. Constraints (16.2) on } \tilde{\mathbf{x}},\tilde{\mathbf{u}} \\
& \qquad\qquad\quad \tilde{x}_0 = x_0 \\
& \qquad\qquad\quad \tilde{u}_0 = T^i x_0 + t^t
\end{aligned}$$

One can see from (16.11) that the conditions of Theorem 16.5.1 are met if and only if $\max\{\xi_i\}$ is negative.

Corollary 16.5.1. *$J^\star$ is a Lyapunov function for the system $x^+ = Ax + B\mu_{T_V}(x)$ if $\max\{\xi_i\} \le 0$.*

A Lyapunov function is insufficient to prove stability for a constrained system; the system must also be invariant. As discussed in [12], since level sets of Lyapunov functions are invariant, it is possible to determine an invariant subset of $\text{conv}(V)$ given the vertices of each region S_i without further processing.

16.5.1 Computation of Stability Criterion

The bilevel optimization problem (16.11) differs from that tackled in the previous section in that the upper level is an indefinite quadratic program, while the lower is a convex QP. In the following, we demonstrate that this class of problems can also be re-formulated as a disjunctive LP and hence solved using standard MILP software.

We begin with the following lemma, which demonstrates that an indefinite QP can be written as a mixed-integer LP.

Lemma 16.5.1. *Consider the following indefinite QP*

$$J^\star := \min_z \quad \frac{1}{2} z^T D z \qquad \text{s.t.} \quad Bz \le b \ , \quad Cz = c \ , \tag{16.12}$$

where $B \in \mathbb{R}^{m \times n}$, $C \in \mathbb{R}^{l \times n}$ *and assume that Slater's condition holds. If* $(z^\star, \lambda, \gamma)$ *is an optimal solution of the MILP* (16.13)*, then* $z^\star$ *is an optimizer of* (16.12) *and* $\check{J} = J^\star$.

$$\begin{aligned} \check{J} = \min_{x,\lambda,\gamma} \quad & -\frac{1}{2}\left(b^T\lambda + c^T\gamma\right) && \\ \text{s.t.} \quad & Bz \le b \ , \quad Cz = c \ , && \textit{Primal feasibility} \\ & \nabla_z \mathscr{L} = Dz + B^T\lambda + C^T\gamma = 0 \ , && \textit{Stationarity} \\ & \lambda \ge 0 \ , \quad \gamma \ \textit{free} \ , && \textit{Dual feasibility} \\ & \lambda_i = 0 \quad \text{or} \quad B_i z = b_i \ , && \textit{Complementarity} \end{aligned} \tag{16.13}$$

Proof. The constraints of (16.13) are precisely the KKT conditions of (16.12), which are necessary but not sufficient because the problem is indefinite. We gain sufficiency by minimizing the cost function of (16.12) while enforcing the necessary optimality conditions, which leads to an optimal solution. We have now to show that the linear cost function of (16.13) is in fact equivalent to the indefinite cost of (16.12).

We begin by taking the inner product of the stationarity condition and the primal optimization variable z

$$z^T \nabla_z \mathscr{L} = 0 = z^T D z + z^T B^T \lambda + z^T C^T \gamma$$

The complementarity conditions $\lambda^T(Bz - b) = 0$ and $\gamma^T(Cz - c) = 0$ then give the result

$$z^T D z = -b^T \lambda - c^T \gamma$$

□

We can now show that an optimizer of (16.11) can be computed by solving a mixed-integer linear program.

Theorem 16.5.2. *Consider the following quadratic bilevel optimization problem:*

$$
\begin{aligned}
J^\star = \min \;& \frac{1}{2} y^T S y + \frac{1}{2} z^{\star T} T z^\star \\
\text{s.t.} \;& Ay \le b \\
& z^\star = \arg\min \left\{ \frac{1}{2} z^T D z \mid Ez + Fy \le g \right\}
\end{aligned}
\tag{16.14}
$$

where D is positive definite and S and T are indefinite matrices. If $(y, z, \beta^L, \beta^U, \lambda^L, \lambda^U, \lambda^{UL}, l)$ is an optimal solution of (16.15)*, then (y, z) is an optimal solution to* (16.14).

$$
\min \; -\frac{1}{2}\left(b^T \lambda^U + g^T \lambda^{UL}\right) \tag{16.15a}
$$

$$
\begin{aligned}
\text{s.t.}\;& \beta_i^L \in \{0,1\}\,,\; \beta_i^U \in \{0,1\} \\
& \textit{Upper level} \\
& \left|\begin{array}{l}
\textit{Primal and dual feasibility} \\
\lfloor Ay \le b\,, \quad \lambda^U \ge 0 \\
\textit{Stationarity} \\
\left|\begin{array}{l}
\nabla_y \mathscr{L}^U = 0 = Sy + A^T \lambda^U + F^T \lambda^{UL} \\
\nabla_z \mathscr{L}^U = 0 = Tz + D^T \gamma^U + E^T \lambda^{UL} \\
\nabla_{\lambda^L} \mathscr{L}^U = 0 = E\gamma^U - l
\end{array}\right. \\
\textit{Complementarity} \\
\left|\begin{array}{ll}
\beta_i^L = 1 \Rightarrow \lambda_i^{UL} = 0\,, & \beta_i^L = 0 \Rightarrow l_i = 0 \\
\beta_i^U = 1 \Rightarrow \lambda_i^U = 0\,, & \beta_i^U = 0 \Rightarrow A_i y = b_i
\end{array}\right.
\end{array}\right. \\
& \textit{Lower level} \\
& \left|\begin{array}{l}
\textit{Primal and dual feasibility} \\
\lfloor Ez + Fy \le g\,, \quad \lambda^L \ge 0 \\
\textit{Stationarity} \\
\lfloor \nabla_z \mathscr{L}^L = 0 = Dz + E^T \lambda^L = 0 \\
\textit{Complementarity} \\
\lfloor \beta_i^L = 1 \Rightarrow \lambda_i^L = 0\,, \quad \beta_i^L = 0 \Rightarrow E_i z + F_i y = g_i
\end{array}\right.
\end{aligned}
\tag{16.15b}
$$

Proof. The matrix D is positive definite and so its KKT conditions are both necessary and sufficient for optimality of the lower level problem. As a result, we can replace the lower level problem with these conditions in order to get an equivalent single level problem with mixed-integer constraints. We introduce the binary variable β^L, which encodes the complementarity conditions of the lower level problem and define the following optimization problem as a function of this variable:

$$J(\beta^L) := \min_{y,z,\lambda^L} \frac{1}{2}y^T Sy + \frac{1}{2}z^T Tz \tag{16.16}$$
$$\text{s.t. } Ay \le b$$
$$\text{Lower level optimality conditions (16.15b)}$$

For each β^L, (16.16) is a single-level indefinite quadratic program, which can be written as a MILP using Lemma 16.5.1. This gives the optimization problem (16.15), where we introduce appropriate dual variables λ^U, λ^{UL}, l and binaries β^U to represent the upper-level complementarity conditions.

Finally, we have $J^\star = \min_{\beta^L} \{J(\beta^L) | (16.16) \text{ feasible}\}$, which gives the desired result. □

Remark 16.5.2. Note that the structure of the problem (16.14) differs slightly from that required to solve the stability criteria (16.11) since it does not include any equality constraints. These constraints were left out for clarity and space restrictions, but the proof can be readily extended to this case.

Stability through Robust MPC

In the previous section we demonstrated that a stability certificate can be obtained if the approximation error is sufficiently small when compared to the optimal cost function. We here introduce a second approach based on a robust MPC formulation. The benefit is a reduction in offline computational effort, since no PWQ cost functions need to be compared, although this comes at the cost of an increase in the conservativeness of the control law.

We propose to model the approximation error as a bounded additive disturbance to the optimal control law. Let $u_0^\star(x)$ be the optimal control law, and $\tilde{\mu}^\star(x)$ an approximate controller such that $\|u_0^\star(x) - \tilde{\mu}^\star(x)\|_\infty \le \alpha$ for some $\alpha \ge 0$. Consider the closed-loop system subject to the approximate control law (16.1)

$$x^+ = Ax + B\tilde{\mu}^\star(x) \ . \tag{16.17}$$

The behaviour of this system is contained within the set of possible state evolutions of the following uncertain system

$$x^+ = Ax + B\mu^\star(x) + Bw \ , \qquad w \in \mathbb{W}_\alpha, \tag{16.18}$$

where $\mathbb{W}_\alpha$ is the bounded set $\mathbb{W}_\alpha := \{w | \|w\|_\infty \le \alpha\}$.

Several robust MPC schemes have been proposed to control the system (16.18) that are able to guarantee stability of the uncertain system, as well as satisfaction of the constraints (16.2) for all disturbances $w \in \mathbb{W}_\alpha$. The reader is referred to, e.g. [16], for a survey of common approaches. Many of these methods maintain the structure of the optimal control problem (16.2), defining the robust control law implicitly as the optimal solution to a quadratic optimization problem of the form (16.2).

This leads to a simple, potentially conservative procedure for generating stabilizing approximate controllers that guarantee satisfaction of the constraints (16.2). We first pose a robust MPC control problem for (16.18), whose optimal control law $u_0^\star(x)$ satisfies the constraints (16.2) for all disturbances $w \in \mathbb{W}_\alpha$. We then compute an approximate controller $\tilde{\mu}(x)$ that satisfies the condition $\|u_0^\star(x) - \tilde{\mu}(x)\|_\infty \leq \alpha$, as verified by (16.9a). The resulting sub-optimal controller will be stabilizing and will satisfy the system constraints (16.2).

16.6 Example

Consider the following simple two-state, two-input example:

$$x^+ = \begin{bmatrix} 1 & 1 \\ 0 & 1 \end{bmatrix} x + \begin{bmatrix} 0.42 & 0.90 \\ 0.38 & 0.67 \end{bmatrix} u \ ,$$

with the input and state constraints $\|u\|_\infty \leq 0.1$, $|x_1| \leq 40$, $|x_2| \leq 10$ and a horizon N of length 10 with the stage cost taken to be $l(x,u) := x'x + 30u'u$. The optimal control law in this case requires $1,155$ regions and can be seen in Figure 16.1a.

16.6.1 Bilevel Approximation

We here approximate this optimal control law with one interpolated from 36 points, which results in 52 simplical regions (Figure 16.1b). The error γ between the optimal and approximate control laws is only 0.06 and the maximum error between the optimal and suboptimal cost functions is 0.75. The approximate system is stable, as verified by solving MILP (16.15).

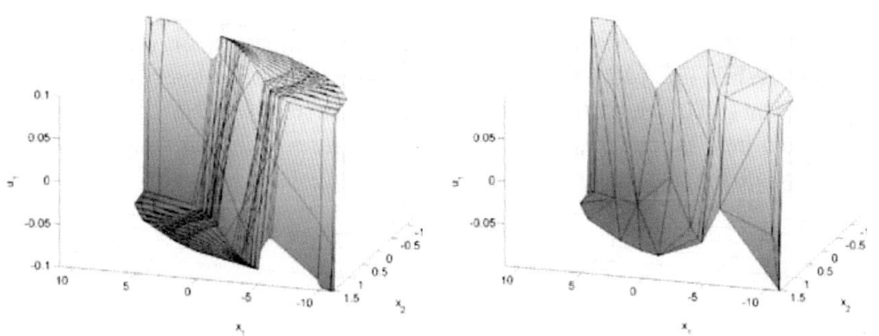

(a). Optimal control law $u_1(x)$; $1,155$ regions.

(b). Approximate control law $u_1(x)$; 52 regions.

Fig. 16.1. Optimal and approximated control law $u_1(x)$ for Example 16.6

16.6.2 Stability Certificate

We now use the approach introduced in Section 16.5.1 to provide a certificate of stability for an approximate PWA controller. In order to highlight the benefits of the proposed method when compared to those existing in the literature, we take an extremely rough approximation of the optimal control law given by the MPC problem (16.2) for the system and setup of Section 16.6.1. As before, the optimal controller consists of $1,155$ regions, but now the approximate control law has been drastically simplified to only 7 simplical regions.

Several proposals have been put forward in the literature that upper-bound the optimal cost function $J^\star$ with a convex piecewise affine function which is the interpolation of the optimal cost at the vertices of a simplical partition (see [1] and [13] for surveys). This conservative upper bound can then be used to generate a stability certificate by testing if this upper bound satisfies the conditions of Theorem 16.5.1. The resulting piecewise affine cost function and the stability test is shown in Figure 16.2. It can be seen from Figure 16.2b that the conditions of Theorem 16.5.1 are not satisfied, since the error between the optimal and suboptimal costs (blue surface) lies clearly above the stage cost $l(x,u)$. As a result, one would need to add significant additional complexity to this approximate control law before this conservative condition would provide a certificate of stability.

Figure 16.3 shows the same condition, with the approximate cost function now taken to be that proposed in this paper:

$$\tilde{J}(x) := \min\{J(\mathbf{x},\mathbf{u})|u_0 = \mu(x),\ \text{Conditions (16.2)}\} \tag{16.19}$$

where $\mu(x)$ is the approximate interpolated control law. Note the key difference that (16.19) is only interpolating over the first step of the control law and leaving the remainder implicitly defined, whereas previous proposals must interpolate over the entire control horizon. The result is a significant reduction in the conservatism of the stability criteria given in Theorem 16.5.1. This can be seen in Figure 16.3b where it is clear that this simple approximate control law defined over only 7 regions is in fact stable. While one can see this from the figure, the mixed-integer optimization problem given in Theorem 16.5.2 was used to confirm this fact. Note that the control law certified here is identical to that shown in Figure 16.2; it is only the stability proof approach that has changed.

References

1. Alessio, A., Bemporad, A.: A survey of explicit model predictive control. In: Proc. of the Int. Workshop on Assessment and Future Directions of NMPC (2008)
2. Bard, J.F., Moore, J.T.: A branch and bound algorithm for the bilevel programming problem. SIAM Journal of Scientific and Statistical Computing 11(2), 281–292 (1990)
3. Bemporad, A., Filippi, C.: An algorithm for approximate multiparametric convex programming. Computational Optimization and Applications 35(1), 87–108 (2006)
4. Bemporad, A., Morari, M., Dua, V., Pistikopoulos, E.N.: The explicit linear quadratic regulator for constrained systems. Automatica 38(1), 3–20 (2002)

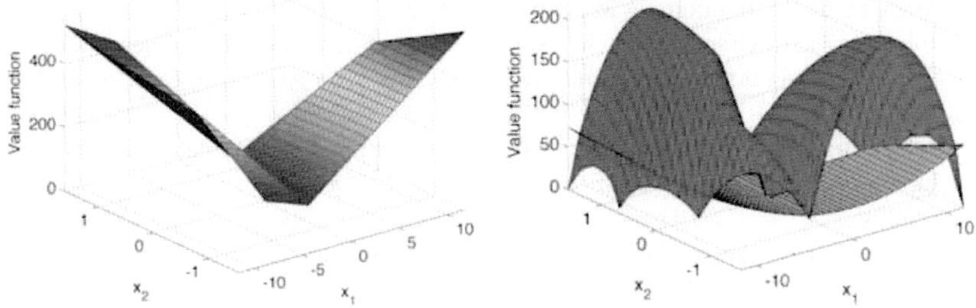

(a). Cost function interpolated linearly across the 7 simplical regions of Example 16.6.2.

(b). Blue: Approximation error between sub-optimal PWA cost function and optimal cost function. Orange: Stage cost $l(x,u)$

Fig. 16.2. Linearly interpolated cost function for Example 16.6.2. Conditions of stability Theorem 16.5.1 are not satisfied since the blue surface does not lie below the orange surface in 16.3b.

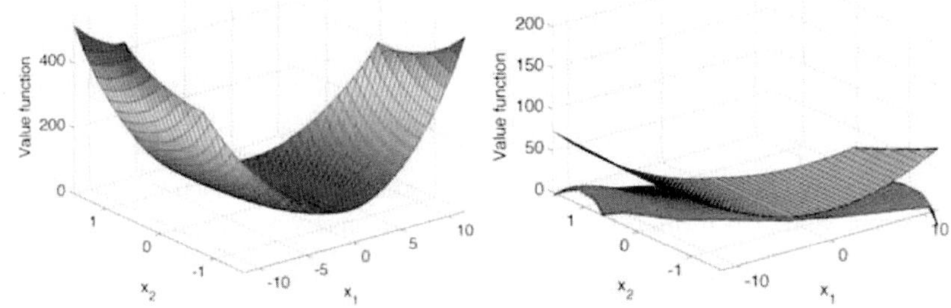

(a). Implicitly defined sub-optimal piecewise-quadratic cost function $\tilde{J}(x)$ (16.19).

(b). Blue: Approximation error between sub-optimal PWQ cost function (16.19) and optimal cost function. Orange: Stage cost $l(x,u)$.

Fig. 16.3. Implicitly defined piecewise quadratic cost function (16.19) for Example 16.6.2. Conditions of stability Theorem 16.5.1 are met, since the blue surface lies entirely below the orange in 16.3b.

5. Colson, B., Marcotte, P., Savard, G.: Bilevel programming: A survey. 4OR: A Quarterly Journal of Operations Research 3(2), 87–107 (2005)
6. De La Pena, D.M., Bemporad, A., Filippi, C.: Robust explicit MPC based on approximate multiparametric convex programming. IEEE Trans. on Automatic Control 51(8), 1399–1403 (2006)
7. Fortune, S.: Voronoi diagrams and delaunay triangulations. In: Goodman, J.E., O'Rourke, J. (eds.) Handbook of Discrete and Computational Geometry, 2nd edn., pp. 513–528. Chapman and Hall, Boca Raton (2004)

8. Jeroslow, R.G.: The polynomial hierarchy and a simple model for competitive analysis. Mathematical Programming 32, 146–164 (1985)
9. Johansen, T.A., Grancharova, A.: Approximate explicit constrained linear model predictive control via orthogonal search tree. IEEE Trans. on Automatic Control 48, 810–815 (2003)
10. Johansen, T.A., Petersen, I., Slupphaug, O.: On explicit suboptimal LQR with state and input constraints. In: Proc. of the IEEE Conf. on Decision and Control, pp. 662–667 (2000)
11. Johansen, T.A.: Approximate explicit receding horizon control of constrained nonlinear systems. Automatica 40(2), 293–300 (2004)
12. Jones, C.N., Barić, M., Morari, M.: Multiparametric linear programming with applications to control. European Journal of Control 13(2–3), 152–170 (2007)
13. Jones, C.N., Kerrigan, E.C., Maciejowski, J.M.: On polyhedral projection and parametric programming. Journal of Optimization Theory and Applications 137(3) (2008)
14. Jones, C.N., Morari, M.: The double description method for the approximation of explicit MPC control laws. In: Proc. of the IEEE Conf. on Decision and Control (December 2008)
15. Jones, C.N., Morari, M.: Approximate explicit MPC using bilevel optimization. In: Proc. of the European Control Conference, 2009 (To appear)
16. Limon, D., Alamo, T., Raimondo, D., de la Pena, D.M., Bravo, J., Ferramosca, A., Camacho, E.: Input-to-state stability: A unifying framework for robust model predictive control. In: Assessment and Future Directions of NMPC, Springer, Heidelberg (2009)
17. Löfberg, J.: YALMIP: A toolbox for modeling and optimization in MATLAB. In: Proceedings of the CACSD Conference, Taipei, Taiwan (2004)
18. Mayne, D.Q., Rawlings, J.B., Rao, C.V., Scokaert, P.O.M.: Constrained model predictive control: Stability and optimality. Automatica 36(6), 789–814 (2000)
19. Scokaert, P.O.M., Mayne, D.Q., Rawlings, J.B.: Suboptimal model predictive control (feasibility implies stability). IEEE Trans. on Automatic Control 44(3), 648–654 (1999)
20. Seron, M.M., Goodwin, G.C., De Doná, J.A.: Geometry of MPC for constrained linear systems. Technical Report EE0031, The University of Newcastle, Australia (2000)

17

Reflex-Type Regulation of Biped Robots*

Hidenori Kimura and Shingo Shimoda

RIKEN, BSI-TOYOTA Collaboration Center, 2-1 Hirosawa, Wako, Saitama 351-0198, Japan

Summary. A new type of neural controller is proposed to circumvent the difficulties associated with biped locomotion, or other robot tasks. The controller works only in those plants in which the actuation management to reduce the sensor input is easily carried out. This assumption is motivated by our observation that some fundamental reflex is innate in our control systems in brain. Our controller can deal with state discontinuity, model multiplicity and control/learning parallelism.

After some toy experiments that demonstrated some novel properties of our controller, we challenged biped locomotion. Very rude indications for locomotion targets are given, and our robot, after a couple of minutes learning, was able to walk biped steadily. The balance which was not explicitly taken care of emerged spontaneously. Some theoretical issues are also addressed.

17.1 Introduction

Human being is the only animal that walks biped in daily life. To mimic biped locomotion in robots has attracted many researchers in the area of robotics, brain sciences and control. There have been considerable amount of work concerning biped robot using variety of approaches [1][2][3]. Theoretical studies have also been done from many points of view (e.g. [4][5]). It is of great interest specially from control point of view for many reasons.

First, it is a combination of postures (statics) and their transitions (dynamics). Second, it needs a delicate integration of locomotion and balance. Third, it embodies both innate ability and acquired skill through development and learning.

There are some essential difficulties for *conventional* control theory to deal with biped locomotion in its traditional framework of model-based control. First, the process of biped locomotion cannot be described by a single model. Each posture taken by humans during biped walking should be described in different coordinate frames, due to alternate change of the supporting leg. So, traditional model-based control

*We deeply acknowledge Drs. I. Maeda and H. Yamamoto of Toyota Motor Co. for their substantial support and encouragement to this work.

X. Hu et al. (Eds.): Three Decades of Progress in Control Sciences, pp. 249–264, 2010.

must be generalized to *multi-model* framework. This is not at all an easy thing to do. Second, some state variables, specially velocities of the links, are discontinuously changed at the time of touching down of each leg. Also, the reacting forces from the floor might change discontinuously. Therefore, we need to extend control theory to allow discontinuity of state variables. Third, the notion of stability must be modified. Of course, stability is concerned with the infinite-time behaviors of the system in its original sense. If we confine our time horizon to be finite, the issue of stability does not come to the stage. The stability issue in biped locomotion lies in balance keeping against gravity force and nothing to do with infinite time nor long time behaviors. The notion of stability must be re-defined to deal with the stability issues in biped locomotion properly. Some attemps have been done to deal with the stability of biped locomotion subject to several restricting assumptions (see, e.g. [5][7]). But it does not seem to be adequately treated. Fourth, biped locomotion is extremely robust. We can walk on an irregular terrain easily without any special care, which is very difficult for humanoid robots because the disturbance due to irregularities of the walking surface comes from the foot and gives big impact to the body because foot is far distant from the center of mass of the body and creates a big disturbing torque around the center of mass. The brain motor system must have an extremely robust control scheme for biped walking which we do not know yet [8]. Finally, biped locomotion can be regarded at least partially as an acquired skill through learning. We do not know how the skill for walking is acquired, and seems to be very difficult to embody these learning processes in the artifacts.

In order to deal with the above difficult issues properly, we must create a new framework of control theory that addresses properly the model multiplicity, discontinuity of state variables and the plasticity of control actions. The well-known *hybrid control* may be thought to be an appropriate framework for such purpose [5][10], but it seems, at least for us, to be too general to address some relevant results to our problem. We must abandon the idea of *general* theory. We must confine the class of plants that includes human body to be controlled. It is our premise that human body that has been evolved for long time essentially has some beneficial structure that is suitable for biped locomotion. We assume that we know the way of manipulating actuators to decrease the sensor input, just like a new-born baby can withdraw his/her hand from a hot plate. In other words, we assume that our controller is equipped with some *innate* capability of simple but effective control rules. Since it is difficult to identify such classes of plants in mathematical terms, we just try to mimic human body and brain in the simplest form [16]. Our controller is made of classical McCulloch Pitts neurons [11] with Hebbian plasticity rule [12], with slight modifications. We use the clusters or aggregates of neurons directly in all computations directly. We didn't use any type of simbols in our computations based on the representation of numeric values by the number of firing neurons in the cluster. This idea comes from our approach to mimic brain functions as faithfully as possible, because we believe that the brain does not use any symbolism for execution of lower functions like motor control. The innate rules of motor control built in human body are called *reflex*. Therefore, we call our method *reflex-type* regulation.

We applied our reflex-type regulation to some simple control problems of robots. The attempts produced unexpectedly good results, to our surprise. Then, we extended our experimental ideas to biped locomotion. Again, it gave very good results. The remarkable thing is that the balance which is absolutely necessary for keeping biped walk *emerged spontaneously* in our experiments. We used no model, no trajectory planing, no stabilization at all. After some trials, however, our robot began to walk, first slowly in awkward way, but after a couple of minutes it began to walk steadily and smoothly with its own gait and pace.

The notion of reflex-type regulation was born as an outcome of our long-term challenge to find a universal control principal of biological control [13][14]. In this sense, reflex-type regulator is an example of *compound control* that embodies biological control principles in artifacts [15]. Our theoretical analysis is still behind the experimental advance; in other words, we do not know why balance has emerged. In Section 17.2, we briefly introduce our neurons, the basic computational unit, of our controller. Our controller is introduced in Section 17.3, where three types of controllers are proposed, cluster, output regulator and self-reference regulator. Section 17.4 reports some simple experiments, which is extended to biped locomotion in Section 17.5. Section 17.6 concludes the paper.

17.2 Neurons and Its Cluster

Our computational elements are composed of traditional McCulloch-Pitts type neurons [11] with slight modifications, as shown in Figure 17.1. The inputs signals from neighboring neurons and/or sensors are summed over and compared with the changing threshold. If the summed inputs are greater than the present threshold, the neuron fires. If the summed inputs are not greater than the present threshold, it is silent. This rule is given by

$$x(t) = \begin{cases} 1 & if\ \Sigma_{i=1}^{N} u_i(t) \geq \theta(t), \\ 0 & otherwise, \end{cases} \tag{17.1}$$

where $u_i(t), \theta(t)$ and $x(t)$, denote the i-th input, threshold and output, respectively. In addition to the rule (17.1) which is nothing but a celebrated McCulloch-Pitts rule, we assume that the threshold is tuned according to the output. It is decreased by $\overline{\Delta\theta}$ if it fires, in order to decrease the possibility of firing next time, and it is reduced by $\underline{\Delta\theta}$ if it does not fired, in order to increase the possibility of firing next time. This rule is described simply by the relation

$$\theta(t+1) = \theta(t) + \overline{\Delta\theta}x(t) + \underline{\Delta\theta}\big(x(t) - 1\big). \tag{17.2}$$

In the continuous time frame, it is written in the form

$$\dot{\theta}(t) = \overline{\Delta\theta}x(t) + \underline{\Delta\theta}\big(x(t) - 1\big). \tag{17.3}$$

We introduce the notion of changing connection weight in the spirit of Hebb [12] with slight modification. Let $w(t)$ be the connection weight between input neuron

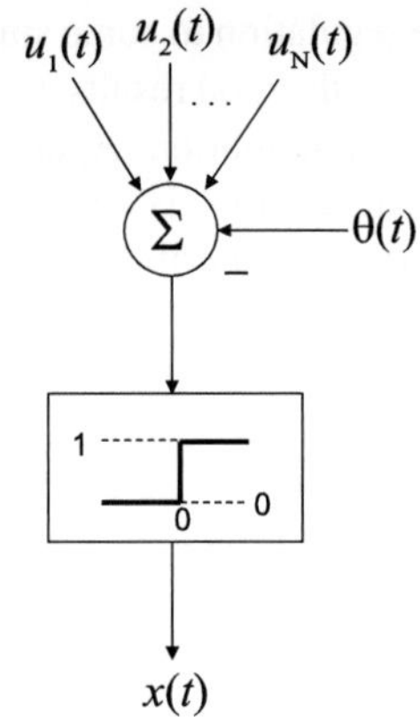

Fig. 17.1. McCulloch-Pitts Type Neuron: The inputs are summed up and compared to the threshold $\theta(t)$.

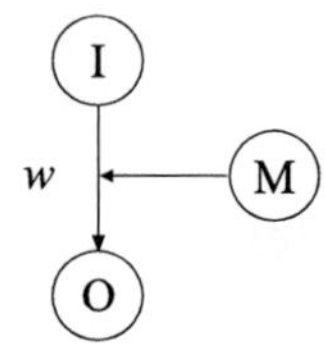

Fig. 17.2. Extended Hebbian Rule: I; input neuron, O; output neuron, M; mediator neuron

I and output neuron O, at time t. We introduce the third neuron M which we call *mediator* as shown in Fig. 17.2. The changing rule of $w(t)$ is as follows. If I, O and M fire simultaneously, it is increased by $\overline{\Delta w}$ in order to strengthen the connectivity of I and O. Otherwise, it is reduced by $\underline{\Delta w}$. The rule is described in the form

$$w(t+1) = w(t) + \overline{\Delta w} x(t) y(t) z(t) + \underline{\Delta w}\big(x(t)y(t)z(t) - 1\big), \tag{17.4}$$

where $x(t), y(t)$ and $z(t)$ are the output of I, O and M, respectively. We call (17.4) the *extended Hebbian rule*, because we introduced mediator which was not found in Hebb's original work.

We frequently use the cluster of neurons represented in Fig. 17.3. The input $u(t)$ is fed commonly to each neuron of the cluster and the output is the sum of the outputs of cluster neurons. Since the output of each neuron is zero or one, the sum of the outputs is always equal to the number of firing neurons in the cluster. If the input weights are all unity and the threshold θ_i of the i-th neuron is set equal to i, then the output is equal to the maximum integer that is not greater than u, which is usually denoted by $[u]$, the Gauss convention, because

$$[u] = \Sigma_{i=1}^{N} sig[u - i], \tag{17.5}$$

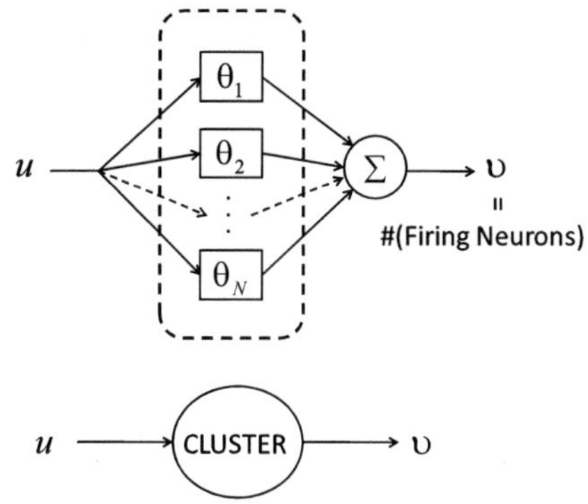

Fig. 17.3. Cluster – Fundamental Module of our Neural Network: The simplified representation is also given.

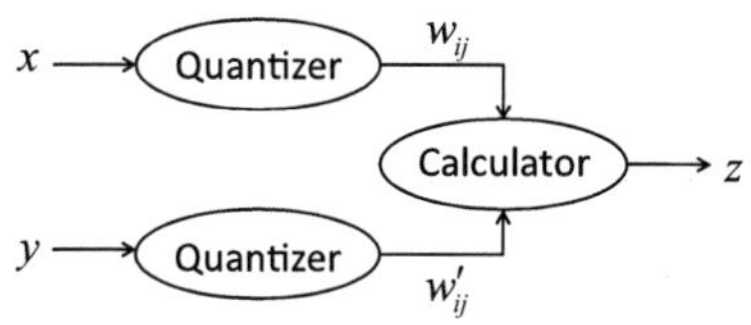

Fig. 17.4. Non-symbolic Calculations

where $sig[a]$ is the discrete sigmoid fundation denoted by

$$sig[a] = \begin{cases} 1 & a > 0, \\ 0 & a \leq 0. \end{cases} \tag{17.6}$$

We call this cluster to be a *quantizer*. We call the cluster with $\theta_i = i$ a *natural order (N.O.) cluster*, which will be used extensively in the sequel.

Let us think of the cluster configuration with two inputs as shown in Fig. 17.4. Actually, by choosing appropriate set of connection weights and thresholds of the cluster, we can impliment elementary arithmetics, addition, subtraction, multiplication and division by the cluster of Fig. 17.4. Examples of weights and thresholds for implimenting the basic arithmetics are shown in Table 17.1.
The multiplication may need explanation. The calculator needs the neurons of the squared number N^2 of that of quantizers N. It consists of N aggregates of N neurons and w_{kij} in Table 17.1 denotes the weight connecting the k-th neuron of a quantizer and the i-th neuron in the j-th aggregate of calculator.

We can also construct a *function generator* for any analytic function $f(u)$ by connecting the two N.O. clusters with weights

$$\begin{aligned} w_{0j} &= f(0), \\ w_{ij} &= f(i) - f(i-1), \quad i = 1, 2, \cdots, N. \end{aligned} \tag{17.7}$$

Table 17.1. Four Arithmetic Operations

	w_{ij}	w'_{ij}	θ_j	#C-cluster
Addition	1	1	j	N
Subtraction	1	-1	0	N
Multiplication	$w_{kij} = \begin{cases} 1, & k=i \\ 0, & k \neq i \end{cases}$	$w'_{kij} = \begin{cases} 1, & k=j \\ 0, & k \neq j \end{cases}$	2	N^2
Subtraction	1	-i	0	N

N: number of neurons in input clusters

where w_{ij} denotes the connection weight between the i-th neuron in the quantizer and the j-th neurons in the function generator (an N.O. cluster) as shown in Fig. 17.5. Since w_{ij} does not depend on j and the number of firing neurons in the upper cluster is $\lfloor u \rfloor$, the inputs to the neurons of the lower cluster become equal to $f(\lfloor u \rfloor)$. Thus, the number of firing neurons in the lower cluster which is actually the output of the network is equal to

$$\Sigma_{i=1}^{f[u]} sig.\lfloor f(\lfloor u \rfloor) - i \rfloor = \lfloor f(\lfloor u \rfloor) \rfloor .$$

Thus, the approximate values of $f(u)$ is computed. The reader may have already noticed that we have dealt with numeric values as the form of the number of firing neurons in the cluster. This gives us the possibility of *non-symbolic* computations of numerical values. The most elementary way of dealing with numerical values is the enumeration by fingers, which is totally the non-symbolic method. Our approach is similar to the enumeration by fingers. It is a good reason to suppose that brain may take this sort of non-symbolic manipulation of the numerical values for lower-level functions like motor control. The method is expected to be extremely robust compared with manipulations using digital representations of the numeric values because malfunction of a small portion of the cluster does not affect the manipulations significantly.

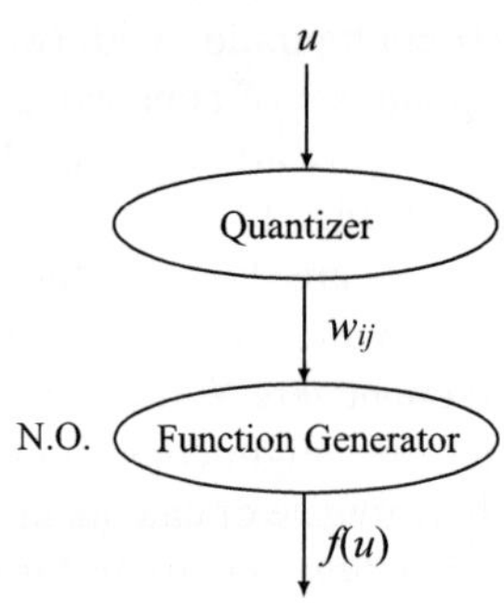

Fig. 17.5. Function Generator

The basic drawback of our non-symbolic manipulation lies in the fact that only positive values can be dealt with. However, we can extend our method to deal with negative values by duplicating the cluster, one for positive values and the other for negative values. See [16] for detail.

Now, we consider macroscopic behaviors of the input/output relations of a cluster under the assumption that the tuning parameters $\overline{\Delta\theta}$ and $\underline{\Delta\theta}$ of the thresholds given in (17.3) are all identical in each neuron of the cluster. Thus, we assume that each neuron is subject to the rule

$$\theta(t+1) = \theta(t) + \alpha x(t) + \beta\left(x(t) - 1\right) \tag{17.8}$$

or

$$\dot{\theta}(t) = \alpha x(t) + \beta\left(x(t) - 1\right), \tag{17.9}$$

where α and β are positive numbers. We also assume that the initial values of the thresholds of the neurons in the cluster are equally distributed in some band with length $\alpha+\beta$. In other words, it is equally distributed in $[\sigma, \sigma+\alpha+\beta]$ for some σ. It is not difficult to see that the thresholds are equally distributed in the band with length $\alpha+\beta$ at each time t. The following theorem which was proven in [10] represents a stabilizing characteristics of the threshold tuning rule (17.8).

Theorem 17.2.1. *Under the assumptions that the neurons are subject to the identical tuning rule of (17.8) and the initial thresholds are distributed identically in a band with width $\alpha+\beta$, then the output $u(t)$ to the input $v(t)$ of the cluster is given by*

$$u(t) = \begin{cases} N, & if \quad v(t) \geq \Theta(t) + \frac{\gamma}{2} \\ \frac{N}{\gamma}\left(v(t) - \Theta(t)\right) + \frac{N}{2}, & if \quad \Theta(t) \leq v(t) < \Theta(t) + \frac{\gamma}{2} \\ 0 & if \quad v(t) < \Theta(t), \end{cases} \tag{17.10}$$

where N denotes the number of neurons in the cluster, $\Theta(t)$ the average threshold at time t given by

$$\Theta(t) = \frac{1}{N}\Sigma_{i=1}^{N}\theta_i(t), \tag{17.11}$$

and finally $\gamma = \alpha+\beta$. The average $\Theta(t)$ follows the dynamics

$$\Theta(t+1) = \Theta(t) + \frac{\gamma}{N}u(t) - \beta \tag{17.12}$$

or

$$\dot{\Theta}(t) = \frac{\gamma}{N}u(t) - \beta.$$

The proof is found in [10].

The representation (17.10) can be simplified by introducing the transformation

$$u'(t) = \frac{\gamma}{N}u(t) - \beta. \tag{17.13}$$

Then, the new output is described by

$$u'(t) = SAT_{\alpha,\beta}\left(v(t) - \Theta(t)\right)$$

and the dynamics of the threshold is given by

$$\Theta(t+1) = \Theta(t) + u'(t),$$

or

$$\dot{\Theta}(t) = u'(t),$$

where $SAT_{\alpha,\beta}(x)$ is a saturation function given by

$$SAT_{\alpha,\beta}(x) = \begin{cases} \alpha, & x \geq \frac{\alpha+\beta}{2} \\ x + \frac{\alpha-\beta}{2} & |x| \leq \frac{\alpha+\beta}{2} \\ -\beta & x \leq -\frac{\alpha+\beta}{2}. \end{cases} \tag{17.14}$$

The profile of the function $SAT_{\alpha,\beta}(x)$ is depicted in Fig. 17.6.

17.3 Control Systems Using Neuron Clusters

Now, we use the clusters described in the preceding section as a feedback controller of the plant. The basic configuration is given in Fig. 17.7, where the dynamics of the threshold $\theta(t)$ acts as an integrator of the controller output (plant input). It is interesting to notice that the dynamics of $\theta(t)$ plays a role of *anti-windup controller* with integral action. Due to this action, the equilibrium point of the closed system is determined solely by the cluster itself, irrespective of the plant, provided that the closed-loop system is stable. To see this, assume that the $\Theta(t)$ converges to Θ_∞ as t goes to infinity. Then, due to (17.12), $u(t)$ must converge to

$$u_\infty = \frac{\beta}{\alpha+\beta}N, \tag{17.15}$$

which means that the equilibrium point is always located in the linear part of the saturation function (17.14). The plant output $y(t)$ converges to the value corresponding to the input (17.15). If the plant is described by a linear state-space model

$$\begin{bmatrix} \dot{x} \\ y \end{bmatrix} = \begin{bmatrix} A & B \\ C & 0 \end{bmatrix} \begin{bmatrix} x \\ u \end{bmatrix}, \tag{17.16}$$

then, the output converges to

$$y_\infty = CA^{-1}Bu_\infty. \tag{17.17}$$

Stability analysis of the closed-loop system can be carried out via multivariable circle criteria or linear matrix inequality which will be discussed elsewhere.

The feedback control scheme of Fig. 17.7 that uses only a single cluster as a feedback controller cannot assign the converging output freely. In order to regulate the plant output freely, the scheme of Fig. 17.8 is considered. In Fig. 17.8, $\otimes$ denotes the multiplication. The controller is represented as

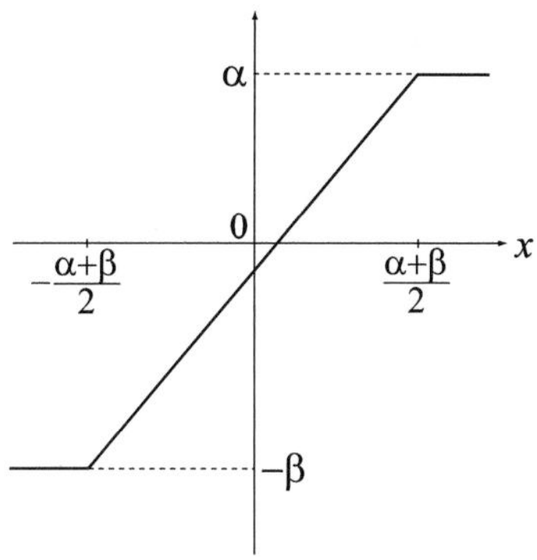

Fig. 17.6. The Function $SAT_{\alpha,\beta}(x)$

$$\begin{aligned} u(t) &= Kg(t)\int_{-\infty}^{t}\big(r-y(\tau)\big)d\tau, \\ g(t) &= SAT_{\alpha,\beta}\big(y(t)-\Theta(t)\big), \\ \dot{\Theta}(t) &= \frac{\alpha+\beta}{N}u(t)-\beta, \end{aligned} \tag{17.18}$$

The above scheme represents a *variable gain servo controller with saturation*. The output of the plant follows the command signal provided that the feedback system is stable. This is a usual servo system based on the integral action with the neuron cluster as its stabilizer.

The output regulation is extended to the so-called *self-reference controller* where the command signal is generated autonomously, as shown in Fig. 17.9. The command r in Fig. 17.8 is replaced by the integration of the plant input $u(t)$ in Fig. 17.9. In this scheme, it is expected that the input $u(t)$ acts as an *inhibitor* to decrease the plant output $y(t)$ in some way due to the basic premise of reflex type regulation.

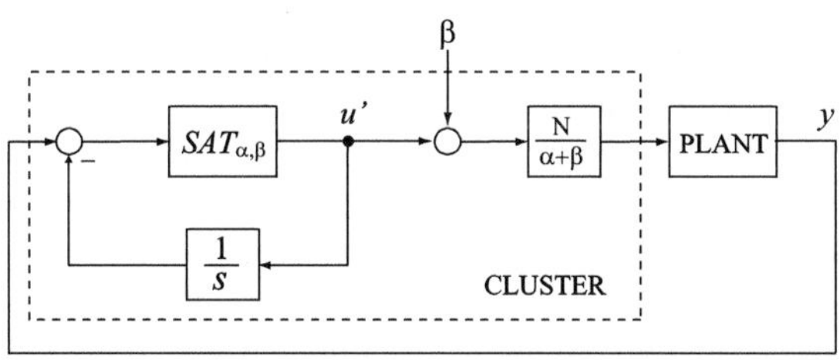

Fig. 17.7. Closed-loop System using Cluster as a Feedback Controller

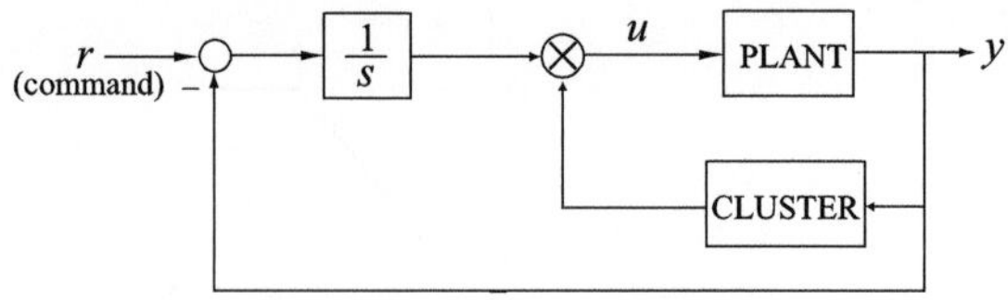

Fig. 17.8. Output Regulation with Cluster as a Feedback Controller
⊗ denotes multiplication

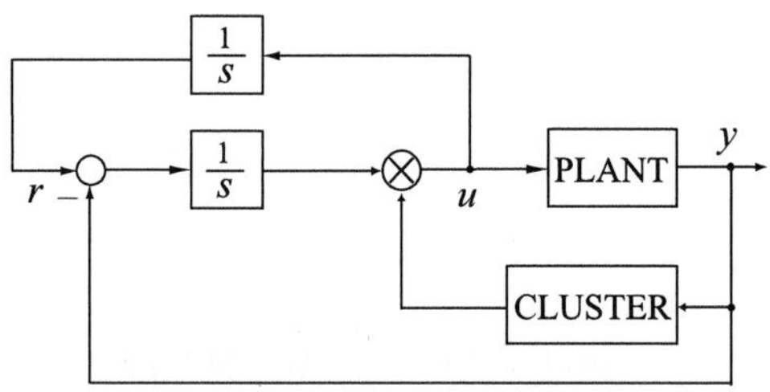

Fig. 17.9. Self-Reference Controller

17.4 Simple Experiments

We show two simple experiments that exhibit superior performances of our neural controller. The first experiment demonstrates that our output regulator is able to find an appropriate trajectory to reach the target. The second one shows that our self-reference regulator is able to find an appropriate posture that minimizes energy consumption.

17.4.1 Trajectory Shaping Experiment

We constructed a single arm to lift a tip load from a lower position to a higher one based on our output regulator. The upper target position is given as the reference. The trajectory connecting the initial location and the target is not specified and left for the *selection* by the regulator. Figure 17.10 shows the trajectories for changing weight of the tip load. It is observed that the horizontal part of the trajectory increases as the tip load weight increases. This is exactly the way we do when we are asked to lift a weight against gravity (Fig. 17.11). The experiment has shown that our output regulator may have capability to mimic the human way of doing task. In other words, our output regulator is able to find an efficient way of carrying out the task [16]. We have not yet been successful to explain why the arm can generate such an efficient trajectory.

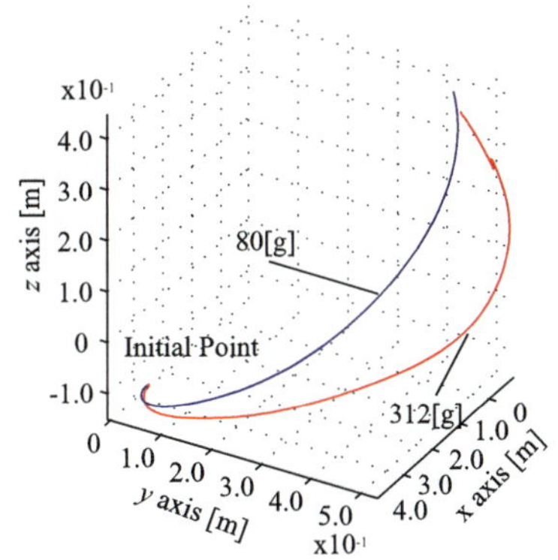

Fig. 17.10. Experimental Result: Trajectories with 80 g and 312 g weights on the end-effector

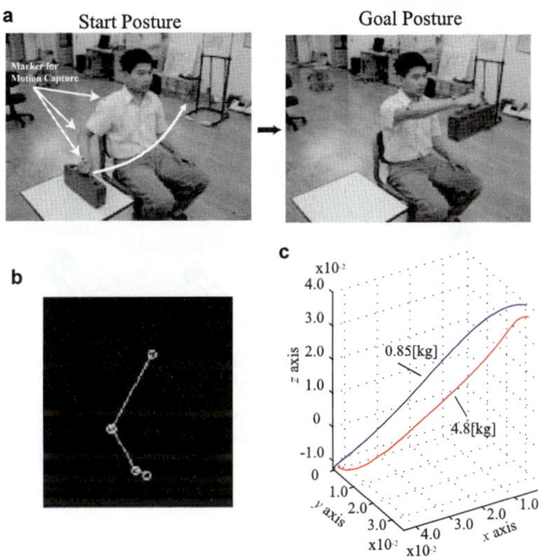

Fig. 17.11. Trajectories of Human's natural motions: **a** Overview of Experiments. **b** Motion capture data. **c** Trajectories with 0.85 g and 4.8 kg weight.

17.4.2 Convergence to Minimum Energy Posture

We constructed a two-degree-of-freedom arm shown in Fig. 17.12. The task of this arm is to take a posture in which the angle between the links is specified. The basement angle (the angle of the lower arm against the horizontal line) is left free. We implement our output regulator to the upper joint to accomplish the task of having a desired angle between the links (Fig. 17.13). To the base joint we implement our self-reference regulator to choose angle against the horizontal line freely "by itself"

(Fig. 17.14). The experiments demonstrated that after a few trials the base joint chose the unique posture as its target which does not require any torque of the base joint [16]. We call this posture as *zero-torque posture*.

Actually, the zero torque posture is characterized as the posture where the mass center of the arm is on the vertical line passing through the base joint. The zero torque posture changes as the tip mass changes, as shown in Fig. 17.13.

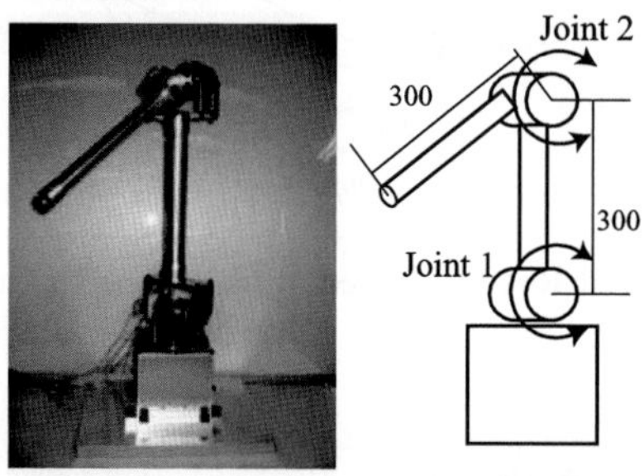

Fig. 17.12. A Two-Degree-of-Freedom arm

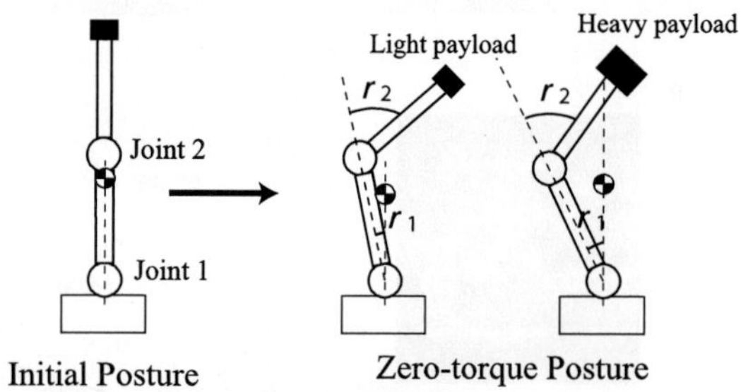

Fig. 17.13. Change of the Center of Mass

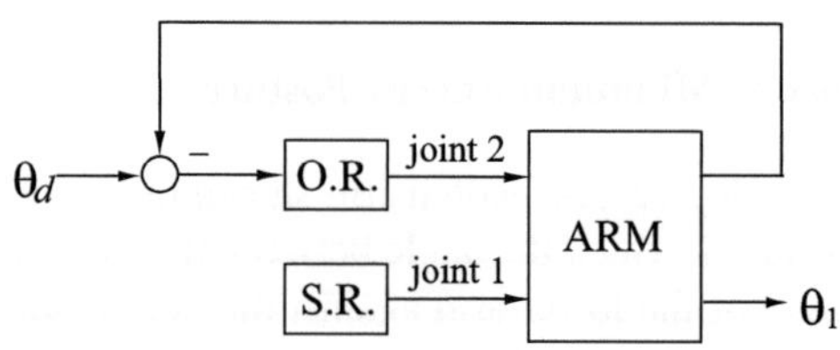

Fig. 17.14. Control Systems for 2DOF Arm

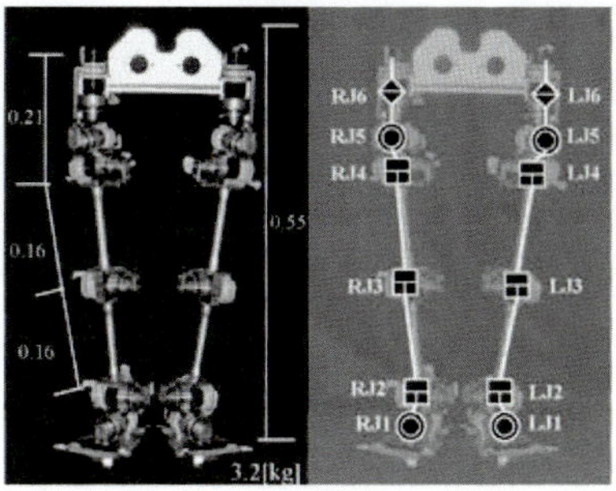

Fig. 17.15. Overview of 12DOF Biped Robot

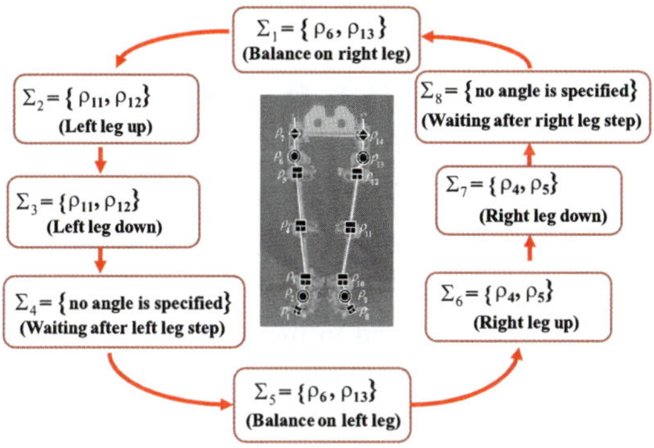

Fig. 17.16. Constraint Conditions for Bipedal Walking: $\Sigma_i = \{$Constraint Conditions (specified angle in Task i)$\}$

17.5 Biped Locomotion

The robot we have constructed is shown in Fig. 17.15. It has 14 DOF with height 0.5m and weight approximately 3.5kg. Each joint is controlled by independent drivers. The locomotion is carried out by control commands to shift from the present posture to the next one. We picked up eight postures (four postures to each leg) described in Fig. 17.16, where ρ_i denotes the i-th joint.

To specify each posture as a *snapshot*, we only use a few joints as is shown in Fig. 17.16 which were controlled by output regulators described in Section 17.3.

No specific command were given to the rest of the joints. Instead, they were asked to *choose their own command signals* by themselves through the self-reference controller discussed in Section 17.3. We did not use any model for control, nor trajectory design for guaranteeing stability. We just impliment output regulators and self-reference regulators to each joint and switched from output regulation to self-reference regulation and vise versa, depending upon what sort of posture transitions were required at each time. It is worth nothing that at each trial, the states of the

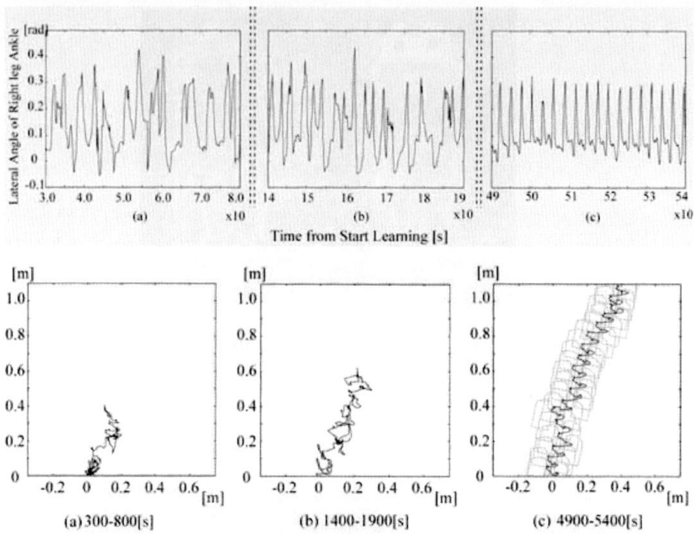

Fig. 17.17. Experimental Results

integrators were memorized and used in the next trial when the robot came to the same posture. This procedure seems to act as a sort of learning.

We summarize our design principles:

1. Snapshot postures are assigned.
2. Each time when a snapshot is fulfilled, the transition to the next snapshot is ordered through the controllers.
3. Those joints which are involved in the specification of the next snapshot are given desired commands, and all the rest of the joints are set free and subject to the selections through self-regulations.
4. If the robot falled down, it was raised up manually by human being, but the controllers were still working during this *salvation*.
5. No model nor trajectory was given.

Fig. 17.17 shows a process of learning through the time profile of leg angle and the trajectory of center of mass. It should be noted that the robot chose its own pace and gate through learning that fit his/her body structure. The pace will change if the robot carries a tip load.

It is surprizing to notice that our robot is also very energy efficient. Fig. 17.18 shows that after learning our robot walked very efficiently under the efficiency index

$$I = \frac{(Consumed\ Energy)}{(Mass) \cdot (Walking\ Distance)}$$

compared with existing humanoid robots. We do not know where this efficiency comes from.

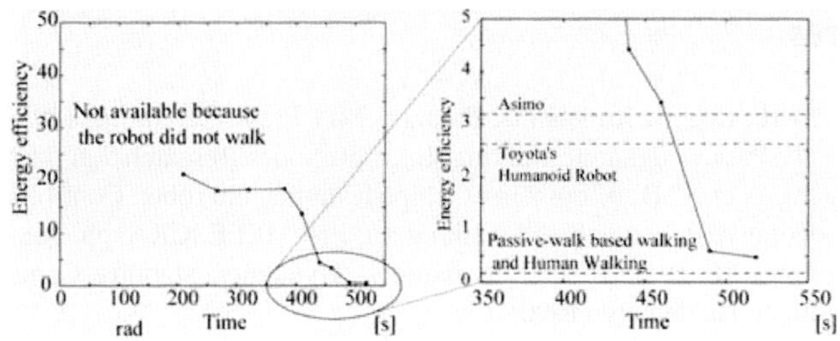

Fig. 17.18. Energy Efficiency

17.6 Concluding Remark

A neuron-based control scheme is proposed that is tailored to robot control, specially to biped locomotion. It has a large number of elementary neurons that obey some simple local rules which date back to McCulloch-Pitts and Hebb. Our controller deals only with such plants that are so simple that the control action can be designed to reduce the input to the sensors during interactions with the environment. Since this knowledge is similar to our innate knowledge from posture reflex, we call this controller as reflex-type regulation.

The reflex-type regulation has a number of distinct characteristic features that enable us to deal with some difficult problems such as model multiplicity, state-discontinuity and control/learning parallelism. These capabilities seem to be given by versatility of our neural schemes, especially of self-reference regulation scheme.

We implemented our reflex-type controller to control simple manipulators. The results were very encouraging. We installed our reflex-type controller for our biped robot. We indicated control actions to the robot for locomotion only, by assigning some snapshots and transition commands. Then, the balance (stability) actually emerged. Our strategy of controling biped locomotion is totally different from existing humanoid robots which are based on precise modeling, fine manufacturing and intensive computations. We do not need any model, nor trajectory planning at all. We just impliment two types of regulators (output regulator and self-reference regulator) to each joint of the robot and just *let it go*. Then, balance soon emerged and is combined locomotion delicately.

There are many things to be done to theoretize our approach. We have not yet been able to identify the class of plants in which our reflex-type regulation works. We have not yet been successful in clarifying the mechanism of how balance emerges through learning. We have not yet captured the essential feature of our learning process in which the integrator state of self-regulators plays a fundamental role.

References

1. Raibert, M.H.: Legged Robots that Balance. MIT Press, Cambridge (1986)
2. McGeer, T.: Passive dynamic walking. Int. J. Robotics Research 9, 62–82 (1990)
3. Yamaguchi, J., et al.: Development of a bipedal humanoid robot: Control method of whole body cooperative dynamic biped walking. In: Proc. IEEE ICRA, pp. 368–374 (1999)
4. Vukobratovic, M., et al.: Biped Locomotion – Dynamics, Stability, Control and Application. Springer, Heidelberg (1990)
5. Grizzel, J.W., et al.: Asymptotically stable walking for biped robots: analysis viasystems with impulse effects. IEEE Trans. Automat. Contr. 46, 51–64 (2001)
6. Kato, R., Mori, M.: Control method of biped locomotion giving asymptotic stability of trajectory. Automatica 20, 405–414 (1984)
7. Gobel, R., et al.: Hybrid systems. IEEE Control Systems Magazine 29, 28–93 (2009)
8. Taga, G., et al.: Self-organized control of bipedal locomotion by neural oscillators in unpredictable environment. Biological Cybernetics 65, 147–159 (1991)
9. Tabuada, P.: Event-triggered real-time scheduling of stabilizing control tasks. IEEE Trans. Automat. Contr. 52, 1680–1685 (2007)
10. Shimoda, S., Kimura, H.: Bio-mimetic approach to tacit learning based on compound control. IEEE Trans. Systems, Man and Cybernetics, B (2009)
11. McCulloch, W.S., Pitts, W.: A logical calculus of the idea immanent in nervous activity. Bull. Math. Biophys. 15, 115–133 (1943)
12. Hebb, D.O.: The Organaization of Behaviors. Wiley, New York (1949)
13. Tanaka, R.J., et al.: Mathematical description of gene regulatory unit. Biophysical J. 91, 1235–1247 (2006)
14. Tanaka, R.J., Kimura, H.: Mathematical classification of regulatory logics for compound environmental changes. J. Theoretical Biology 251, 363–379 (2008)
15. Tanaka, R.J., et al.: Compound control – adaptation to multiple environmental change. IEEE, CDC (2009)
16. Shimoda, S., Kimura, H.: Neural computation scheme of compound control: Tacit learning for bipedal locomotion. SICE J. Control 1(4), 275–283 (2008)

18

Principal Tangent Sytem Reduction[*,†]

Arthur J. Krener[1,‡] and Thomas Hunt[2]

[1] Department of Applied Mathematics, Naval Postgraduate School, Monterey, CA 93943-5216, USA
[2] Department of Mathematics, University of California Davis, CA 95616-8633, USA

Summary. We have outlined a new method of model reduction for nonlinear control systems. The advantages of this new approach are that it is not a local method and it is not an analytic method. All it requires is numerical code that can simulate the system. Further development of this method is needed and we will report on that in future publications.

18.1 Introduction

The paper is a preliminary description of a computationally-based algorithm for obtaining a reduced order nonlinear model from a high order nonlinear model of a *control system*. We call this algorithm Principal Tangent System Reduction (PTSR). By a control system we mean a dynamical process with inputs u and outputs y. The inputs are the way that the external world affects the system and the outputs are the way the system affects the external world. In the model the inputs appear as forcing terms in the dynamics and the outputs are functions of the state x of the dynamics.

We assume that the high order model is encompassed in a large computer code obtained by discretizing a high or infinite dimensional differential (or difference) equation of the form

$$\dot{x} = f(x,u), \qquad y = h(x) \tag{18.1}$$

In the case of an infinite dimensional system, f is a partial differential (difference) operator or a delayed differential operator. We further assume the dimension m of the input u and the dimension p of the output y are small but the dimension n of the state x of the differential equation is high. The control system defines a mapping from input time trajectories to output time trajectories. The computer code allows us to simulate its input to output behavior. The code may be extremely complex consisting

*The authors would like to thank Wei Kang, Francis Giraldo, Kayo Ide, Andrzej Banaszuk and Stephanie Taylor for their thoughtful comments and suggestions.

†This paper is dedicated to our esteemed colleagues and good friends, Christopher Byrnes and Anders Lindquist.

‡Research supported in part by NSF DMS-0505677.

X. Hu et al. (Eds.): Three Decades of Progress in Control Sciences, pp. 265–274, 2010.

of several modules developed over many work years. Our goal is to develop a low order model with approximately the same input to output mapping as the full order model.

There are several reasons for wanting a reduced order model. One is that it can lead to greater understanding of the essential features of the dynamic process. It may be very difficult to develop an intuition and understanding using a large model. And such an understanding is particularly important if one is trying to control the system using feedback.

Another is that it may be very expensive to run the high order model. It may be impossible to do so in real time on existing machines and this may be critical in controlling a plant in real time.

18.2 Principal Tangent System Reduction

Let us now describe the basic elements of our approach. Following Moore [11], Scherpen [12] and Krener [6] the state space of this reduced order model should be the directions that are easiest to control and that affect the output the most. Moore solved this problem for linear systems and Scherpen and Krener extended it to non-linear systems but their approaches were local around the origin. In contrast to these analytic approaches, we would like to develop a more global approach that can be implemented by computational rather than analytic techniques. It is closer to the POD like approach of Lall, Marsden and Glavski [10] but it differs from theirs in two important ways. Their approach is local around an equilibrium point of the system and the reduced state space is a linear subspace of the full state space. Our approach is more global and the reduced state space is a submanifold of the full state space.

The approach will make it possible for reduction of system models defined by computer codes. And the reduced order model will be realized by a simpler computer code. In some cases when the large model is given analytically, so will the reduced model. This will revolutionize the way knowledge is extracted from complicated processes in a multitude of disciplines.

To perform this model reduction we need local measures of relative controllability and relative observability of the various state directions. We discuss controllability first. Consider a state x^0 of the system (18.1), a nominal control $u^0(t)$ and the resulting state $x^0(t)$ and output $y^0(t)$ trajectories through this point satisfying $x^0(0) = x^0$. The controllability question is if we had excited the system with a different control on the time interval $[-T, 0]$ where T is a characteristic time of the system, how easy would it have been to reach states $x = x(0)$ near x^0. To address this question we turn to the linear approximating system around the state trajectory,

$$\dot{\delta x} = F(t)\delta x + G(t)\delta u, \tag{18.2}$$

$$\delta y = H(t)\delta x \tag{18.3}$$

$$
\begin{aligned}
F(t) &= \frac{\partial f}{\partial x}(x^0(t), u^0(t)) \\
G(t) &= \frac{\partial f}{\partial u}(x^0(t), u^0(t)) \\
H(t) &= \frac{\partial h}{\partial x}(x^0(t)) \\
\delta u &\approx u - u^0(t) \\
\delta x &\approx x - x^0(t) \\
\delta y &\approx y - y^0(t)
\end{aligned}
$$

Let $\Phi(t)$ be the fundamental matrix solution of this linear dynamics

$$\frac{\mathrm{d}}{\mathrm{dt}}\Phi(t) = F(t)\Phi(t), \qquad \Phi(0) = I$$

then the controllability gramian at x^0 of the linear approximating system (18.2) is given by

$$P_c(x^0) = \int_{-T}^{0} \Phi^{-1}(t)G(t)G'(t)\Phi^{-1}(t)\,\mathrm{dt} \tag{18.4}$$

The square roots of the eigenvalues of $P_c(x^0)$ are the singular values of the mapping defined by the local linear system (18.2) from inputs $\delta u(t)$ on the interval $[-T, 0]$ to $\delta x(0)$. If any of these are very small then it is difficult for an input to change the state in some direction and such directions may be ignored in a reduced order model.

If we excite the linear approximating system (18.2) starting from $\delta x(-T) = 0$ with $u(t)$ being standard white noise then the covariance of its state at time 0 is $P_c(x^0)$. Looked at from another point of view, if it is invertible then the local controllability gramian measures the minimum control energy needed to excite the linear system from $\delta x(-T) = 0$ to $\delta x(0)$, that energy being $(\delta x^0)'P_c^{-1}(x^0)\delta x^0$. In other words the inverse of the local controllability gramian defines a Riemannian metric on the state space which we call the *controllability metric*.

It would be very difficult to compute the local controllability gramian from (18.4) and impossible for a system defined by computer routines. Therefore we compute it empirically using these computer routines to approximate it. There are several ways of doing this, our preferred method is to excite the nonlinear system (18.1) starting from $x^0(-T)$ with $u(t)$ scaled white noise and take $P_c(x^0)$ to be the inverse scaled covariance of the ensemble of endpoints $x(0)$, more precisely the empirical expectation of $(x(0) - x^0)(x(0) - x^0)'$. The use of white noise ensures a certain degree of robustness in the calculation as the noise driven trajectories will explore the dynamics in the nearby region of the state space. There are subtleties about simulating nonlinear stochastic differential equations that must be taken into account [4].

If the state dimension n is large then this Monte Carlo approach to computing the partial local controllability gramian can be much more efficient because computations of the samples can be done in parallel and the sample size can be much smaller than n. Of course if $l < n$ simulations are done then the resulting covariance can have rank at most l so we are computing only a partial local controllability gramian. Since it is not invertible its inverse does not exist, but we can define a generalized inverse by setting $(\delta x^0)' P_c^{-1}(x^0)\delta x^0 = b' P_c(x^0) b$ if $\delta x^0 = P_c(x^0) b$ and ∞ otherwise. Then it defines a sub Riemannian metric on the state space, with the length of some tangent vectors being infinite.

Even if it is only a partial local controllability gramian, the eigenvectors of the $P_c(x^0)$ corresponding to the largest eigenvalues are the directions that are most likely to be excited by the control. We call these the principal eigenvectors. The other directions are difficult to excite and may be ignored in a reduced order model.

Of course the state coordinates must be appropriately scaled before computing the local controllability gramian but scaling can be a subjective process. In a moment we shall introduce a second Riemannian metric on the state space, the observability metric, and we can use these two metrics to scale each other. The number l of samples used in empirically computing the local controllability gramian should be large compared to the expected state dimension k of the reduced order model so that the impact of the subjective scaling is minimized but it can be small relative to the dimension of the full order model.

Besides its use in model reduction, the local controllability gramian can be used to compare the effectiveness of different forms of actuation for a system. If one seeks to effectively control a plant then one should choose the form of actuation that maximizes its minimum eigenvalues.

Next we discuss the observability metric. Given a state we can simulate the system from nearby states and see how small changes in the state cause small or large changes in the output trajectory. From the output point of view the most important directions are those where there are relatively large changes in the output trajectories for relatively small changes in the initial state. By repeated simulations of the system we shall compute a symmetric matrix field that encodes the sensitivity of the output trajectory to changes in the initial state. If we could compute this matrix field everywhere and if it were everywhere positive definite, it would define a Riemannian metric on the state space called the *observability metric*. But of course we cannot compute it everywhere, fortunately we do not have to as we shall explain below.

The observability metric at x^0 is defined by the local observability gramian $P_o(x^0)$ of the linear approximating system (18.2) over some time interval $[0,T]$ where again T is a time characteristic of the system. The observability gramian of the linear approximating system at x^0 is

$$P_o(x^0) = \int_0^T \Phi'(t) H'(t) H(t) \Phi(t)\, \mathrm{dt} \tag{18.5}$$

We call $P_o(x^0)$ the local observability gramian of the the nonlinear system (18.1). If the system is observable then the local observability gramian is positive definite and defines the observability Riemannian metric. The square roots of the eigenvalues of $P_o(x^0)$ are the singular values of the mapping from initial state δx^0 to output trajectory $\delta y^0(t)$ of the linear approximating system (18.2). We call them the local singular values of the mapping from initial state $x(0)$ to output trajectory $y(t)$ of the nonlinear system (18.2). If any of these singular values are small then some changes in the initial state $x(0)$ have little effect on the output and those state directions can be ignored to obtain a reduced order model with substantially the same input to output behavior.

Note also that if any of these singular values are small then it is difficult to estimate some state coordinates from the output trajectory. The reciprocal of the smallest local singular value is called the local unobservability index. It is a measure of the signal to noise ratio that is necessary to effectively estimate the state of the system from its output trajectory. Any estimation scheme that is exact when no noise is present must have a gain no smaller than the local unobservability index. If an estimation scheme has high gain then it is sensitive to model errors and observation noise.

But computing the observability gramian (18.5) is difficult for all but the simplest systems and it is impossible for systems that are defined by numerical routines. Fortunately it is possible to approximate it empirically, through simulations of the system. Given a frame of directions $V(x^0) = [v^1(x^0), \ldots, v^r(x^0)]$ which is orthonormal,$V'V = I$, but not necessarily complete $r \leq n$, and the length of a small state displacement $\varepsilon > 0$, let $x^{\pm i} = x^0 \pm \varepsilon v^i$ and $y^{\pm i}(t)$ be the corresponding output. If the state dimension n is moderate then the frame V might span all the tangent directions at x^0, $r = n$ but if the state dimension is large then we choose the frame V to span the most important directions from the point of view of controllability. If we have computed the local controllability gramian $P_c(x^0)$ by Monte Carlo simulation then we can take the frame V to be its principal eigenvectors. Notice that the computation of the $2r$ output trajectories $y^{\pm i}(t)$ can be done in parallel.

The empirical local observability gramian at x^0 is the $n \times n$ matrix

$$P_o(x^0) = V(x^0)Q_o(x^0)V'(x^0)$$

whose (i, j) component of $Q_o(x^0)$ is

$$\frac{1}{4\varepsilon^2} \int_0^T (y^{+i}(t) - y^{-i}(t))'(y^{+j}(t) - y^{-j}(t))\, dt \tag{18.6}$$

If $l < n$ then $P_o(x^0)$ is a partial gramian which measures changes in the output due to changes in the state in the directions of the frame. It defines a pseudo Riemannian structure on the state space, some tangent vectors have zero length.

We shall explain how it is used in model reduction below. But before we do that we note that the local observability gramian has uses other than model reduction. It can be used to compare the effectiveness of various observation schemes [9]. In the left side of Figure 18.1 we show a contour plot of the local unobservabity index as a function of the location of an Eulerian observation of the flow of two point vortices. An Eulerian observation is the fluid velocity at a fixed location in the fluid domain. The right side shows the local unobservabity index as a function of the starting location of a Lagrangian observation. A Lagrangian observation is the velocity of a fluid particle. The scale is logarithmic and the local unobservability index varies over seven orders of magnitude. There are some very good places to make an observation and some very poor places.

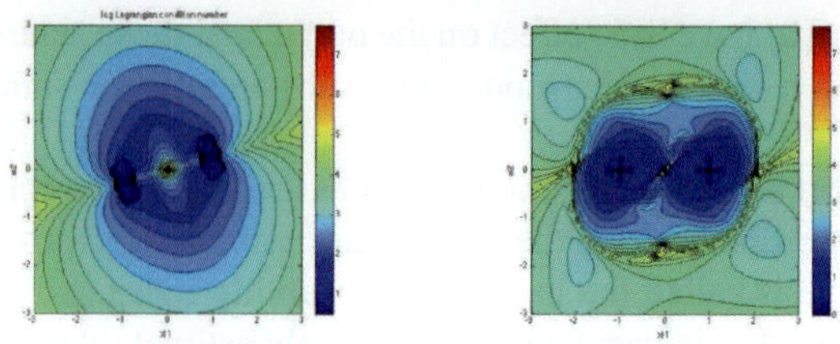

Fig. 18.1. Log of Eulerian and Lagrangian Unobservability Indices

The two (pseudo and sub) metrics on the state space allow us to scale one with the other. The most important directions are those that maximize changes in the output trajectories for given changes in the input trajectories. These are the principal generalized eigenvectors of the observability metric relative to the controllability metric. (By principal generalized eigenvectors we mean the ones corresponding to the largest generalized eigenvalues of the observability metric relative to the controllability metric.) If we could compute them everywhere then these principal eigevectors would span a subbundle of the tangent bundle of the state space. We call this subbundle the principal subbundle. The desired reduced order state space is a submanifold everywhere tangent to the principal subbundle called the principal submanifold. We need to numerically compute the principal subbundle and the principal submanifold. We assume that there is a distinguished point of the state space, perhaps an asymptotically stable equilibrium point of the dynamics to which the system returns in the absence of excitation by the input. (In fact there may be several such points in which case we do the following at each.) Starting at the distinguished point we compute the observability and controllability metrics and the principal eigenvectors of the former relative to the latter. Assume that there are k dominant eigenvalues and the rest are relatively small. Around the distinguished point we construct k simplices in the directions of the principal eigenvectors. We repeat this process a number of times to obtain a piecewise linear k dimensional submanifold. We shall discuss how this is done in more detail when we discuss Principal Tangent Data Reduction below.

After the simplicial complex is constructed, at each of its nodes we project the dynamics onto this submanifold along the directions of the nonprincipal eigevectors to get a reduced dynamics at that node. Using barycentric coordinates on each k simplex, we get the dynamics at states in a simplex by forming the barycentric average of the dynamics at its $k+1$ nodes. In a similar fashion the output mapping is evaluated at each vertex and the output value at a point in a simplex is the barycentic average of the output at its $k+1$ nodes. In this way we construct a reduced order model that is piecewise linear.

It will not always possible to describe the reduced order model analytically but it will be given numerically as routines that compute the reduced dynamics and output at any state on the simplicial complex and any input. To simplify the calculation for any input we shall assume at first that the dynamics is affine in the input

$$\dot{x} = f(x,u) = g_0(x) + \sum g_i(x)u_i$$

Computing the local observability and controllability gramians for a high dimensional system is a daunting task and it might have been impossible a few years ago. That is why Principal Tangent System Reduction is a cyber enabled algorithm. Moreover it is not neccessary to compute the full local observability and controllability gramians, one need only compute it with respect to the most important state directions. Notice also that it is not necessary to compute them everywhere. If we have partially computed the submanifold then all we need to do is compute at the faces along its boundary. Notice also that it is not necessary for k to remain constant so the resulting reduced order state is better thought as a simplicial complex rather than a piecewise linear manifold.

Principal Tangent System Reduction can be viewed as a nonlinear extension of Proper Orthogonal Decomposition (POD) but with several very important differences. In POD one has an empirical covariance and the inverse of this defines an inner product on the linear state space. To normalize this metric another one is needed and this usually is taken to be the standard inner product in the state coordinates in which the system is given. The resulting reduced order system is obtained by Galerkin projection onto the linear subspace spanned by the principal directions.

In PTSR the two metrics arise naturally from the controllability and observability properties of the system. They do depend on the metrics chosen on the input and output spaces but these are smaller spaces for which it is easier to define reasonable metrics. We ignored this point in the discussion above because we implicitly assumed that the standard metrics on the input and output spaces was appropriate. If not then changes of coordinates should be made on the input and output spaces so that the standard metrics are appropriate.

The other way PTSR differs from POD is that the reduced order state space is a piecewise linear submanifold and reduced order dynamics is obtained by Petrov Galerkin projections at each vertex.

18.3 The Local Hankel Point of View

The Hankel map of a control system (18.1) is closely related to its input to output mapping. It is the mapping from past inputs to future outputs. For asymptotically stable systems it can be defined over semi-infinite time intervals. The past control $\{u(t) : t \in (-\infty, 0]\}$ is mapped to the current state $x^0 = x(0)$ where $x(t)$ satisfies

$$\begin{aligned} \dot{x} &= f(x,u) \\ x(-\infty) &= 0 \end{aligned}$$

Then the current state x^0 is mapped to the future output $\{y(t) : t \in [0,\infty)\}$ by

$$\begin{aligned} \dot{x} &= f(x,0) \\ y &= h(x) \\ x(0) &= x^0 \end{aligned}$$

The advantage of the Hankel map over the input to output mapping is that it factors through the current state space at time $t = 0$. The hope is that if the Hankel maps of two systems are close then so are their input to output mappings.

For exponentially stable linear systems both the Hankel and input to output mappings are linear. If the state space is finite dimensional then the Hankel map is of finite rank while the input to output map need not even be compact. B. C. Moore [11] showed that the square roots of the eigenvectors of the infinite time observability gramian relative to the inverse of infinite time controllability gramian are the singular values of its Hankel map. Moore's balanced reduction is the Petrov Galerkin projection of the linear system onto the principal singular value vectors of the Hankel map.

When trying to extend this point of view to nonlinear systems two problems arise. The first is that the system and its associated mappings are nonlinear and so only a local analysis is possible. The second problem is that the nonlinear system may not be globally asymptotically stable so one must compute the gramians over a finite time interval. With this in mind we define the local Hankel map of (18.1) around x^0 with time constant T. For convenience we assume that the reference control is $u^0(t) = 0$. The local Hankel map is the mapping from past inputs $\{\delta u(t) : t \in [-T, 0]\}$ to the current state $\delta x(0)$ defined by the linear approximating system (18.1) with initial condition $\delta x(-T) = 0$ followed by the mapping from current state $\delta x(0)$ to future output $\{y(t) : t \in [0, T\}$ defined by the same linear system with future input $\delta u(t) = 0,\ t \in [0, T]$. The local Hankel singular values of the nonlinear system (18.1) at x^0 are the singular values of the local Hankel map at x^0. These can be computed as the square roots of the eigenvalues of the exact local observability gramian (18.5) relative to the inverse of the exact local controllability gramian (18.4) and approximated from their empirical approximations. Therefore PTSR is a local balanced truncation scheme.

18.4 Principal Tangent Data Reduction

With similar thinking we get a method for nonlinear principal component analysis which we call Principal Tangent Data Reduction (PTDR) [2]. One has a large ensemble of data points in a high-dimensional space. The goal is to fit a low-dimensional submanifold through or near the data points. The first step is to define the concept of a neighborhood of a data point. It may be a ball of fixed radius or it may be a fixed number of closest points. Then one computes the empirical covariance of the directions of the neighboring data points. At each point most of the variability in the data is in the directions of the principal eigenvectors of the local covariance. These local covariances define a symmetric matrix field on the high dimensional space and their principal eigenvectors span a subbbundle of the high dimensional data space.

For simplicity of exposition we assume that there are two principal eigenvalues, the remaining ones are significantly smaller. Starting from a distinguished point in the data (e.g., a point with many neighbors) we construct four triangles in the plane spanned by the two principal eigenvectors. On the outside edges of these triangles, new triangles are constructed as follows. Suppose $[x^1, x^2]$ is an outside edge of the triangle $[x^1, x^2, x^3]$. To find a vertex to form a new triangle, we solve an optimization problem. Let P^1 and P^2 be the local covariances at x^1 and x^2 and $x^m = (x^1 + x^2)/2$. We seek x to maximize $\sqrt{(x - x^1)' P^1 (x - x^1)} + \sqrt{(x - x^2)' P^2 (x - x^2)}$ subject to $|x - x^m| = s$ and $(x - x^m)'(x^3 - x^m) < 0$. Maximizing the objective tends to force x to be near the tangent planes spanned by the principal eigenvectors at x^1 and x^2. Maximizing the sum of square roots rather than the sum of squares tends to make the new triangle close to isosceles. The first constraint ensures that the resulting triangle is neither too big or too small, the parameter s is adjusted based on the length of $[x^1, x^2]$ so as to make the new triangle close to equilateral. The second constraint ensures that the new triangle does not overlap the previous ones.

First we search the nearby outside vertices to see if there is a suitable one to form a new triangle that is a near optimal solution that nearly satisfies the first constraint. If so then we construct the new triangle with this vertex along with x^1 and x^2. If there no such existing vertex then we find the nearest data point and make it the vertex of the new triangle with other vertices x^1 and x^2.

In Figure 18.2 we show an example using 5,000 data points in $\mathbb{R}^3$ approximately uniformly distributed on a torus of radii 4 and 1. Each of the data points is corrupted by a Gaussian random vector of magnitude 0.05. There are 451 triangles.

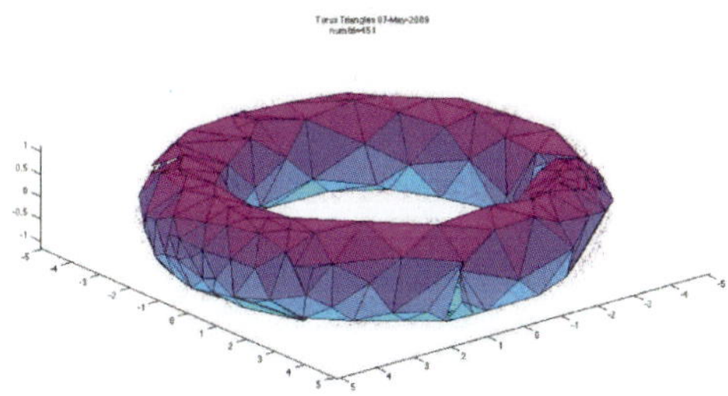

Fig. 18.2. Example of Principal Tangent Data Reduction

References

1. Hermann, R., Krener, A.J.: Nonlinear controllability and observability. IEEE Trans. Automat. Control 22, 728–740 (1977)
2. Hunt, T., Krener, A.J.: Principal Tangent Data Reduction (submitted)
3. Ide, K., Kuznetsov, L., Jones, C.K.R.T.: Lagrangian data assimilation for point-vortex systems. Journal of Turbulence, 053 (2002)
4. Kloden, P.E., Platen, E.: Numerical Solution of Stochastic Differential Equations. Springer, Berlin (1992)

5. Krener, A.J.: The Importance of State Coordinates of a Nonlinear System. In: Bonivento, C., Isisdori, A., Marconi, L., Rossi, C. (eds.) Advances in Control Theory and Applications. LNCIS, vol. 353, pp. 161–170. Springer, Heidelberg (2006)
6. Krener, A.J.: Reduced Order Models for Nonlinear Control Systems. In: Astolfi, A., Marconi, L. (eds.) Analysis and Design of Nonlinear Control Systems, In Honor of Alberto Isidori, Springer, Heidelberg (2008)
7. Krener, A.J.: Observability of Vortex Flows. In: Fourty Seventh Conference on Decision and Control, Cancun, Mexico (2008)
8. Krener, A.J.: Eulerian and Lagrangian Observability of Point Vortex Flows. Tellus A 60, 1089–1102 (2008)
9. Krener, A.J., Ide, K.: Measures of Unobservability (submitted)
10. Lall, S., Marsden, J.E., Glavski, S.: A subspace approach to balanced truncation for model reduction of nonlinear control systems. Int. J. Robust Nonlinear Control 12, 519–535 (2002)
11. Moore, B.C.: Principle Component Analysis in Linear Systems: Controllability, Observability and Model Reduction. IEEE Trans. Auto. Con. 26, 17–32 (1981)
12. Scherpen, J.M.A.: Balancing for Nonlinear Systems. Systems and Control Letters 21, 143–153 (1993)

19

The Contraction Coefficient of a Complete Gossip Sequence*,†

J. Liu[1], A.S. Morse[1], B.D.O. Anderson[2], and C. Yu[2]

[1] Yale University, New Haven, CT 06511, USA
[2] Australian National University and National ICT Australia, Canberra ACT 0200, Australia

Summary. A sequence of allowable gossips between pairs of agents in a group is *complete* if the gossip graph which the sequence generates contains a tree spanning the graph of all allowable gossip pairs. The stochastic matrix of a sequence of allowable gossips is shown to be a contraction for an appropriately defined Euclidian seminorm, if and only if the gossip sequence is complete. The significance of this result in determining the convergence rate of an infinite aperiodic sequence of gossips is explained.

19.1 Introduction

There has been considerable interest recently in developing algorithms for distributing information among the members of a group of sensors or mobile autonomous agents via local interactions. Notable among these are those algorithms intended to cause such a group to reach a consensus in a distributed manner [9, 12, 2, 11, 13]. In a typical consensus seeking process, the agents in a given group are all trying to agree on a specific value of some quantity. Each agent initially has only limited information available. The agents then try to reach a consensus by communicating what they know to their neighbors either just once or repeatedly, depending on the specific problem of interest. One particular consensus problem which has received much attention is called "gossiping." In a typical gossiping problem each agent has control over a real-valued scalar "gossiping" variable. What distinguishes gossiping

*The authors wish to thank Shaoshuai Mou, Ming Cao and Sekhar Tatikonda for useful discussions which have contributed to this work. The research of the first two authors is supported by the US Army Research Office, the US Air Force Office of Scientific, and the National Science Foundation. B.D.O. Anderson is supported by Australian Research Council's Discovery Project DP-0877562 and National ICT Australia-NICTA. NICTA is funded by the Australian Government as represented by the Department of Broadband, Communications and the Digital Economy and the Australian Research Council through the ICT Centre of Excellence program. C. Yu is supported by the Australian Research Council through an Australian Postdoctoral Fellowship under DP-0877562.

†This paper is dedicated to Chris Byrnes and Anders Lindquist.

X. Hu et al. (Eds.): Three Decades of Progress in Control Sciences, pp. 275–289, 2010.

from more general consensus seeking is that during a gossiping process only two agents are allowed to communicate with each other at any one clock time. A pair of agents gossip by updating the current values of their gossip variables to new values which are both equal to the average of their current values. Generally not every pair of agents is allowed to gossip. The edges of a given, undirected "allowable gossip graph" specify which gossip pairs are allowable. The actual sequence of gossip pairs which occurs during a specific gossip sequence might be determined either probabilistically [4] or deterministically, depending on the problem of interest. It is the latter type of problem to which this paper is addressed.

Of particular interest is the rate at which a sequence of agent gossip variables converge to a common value. The convergence rate question for more general deterministic consensus problems has been studied recently in [6, 1]. In [4, 3, 15] the convergence rate question is addressed for gossiping algorithms in which the sequence of gossip pairs under consideration is determined probabilistically. A modified gossiping algorithm intended to speed up convergence is proposed in [7] without proof of correctness, but with convincing experimental results. The algorithm has recently been analyzed in [10].

A typical gossiping process can be modeled as a discrete time linear system of the form $x(t+1) = M(t)x(t), \quad t = 0,1,\dots$ where x is a vector of agent gossip variables and each value of $M(t)$ is a specially structured doubly stochastic matrix (see section 19.2). A sequence of allowable gossips pairs is *complete* if the gossip graph which the sequence generates contains a tree spanning the graph of all allowable gossip pairs. (see section 19.6). The specific goal of this paper is to find a seminorm with respect to which the state transition matrix of any such complete sequence is a contraction. The role played by seminorms in characterizing convergence rate is explained in section 19.5. Three different types of seminorms are considered in section 19.5. Each is compared to the well known coefficient of ergodicity which plays a central role in the study of convergence rates for nonhomogeneous Markov chains [14]. Somewhat surprisingly, it turns out that a particular Euclidean seminorm on $\mathbb{R}^{n\times n}$ has the required property - namely that in this seminorm, the state transition matrix of any complete gossip sequence is a contraction. The value of the seminorm for a given gossip sequence is what is meant by the "contraction coefficient" of the sequence.

19.2 Gossiping

The type of gossiping we want to consider involves a group of n agents labeled 1 to n. Each agent i has control over a real-valued scalar quantity x_i called a *gossip variable* which the agent is able to update at discrete clock times $t = 1,2,\dots$. A *gossip* occurs at time t between agents i and j if the values of both agents variables at time $t+1$ equal the average of their values at time t. In other words $x_i(t+1) = x_j(t+1) = \frac{1}{2}(x_i(t) + x_j(t))$. Generally not every pair of agents are allowed to gossip. The edges of a given simple directed n-vertex graph $\mathbb{A}$, called an *allowable gossip graph*, specify which gossip pairs are allowable. In other words a gossip between agents i and j is *allowable* if (i,j) is an edge in $\mathbb{A}$. In this paper we will stipulate that at most one allowable gossip can occur at each clock time; the value of the gossip variable of any agent which does not gossip at time t, does not change at time $t+1$. The goal of gossiping is for the n agents to reach a consensus in the sense that all n gossip variable ultimately reach the same value in the limit as $t \to \infty$. For this to be possible, no matter what the initial values of the gossiping variables are, it is clearly necessary that $\mathbb{A}$ be a connected graph, an assumption we henceforth make.

A gossiping process can be conveniently modeled as a discrete time linear system of the form $x(t+1) = M(t)x(t), \quad t = 0,1,\ldots,$ where $x \in \mathbb{R}^n$ is a state vector of gossiping variable and $M(t)$ is a matrix characterizing how x changes as the result of the gossip between two given agents at time t. Because there are only a finite number of allowable gossips, there are only a finite number of values that each $M(t)$ can take on. In particular, if agents i and j gossip at time t, then $M(t) = S_{ij}$ where S_{ij} is the $n \times n$ matrix for which $s_{ii} = s_{ij} = s_{ji} = s_{jj} = \frac{1}{2}$, $s_{kk} = 1, k \notin \{i,j\}$ and all remaining entries equal zero. Thus S_{ij} is a nonnegative matrix whose row sums and column sums all equal one. Matrices with these two properties are called *doubly stochastic*. Note that the type of doubly stochastic matrix which characterizes a gossip {i.e., a *gossip matrix*} has two additional properties - it is symmetric and its diagonal entries are all positive. Mathematically, reaching a consensus by means of an infinite sequence of gossips modeled by a corresponding infinite sequence of gossip matrices $M(1), M(2), \ldots,$ means that the sequence of matrix products $M(1)$, $M(2)M(1)$, $M(3)M(2)M(1)$, ... converges to a matrix of the form $\mathbf{1}c$ where $\mathbf{1} \in \mathbb{R}^n$ is a vector whose entries are all ones. It turns out that if convergence occurs, the limit matrix $\mathbf{1}c$ is also a doubly stochastic matrix which means that $c = \frac{1}{n}\mathbf{1}'$.

19.3 Stochastic Matrices

Doubly stochastic matrices are special types of "stochastic matrices" where by a *stochastic matrix* is meant a nonnegative $n \times n$ matrix whose row sums all equal one. It is easy to see that a nonnegative matrix S is stochastic if and only if $S\mathbf{1} = \mathbf{1}$. Similarly a nonnegative matrix S is doubly stochastic if and only if $S\mathbf{1} = \mathbf{1}$ and $S'\mathbf{1} = \mathbf{1}$. Using these characterizations it is easy to prove that the class of stochastic matrices in $\mathbb{R}^{n \times n}$ is closed under multiplication as is the class of doubly stochastic matrices in $\mathbb{R}^{n \times n}$. It is also true that the class of nonnegative matrices in $\mathbb{R}^{n \times n}$ with positive diagonals is closed under multiplication.

19.3.1 Graph of a Stochastic Matrix

Many properties of a stochastic matrix can be usefully described in terms of an associated directed graph determined by the matrix. The *graph* of non-negative matrix $M \in \mathbb{R}^{n \times n}$, written $\gamma(M)$, is a directed graph on n vertices with an arc from vertex i to vertex j just in case $m_{ji} \neq 0$; if (i,j) is such an arc, we say that i is a *neighbor* of j and that j is an *observer* of i. Thus $\gamma(M)$ is that directed graph whose adjacency matrix is the transpose of the matrix obtained by replacing all nonzero entries in M with ones.

19.3.2 Connectivity

There are various notions of connectivity which are useful in the study of the convergence of products of stochastic matrices. Perhaps the most familiar of these is the idea of "strong connectivity." A directed graph is *strongly connected* if there is a directed path between each pair of distinct vertices. A directed graph is *weakly connected* if there is an undirected path between each pair of distinct vertices. There are other notions of connectivity which are also useful in this context. To define several of them, let us agree to call a vertex i of a directed graph $\mathbb{G}$, a *root* of $\mathbb{G}$ if for each other vertex j of $\mathbb{G}$, there is a directed path from i to j. Thus i is a root of $\mathbb{G}$, if it is the root of a directed spanning tree of $\mathbb{G}$. We will say that $\mathbb{G}$ is *rooted at* i if i is in fact a root. Thus $\mathbb{G}$ is rooted at i just in case each other vertex of $\mathbb{G}$ is *reachable* from vertex i along a directed path within the graph. $\mathbb{G}$ is *strongly rooted at* i if each other vertex of $\mathbb{G}$ is reachable from vertex i along a directed path of length 1. Thus $\mathbb{G}$ is strongly rooted at i if i is a neighbor of every other vertex in the graph. By a *rooted graph* $\mathbb{G}$ is meant a directed graph which possesses at least one root. A *strongly rooted graph* is a graph which has at least one vertex at which it is strongly rooted. Note that a nonnegative matrix $M \in \mathbb{R}^{n\times n}$ has a strongly rooted graph if and only if it has a positive column. Note that every strongly connected graph is rooted and every rooted graph is weakly connected. The converse statements are false. In particular there are weakly connected graphs which are not rooted and rooted graphs which are not strongly connected.

19.3.3 Composition

Since we will be interested in products of stochastic matrices, we will be interested in graphs of such products and how the are related to the graphs of the matrices comprising the products. For this we need the idea of "composition" of graphs. Let $\mathbb{G}_p$ and $\mathbb{G}_q$ be two directed graphs with vertex set $\mathcal{V}$. By the *composition* of $\mathbb{G}_p$ with $\mathbb{G}_q$, written $\mathbb{G}_q \circ \mathbb{G}_p$, is meant the directed graph with vertex set $\mathcal{V}$ and arc set defined in such a way so that (i,j) is an arc of the composition just in case there is a vertex k such that (i,k) is an arc of $\mathbb{G}_p$ and (k,j) is an arc of $\mathbb{G}_q$. Thus (i,j) is an arc in $\mathbb{G}_q \circ \mathbb{G}_p$ if and only if i has an observer in $\mathbb{G}_p$ which is also a neighbor of j in $\mathbb{G}_q$. Note that composition is an associative binary operation; because of this, the definition extends unambiguously to any finite sequence of directed graphs $\mathbb{G}_1, \mathbb{G}_2, \dots, \mathbb{G}_k$ with the same vertex set.

Composition and matrix multiplication are closely related. In particular, the graph of the product of two nonnegative matrices $M_1, M_2 \in \mathbb{R}^{n\times n}$, is equal to the composition of the graphs of the two matrices comprising the product. In other words, $\gamma(M_2M_1) = \gamma(M_2) \circ \gamma(M_1)$.

If we focus exclusively on graphs with self-arcs at all vertices, more can be said. In this case the definition of composition implies that the arcs of both $\mathbb{G}_p$ and $\mathbb{G}_q$ are arcs of $\mathbb{G}_q \circ \mathbb{G}_p$; the converse is false. The definition of composition also implies that if $\mathbb{G}_p$ has a directed path from i to k and $\mathbb{G}_q$ has a directed path from k to j, then $\mathbb{G}_q \circ \mathbb{G}_p$ has a directed path from i to j. These implications are consequences of the requirement that the vertices of the graphs in question have self arcs at all vertices. It is worth emphasizing that the union of the arc sets of a sequence of graphs $\mathbb{G}_1, \mathbb{G}_2, \ldots, \mathbb{G}_k$ with self-arcs must be contained in the arc set of their composition. However the converse is not true in general and it is for this reason that composition rather than union proves to be the more useful concept for our purposes.

19.4 Convergability

It is of obvious interest to have a clear understanding of what kinds of stochastic matrices within an infinite product guarantee that the infinite product converges. There are many ways to address this issue and many existing results. Here we focus on just one issue.

Let $\mathscr{S}$ denote the set of all stochastic matrices in $\mathbb{R}^{n\times n}$ with positive diagonals. Call a compact subset $\mathscr{M} \subset \mathscr{S}$ *convergable* if for each infinite sequence of matrices $M_1, M_2, \ldots, M_i, \ldots$ from $\mathscr{M}$, the sequence of products $M_1M_1, M_3M_2M_1 \cdots$ converges exponentially fast to a matrix of the form $\mathbf{1}c$. Convergability can be characterized as follows.

Theorem 19.4.1. *Let $\mathscr{R}$ denote set of all matrices in $\mathscr{S}$ with rooted graphs. Then a compact subset $\mathscr{M} \subset \mathscr{S}$ is convergable if and only if $\mathscr{M} \subset \mathscr{R}$.*

The theorem implies that $\mathscr{R}$ is the largest subset of $n \times n$ stochastic matrices with positive diagonals whose compact subsets are all is convergable. $\mathscr{R}$ itself is not convergable because it is not closed and thus not compact.

Proof of Theorem 19.4.1: The fact that any compact subset of $\mathscr{R}$ is convergable is an immediate consequence of Proposition 11 of [5]. To prove the converse, suppose that $\mathscr{M} \subset \mathscr{S}$ is convergable. Then by continuity, every sufficiently long product of matrices from $\mathscr{M}$ must be a matrix with a positive column. Therefore, the graph of every sufficiently long product of matrices from $\mathscr{M}$ must be strongly rooted. It follows from Proposition 5 of [5], that $\mathscr{M}$ must be a subset of $\mathscr{R}$. $\square$

Although doubly stochastic matrices are stochastic, convergability for classes of doubly stochastic matrices has a different characterization than it does for classes of stochastic matrices. Let $\mathscr{D}$ denote the set of all doubly stochastic matrices in $\mathscr{S}$. In the sequel we will prove the following theorem.

Theorem 19.4.2. *Let $\mathscr{W}$ denote set of all matrices in $\mathscr{D}$ with weakly connected graphs. Then a compact subset $\mathscr{M} \subset \mathscr{D}$ is convergable if and only if $\mathscr{M} \subset \mathscr{W}$.*

The theorem implies that $\mathscr{W}$ is the largest subset of $n \times n$ doubly stochastic matrices with positive diagonals whose compact subsets are all is convergable. Like $\mathscr{R}$, $\mathscr{W}$ is not convergable because it is not compact.

An interesting set of stochastic matrices in $\mathscr{S}$ whose compact subsets are known to be convergable, is the set of all "scrambling matrices." A matrix $S \in \mathscr{S}$ is *scrambling* if for each distinct pair of integers i, j, there is a column k of S for which s_{ik} and s_{jk} are both nonzero [14]. In graph theoretic terms S is a scrambling matrix just in case its graph is "neighbor shared" where by *neighbor shared* we mean that each distinct pair of vertices in the graph share a common neighbor [5]. Convergability of compact subsets of scrambling matrices is tied up with the concept of the coefficient of ergodicity [14] which for a given stochastic matrix $S \in \mathscr{S}$ is defined by the formula

$$\tau(S) = \frac{1}{2} \max_{i,j} \sum_{k=1}^{n} |s_{ik} - s_{jk}|$$

It is known that $0 \leq \tau(S) \leq 1$ for all $S \in \mathscr{S}$ and that

$$\tau(S) < 1 \tag{19.1}$$

if and only if S is a scrambling matrix. It is also known that

$$\tau(S_2 S_1) \leq \tau(S_2)\tau(S_1), \quad S_1, S_2 \in \mathscr{S} \tag{19.2}$$

It can be shown that 19.1 and 19.2 are sufficient conditions to ensure that any compact subset of scrambling matrices is convergable. But $\tau(\cdot)$ has another role. It provides a worst case convergence rate for any infinite product of scrambling matrices from a given compact set $\mathscr{C} \subset \mathscr{S}$. In particular, it can be easily shown that as $i \to \infty$, any product $S_i S_{i-1} \cdots S_2 S_1$ of scrambling matrices $S_i \in \mathscr{C}$ converges as to a matrix of the form $\mathbf{1}c$ as fast as λ^i where

$$\lambda = \max_{S \in \mathscr{C}} \tau(S)$$

This preceding discussion suggests the following question. Can analogs of the coefficient of ergodicity satisfying formulas like 19.1 and 19.2 be found for the set of stochastic matrices with rooted graphs or perhaps for the set of doubly stochastic matrices with weakly connected graphs? In the sequel we will provide a partial answer to this question for the case of stochastic matrices and a complete answer for the case of doubly stochastic matrices. Our approach will be to appeal to certain types of seminorms of stochastic matrices.

19.5 Seminorms

Let $||\cdot||_p$ be the induced p norm on $\mathbb{R}^{m\times n}$. We will be interested in $p = 1, 2, \infty$. Note that

$$||A||_1 = \max \text{column sum } A \qquad ||A||_2 = \sqrt{\mu(A'A)} \qquad ||A||_\infty = \max \text{row sum } A$$

where $\mu(A'A)$ is the largest eigenvalue of $A'A$; that is, the square of the largest singular value of A. For $M \in \mathbb{R}^{m\times n}$ define

$$|M|_p = \min_{c \in \mathbb{R}^{1\times n}} ||M - \mathbf{1}c||_p$$

As defined, $|\cdot|_p$ is nonnegative and $|M|_p \leq ||M||_p$. Let M_1 and M_2 be matrices in $\mathbb{R}^{m\times n}$ and let c_0, c_1 and c_2 denote values of c which minimize $||M_1 + M_2 - \mathbf{1}c||_p$, $||M_1 - \mathbf{1}c||_p$ and $||M_2 - \mathbf{1}c||_p$ respectively. Note that

$$\begin{aligned}|M_1 + M_2|_p = ||M_1 + M_2 - \mathbf{1}c_0||_p &\leq ||M_1 + M_2 - \mathbf{1}(c_1 + c_2)||_p \\ &\leq ||M_1 - \mathbf{1}c_1||_p + ||M_2 - \mathbf{1}c_2||_p \\ &= |M_1|_p + |M_2|_p.\end{aligned}$$

Thus the triangle inequality holds which makes $|\cdot|_p$ a *seminorm.* $|\cdot|_p$ behaves much like a norm. For example, if N is a sub-matrix of M, then $|N|_p \leq |M|_p$. However $|\cdot|_p$ is not a norm because $|M|_p = 0$ does not imply $M = 0$; rather it implies that $M = \mathbf{1}c$ for some row vector c which minimizes $||M - \mathbf{1}c||_p$. For our purposes, $|\cdot|_p$ has a particularly important property:

Lemma 19.5.1. *Suppose $\mathscr{M}$ is a subset of $\mathbb{R}^{n\times n}$ such that $M\mathbf{1} = \mathbf{1}$ for all $M \in \mathscr{M}$. Then*

$$|M_2 M_1|_p \leq |M_2|_p |M_1|_p \tag{19.3}$$

We say that $|\cdot|_p$ is *sub-multiplicative* on $\mathscr{M}$.

Proof of Lemma 19.5.1: Let c_0, c_1 and c_2 denote values of c which minimize $||M_2M_1 - \mathbf{1}c||_p, ||M_1 - \mathbf{1}c||_p$ and $||M_2 - \mathbf{1}c||_p$ respectively. Then

$$\begin{aligned}|M_2M_1|_p &= ||M_2M_1 - \mathbf{1}c_0||_p \\ &\leq ||M_2M_1 - \mathbf{1}(c_2M_1 + c_1 - c_2\mathbf{1}c_1)||_p \\ &= ||M_2M_1 - \mathbf{1}c_2M_1 - M_2\mathbf{1}c_1 + \mathbf{1}c_2\mathbf{1}c_1)||_p \\ &= ||(M_2 - \mathbf{1}c_2)(M_1 - \mathbf{1}c_1)||_p \\ &\leq ||(M_2 - \mathbf{1}c_2)||_p||(M_1 - \mathbf{1}c_1)||_p \\ &= |M_2|_p|M_1|_p\end{aligned}$$

Thus 19.3 is true. □

We say that $M \in \mathbb{R}^{n\times n}$ is *semi-contractive* in the p-norm if $|M|_p < 1$. In view of Lemma 19.5.1, the product of semi-contractive matrices in $\mathscr{M}$ is thus semi-contractive. The importance of these ideas lies in the following fact.

Proposition 19.5.1. *Suppose $\mathscr{M}$ is a subset of $\mathbb{R}^{n\times n}$ such that $M\mathbf{1} = \mathbf{1}$ for all $M \in \mathscr{M}$. Let p be fixed and let $\mathscr{M}$ be a compact set of semi-contractive matrices in $\mathscr{M}$. Let*

$$\lambda = \sup_{\mathscr{M}} |M|_p$$

Then for each infinite sequence of matrices $M_i \in \bar{\mathscr{M}}$, $i \in \{1, 2, \ldots\}$, the matrix product

$$M_iM_{i-1}\cdots M_1$$

converges as fast as λ^i to a rank one matrix of the form $\mathbf{1}\bar{c}$.

Proof of Proposition 19.5.1: To be given in the full length version of this paper.

19.5.1 The Case $p = 1$

We now consider in more detail the case when $p = 1$. For this case it is possible to derive an explicit formula for the seminorm $|M|_1$ of a nonnegative matrix $M \in \mathbb{R}^{n\times n}$.

Proposition 19.5.2. *Let q be the unique integer quotient of n divided by 2. Let $M \in \mathbb{R}^{n\times m}$ be a non-negative matrix. Then*

$$|M|_1 = \max_{j\in\{1,2,\ldots,n\}} \left\{ \sum_{i\in\mathscr{L}_j} m_{ij} - \sum_{i\in\mathscr{S}_j} m_{ij} \right\}$$

where $\mathscr{L}_j$ and $\mathscr{S}_j$ are respectively the row indices of the q largest and q smallest entries in the jth column of M.

This result is a direct consequence of the following lemma and the definition of $|\cdot|_1$.

Lemma 19.5.2. *Let q denote the unique integer quotient of n divided by 2. Let y be a non-negative n vector and write $\mathscr{L}$ and $\mathscr{S}$ for the row indices of the q largest and q smallest entries in y respectively. Then*

$$|y|_1 = \sum_{i\in\mathscr{L}} y_i - \sum_{i\in\mathscr{S}} y_i \qquad (19.4)$$

where y_i is the ith entry in y.

Proof of Lemma 19.5.2: To be given in the full length version of this paper.

Consider now the case when M is a doubly stochastic matrix S. Then the column sums of S are all equal to 1. This implies that $|S|_1 \leq 1$ because $|S|_1 \leq ||S||_1 \leq 1$. The column sums all equaling one also implies that

$$\sum_{i\in\mathscr{L}_j} s_{ij} + rs_{(q+r)j} + \sum_{i\in\mathscr{S}_j} a_{ij} = 1, \quad j \in \{1,2,\ldots,n\}$$

Therefore

$$|S|_1 = \max_{i\in\{1,2,\ldots,n\}} \left\{ 2\sum_{i\in\mathscr{L}_j} s_{ij} + rs_{(q+r)j} - 1 \right\}$$

This means that S is semi-contractive in the one-norm just in case

$$\sum_{i\in\mathscr{L}_j} s_{ij} + \frac{r}{2} s_{(q+r)j} < 1, \quad j \in \{1,2,\ldots,n\}$$

We are led to the following result.

Theorem 19.5.1. *Let q be the unique integer quotient of n divided by 2. Let $S \in \mathbb{R}^{n\times n}$ be a doubly stochastic matrix. A doubly stochastic matrix. Then $|S| \leq 1$. Moreover S is a semicontraction in the one norm if and only if the number of nonzero entries in each column of S exceeds q.*

19.5.2 The Case $p = 2$

For the case when $p = 2$ it is also possible to derive an explicit formula for the seminorm $|M|_2$ of a nonnegative matrix $M \in \mathbb{R}^{n\times n}$. Towards this end note that for any $x \in \mathbb{R}^n$, the function

$$g(x,c) = x'(M - \mathbf{1}c)'((M - \mathbf{1}c)x = x'M'Mx - 2x'M'\mathbf{1}cx + n(cx)^2$$

attains its minimum with respect to c at

$$c = \frac{1}{n}\mathbf{1}'M$$

This implies that

$$|M|_2 = ||M - \frac{1}{n}\mathbf{1}\mathbf{1}'M||_2 = \sqrt{\mu\{(M - \frac{1}{n}\mathbf{1}\mathbf{1}'M)'(M - \frac{1}{n}\mathbf{1}\mathbf{1}'M)\}}$$

where for any symmetric matrix T, $\mu\{T\}$ is the largest eigenvalue of T. We are led to the following result.

Proposition 19.5.3. *Let $M \in \mathbb{R}^{n\times n}$ be a nonnegative matrix.*

$$|M|_2 = \sqrt{\mu\{M'(I - \frac{1}{n}\mathbf{1}\mathbf{1}')M\}} \tag{19.5}$$

where $I - \frac{1}{n}\mathbf{1}\mathbf{1}'$ is the orthogonal projection on the orthogonal complement of the span of $\mathbf{1}$.

Now suppose that M is a doubly stochastic matrix S. Then $S'S$ is also doubly stochastic and $\mathbf{1}'S = \mathbf{1}'$. The latter and 19.5 imply that

$$|S|_2 = \sqrt{\mu\{S'S - \frac{1}{n}\mathbf{1}\mathbf{1}'\}} \tag{19.6}$$

More can be said:

Lemma 19.5.3. *If S is doubly stochastic, then $\mu\{S'S - \frac{1}{n}\mathbf{1}\mathbf{1}'\}$ is the second largest eigenvalue of $S'S$.*

Proof: Since $S'S$ is symmetric it has orthogonal eigenvectors one of which is $\mathbf{1}$. Let $\mathbf{1}, x_2, \ldots, x_n$ be such a set of eigenvectors with eigenvalues $1, \lambda_2, \ldots, \lambda_n$. Then $S'S\mathbf{1} = 1$ and $S'Sx_i = \lambda_i x_i$, $i \in \{2,3,\ldots,n\}$. Clearly $(S'S - \frac{1}{n}\mathbf{1}\mathbf{1}')\mathbf{1} = 0$ and $(S'S - \frac{1}{n}\mathbf{1}\mathbf{1}')x_i = \lambda_i x_i$, $i \in \{2,3,\ldots,n\}$. Since 1 is the largest eigenvalue of $S'S$ it must therefore be true that the second largest eigenvalue $S'S$ is the largest eigenvalue of $S'S - \frac{1}{n}\mathbf{1}\mathbf{1}'$. □

We summarize:

Theorem 19.5.2. *For $p = 2$, the seminorm of a doubly stochastic matrix S is the second largest singular value of S*

We are now in a position to characterize in graph theoretic terms those doubly stochastic matrices with positive diagonals which are semicontractions for $p = 2$

Theorem 19.5.3. *Let S be a doubly stochastic matrix with positive diagonals. Then $|S|_2 \leq 1$. Moreover S is a semicontraction in the 2-norm if and only if the graph of S is weakly connected.*

To prove this theorem we need several concepts and results. Let $\mathbb{G}$ denote a directed graph and write $\mathbb{G}'$ for that graph which results when the arcs in $\mathbb{G}$ are reversed; i.e., the *dual graph*. Call a graph *symmetric* if it is equal to its dual. Note that in the case of a symmetric graph, rooted, strongly connected and weakly connected are equivalent properties. Note also that if $\mathbb{G}$ be the graph of a nonnegative matrix M with positive diagonals, then $\mathbb{G}'$ is the graph of M' and $\mathbb{G}' \circ \mathbb{G}$ is the graph of $M'M$.

Lemma 19.5.4. *A directed graph $\mathbb{G}$ with self arcs at all vertices is weakly connected if and only if $\mathbb{G}' \circ \mathbb{G}$ is strongly connected.*

Proof of Lemma 19.5.4: Since $\mathbb{G}$ has self arcs at all vertices so does $\mathbb{G}'$. This implies that the arc set of $\mathbb{G}' \circ \mathbb{G}$ contains the arc sets of $\mathbb{G}$ and $\mathbb{G}'$. Thus for any undirected path in $\mathbb{G}$ between vertices i and j there must be a corresponding directed path in $\mathbb{G}' \circ \mathbb{G}$ between the same two vertices. Thus if $\mathbb{G}$ is weakly connected, $\mathbb{G}' \circ \mathbb{G}$ must be strongly connected.

Now suppose that (i, j) is an arc in $\mathbb{G}' \circ \mathbb{G}$. Then because of the definition of composition, there must be a vertex k such that (i,k) is an arc in $\mathbb{G}$ and (k, j) is an arc in $\mathbb{G}'$. This implies that (i,k) and (j,k) are arcs in $\mathbb{G}$. Thus $\mathbb{G}$ has an undirected path from i to j. Now suppose that $(i, v_1), (v_1, v_2), \ldots, (v_q, j)$ is a directed path in $\mathbb{G}' \circ \mathbb{G}$ between i and j. Between each pair of successive vertices along this path there must therefore be an undirected path in $\mathbb{G}$. Thus there must be an undirected path in $\mathbb{G}$ between i and j. It follows that if $\mathbb{G}' \circ \mathbb{G}$ is strongly connected, then $\mathbb{G}$ is weakly connected. □

Lemma 19.5.5. *Let T be a stochastic matrix with positive diagonals. If T has a strongly connected graph then the magnitude of its second largest eigenvalue is less than 1. If, on the other hand the magnitude of the second largest eigenvalue of T is less than one, then the graph of T is weakly connected.*

Proof of Lemma 19.5.5: Suppose that the graph of T is strongly connected. Then via Theorem 6.2.24 of [8], T is irreducible. Thus there is an integer k such that $(I + T)^k > 0$. Since T has positive diagonals, this implies that $T^k > 0$. Therefore T is primitive [8]. Thus by the Perron-Frobenius theorem [14], T can have only one eigenvalue of maximum modulus. Since the spectral radius of T is 1 and 1 is an eigenvalue, the magnitude of the second largest eigenvalue of T must be less than 1.

To prove the converse, suppose that T is a stochastic matrix whose second largest eigenvalue in magnitude is less than 1. Then

$$\lim_{i \to \infty} T^i = \mathbf{1}c \tag{19.7}$$

for some row vector c. Suppose that the graph of T is not weakly connected. Therefore if q denotes the number of weakly connected components of the graph, then $q > 1$. This implies that $T = P^{-1}DP$ for some permutation matrix P and block diagonal matrix D with q blocks. Since $D = PTP'$, D is also stochastic. Thus each of its q diagonal blocks is stochastic. Since T^i converges to $\mathbf{1}c$, D^i must converge to a matrix of the form $\mathbf{1}\bar{c}$. But this is clearly impossible because $\mathbf{1}\bar{c}$ cannot have $q > 1$ diagonal blocks. □

Proof of Theorem 19.5.3: Let S be a doubly stochastic matrix with positive diagonals. Then 1 is the largest singular value of S because $S'S$ is doubly stochastic. From this and theorem 19.5.2 it follows that $|S|_2 \leq 1$.

Suppose S is a semicontraction. Then in view of Theorem 19.5.2, the second largest eigenvalue of $S'S$ is less than 1. Thus by Lemma 19.5.5, the graph of $S'S$ is weakly connected. But $S'S$ is symmetric so its graph must be strongly connected. Therefore by Lemma 19.5.4, the graph of S is weakly connected.

Now suppose that the graph of S is weakly connected. Then the graph of $S'S$ is strongly connected because of Lemma 19.5.4. Thus by Lemma 19.5.5, the magnitude of the second largest eigenvalue of $S'S$ is less than 1. From this and Theorem 19.5.2 it follows that S is a semicontraction. □

Proof of Theorem 19.4.2: Let $\mathscr{M}$ be any compact subset of $\mathscr{W}$. In view of Theorem 19.5.3, each matrix in $\mathscr{M}$ is a semicontraction in the two-norm. From this and Proposition 19.5.1, it follows that $\mathscr{M}$ is convergable.

Now suppose that $\mathscr{M}$ is convergable. and let S be a matrix in $\mathscr{M}$. Then S^i converges to a matrix of the form $\mathbf{1}c$ as $i \to \infty$. This means that the second largest eigenvalue of S must be less than 1 in magnitude. Thus by Lemma 19.5.5, S must have a weakly connected graph. □

The importance of Theorem 19.5.3 lies in the fact that the matrices in every convergable set of doubly stochastic matrices are contractions in the 2-norm. In view of Proposition 19.5.1, this enables one to immediately compute a rate of convergence for any infinite product of matrices from any given convergable set. The coefficient of ergodicity mentioned earlier does not have this property. If it did, then every doubly stochastic matrix with a weakly connected graph would have to be a scrambling matrix. The following counterexample shows that this is not the case.

$$S = \begin{bmatrix} .5 & .25 & 0 & 0 & 0 & .25 \\ .25 & .5 & 0 & 0 & 0 & .25 \\ 0 & 0 & .5 & .5 & 0 & 0 \\ 0 & 0 & .5 & .25 & 0 & .25 \\ 0 & 0 & 0 & 0 & .875 & .125 \\ .25 & .25 & 0 & .25 & .125 & .125 \end{bmatrix}$$

In particular, S doubly stochastic matrix with a weakly connected graph but it is not a scrambling matrix.

19.5.3 The Case $p = \infty$

Note that in this case $|S|_\infty \leq 1$ for any stochastic matrix because $|S|_\infty \leq ||S||_\infty \leq 1$. Despite this, the derivation of an explicit formula for the seminorm $|S|_\infty$ has so far eluded us. We suspect that $|S|_\infty$ is the coefficient of ergodicity of S. This conjecture is prompted by the following.

Proposition 19.5.4. *For any stochastic matrix $S \in \mathbb{R}^{n\times n}$,*

$$\tau(S) \leq |S|_\infty \leq 2\tau(S) \tag{19.8}$$

where $\tau(S)$ is the coefficient of ergodicity

$$\tau(S) = \frac{1}{2}\max_{i,j}\sum_{k=1}^{n}|s_{ik} - s_{jk}| \tag{19.9}$$

One implication of Proposition 19.5.4 is that M cannot be a contraction in ∞ seminorm if M is not a scrambling matrix. But as noted at the end of the last section, there are doubly stochastic matrices with weakly connected graphs which are not scrambling matrices. There are also plenty of stochastic matrices with rooted graphs which are not scrambling matrices thus there are matrices in both $\mathscr{R}$ and $\mathscr{W}$ which are not contractions in this norm. For this reason, the ∞ seminorm does not have the property we seek; i.e., providing a seminorm with value less than one for either stochastic matrices with rooted graphs or doubly stochastic matrices with weakly connected graphs.

Proof of Proposition 19.5.4: To be given in the full length version of this paper.

19.6 Complete Gossip Sequences

Let $(i_1, j_1), (i_2, j_2), \ldots, (i_m, j_m)$ be a given sequence of pairs of indices of agent pairs which have gossiped. Let $S = S_m S_{m-1} \cdots S_1$ where S_i is the stochastic matrix associated with the ith gossip in the sequence. By the sequence's *contraction coefficient* is meant the seminorm $|S|_2$. By the *gossip graph* of the sequence is meant that *undirected* graph with vertex set $\{1, 2, \ldots, n\}$ and edge set equal to the union of all pairs (i_k, j_k) in the sequence. This graph, which does not take into account the sequence's order, is clearly a subgraph of the allowable gossip graph $\mathbb{A}$ for the gossiping process under consideration. We call such a gossiping sequence *complete* if its corresponding gossip graph contains a tree spanning $\mathbb{A}$. Call the directed graph of S, the sequence's *composed* gossip graph. In general the gossip graph of a given sequence must be a subgraph of the undirected version of the sequence's composed gossip graph.

The main result we want to prove is as follows.

Theorem 19.6.1. *A gossip sequence $(i_1, j_1), (i_2, j_2), \ldots, (i_m, j_m)$ is complete if and only if the sequence's contraction coefficient is less than 1.*

To prove Theorem 19.6.1 we need several preliminary results.

Lemma 19.6.1. *Let $\mathbb{G}$ and $\mathbb{H}$ be directed graphs on the same n vertices. Suppose that both graphs have self arcs at all vertices. If there is an undirected path from i to j in $\mathbb{H} \circ \mathbb{G}$, then there is an undirected path from i to j in the union of $\mathbb{H}$ and $\mathbb{G}$.*

Proof of Lemma 19.6.1: First suppose that there is an undirected path of length one between vertices i and j in $\mathbb{H} \circ \mathbb{G}$. Then either (i,j) or (j,i) must be an arc in $\mathbb{H} \circ \mathbb{G}$. Without loss of generality suppose that (i,j) is an arc in $\mathbb{H} \circ \mathbb{G}$. Then because of the definition of composition, there must be a vertex k such that (i,k) is an arc in $\mathbb{G}$ and (k,j) is an arc in $\mathbb{H}$. This implies that (i,k) and (k,j) are arcs in $\mathbb{H} \cup \mathbb{G}$. Thus $\mathbb{H} \cup \mathbb{G}$ has a directed path from i to j. Therefore $\mathbb{H} \cup \mathbb{G}$ has an undirected path from i to j. Now suppose that $(i,v_1),(v_1,v_2),\ldots,(v_q,j)$ is a undirected path in $\mathbb{H} \circ \mathbb{G}$ between i and j. Between each pair of successive vertices along this path there must therefore be an undirected arc in $\mathbb{H} \circ \mathbb{G}$. Therefore between each pair of successive vertices along this path there must therefore be an undirected arc in $\mathbb{H} \cup \mathbb{G}$. Thus there must be an undirected path in $\mathbb{H} \cup \mathbb{G}$ between i and j. □

It is obvious that the preceding lemma applies to a finite set of directed graphs. In the sequel we appeal to this extension without special mention.

Lemma 19.6.2. *A gossip sequence is complete if and only if its composed gossip graph is weakly connected.*

Proof of Lemma 19.6.2: Because all vertices of all directed graphs under consideration have self-arcs, a gossip graph is always a subgraph of the undirected version of the composed gossip graph. Thus if the gossip sequence is complete then the undirected version of the composed gossip graph must contain a subgraph which is a spanning tree in $\mathbb{A}$. Thus the composed gossip graph must be weakly connected.

Now suppose that the composed gossip graph is weakly connected. Then the union of the one-gossip graphs comprising the composition must be weakly connected because of Lemma 19.6.1. But the union of the undirected versions of one-gossip graphs is the gossip graph of the sequence. Thus gossip graph of the sequence must be connected and therefore must contain a subgraph which is a spanning tree of $\mathbb{A}$. Thus the gossip sequence must be complete.□

Proof of Theorem 19.6.1: Suppose that the gossip sequence generating the sequence of matrices $S_1, S_2, \ldots, S_m$ is complete. In view of Lemma 19.6.2, the composed gossip graph is weakly connected. Therefore by Theorem 19.5.3, the matrix $S_m \cdots S_1$ is a semicontraction so the contraction coefficient of the sequence is less than one.

Now suppose that the contraction coefficient is less than one. Therefore matrix $S = S_m \cdots S_1$ is a semicontraction. Then by Theorem 19.5.3, the composed gossip graph is weakly connected. Thus by Lemma 19.6.2, the gossip sequence must be complete. □

19.7 Concluding Remarks

Let $\mathbb{A}$ be a given allowable gossip graph. Let us call a complete gossip sequence *minimal* if there is no shorter sequence of allowable gossips which is compete. It is easy to see that a gossip sequence will be minimal if and only if its gossip graph is a minimal spanning tree of $\mathbb{A}$. For a given allowable gossip graph there can be many complete minimal sequences. Moreover, there can be differing largest singular values for the different doubly stochastic matrices associated with different complete minimal sequences. A useful challenge then would be to determine those complete minimal sequences whose associated singular values are as small as possible. This issue will be addressed in a future paper.

The definition of a contraction coefficient proposed in this paper is appropriate for aperiodic gossip sequences. For example, suppose $(i_1, j_1), (i_2, j_2), \ldots$ is an infinite gossip sequence composed of successive complete subsequences which are each of length at most m. Suppose in addition that $\lambda < 1$ is a positive constant which bounds from above the contraction coefficients for each successive subsequence. Then it is easy to see that the entire sequence converges at a rate no slower than $\lambda^{\frac{1}{m}}$.

For certain applications it is useful to consider gossip sequences $(i_1, j_1), (i_2, j_2)$, $\ldots$ which are periodic in the sense that for some integer m subsequence (i_1, j_1), $(i_2, j_2), \ldots, (i_m, j_m)$ repeats itself over and over. In this case a more useful notion of a contraction coefficient might be the second largest eigenvalue of the doubly stochastic matrix S defined by the product of the m stochastic matrices corresponding to the m gossips in the sequence. As with the notion of a contraction coefficient for aperiodic sequences considered in this paper, it would be interesting to determine those complete minimal sequences whose associated contraction coefficients are as small as possible.

Results similar to the main results of this paper for gossip sequences have recently appeared in [16]. A careful comparison of findings will be made in a future paper on deterministic gossiping.

References

1. Olshevsky, A., Tsitsiklis, L.: Disturbance rates in distributed consensus and averaging. In: Proceedings IEEE CDC, pp. 3387–3392 (2006)
2. Blondel, V.D., Hendrichkx, J.M., Olshevsky, A., Tsitsiklis, J.N.: Convergence in multi-agent coordination, consensus, and flocking. In: Proc. of the 44th IEEE Conference on Decision and control, pp. 2996–3000 (2005)
3. Boyd, S., Ghosh, A., Prabhakar, B., Shah, D.: Gossip algorithms: Design, analysis and applications. In: Proceedings INFOCOM (2005)
4. Boyd, S., Ghosh, A., Prabhakar, B., Shah, D.: Randomized gossip algorithms. IEEE Transactions on Information Theory (June 2005)
5. Cao, M., Morse, A.S., Anderson, B.D.O.: Reaching a consensus in a dynamically changing environment – a graphical approach. SIAM J. on Control and Optimization, 575–600 (February 2008)
6. Cao, M., Spielman, D.A., Morse, A.S.: A lower bound on convergence of a distributed network consencus algorithm. In: Proc. 2005 IEEE CDC, pp. 2356–2361 (2005)
7. Cao, M., Spielman, D.A., Yeh, E.M.: Accelerated gossip algorithms for distributed computation. In: Proceedings of the Allerton Converence (2006)
8. Horn, R.C., Johnson, C.R.: Matrix Analysis. Cambridge University Press, New York (1985)

9. Jadbabaie, A., Lin, J., Morse, A.S.: Coordination of groups of mobile autonomous agents using nearest neighbor rules. IEEE Transactions on Automatic Control, 988–1001 (June 2003), also in Proc. 2002 IEEE CDC, pp. 2953–2958.
10. Liu, J., Cao, M., Anderson, B.D.O., Morse, A.S.: Analysis of accelerated gossip algorithms. In: Proceedings 2009 CDC, submitted (2009)
11. Moreau, L.: Stability of multi-agent systems with time-dependent communication links. IEEE Transactions on Automatic Control 50, 169–182 (2005)
12. Olfati-Saber, R., Murray, R.M.: Consensus seeking in networks of agents with switching topology and time-delays. IEEE Transactions on Automatic Control 49, 1520–1533 (2004)
13. Ren, W., Beard, R.: Consensus seeking in multiagent systems under dynamically changing interaction topologies. IEEE Transactions on Automatic Control 50, 655–661 (2005)
14. Seneta, E.: Non-negative Matrices and Markov Chains. Springer, Heidelberg (2006)
15. Xiao, L., Boyd, S.: Fast linear iterations for distributed averaging. Systems and Control Letters 53, 65–78 (2004)
16. Nedić, A., Olshevsky, A., Ozdaglar, A., Tsitsiklis, J.N.: On distributed averaging algorithms and quantization effects. IEEE Transactions on Automatic Control 54(11), 2506–2517 (2009)

20

Covariance Extension Approach to Nevanlinna-Pick Interpolation: Kimura-Georgiou Parameterization and Regular Solutions of Sylvester Equations*

György Michaletzky

Department of Probability Theory and Statistics, Eötvös Loránd University, Pázmány Péter sétány 1/C, H-1117, Hungary

Summary. In this paper – using the equivalence of the Nevanlinna-Pick-interpolation problem formulated for positive real (possibly matrix-valued) functions and the general form of the Covariance Extension Problem – all solutions of the interpolation problem are derived via a stochastic (geometric) argument. We show how this problem is connected with a special form of Beurling's problem, i.e. characterization of shift invariant subspaces of special function spaces. We also generalize the Kimura-Georgiou-parameterization of interpolants with low rank to the multivariate case and derive a short state-space form of this parameterization. Finally, we point out the connection between the regular solutions of a Sylvester-equation and the solutions of the interpolation problem described by the Kimura-Georgiou-parameterization.

20.1 Introduction

The problem of interpolation with bounded, analytic or equivalently with positive real functions is quite classical in mathematics and is known as the Nevanlinna-Pick-problem. The case of a finite interpolation set was treated by Pick [21], who obtained an algebraic solvability criterion. Later on, Nevanlinna [20] solved the problem for countable sets, he gave a constructive algorithm yielding both a parameterization of all solutions and a criterion for uniqueness.

There are several alternative approaches giving a unified treatment of a wide range of interpolation and approximation problems. A powerful approach is based on the commutant lifting [22] and contractive intertwining dilations [12]. An elegant approach in Ball [4], Ball-Helton [6], Ball-Gohberg-Rodman [5] is based on the theory of spaces with indefinite inner product, while a more function-theoretic approach, using Hankel-operators, stems form Adamyan et al [2]. A recursive solution of the extension problem was constructed by Schur [23].

In the case when all interpolation nodes are at the origin the Nevanlinna-Pick interpolation problem formulated for positive real functions coincides with the

*Dedicated to Christopher I. Byrnes and Anders Lindquist.

X. Hu et al. (Eds.): Three Decades of Progress in Control Sciences, pp. 291–311, 2010.

Carathéodory-extension problem [10] of describing all positive semidefinite extensions of a finite Toeplitz-matrix. This can be considered as a *covariance extension* problem or the reconstruction of the spectral density. The theory of orthogonal polynomials plays an important role in the classical analysis of this problem. Szegő [24], Akhiezer [3], Geronimus [16] and Grenander-Szegő [17] give a comprehensive introduction to the scalar case. Concerning a matrix extension see Fuhrmann [13].

Dewilde and Dym in [11] showed that the solution of the Nevanlinna-Pick-problem for analytic functions with positive real part can be considered as solutions of a *covariance extension problem* in the sense that the Nevanlinna-Pick data set determines the covariances inside a subspace while this latter is defined using the interpolations nodes and values. If all interpolation nodes are at the origin then some of the first auto-covariances are prescribed.

T. Georgiou [14] studies – for scalar valued functions (although his unpublished dissertation contains results in the multivariate case, as well) – the set of all solutions of McMillan-degree equal to the length of the partial covariance sequence. The so-called maximum entropy extension – producing an AR process – is in this set. While Georgiou gives a parameterization of this set in terms of orthogonal and conjugate orthogonal polynomials H. Kimura [18] [19] provides a parameterization in state space form. The maximum entropy solution gives the basis of this parameterization. The free parameters are represented as the gains of additional feedback loops attached to the lattice configuration in this parameterization. Later Byrnes et al [9] proved that prescribing the zero structure in advance there exists a unique extension belonging to the set specified above. See also Byrnes-Lindquist [7]. Their proof uses an elegant degree-theoretic argument. In Byrnes et al. [8] a convex optimization approach was applied to this problem.

Section 20.2 specifies the notations used in the present paper. Section 20.3 introduces the Nevalinna–Pick-interpolation problem (Problem 20.3.2) and the Covariance Extension Problem (Problem 20.3.3) with recalling a theorem proven by Dewilde and Dym [11] stating the equivalence of these two problems.

In Section 20.4 we show that Covariance Extension Problem i.e. characterizing the weakly stationary processes producing a *prescribed covariance structure* (scalar-product structure) inside a given subspace is equivalent to characterizing those weakly stationary processes for which two appropriate *complementary subspaces* in that given subspace are *orthogonal* to each other.

In section 20.5 we generalize the Kimura-Georgiou parameterization. As we have already pointed out in this case the extensions are rational functions for which the number of poles is equal to the length of the interpolation data. In other words these functions can be written as a ratio of two polynomials of given degree. The natural generalization is to consider extensions of the form $F^{-1}G$, where F and G are rational functions with a fixed pole structure determined by the interpolation nodes and directions. We show that this form of the interpolants is strongly connected to regular solutions of an appropriate Sylvester-equation. This Sylvester-equation is based on the interpolation nodes/directions and the left-pole structure of the interpolant. The *interpolation values* do not play any role in this equation.

In the parameterization form given by T. Georgiou the numerator and denominator polynomials are expressed as sums of the orthogonal polynomials of the first and the second kind using the same set of coefficients. In the present paper (see Lemma 20.5.1) we show that any two sets of functions can be used here assuming that the jth pair provides a solution of the interpolation problem determined by the first j interpolation nodes. Together with this parameterization which is similar to the one developed by Kimura and Georgiou we present it in state-space form, as well (see Section 20.5.2).

Note that the parameterization of the rational functions with a given McMillan-degree satisfying the interpolation conditions but without imposing the positive real condition was considered and solved by Antoulas et al. in [1].

20.2 Notation

A complex function will be called *stable* if it is analytic on the open unit disk, while if its poles are inside the closed unit disk then it is a so-called *anti-stable* function. (Sometimes instead of using the notion stable we are going to say *analytic*. When the region of analyticity is not specified explicitly, it is assumed implicitly that it refers to the open unit disk. In some cases instead of the notion anti-stable we write *co-analytic*.)

The Hilbert-space $H_2 = H_2^p$ will be identified by the set of functions having the form $f(z) = f_0 + f_1 z + f_2 z^2 + \dots$, where f_j, $j \geq 0$ is a sequence of p-dimensional **row-vectors** satisfying the condition $\sum_{j=0}^{\infty} \|f_j\|^2 < \infty$. These vector-valued functions are analytic inside the open unit circle and have square-integrable radial limits on the boundary.

For a complex valued matrix A its adjoint is denoted by A^*. If $f(z), z \in \mathbb{C}$ is an $m \times p$ complex function, then f^* denotes its para-hermitian conjugate function, $f^*(z) = \{f(\bar{z}^{-1})\}^*$.

If Ξ is a $p \times p$ valued inner function, (i.e. $\Xi\Xi^* = I$ almost surely), then $H_2(\Xi)$ denotes the set of functions $f \in H_2$ for which $f\Xi^*$ is analytic outside of the closed unit circle.

For a p-dimensional stationary process with zero mean $y(k),\ k \in \mathbb{Z}$, $R_y(k) = R(k) = E(y(k)y(0)^T)$, $k \in \mathbb{Z}$, denotes its covariance function.

Consider the Hilbert space $H(y)$ generated by the scalar-valued random variables

$$\{y^{(i)}(k) \mid i = 1,2,\dots,p; k \in \mathbb{Z}\},$$

where $y^{(i)}(k)$ denotes the i-th coordinate of $y(k)$ with the inner product $\langle \xi, \eta \rangle = \mathrm{E}\{\xi \cdot \overline{\eta}\}$. The closed subspace generated by the coordinates of $y(k), y(k-1), \dots$ is denoted by $H_k^-(y)$, while the closed subspace $H_k^+(y)$ is spanned by the coordinate variables of $y(k), y(k+1), \dots$.

The unitary operator $U : H(y) \to H(y)$ is the shift determined by

$$Uy^{(i)}(k) = y^{(i)}(k+1).$$

Together with the operator U we shall consider analytic functions of it. Namely, if f is a vector-valued analytic function on the closed unit disk then with coordinates f_i, $i = 1, \ldots, p$, we can consider the operator $f(U^{-1})$. The shorthand notation $f(z)y(k)$ or $f y(k)$ will be used for the random variable $\sum_{i=1}^{p} f_i(U^{-1})U^k y^{(i)}(0)$. Note, that in this notation the multiplication by the function z is identified by the application of the operator U^{-1}.

20.3 Covariance Extension Formulation of the Nevanlinna-Pick Interpolation Problem

Let us consider a special form of the Nevanlinna-Pick interpolation problem on positive real functions assuming that the interpolation conditions fix the value of the function at zero, as well. Namely, let us assume that the spectrum of the matrix $\mathscr{A}$ of size $N \times N$ is in the open unit disk, $R(0)$ is a positive definite matrix of size $p \times p$, while $\mathscr{B}$, and $\mathscr{C}$ determine the "extra" interpolation direction and values, respectively. Now we might formulate the Interpolation Problem 20.3.1:

Problem 20.3.1 (Nevanlinna-Pick interpolation problem). Characterize all analytic functions Z° with positive real part for which the functions

$$\left(Z^\circ - \frac{R(0)}{2}\right) z^{-1}, \quad \left(z^{-1}\left(Z^\circ - \frac{R(0)}{2}\right)\mathscr{B}^* - \mathscr{C}^*\right)(zI - \mathscr{A}^*)^{-1} \tag{20.1}$$

are analytic inside the complex unit circle.

Note that here the "extra" interpolation conditions are formulated on the function without the constant term. This formulation differs a little from the usual ones but with some calculations it can be transformed into it.

Let us assume that the Pick-matrix corresponding to this interpolation problem is positive definite. It is well-known that in this case there exists a solution of the problem above, moreover from this is follows that the pair $\mathscr{A}, \mathscr{B}$ is reachable.

If Z is a special solution of the interpolation problem, then obviously $z^{-1}(Z^\circ - Z)$ and $(Z^\circ - Z)\mathscr{B}^* z^{-1}(zI - \mathscr{A}^*)^{-1}$ are analytic inside the unit circle. Introducing the inner function Ξ generated by the pair $\mathscr{A}, \mathscr{B}$ (i.e. $\Xi(z) = \mathscr{D}_\Xi + \mathscr{C}_\Xi\left(z^{-1}I - \mathscr{A}\right)^{-1}\mathscr{B}$ we obtain that the interpolation Problem 20.3.2 can be formulated as follows:

Problem 20.3.2 (Nevanlinna-Pick interpolation problem). Characterize all analytic functions Z° with positive real part for which the function

$$z^{-1}(Z - Z^\circ)\,\Xi^* \tag{20.2}$$

is analytic inside the complex unit circle. (See Dewilde and Dym [11].)

For later use (see Section 20.5 we assume that the inner function Ξ is written in the form of products of elementary inner functions, i.e. $\Xi = \xi_n \times \xi_{n-1} \times \cdots \times \xi_1$ where each factor has the form

$$\xi_j(z) = \left(I - U_j U_j^*\right) + \zeta_j(z) U_j U_j^* \,. \tag{20.3}$$

Here U_j is an arbitrary matrix of size $p \times m_j$ satisfying the equation

$$U_j^* U_j = I_{m_j} \,,$$

i.e. the columns of U_j are orthogonal to each other and they are of unit length, and the function ζ_j is defined as follows

$$\zeta_j(z) = \begin{cases} -\frac{\bar{a}_j}{|a_j|} \frac{z-a_j}{1-z\bar{a}_j} \,, & \text{if } a_j \neq 0 \\ z \,, & \text{if } a_j = 0 \end{cases} \tag{20.4}$$

Note that the matrix $\xi_j(0)$ is self-adjoint and ζ_j is normalized in such a way that $\zeta_j(0) > 0$, especially, it is a real number.

Now the numbers $a_1, \ldots, a_n$ are the *nodes* of the covariance-extension problem. and

$$N = \sum_{j=1}^{n} m_j \tag{20.5}$$

Because of this assumed structure instead of Ξ we shall use the notation Ξ_n.

Furthermore, following Dewilde and Dym [11], this problem can be considered as a covariance extension problem. Namely, let $y(k),\ k \in \mathbb{Z}$ be a p-dimensional (wide-sense) stationary process of full rank with zero mean and spectral density $\Phi = Z + Z^*$. Assume that $\log\det\Phi$ is integrable.

Using the inner function Ξ_n let us define a subspace $\mathcal{M} \subset H_0^-(y)$ as follows:

$$\mathcal{M}_n = \{f(z)y(0) \mid f \in H_2(\Xi_n)\} \,. \tag{20.6}$$

For a given vector $\zeta \in \mathcal{M}_n$ let us denote by T_ζ the corresponding function in $H_2(\Xi_n)$ for which $\zeta = T_\zeta(z)y(0)$.

Let us observe that the subspace $\mathcal{M}_n$ is algebraically isomorphic to the coinvariant subspace $H_2(\Xi_n)$, but the scalar-product (covariance-structure) in $\mathcal{M}_n$ is determined by the given process y.

If it is needed we use the notation $\mathcal{M}_n(y)$ to emphasize that it was constructed from the process y.

In the special case, when $\Xi_n(z) = z^n I_p$ then $H_2(\Xi_n)$ contains all the p-dimensional row vectors with polynomial entries of degree no greater than n, consequently we have that $\mathcal{M}_n = \langle y(0), y(-1), \ldots, y(-n) \rangle$.

Problem 20.3.3 (Covariance extension problem). Characterize all stationary processes y° for which

$$\langle f(z)y(0), g(z)y(0) \rangle = \langle f(z)y^\circ(0), g(z)y^\circ(0) \rangle \,,$$

for every $f, g \in H_2(\Xi_n)$, i.e. for which the covariance structures of $\mathcal{M}_n(y)$ and $\mathcal{M}_n(y^\circ)$ coincide, where the subspace $\mathcal{M}_n(y^\circ)$ is constructed from the process y° using the same steps as in the definition of $\mathcal{M}_n = \mathcal{M}_n(y)$.

Note that in the special case, when $\Xi_n(z) = z^n I_p$ we obtain the Carathéodory-extension problem.

The following theorem shows that Problem 20.3.3 and Problem 20.3.2 are equivalent.

Theorem 20.3.1 (Dewilde and Dym [11]). *The analytic part Z° of the spectral density defined by the stationary process y° is a solution of the interpolation problem (20.2) if and only if the covariance structures of $\mathscr{M}_n(y)$ and $\mathscr{M}_n(y^\circ)$ coincide, where the subspace $\mathscr{M}_n(y^\circ)$ is constructed from the process y° using the same steps as in the definition of $\mathscr{M}_n = \mathscr{M}_n(y)$.*

The theorem above plays a crucial role in the present paper.

To construct state-space equation we start with the analysis of the structure of the subspace $\mathscr{M}_n$. Consider first the subspace $zH_2 \cap H_2(\Xi_n) \subset H_2(\Xi_n)$. Let us observe that a function f belongs to this subspace if and only if $f \in H_2(\Xi_n)$ and $f(0) = 0$. Note that in this case $z^{-1}f(z) \in H_2(\Xi_n)$, as well.

The "corresponding" subspaces of $H_0^-(y)$ will be denoted by $\mathscr{M}_n'$ and $\mathscr{M}_n''$, respectively. I.e.

$$\begin{aligned} \mathscr{M}_n' &= \{f(z)y(0) \mid f \in (zH_2 \cap H_2(\Xi_n))\} \, , \\ \mathscr{M}_n'' &= \{f(z)y(0) \mid zf \in (zH_2 \cap H_2(\Xi_n))\} \, . \end{aligned}$$

These subspaces can also be defined using the process

$$y_n(t) = \Xi(z)y(t) \, . \tag{20.7}$$

Especially, in this case

$$\mathscr{M}_n = H_0^-(y) \cap H_0^+(y_n) \, . \tag{20.8}$$

20.4 State-Space Equations

In this section we give a stochastic-geometric characterization of the solutions of the covariance extension Problem 20.3.3 (or in other words of the equivalent Nevanlinna-Pick interpolation Problem 20.3.2). This characterization is based on a state-space equation reformulation of the quantities considered in the previous section.

Denote the normalized projection error of $y(0)$ onto $\mathscr{M}_n'$ by $e_n(0)$. Since y is assumed to be of full rank and $\mathscr{M}_n' \subset H_{-1}^-(y)$ the covariance matrix of the projection error above is nonsingular, thus it can be normalized. Using the shift operator we define the sequence $e_n(t) = U^t e_n(0)$.

Next let us choose an *arbitrary* basis in $\mathscr{M}_n'$. To this aim we might consider a basis $T_1, \dots, T_N \in (zH_2 \cap H_2(\Xi_n))$ and define $x_j(-1) = T_j(z)y(0)$. Form the vector $x(-1)$ from $x_1(-1), \dots, x_N(-1)$ and apply again the shift operator to construct the stationary sequence $x(t)$, $t \in \mathbb{Z}$. Let us emphasize that the functions $T_1, \dots, T_N$ can be fixed independently of the process y, depending only on the inner function Ξ_n.

Obviously, the coordinates $e_n(0)$ and $x(-1)$ generate the subspace $\mathscr{M}_n$. On the other hand the coordinates of $x(0)$ are in $\mathscr{M}_n^{''} \subset \mathscr{M}_n$, due to the fact that $U\mathscr{M}_n^{'} = \mathscr{M}_n^{''} \subset \mathscr{M}_n$. Thus

$$x(0) = Ax(-1) + Be_n(0)$$

for some matrices A and B.

Moreover, since $e_n(0)$ is the normalized projection error of $y(0)$ onto $\mathscr{M}_n^{'}$ we obtain that

$$y(0) = Cx(-1) + De_n(0)$$

for appropriate C and D matrices, where D is nonsingular by the construction.

Applying the shift operator we obtain the following representation

$$\begin{cases} x(t) = Ax(t-1) + Be_n(t) \\ y(t) = Cx(t-1) + De_n(t) \end{cases} \tag{20.9}$$

Note, that here the process $e_n(t)$ is – in general – not an uncorrelated sequence. Also note that after chosing the basis-functions $T_1, \ldots, T_N$ the matrices A, B, C and D are already uniquely determined. Thus their value – up to the usual nonsingular state-space transformation – depends on the inner function Ξ_n and on the covariance structure inside $\mathscr{M}_n$ induced by the process y, as well.

Denote by $W(e)_n$ the transfer function giving the process $y(t)$ as the output if $e_n(t)$ is used for the input. I.e.

$$y(t) = W(e)_n e_n(t)\,. \tag{20.10}$$

Its inverse is given by the equations

$$\begin{cases} x(t) = \left(A - BD^{-1}C\right) x(t-1) + BD^{-1}y(t) \\ e_n(t) = -D^{-1}Cx(t-1) + D^{-1}y(t) \end{cases} \tag{20.11}$$

Let us introduce the notation $A_{\text{int}} = A - BD^{-1}C$, $B_{\text{int}} = BD^{-1}$. The matrices $A_{\text{int}}, B_{\text{int}}$ are determined – up to a state-space equivalence – by the inner function Ξ_n, the covariance structure in $\mathscr{M}_n$ plays some role only in the determination of $D^{-1}C, D^{-1}$.

Proposition 20.4.1. *The matrix A is stable (in discrete-time sense), thus its eigenvalues are strictly inside the complex unit circle. Consequently, the transfer function*

$$W(e)_n(z) = D + C\left(z^{-1}I - A\right)^{-1} B$$

is stable, i.e. analytic inside the unit circle. The function $W(e)_n^{-1}$ is stable, as well.

Proof. Since the random variables $e_n(0)$ and $x(-1)$ are uncorrelated the Lyapunov equation

$$P = APA^* + BB^* \tag{20.12}$$

holds, where P is the state covariance matrix.

$$P = \mathrm{cov}(x(0)) > 0\,.$$

So if β is a left eigenvector of A, i.e. $\beta^* A = \lambda \beta^*$, then (20.12) implies that $|\lambda| \le 1$. But if $|\lambda| = 1$, then $\beta^* B = 0$, thus the equation $\beta^* x(t) = \lambda \beta^* x(t-1)$ should hold. Invoking the regularity of the process y we obtain that the corresponding backward shift operator has no (nontrivial) finite dimensional invariant subspaces, thus λ should be of modulus strictly less than one.

The inverse transfer function $W(e)_n^{-1}$ maps $y(0)$ into $e_n(0) \in H_0^-(y)$, so it is obviously stable. Thus $W(e)_n$ is a stable, minimum phase transfer function. □

The following theorem gives a simple stochastic-geometric characterization of the solutions of the Nevanlinna-Pick interpolation problems.

Theorem 20.4.1. *Suppose that y° is another stationary process. Construct the process x° similarly as for the original process y, i.e. using the same basis-functions $T_1, \dots, T_N$. Define the sequence $e_n^\circ(t)$ via the relation*

$$e_n^\circ(t) = W(e)_n^{-1} y^\circ(t)\,. \tag{20.13}$$

Then the analytic part Z° of the spectral density of y° is a solution of the Covariance Extension Problem 20.3.2 if and only if the random variables $x^\circ(-1)$ and $e_n^\circ(0)$ are uncorrelated.

Proof. Consider first an arbitrary stationary process y° and construct x° and e_n° according to the theorem. Since the process x° is created using the same steps as for the case of the process y and the process and e_n° is defined using the transfer function $W(e)_n$ the same system equations (20.9) hold.

If y° provides a solution of the approximation problem then according the Theorem 20.3.1 the covariance structure inside $\mathscr{M}_n(y)$ and $\mathscr{M}_n(y^\circ)$ coincide. Because of the construction the random variables $x(-1)$ and $e_n(0)$ are uncorrelated, their coordinates are in the subspace $\mathscr{M}_n$, thus $x^\circ(-1)$ and $e_n^\circ(0)$ should be uncorrelated, as well.

Conversely, if $x^\circ(-1)$ and $e_n^\circ(0)$ are uncorrelated then the Lyapunov-equation

$$P^\circ = AP^\circ A^* + BB^*$$

should hold true, where P° denotes the covariance matrix of $x^\circ(0)$. Since this equation has a unique solution in view of the stability of A (see Proposition 20.4.1) we obtain that

$$P = P^\circ\,.$$

Now computing the covariance of $y^\circ(0)$ and the cross-covariance between $y^\circ(0)$ and $x^\circ(-1)$ using the second equation in (20.9) for the new process, we obtain they should coincide with those of for the original process y. But the coordinates of $x^\circ(-1)$ and $y^\circ(0)$ generate $\mathscr{M}_n(y^\circ)$ proving that the spaces $\mathscr{M}_n(y^\circ)$ and $\mathscr{M}_n(y)$ are isometric, concluding the proof of the theorem. □

The next immediate but useful corollary shows that the orthogonality condition required by the previous Theorem can be formulated as a set of matrix equations.

Corollary 20.4.1. *Suppose that y° is another stationary process with a – possibly non-minimal – realization*

$$\begin{aligned} x_r^\circ(t) &= A_r x_r^\circ(t-1) + B_r e_r^\circ(t) , \\ y^\circ(t) &= C_r x_r^\circ(t-1) + D_r e_r^\circ(t) , \end{aligned} \tag{20.14}$$

where now $e_r^\circ(t)$, $t \in \mathbb{Z}$ form an uncorrelated sequence and the matrix A_r is stable.

Then the analytic part Z° of the spectral density of y° is a solution of the Covariance Extension Problem 20.3.2 if and only if the matrix equations

$$\begin{aligned} S_r &= A_{int} S_r A_r^* + B_{int} \overline{C}_r \\ C_r S_r^* &= CP \end{aligned} \tag{20.15}$$

have a solution in S_r, where $\overline{C}_r = cov\,(y^\circ(t), x_r^\circ(t))$.

Proof. The stability of the matrix A_r implies that $e_r^\circ(t)$ is uncorrelated with the vectors $y^\circ(t-1), y^\circ(t-2), \ldots$.

Now let us construct the process x° starting with y° similarly as x for the original process y, i.e. using the same functions $T_1, \ldots, T_N$. Observe that the coordinates of $x^\circ(t-1)$ should be in the subspace $H_{t-1}^-(y^\circ)$. Define the sequence $e_n^\circ(t)$ via the relation

$$e_n^\circ(t) = W(e)_n^{-1} y^\circ(t) . \tag{20.16}$$

Computing the covariance matrix between the vectors $x^\circ(t)$ and $x_r^\circ(t)$ using equations (20.11) and (20.14) the orthogonality of $e_r^\circ(t)$ and $x^\circ(t-1)$ gives that the unique solution of the first equation in (20.15) is $S_r := \mathrm{cov}\,(x^\circ(t), x_r^\circ(t))$. Now

$$y^\circ(0) = Cx^\circ(-1) + De_n^\circ(0) = C_r x_r^\circ(-1) + D_r e_r^\circ(0) .$$

Taking the covariance with $x^\circ(-1)$ we obtain that $x^\circ(-1)$ and $e_n^\circ(0)$ are uncorrelated if and only if the equation

$$C_r S_r^* = CP$$

holds, concluding the proof of the corollary. □

Remark 20.4.1. T. Georgiou in [15] considers the following problem closely related to Theorem 20.4.1. Namely, characterize the possible state covariances using various second order stationary processes $e(t)$,$t \in \mathbb{Z}$, with zero mean as inputs in

$$x(t) = \Gamma x(t-1) + Ge(t)$$

where Γ and G are fixed matrices. Furthermore given a possible state covariance P characterize the input processes producing this prescribed state covariance.

Georgiou proves that $P > 0$ is a so-called admissible state covariance if and only if there exists a solution H of the equation

$$P - \Gamma P \Gamma^* = GH + H^* G^* .$$

Furthermore, for a given H the positive real function Z_e corresponding to the spectral measure of the stationary process e leading to the state covariance P can be written as follows

$$Z_e(z) = HP^{-1}(I - z\Gamma)^{-1}G + Q(z)\left(D_0 + C_0\left(z^{-1}I - \Gamma\right)^{-1}G\right)$$

where the matrices D_0, C_0 are such that $\left(D_0 + C_0\left(z^{-1}I - \Gamma\right)^{-1}G\right)$ be an inner function, while Q is an arbitrary stable function.

20.4.1 Special Solution of the Covariance Extension Problem

Obviously, if we define the sequence $e_n^\circ(t)$ as a *white noise process* and then form

$$y^\circ(t) = W(e)_n e_n^\circ(t)$$

then the criterion considered in Theorem 20.4.1 holds true, because of the stability of the matrix A, thus we obtain a solution of the covariance extension problem.

In this subsection we analyse this special solution, thus assume that e_n° is a white noise process. Observe that in this case

$$\mathrm{Pr}_{\mathscr{M}_n'(y^\circ)} y^\circ(0) = \mathrm{Pr}_{H_{-1}^-(y^\circ)} y^\circ(0)\,,$$

thus DD^* gives the covariance of the error in this projection. On the other hand we obtain the same error covariance projecting $y(0)$ onto $\mathscr{M}_n'$ or more generally performing this projection for any solution of the covariance extension problem. Since $\mathscr{M}_n' \subset H_{-1}^-(y)$ the innovation error for the process y should be smaller than DD^*. In particular, since according to Szegő's theorem $\frac{1}{2\pi}\int \log\det\Phi$ is equal to the logarithm of the trace of the innovation error covariance we obtained that the extension for which e_n° is a white noise maximizes the quantity $\frac{1}{2\pi}\int \log\det\Phi$, i.e. determines the so-called maximum entropy solution.

The previous construction was based on the orthogonal decomposition

$$\mathscr{M}_n = \mathscr{M}_n' \oplus \left(\mathscr{M}_n \ominus \mathscr{M}_n'\right).$$

The coordinates of $e_n(0)$ form a generating set of vectors of $\mathscr{M}_n \ominus \mathscr{M}_n'$, while those of $x(-1)$ generate $\mathscr{M}_n'$. In a similar way instead of $e_n(0)$ generating the subspace $\mathscr{M}_n \ominus \mathscr{M}_n'$ we can choose vectors from $\mathscr{M}_n \ominus \mathscr{M}_n''$. To keep full similarity, this can be achieved via normalizing the projection error of $y_n(0) = \Xi_n y(0)$ onto $\mathscr{M}_n''$. The spectral density of the process y_n is $\Xi_n \Phi \Xi_n^*$ implying that it is a full rank stationary process, thus the covariance matrix of the projection error considered above is nonsingular. Choosing arbitrary normalization we obtain a random vector denoted by $f_n(0)$. Using the shift operator it can be extended into a stationary process $f_n(t) = U^t f_n(0)$, $t \in \mathbb{Z}$.

In this case $x(0)$ and $f_n(0)$ are uncorrelated, their coordinate variables together generate $\mathscr{M}_n$. Consequently, $x(-1)$ and $y(0)$ can be expressed as linear combinations of $x(0)$, $f_n(0)$, yielding a backward state-space representation of y where f_n is the input process. Let us emphasize again, that – in general – $f_n(t)$, $t \in \mathbb{Z}$, is not an uncorrelated sequence.

Denote by $\overline{W}(f)_n$ the transfer function mapping the process f_n into y.

$$y(t) = \overline{W}(f)_n f(t) \tag{20.17}$$

Note that although the process f_n was created using a projection of $y_n(0)$ the transfer function $\overline{W}(f)_n$ is defined in the way that its output is the original process y inputting f_n. We could have defined this transfer function taking y_n as the output process, as well. But in order to fully utilize the similarities to stochastic realization theory of stationary processes it is more appropriate to use here the same output process as before. Also, due to the fact that $y_n(t) = \Xi_n y(t)$, these functions could be transferred into each other.

Proposition 20.4.2. *The transfer function $\overline{W}(f)_n^{-1}$ is stable and maximum phase, i.e. it is analytic inside the unit circle but its inverse is analytic outside of the unit circle.*

Proof. This is an immediate consequence of the inclusions

$$H_0^-(f_n) \subset H_0^-(y) \quad \text{and} \quad H_0^+(f_n) = H_0^+(y_n) \supset H_0^+(y) .$$

□

Introduce the notation:

$$K_n = \overline{W}(f)_n^{-1} W(e)_n . \tag{20.18}$$

Proposition 20.4.3. *Assume that $e_n^\circ(t)$ is a white noise process. Define the process y° as in Proposition 20.4.3. I.e. $y^\circ(t) = W(e)_n e_n^\circ(t)$. Then the process defined by*

$$f_n^\circ(t) = \overline{W}(f)_n^{-1} y^\circ(t)$$

is a white noise process.

Furthermore, the function K_n defined in (20.18) is inner.

Proof. The stability of the matrix A gives that projecting $y(0), y(1), \ldots$ onto $H_{-1}^-(e_n^\circ)$ the projections are in the subspace generated by the coordinates of $x(-1)$. Since $W(e)_n$ is a stable and minimum phase function we have that $H_{-1}^-(y^\circ) = H_{-1}^-(e_n^\circ)$, consequently the previous claim can be expressed as

$$\mathrm{Pr}_{H_{-1}^-(y^\circ)} H_0^+(y^\circ) \subset \mathcal{M}_n'(y^\circ) . \tag{20.19}$$

Furthermore, $H_0^+(y_n^\circ) = \mathcal{M}'(y^\circ) \vee H_0^|(y^\circ)$ thus

$$\mathrm{Pr}_{H_{-1}^-(y^\circ)} H_0^+(y_n^\circ) \subset \mathcal{M}_n'(y^\circ) , \tag{20.20}$$

holds, as well. But obviously, $\mathcal{M}_n'(y^\circ) \subset H_0^+(y_n^\circ) \cap H_{-1}^-(y^\circ)$, so necessarily

$$\mathcal{M}_n^{'}(y^\circ) = H_0^+(y_n^\circ) \cap H_{-1}^-(y^\circ)\,, \tag{20.21}$$

Applying now the forward shift operator we get that the projections of any vectors from $H_0^-(y^\circ)$ onto $H_1^+(y_n^\circ)$ are in $U\mathcal{M}_n^{'}(y^\circ) = \mathcal{M}_n^{''}(y^\circ)$. Especially, projecting the coordinates of $y_n^\circ(0)$ – these are in $H_0^-(y^\circ)$ – onto $H_1^+(y_n^\circ)$ we obtain that the projections are in $\mathcal{M}_n^{''}(y^\circ)$. Consequently, the projection error, in other words the backward innovation process of y_n° is f_n°.

Finally, since $H_0^-(e_n^\circ) \supset \mathcal{M}_n$ the function K_n should be analytic inside the unit circle. On the other hand e_n° and f_n° are both white noise processes, consequently K_n should be inner. □

Since the coordinates of $f_n(0)$ are in the subspace $\mathcal{M}_n(y)$, they can be expressed using $x(-1)$ and $e_n(0)$. This – together with the first equation of (20.9) – gives a realization of K_n. Thus

$$K_n(z) = D_K + C_K\left(z^{-1}I - A\right)^{-1}B. \tag{20.22}$$

On the other hand the projection of the subspace $H_0^+(f_n) = H_0^+(y_n)$ onto $\mathcal{M}_n^{'}$ gives the whole subspace $\mathcal{M}_n^{'}$, the pair (C_K, A) is observable.

Let us point out that if e_n° is a white noise then the spectral density of the corresponding process y° denoted by Φ_n is given as

$$\Phi_n = W(e)_n W(e)_n^* = \overline{W}(f)_n \overline{W}(f)_n^*\,. \tag{20.23}$$

Denote by Z_n the analytic part of Φ_n:

$$\Phi_n = Z_n + Z_n^*.$$

We shall call Z_n as the *maximum entropy solution* of the Nevanlinna-Pick interpolation problem or equivalently of the covariance extension problem.

Later on we are going to use the following observation:

Remark 20.4.2. Note that the identity

$$\overline{W}(f)_n^* = \overline{W}(f)_n^{-1} Z_n + \overline{W}(f)_n^{-1} Z_n^*$$

and Proposition 20.4.2 show that the function $\overline{W}(f)_n^{-1} Z_n^*$ is stable.

Let us mention without proving it that transforming the orthogonality condition formulated in Theorem 20.4.1 in the language of functions, namely for a function $T_\zeta(z) = \sum_{j=1}^\infty \beta_j z^i$. the equations $\int T_\zeta K_n^* z^{-k} = 0 \quad$ for $k \geq 0$.(i.e. the random variable $\zeta = \sum_{j=1}^\infty \beta_j e_n^\circ(-j)$ is in the subspace $\mathcal{M}_n^{'}$) should imply that

$$\int T_\zeta \sum_{k=1}^\infty R_{e_n^\circ}(-k) z^{-k} = 0 \tag{20.24}$$

should be fulfilled, where $R_{e_n^\circ}$ denotes the covariance sequence of the process e_n°, (i.e. ζ is orthogonal to $e_n^\circ(0)$, the usual characterization of all solutions of the Nevanlinna-Pick interpolation problem can be derived. (See Dewilde and Dym [11].)

Theorem 20.4.2. *The positive real function Z° is a solution of the interpolation problem if and only if there exists a contractive, analytic function S, such that $S(0) = 0$ and*

$$Z^\circ = Z_n + W(e)_n S(I - K_n S)^{-1} \overline{W}(f)_n^* \tag{20.25}$$

where Z_n denotes the maximum entropy solution.

20.5 Generalized Kimura-Georgiou Parameterization

In case of the Carathéodory extension problem for scalar-valued functions the so-called Kimura-Georgiou parameterization (see [14], [18], [19]) the solutions with McMillan-degree equal to the number of the given covariances are parameterized in forms of ratio of two polynomials, where these polynomials have special forms. As a generalization of this in the present section we describe the solutions of the the interpolation Problem 20.3.2 of the form

$$(Z^\circ)^* = F^{-1}G\,,$$

where F and G are *analytic* functions of size $p \times p$, but $F\Xi_n^*$ is *co-analytic*. (Note, that if $\Xi_n(z) = z^n I_p$, thus all interpolation nodes are at the origin, then the required property of the function F means that it should be a matrix-valued polynomial of degree no greater than n. As we are going to see, the form of the function G is already determined by the form the F and by the property that the "ratio" is an interpolating function.)

First let us consider various factorizations of the inner function Ξ_n. $\Xi_n = \Theta_j \Xi_j$, $j = 1, \dots k$, where the functions Θ_j, Ξ_j are inner functions $1 \le j \le k$ and consider the interpolation problems formulated by the inner functions Ξ_j, $j = 1, 2, \dots, k$ but always using the same positive real function Z. We shall use the notation Nevanlinna-Pick-interpolation Problem $20.3.2_j$.

Lemma 20.5.1. *Let us assume that the functions*

$$\tilde{Z}_j^* = F_j^{-1} G_j$$

provide solutions of the Nevanlinna-Pick-interpolation Problem $20.3.2_j$, for all $1 \le j \le k$, where the functions F_j, G_j, are stable of size $p \times p$, while $F_j \Xi_j^*$ is anti-stable. Set $F_0 = I$, $G_0 = I$.

Then, if the function Z° provides a solution of the Interpolation Problem 20.3.2 and its parahermitian conjugate function can be written in the form

$$(Z^\circ)^* = F^{-1}G\,, \tag{20.26}$$

where F and G are analytic functions of size $p \times p$, and F is of the form

$$F = \sum_{j=0}^{k} \Gamma_j F_j$$

for appropriate coefficient matrices $\Gamma_0, \ldots, \Gamma_n$, then the function G should be defined as

$$G = \sum_{j=0}^{k} \Gamma_j G_j \,.$$

Proof. Since according to our assumption $(Z^\circ)^*$ is a solution of the interpolation problem the function $\Xi\left(Z^* - (Z^\circ)^*\right)$ and together with this the functions $\Xi_j\left(\tilde{Z}_j^* - (Z^\circ)^*\right) = \Xi_j\left(\tilde{Z}_j^* - Z^*\right) + \Xi_j\left(Z^* - (Z^\circ)^*\right)$, $j = 1, 2, \ldots, k$ should be co-analytic and vanishing at infinity. On the other hand

$$\left(\sum_{j=0}^{k} \Gamma_j F_j\right)(Z^\circ)^* = G \,,$$

so

$$\sum_{j=0}^{k} \Gamma_j F_j \Xi_j^* \Xi_j \left((Z^\circ)^* - \tilde{Z}_j^*\right) = G - \sum_{j=0}^{k} \Gamma_j G_j$$

The right hand side is analytic, while the left hand side is co-analytic and vanishing at infinity. Thus both sides are zero, implying that $G = \sum_{j=0}^{k} \Gamma_j G_j$, proving the lemma. □

In order to prove a general form of the Kimura-Georgiou-parameterization the condition given above – $F = \sum_{j=0}^{k} \Gamma_j F_j$ – should be changed to the following one: F be an analytic function while $F\Xi_n^*$ is co-analytic.

Since $\mathscr{M}_n = H_0^-(y) \cap H_0^+(y_n)$, where $y_n(t) = \Xi_n y(t)$, a vector $\kappa \in \mathscr{M}_n$ if and only if expressing it from the sequence y the corresponding transfer function T_κ (i.e. $\kappa = T_\kappa y(0)$) is analytic and $T_\kappa \Xi^*$ is co-analytic. Thus, to have the representation $F = \sum_{j=0}^{k} \Gamma_j F_j$ for every function F with the required property, an appropriate basis (or generating system) of $\mathscr{M}_n$ should be chosen.

Thus the basic idea behind the Kimura-Georgiou parameterization is a special basis considered in the subspace $\mathscr{M}_n$. There are various possibilities choosing this appropriate basis. For example, in the classical case (Carathéodory extension) the orthogonal polynomials determine this basis. In the present paper – for the more general Nevanlinna-Pick interpolation problem – we follow such an approach which is naturally connected to the possibility of sequentially allocating the interpolation nodes. Namely, as we have assumed, Ξ_n is written in the form of products of elementary inner functions, i.e. $\Xi_n = \xi_n \times \xi_{n-1} \times \cdots \times \xi_1$. Set $\Xi_j = \xi_j \times \xi_{j-1} \times \cdots \times \xi_1$.

Applying the construction above we might define the subspaces $\mathscr{M}_j$ using the inner function Ξ_j and construct the functions $\overline{W}(f)_j^{-1}$, for $1 \leq j \leq n$. Then the coordinates of

$$f_j(0) = \overline{W}(f)_j^{-1} y(0) \,,$$

generate the subspace $\mathscr{M}_j \ominus \mathscr{M}_j''$. Using the special factorization of the function Ξ_n is can be proved that the following lemma holds.

Lemma 20.5.2. *The coordinate random variables of* $f_j(0)$, $j = 1,\dots,n$ *and those of* $y(0)$ *generate the subspace* $\mathscr{M}_n$. *I.e.*

$$\mathscr{M}_n = \langle y(0), f_1(0),\dots,f_n(0)\rangle \ , \tag{20.27}$$

Now the following theorem is an immediate consequence of Lemmata 20.5.1 and 20.5.2.

Theorem 20.5.1 (Cf. Georgiou [14], Kimura [18],[19]). *If the function* Z° *provides a solution of the Interpolation Problem 20.3.2 and its para-hermitian conjugate function is of the form*

$$(Z^\circ)^* = F^{-1}G \,, \tag{20.28}$$

where F *and* G *are analytic functions of size* $p\times p$, *but* $F\Xi_n^*$ *is co-analytic then*

$$(Z^\circ)^* = \left(\sum_{j=0}^{n} \Gamma_j \overline{W}(f)_j^{-1}\right)^{-1} \left(\sum_{j=0}^{n} \Gamma_j \overline{W}(f)_j^{-1} Z_j^*\right) . \tag{20.29}$$

for appropriate coefficient matrices $\Gamma_0,\dots,\Gamma_n$.

Furthermore, in this representation the function F *can be chosen in such a way that* $F\Xi^*$ *be a conjugate outer function such that its value at infinity is the identity matrix.*

Proof. Lemma 20.5.2 gives that the subspace $\mathscr{M}_n$ is generated by the normalized projection errors $f_j(0)$ for $j=1,\dots,n$ and $y(0)$. At the same time $f_j(0) = \overline{W}(f)_j^{-1}y(0)$.

Consequently, the function F with the property $F\Xi_n^*$ is co-analytic can be written in the form

$$F = \sum_{j=0}^{n} \Gamma_j \overline{W}(f)_j^{-1}$$

for some matrices $\Gamma_0,\dots,\Gamma_n$. Now the conditions of Lemma 20.5.1 are fulfilled for the functions $F_j = \overline{W}(f)_j^{-1}$, and $G_j = \overline{W}(f)_j^{-1}Z_j^*$, because Remark 20.4.2 gives that G_j is analytic, and obviously $F_j\Xi_j^*$ is co-analytic.

Consequently, if $F = \sum_{j=0}^n \Gamma_j\overline{W}(f)_j^{-1}$ then G should be $G = \sum_{j=0}^n \Gamma_j\overline{W}(f)_j^{-1}Z_j^*$ proving the first part of the Theorem.

Finally, consider a conjugate inner, conjugate outer factorization of $F\Xi_n^*$ in the form

$$F\Xi_n^* = F_i^*F_o^* \,,$$

where F_i is an inner, F_o is an outer function. (Since F_o is invertible out of the unit disk we might assume that $F_o(\infty) = I$.) Then

$$Z^{'*} = (F_iF)^{-1}(F_iG) \ .$$

Here $F_iF\Xi_n^* = F_o^*$ is co-analytic, moreover it is a conjugate outer function (and the value of $F_iF\Xi_n^*$ at infinity is the identity matrix). This concludes the proof of the Theorem. □

20.5.1 State-Space Equations for the Covariance Extension Problem

Recall that $W(e)_n$ can be written in the form $W(e)_n(z) = D + C\left(z^{-1}I - A\right)^{-1} B$ (cf. equation (20.9)). Note, that the form above is not necessarily a minimal realization of $W(e)_n$, although from the positive definiteness of the matrix P it follows that the pair (A,B) is reachable.

Straightforward calculation gives the following realizations, as well:

(i)

$$\overline{W}(f)_n^{-1}(z) = V - H\left(z^{-1}I - A_{\text{int}}\right)^{-1} B_{\text{int}}, \tag{20.30}$$

for some matrices V, H;

(ii)

$$Z_n(z) = \frac{1}{2}R(0) + C\left(z^{-1}I - A\right)^{-1}\overline{C}^* \tag{20.31}$$

where $\overline{C} = DB^* + CPA^*$;

(iii)

$$\overline{W}(f)_n^{-1}(z)Z_n^*(z) = \frac{1}{2}\overline{W}(f)_n^{-1}R(0) - H\left(I - zA_{\text{int}}\right)^{-1} PC^*\,; \tag{20.32}$$

(iv)

$$\Xi_n(z) = D_\Xi + C_\Xi\left(z^{-1}I - A_{\text{int}}\right)^{-1} B_{\text{int}}, \tag{20.33}$$

for some matrices C_Ξ, D_Ξ.

Remark 20.5.1. Due to the fact that Σ is an inner function and the pair (A,B) is completely controllable there exists a positive definite matrix ρ such that the following equation hold:

$$\begin{bmatrix} A_{\text{int}} & B_{\text{int}} \\ C_\Sigma & D_\Sigma \end{bmatrix}\begin{bmatrix} \rho & 0 \\ 0 & I \end{bmatrix}\begin{bmatrix} A_{\text{int}} & B_{\text{int}} \\ C_\Sigma & D_\Sigma \end{bmatrix}^* = \begin{bmatrix} \rho & 0 \\ 0 & I \end{bmatrix}. \tag{20.34}$$

20.5.2 State-Space Equations for the Kimura-Georgiou Parametrization

In order to find state-space realizations of the solutions provided by the multivariate version of the Kimura-Georgiou parametrization a special state-process should be defined which can be used for all functions $\overline{W}(f)_n^{-1}$. Since the subspaces $\mathcal{M}_j'$ are increasing, i.e. $\mathcal{M}_j' \subset \mathcal{M}_{j+1}'$, $1 \leq j \leq n-1$, we can choose a basis in the largest subspace $\mathcal{M}_n'$ in such a way that its elements sequentially determine bases of $\mathcal{M}_j'$, for all j. Let us point out that for example defining $z_j(t) = \frac{U_j^*}{1 - z\overline{a_j}}\Xi_{j-1}y(t)$, $1 \leq j \leq n$ then the coordinates of $z_1(-1), \ldots, z_j(-1)$ form a basis in the subspace $\mathcal{M}_j'$.

In this case any function $F = \sum_{j=1}^n \Gamma_j \overline{W}(f)_n^{-1}$ can be written in the form

$$F = D_{KM} - C_{KM}\left(z^{-1}I - A_{\text{int}}\right)^{-1} B_{\text{int}} \tag{20.35}$$

for some matrices C_{KM}, D_{KM}. It follows – using equation (20.32) – that now

$$G = \sum_{j=0}^{n} \Gamma_j \overline{W}(f)_j^{-1} Z_j^* = \frac{1}{2} F(z) R(0) - C_{KM} \left(I - z A_{\text{int}} \right)^{-1} P C^* . \tag{20.36}$$

From Theorem 20.5.1 straightforward calculation gives the following proposition.

Proposition 20.5.1. *Assume that a solution $(Z^\circ)^*$ of the Interpolation Problem 20.3.2 is given in the form*

$$(Z^\circ)^* = F^{-1} G ,$$

where F and G are analytic functions of size $p \times p$, $F \Xi_n^*$ is co-analytic.

If $F(0)$ is nonsingular then $(Z^\circ)^*$ has the following state space representation:

$$(Z^\circ)^* = \frac{1}{2} R(0) - D_{KM}^{-1} C_{KM} \left(I - z \left(A_{int} + B_{int} D_{KM}^{-1} C_{KM} \right) \right)^{-1} P C^* \tag{20.37}$$

for some matrices C_{KM}, D_{KM}.

The next lemma gives a matrix-equation counterpart of the relation $F(Z^\circ)^* = G$, when Z° is an interpolating function.

Lemma 20.5.3. *Assume that the function Z° has the following – possibly non-minimal – realization*

$$Z^\circ = \frac{1}{2} R(0) + C_r \left(z^{-1} I - A_r \right)^{-1} \overline{C}_r^* . \tag{20.38}$$

Then, if for some matrices S_r, C_{KM}, D_{KM} the equations

$$A_{int} S_r A_r^* + B_{int} \overline{C}_r = S_r , \tag{20.39}$$

$$CP = C_r S_r^* , \tag{20.40}$$

$$-C_{KM} S_r A_r^* + D_{KM} \overline{C}_r = 0 \tag{20.41}$$

hold, then – defining the functions F and G using formulae (20.35) and (20.36) – the equation

$$F(Z^\circ)^* = G$$

holds, as well.

Conversely, if the pair (C_r, A_r) is observable and S_r is a solution of the equations (20.39), (20.40) and moreover the extra equation $F(Z^\circ)^* = G$ holds, then equation (20.41) is satisfied, as well.

Remark 20.5.2. Corollary 20.4.1 shows that if Z° is a solution of the Covariance Extension Problem 20.3.2 and A_r is a stable matrix, then equations (20.39) and (20.40) have a unique solution S_r.

Proof. Assuming that S_r is a solution of (20.39) and (20.40) let us define the functions F and G by (20.35) and (20.36). Straightforward calculation gives that

$$F(Z^\circ)^* = G + \left(D_{KM} \overline{C}_r - C_{KM} S_r A_r^* \right) (zI - A_r^*) C_r^* .$$

From this identity the statement follows directly. □

The following theorem shows that there is a more or less explicit state-space representation of $(Z^\circ)^*$ in the general case, when $F(0)$ is singular, as well.

Let us recall the notation $N = \sum_{j=1}^n m_j$, i.e. $N = \dim \mathcal{M}_n'$.

Theorem 20.5.2. *Assume that the function Z° of McMillan-degree $N_m \leq N$ with minimal realization*

$$Z^\circ = \frac{1}{2}R(0) + C_m\left(z^{-1}I - A_m\right)^{-1}\overline{C}_m^* , \tag{20.42}$$

provides a solution of the interpolation problem (20.2).

Then the following conditions are equivalent:

(i) its para-hermitian conjugate function can be written in the form

$$(Z^\circ)^* = F^{-1}G , \tag{20.43}$$

where F, G are analytic functions of size $p \times p$ and $F\Xi_n^*$ is co-analytic;

(ii) for the solution S_m of the equation

$$S_m = A_{int}S_mA_m^* + B_{int}\overline{C}_m \tag{20.44}$$

the condition

$$\mathrm{Ker}\,(S_m) = \{0\} \tag{20.45}$$

holds.

Moreover, if these conditions are fulfilled then assuming that the function F is chosen in such a way that $F\Xi_n^*$ is a conjugate outer function (with value I at infinity) a – possibly non-minimal – realization of $(Z^\circ)^*$ is given by the formula

$$(Z^\circ)^* = \frac{1}{2}R(0) + \overline{C}_r\left(zI - A_r^*\right)^{-1} PC^* , \tag{20.46}$$

where the matrices A_r and $\overline{C}_r$ of size $N \times N$ and $p \times N$, respectively, are the solutions of the equation

$$\begin{bmatrix} A_{int} & B_{int} \\ -C_{KM} & D_{KM} \end{bmatrix} \begin{bmatrix} A_r^* \\ \overline{C}_r \end{bmatrix} = \begin{bmatrix} I \\ 0 \end{bmatrix} . \tag{20.47}$$

Furthermore there exists a matrix R such that

$$\begin{bmatrix} A_r^* \\ \overline{C}_r \end{bmatrix} = \begin{bmatrix} \rho A_{int}^*\rho^{-1} \\ B_{int}^*\rho^{-1} \end{bmatrix} - \begin{bmatrix} \rho C_\Xi^* \\ D_\Xi^* \end{bmatrix} R . \tag{20.48}$$

Thus

$$(Z^\circ)^* = \frac{1}{2}R(0) + \left(B_{int}^*\rho^{-1} - D_\Xi^*R\right)\left(zI - \rho A_{int}^*\rho^{-1} + \rho C_\Xi^*R\right)^{-1} PC^* .$$

Proof. (Sketch) Assume that (i) holds. According to Theorem 20.5.1, we might always assume that in the representation $(Z^{\circ})^* = F^{-1}G$ the function $F\Xi_n^*$ is conjugate outer and the function F has the realization $F(z) = D_{KM} - C_{KM}\left(z^{-1}I - A_{\text{int}}\right)^{-1} B_{\text{int}}$. Now invoking Corollary 20.4.1 and Lemma 20.5.3 we get that equation

$$\begin{bmatrix} A - BD^{-1}C & BD^{-1} \\ -C_{KM} & D_{KM} \end{bmatrix} \begin{bmatrix} S_m A_m^* \\ \overline{C}_m \end{bmatrix} = \begin{bmatrix} S_m \\ 0 \end{bmatrix} \tag{20.49}$$

holds. Using "stochastic-geometric" arguments it can be proved that

$$\begin{bmatrix} A - BD^{-1}C & BD^{-1} \\ -C_{KM} & D_{KM} \end{bmatrix} \tag{20.50}$$

is nonsingular implying that $\text{Ker}(S_m)$ is in $\text{Ker}\left(\overline{C}_m\right)$ and it is A_m^*-invariant. The observability of the pair $\left(\overline{C}_m^*, A_m\right)$ implies that $\text{Ker}(S_m)$ should be trivial, proving that (ii) holds, as well.

Conversely, assume that (ii) holds. Consider the subspace $\mathscr{K}$ generated by the $N+p$ dimensional column vectors of $\begin{bmatrix} S_m A_m^* \\ \overline{C}_m \end{bmatrix}$. It is of dimension at most N_m, and the assumption $\text{Ker}(S_m) = \{0\}$ implies that its nonzero elements are not orthogonal to the subspace $\mathscr{L}$ generated by the row vectors of $\left[A - BD^{-1}C, BD^{-1}\right]$. Counting dimensions it follows that there is a subspace $\mathscr{H}$ of dimension at least p which is orthogonal to $\mathscr{K}$ with trivial intersection of $\mathscr{L}$.

Thus we can choose p linearly independent row vectors from $\mathscr{H}$ and form from them the matrix $[-C_{KM}, D_{KM}]$ of size $p \times (N+p)$. Notice, that in this case the row vectors of the matrix

$$\begin{bmatrix} A - BD^{-1}C & BD^{-1} \\ -C_{KM} & D_{KM} \end{bmatrix} \tag{20.51}$$

will be linearly independent, too. Consequently, in addition to (20.44) equation (20.49) holds. Defining the functions F and G using the matrices C_{KM} and D_{KM} obtained above Lemma 20.5.3 gives that $F(Z^{\circ})^* = G$ holds. From the fact that the matrix in (20.51) is nonsingular it follows that $\det(F)$ is not identically zero, so its inverse can be considered. This leads to the equation $(Z^{\circ})^* = F^{-1}G$, where $F\Xi_n^*$ is obviously co-analytic, concluding the proof of the implication (ii) $\to$ (i) and so the proof of the converse statement.

Now, let us assume that (i) (or equivalently (ii)) holds, The invertibility of the matrix (20.50) implies that equation (20.47) – in terms of A_r^* and $\overline{C}_r$ – has a solution. Using again Lemma 20.5.3 we get that

$$F(z)\left(\frac{1}{2}R(0) + \overline{C}_r (zI - A_r^*)^{-1} PC^*\right) = G\,. \tag{20.52}$$

Comparing this to (20.43) we obtain that the representation (20.46) holds.

Finally the normalization of $F\Xi_n^*(\infty) = I$ implies that $D_{KM}D_{\Xi}^* - C_{KM}\rho C_{\Xi}^* = I$.

Together with (20.47) this can be written as

$$\begin{bmatrix} A_{\text{int}} & B_{\text{int}} \\ -C_{KM} & D_{KM} \end{bmatrix} \begin{bmatrix} A_r^* & \rho C_\Xi^* \\ \overline{C}_r & D_\Xi^* \end{bmatrix} = \begin{bmatrix} I & 0 \\ 0 & I \end{bmatrix}.$$

Comparing this to equation (20.34) we obtain that there exists a matrix R such that

$$\begin{bmatrix} A_r^* \\ \overline{C}_r \end{bmatrix} = \begin{bmatrix} \rho A_{\text{int}}^* \rho^{-1} - \rho C_\Xi^* R \\ B_{\text{int}}^* \rho^{-1} - D_\Xi^* R \end{bmatrix}$$

concluding the proof of the theorem. □

References

1. Antoulas, A.C., Ball, J.A., Willems, J.K., On, J.C.: the solutions of the minimal rational interpolation problem. Linear Algebra and Applications 137/138, 511–573 (1990)
2. Adamjan, V.M., Arov, D.Z., Krein, M.G.: Analitical properties of Schmidt pairs of Hankel operators and the generalized Schur-Takagi problem (In Russian). Mat. Sbornik 86, 34–75 (1971)
3. Akhiezer, N.I.: The Classical Moment Problem. Oliver and Boyd, Edinburgh (1965)
4. Ball, J.A.: Interpolation problems of Nevanlinna-Pick and Löewner types for meromorphic matrix functions. Integral Equations and Operator Theory 6, 804–840 (1983)
5. Ball, J.A., Gohberg, I., Rodman, L.: Realization and interpolation of rational matrix functions. Operator Theory, Advances and Applications 33, 1–72 (1988)
6. Ball, J.A., Helton, J.W.: A Beurling-Lax theorem for the Lie group $u(m,n)$ which contains most classical interpolation theory. J. Operator Theory 9, 107–142 (1983)
7. Byrnes, C., Lindquist, A.: On the partial stochastic realization problem. IEEE Trans. Automatic Control 42, 1049–1070 (1997)
8. Byrnes, C., Lindquist, A., Gusev, S.V.: A convex optimization approach to the rational covariance extension problem. TRIA/MAT (1997)
9. Byrnes, C., Lindquist, A., Gusev, S.V., Mateev, A.S.: A complete parametrization of all positive rational extension of a covariance sequence. IEEE Trans. Automatic Control 40, 1841–1857 (1995)
10. Carathéodory, C.: Über die Variabilitätsbereich der Koeffizienten von Potenzreihen, die gegebene Werte nicht annehmen. Math. Ann. 64, 95–115 (1907)
11. Dewilde, P., Dym, H.: Lossless chain scattering matrices and optimum linear prediction: the vector case. Circuit Theory and Applications 9, 135–175 (1981)
12. Foias, C., Frazho, A.E.: The commutant lifting approach to interpolation problems. Operator Theory, vol. 44. Birkhäuser, Basel (1990)
13. Fuhrmann, P.: Ortogonal matrix polynomial and system theory. Rend. Sem. Mat. Univers. Politecn. Torino pp. 68–124 (1988)
14. Georgiou, T.: Partial realization of covariance sequences, Phd. Thesis (1983)
15. Georgiou, T.: The structure of state covariances and its relation to the power spectrum of the input. IEEE Trans. on Automatic Control 47(7), 1056–1066 (2002)
16. Geronimus, Y.L.: Polynomials Orthogonal on a Circle and Interval. Pergamon Press, Oxford (1960)

17. Grenander, U., Szegő, G.: Toeplitz forms and their applications. University of California Press, Berkeley (1958)
18. Kimura, H.: Positive partial realization of covariance sequences. In: Byrnes, C.I., Lindquist, A. (eds.) Modelling, Identification and Robust Control, pp. 499–513. Elsevier Science, Amsterdam (1986)
19. Kimura, H.: Generalized Schwartz form and lattice-ladder realizations of digital filters. IEEE Trans. Circuit Syst. (1987)
20. Nevanlinna, R.: Über beschränkte analytische Funktionen. Ann. Acad. Sci. Fenn. 32, 1–75 (1929)
21. Pick, G.: Über beschränkte analytischer Funktionen, welche durch vorgegebene Funktionswerte bewirkt werden. Math, Ann 77, 7–23 (1916)
22. Rosenblum, M., Rovnyak, J.: An operator-theoretic approach to theorem of the Pick-Nevanlinna and Loewner types. Integral Equations Operator Theory 3, 408–436 (1980)
23. Schur, I.: Über Potenzreihen die im Innerne des Einheitskreises beschränkt sind. J. für die Reine und Angew. Math. 147, 205–232 (1917)
24. Szegő, G.: Orthogonal polynomials. Amer. Math. Soc. Colloq. Publ. XXIII, Providence, Rhode Island (1939)

21

A New Class of Control Systems Based on Non-equilibrium Games*

Yifen Mu and Lei Guo

Key Laboratory of Systems and Control, ISS, Academy of Mathematics and Systems Science, Chinese Academy of Sciences, Beijing 100190, P.R. China

Summary. In this paper, a new class of control systems based on non-equilibrium dynamic games is introduced. Specifically, we consider optimization and identification problems modeled by an infinitely repeated 2×2 generic games between a human and a machine, where the machine takes a fixed and k-step-memory strategy while the human is more intelligent in the sense that she can optimize her strategy. This framework is beyond the frameworks of both the traditional control theory and game theory. By using the concept of state transfer graphs, the optimal strategy of the human will be characterized and the win-lose situation will be discussed. These are carried out for three typical games, i.e., the Prisoner's Dilemma game, the Snowdrift game and the Battle-of-sex game, but with different win-loss results. The problem of strategy identification will also be investigated.

21.1 Introduction

The current theoretical framework for control systems mainly aims at designing control laws for dynamical systems to achieve a certain prescribed performance (e.g. stability, optimality and robustness, etc.). In the control process, the systems (or plants) to be controlled are completely "passive" in an essential way, in the sense that they have no intension to compete with the controllers to achieve their own objectives or "payoffs". This is so even when the structures or parameters of the dynamical systems under control are uncertain and changing in time, because they are again of "passive" character. In these cases, the controllers can be made adaptive by incorporating certain online estimation algorithms, see e.g. [1]-[6].

However, in many practical systems, especially social, economical, biological and ecological systems, which involve adaptation and evolution, people often encounter with the so-called complex adaptive systems (CAS) as described in, e.g. [7]. In a CAS, as summarized in [8], a large number of components, called agents, interact with and adapt to (or learn) each other and their environment actively, leading

*This work was supported by the National Natural Science Foundation of China under grant 60821091 and by the Knowledge Innovation Project of CAS under Grant KJCX3-SYW-S01.

X. Hu et al. (Eds.): Three Decades of Progress in Control Sciences, pp. 313–326, 2010.

to some (possibly unexpected) macro phenomena called emergence. Despite of the flexibility in modeling a wide class of complex systems by CAS, it brings great challenge to understand the evolution of a CAS mathematically, since the traditionally used mathematical tools appear to give limited help in the study of CAS, as pointed out in [8].

As an attempt towards initiating a theoretical investigation for dynamical systems when some parts of the plants to be controlled have intentions to gain their own payoffs, we will, in this paper, consider a dynamic game framework that is somewhat beyond the current control theoretical framework. Intuitively, we will consider a simple scenario where we have two heterogeneous agents (players) in a system playing a repeated noncooperative game [9], where each agent makes its decision based on the previous actions and payoffs of both agents, but the law for generating the actions of one agent is assumed to be fixed. Thus, it may still be referred as a "control system" from the other agent's standpoint.

We would like to point out that, to the best of the authors' knowledge, the above non-equilibrium dynamic game framework seems to be neither contained in the traditional control theory, nor considered in the classical game theory. In fact, in the classical framework of game theory, all agents (players) stand in a symmetric position in rationality, in order to reach some kind of equilibrium; whereas, in our framework, the agents do not share a similar mechanism for decisions making and do not have the same level of rationality. This difference is of fundamental importance, since in many complex systems, such as non-equilibrium economy (e.g. see [10]), the agents are usually heterogenous, and they may indeed differ in either their information obtained or their ability in utilizing it.

We would also like to mention that there have been considerable investigations in game theory in relation to adaptation and learning, which can be roughly divided into two directions. One is called evolutionary game theory (e.g. see [11] and [12]) in which all agents in a large population are programmed to use certain actions to play. An action will spread or diminish according to the value of its corresponding payoff. The other direction is called learning in game theory, see, e.g. [13], [14]], which considers whether the long-run behaviors of individual agents will arrive at some equilibrium [15]-[16]. In both the directions, all the agents in the games are equal in their ability to learn or adapt to the strategies of their opponents. Some recent works and reviews can be found in [17]-[21]. The dynamic game framework to be studied in this paper is partly inspired by the evolutionary game framework in [22], where the best strategy is emerged as a result of evolution, while our optimal strategy to be studied in the next sections will be obtained by optimization and identification.

More specifically, we will consider infinitely repeated games between a human (or "controller") and a machine (or "plan") based on a generic 2×2 game model, which includes standard games such as Prisoners' Dilemma, Snowdrift, and Battle of Sex. The machine's strategy is assumed to be fixed with k-step memory, which may be unknown to the human.

To this end, we need to analyze the state transfer graph for machine strategy with k-memory. We will show that, similar to the Prisoners' Dilemma game as studied recently in [25], the optimal strategy for the current generic games that maximizes the human's averaged payoff is also periodic after finite steps. However, different from the result of [25], even for the case of $k = 1$, the human may lose to the machine while optimizing his own averaged payoff in the Snowdrift game, and the similar conclusion will depend on more conditions on the parameters of the Battle of Sex game. When the machine's strategy is unknown to the human, we will give a necessary and sufficient condition for identifiability, and will investigate the consequences of identification in our non-equilibrium dynamic game problem. Finally, we will discuss possible extensions to games with 2-players but with 3-actions.

The remainder of this paper is organized as follows. In Section 21.2, the main theorems will be stated following the problem formulation; In Section 21.3, the state transfer graph (STG) will be described and some useful properties will be studied. The proofs of some theorems will be given in Section 21.4, and Section 21.5 extends the modeling to games of 2-players with 3-actions. Finally, Section 21.6 will conclude the paper with some remarks.

21.2 Problem Statement and Main Results

Consider a generic 2×2 game with its payoff matrix described in figure 21.1. This matrix can be used to describe many standard games, either symmetric or asymmetric, when the parameters satisfy certain conditions.In the symmetric case, the payoff matrix can be specified as in figure 21.2.

		Player II	
		A	B
Player I	A	(a_{11}, b_{11})	(a_{12}, b_{12})
	B	(a_{21}, b_{21})	(a_{22}, b_{22})

Fig. 21.1. The payoff matrix of the generic 2×2 game

		Player II	
		A	B
Player I	A	(a, a)	(c, b)
	B	(b, c)	(d, d)

Fig. 21.2. The payoff matrix of the symmetric 2×2 game

One well-known example is the Prisoners' Dilemma game where the parameters satisfy $c > a > b > d$ and $2 \cdot a > b + c$, while the actions "A" and "B" mean "Cooperate" and "Defect" respectively.

Another typical example is the Snow Drift game. In this game, two players, called Player 1 and 2, can be two drivers who are on their way home, caught by the snowdrift and thus must decide whether or not to shovel it. They simultaneously choose their actions A or B, where "A" means the player will shovel the snow on the road, and "B" means the player will not. Different action profiles will result in different payoffs for the players. The parameters in the payoff matrix of this game satisfy $d = 0 < c < a < b$.

As for the asymmetric case, the game of Battle of Sex is a typical example (see figure 21.3). Here, the Player 1 can be assumed to be the wife while the Player 2 be the husband, with the action A may stand for watching the ballet while the action B may stand for watching the football. The parameters are assumed to satisfy $a_{21} = b_{21} = 0, a_{11} > a_{12} > 0, a_{11} > a_{22} > 0$, and $b_{22} > b_{11} > 0, b_{22} > b_{12} > 0$. Without loss of generality, we may specify the matrix as follows where $a > b > 0, a > c > 0$:

		Player II	
		A	B
Player I	A	(a, b)	(c, c)
	B	(0, 0)	(b, a)

Fig. 21.3. The payoff matrix of the Battle of Sex game

From the parameter inequalities, it is easy to compute the Nash Equilibria of these games. Our purpose is, however, not to investigate the Nash equilibrium in the game theory. Instead, we will consider the scenario where Player 1 has the ability to search for the best strategy so as to optimize his payoff, while Player 2 acts according to a given strategy. Clearly, this non-equilibrium dynamic game problem is different from either the standard control problem or the classical game problem, and thus may be regarded as a new class of "control systems". A preliminary study was initiated recently for the Prisoners' Dilemma game in [25], where some basic notations and ideas will be adopted in what follows.

Vividly, let Player 1 be a human (we say it is a "he" henceforth) while his opponent Player 2 is a machine. Assume they both know the payoff matrix. The action set of both players is denoted as $\mathscr{A} = \{A, B\}$, and the time set is discrete, $t = 0, 1, 2, \ldots$. At time t, both players will choose their actions and get their payoffs simultaneously. Let $h(t)$ denote the human's action at t and $m(t)$ the machine's.

Define the history of time t, H_t, as the sequence of two players' action profiles before time t i.e.

$$H_t \triangleq (m(0),h(0);m(1),h(1);...;m(t-1),h(t-1)).$$

Denote the set of all histories for all time t as $H = \bigcup_t H_t$.

As a start, we consider the case of pure strategy and define the **strategy** of either player as a function $f: H \to \mathscr{A}$. In this paper, we will further confine the machine's strategy with finite k-memory as follows:

$$m(t+1) = f(m(t-k+1),h(t-k+1);...;m(t),h(t)) \tag{21.1}$$

which, obviously, is a discrete function from $\{0,1\}^{2k}$ to $\{0,1\}$, where and hereafter, 0 and 1 stands for A and B respectively. Moreover, the following mapping can establish a one-to-one correspondence between the vector set $\{0,1\}^{2k}$ and the integer set $\{1,2,......2^{2k}\}$:

$$s(t) = \sum_{l=0}^{k-1} \{2^{2l+1} \cdot m(t-l) + 2^{2l} \cdot h(t-l)\} + 1 \tag{21.2}$$

For convenience, in what follows we will denote $s_i = i$ and call it a **state** of the game under the given strategies.

In the simplest case where $k = 1$, the above mapping reduces to

$$s(t) = 2 \cdot m(t) + h(t) + 1, \tag{21.3}$$

which establishes a one-to-one correspondence between the value set $s(t) \in \{s_1, s_2, s_3, s_4\}$ with $s_i = i$ and $(m(t),h(t))$:

s(t)	(m(t),h(t))
S_1	(0,0)
S_2	(0,1)
S_3	(1,0)
S_4	(1,1)

and the machine strategy (21.1) can be written as

$$\begin{aligned} m(t+1) &= f(m(t),h(t)) \\ &= a_1 I_{\{s(t)=s_1\}} + ... + a_4 I_{\{s(t)=s_4\}} \\ &= \sum_{i=1}^{4} a_i I_{\{s(t)=s_i\}} \end{aligned} \tag{21.4}$$

which can be simply denoted as a vector $A = (a_1, a_2, a_3, a_4)$ with a_i being 0 or 1.

Given any strategies of both players together with any initial state, the game will be carried on and a unique sequence of states $\{s(1), s(2), ...\}$ will be produced. Such a sequence will be called a **realization** [15].

Obviously, each state $s(t)$ corresponds to a pair $(m(t), h(t))$, and so by the definition of the payoff matrix, the human and the machine will obtain their payoffs, denoted by $p(s(t))$ and $p_m(s(t))$, respectively. Let us further define the extended payoff vector for the human as **P(s(t))** $\triangleq (p(s(t)), w(s(t)))$, where $w(s(t))$ indicates the relative payoff to the machine at t, i.e.,

$$w(s(t)) = sgn\{p(s(t)) - p_m(s(t))\} \triangleq w(t), \tag{21.5}$$

where $sgn(\cdot)$ is the sign function and $sgn\{0\} = 0$.

For the above infinitely repeated games, the human may only observe the payoff vector **P(s(t))**, but since there is an obvious one-to-one correspondence between **P(s(t))** and $s(t)$, we will assume that $s(t)$ is observable to the human at each time t throughout the paper.

Now, for any given human and machine strategies with their corresponding realization, the **averaged payoff** (or ergodic payoff) [23] of the human can be defined as

$$P_\infty^+ = \overline{\lim_{T\to\infty}} \frac{1}{T}\sum_{t=1}^{T} p(t). \tag{21.6}$$

In the case where the limit actually exists, we may simply write $P_\infty^+ = P_\infty$. Similarly, W_∞^+ can be defined.

The basic questions that we are going to address are as follows:

1. How can the human choose his strategy g so as to obtain an optimal averaged payoff?
2. Is the human's optimal strategy necessarily gives a payoff that is better than the machine's?
3. Can the human still obtain an optimal payoff when the machine's strategy is unknown to him?

The following theorems and proposition will give some answers to these questions.

Theorem 21.2.1. *Consider the generic* 2×2 *game described in Fig 1, and any machine strategy with finite k-memory. Then, there always exists a human strategy also with k-memory, such that the human's payoff is maximized and the resulting system state sequence* $\{s(t)\}$ *will become periodic after some finite time.*

The proof of Theorem 21.2.1 is just the same as that in [25] for the case of the Prisoners' Dilemma game, so we refer the readers to [25] for the proof details. Also, One can see from the proof that the optimal payoff values will remain the same for different initial values of the state transfer graph (STG), as long as they share the same reachable set. In particular, this observation is true when the STG is strongly connected, see Section 21.3 for the definition of STG.

Moreover, as will be illustrated by Example 21.3.1, Theorem 21.2.1 will enable us to find the optimal human strategy by searching on the STG with considerably reduced computational complexity. Furthermore, since Theorem 21.2.1 only concerns with the properties of the optimal human trajectory, a natural question is: whether or not the human's optimal averaged payoff value is better than that of the machine's. This is a subtle question, and will be addressed in the following theorem.

Theorem 21.2.2.

1. *For the standard Prisoners' Dilemma game, the optimal strategy of the human will not lose to any machine whose strategy is of 1-memory. However, when* $k > 1$, *there exists such machine strategies, that the human's optimal strategy will lose to them.*
2. *For the Snowdrift game, there exists such a machine strategy with 1-memory, that the optimal strategy of the human will lose to the machine.*
3. *For the game of Battle of Sex, whether or not the human will always win the machine with 1-memory is indefinite, i.e., it depends on more conditions on the payoff parameters.*

Remark 21.2.1.

(1) For the Prisoners' Dilemma game, when $k \geq 2$, the game becomes more complicated and subtle. As demonstrated in Section 21.4 of [25], whether the human can win while getting his optimal payoff depends on delicate relationships among s, p, r, t.

(2) Theorem 21.2.2 (2) will remain valid for the machine strategy with k-memory in general, since $k = 1$ is a special case.

Remark 21.2.2. As has been noted in [25], it is the game structure that brings about a somewhat unexpected win-loss phenomenon: such an one-sided optimization problem (for the human) may not always win even if the opponent has a fixed strategy. Similar phenomena do exist practically, but, of course, cannot be observed in the traditional framework of optimal control. We would also like to note that the differences among the results of the three games can be attributed to the differences in the game structures.

As will be shown in Section 21.3, when the machine strategy is known to the human, the human can find the optimal strategy with the best payoff. A natural question is: What if the machine strategy is unknown to the human?

One may hope to identify the machine strategy within finite steps before making optimal decision. A machine strategy which is parameterized by a vector A (like in (21.4) for the case of $k = 1$), is called **identifiable** if there exists a human strategy such that the vector A can be constructed from the corresponding realization and the initial state.

Proposition 21.2.1. *A machine strategy with k-memory is identifiable if and only if its corresponding STG is strongly connected.*

Proposition 21.2.1 is somewhat intuitive, which can be used to identify non-identifiable machine strategies. Consider the simple case where $k = 1$. Then it is easy to see that the STG corresponding to the machine strategy $A = (0, 0, *, *)$ or $A = (*, *, 1, 1)$

will not be strongly connected, and so will not be identifiable by Proposition 21.2.1. In fact, as can be easily seen, only part of the entries of such $A = (a_1, a_2, a_3, a_4)$ can be identified from any given initial state.

If the machine makes mistakes with a tiny possibility, however, the machine strategy may become identifiable. For example, if it changes its planed decision with a small positive probability to any other decisions, then the corresponding STG will be a Markovian transfer graph which is strongly connected. Hence, all strategies will be identifiable.

To illustrate how to identify the machine strategy, let us again consider the case of $k = 1$. In this case, one effective way for the human to identify the machine strategy is to randomly choose his action at each time. One can also use the following method to identify the parameters:

$$h(t+1) = \begin{cases} 0 & a_{s(t)} \textit{ is not known at time } t, \\ & \textit{or } a_{s(t)} \textit{ is known, but } a_{2 \cdot a_{s(t)}+1} \textit{ is not}; \\ 1 & \textit{otherwise}. \end{cases} \tag{21.7}$$

Theorem 21.2.3. *For any identifiable machine strategy with $k = 1$, it can be identified using the above human strategy with at most 7 steps from any initial state.*

Remark 21.2.3. For non-identifiable machine strategies, one may be surprised by the possibility that identification may lead to a worse human's payoff. We have shown that this is true for the PD game [25]. It is true for the Snow drift game too. For example, if the machine takes the non-identifiable strategy $A = (0, 1, 1, 1)$, then by acting with "A" blindly, the human can get a payoff a by the payoff matrix at each time. However, once he tries to identify the machine's strategy, he may use the "B" to probe it. Then the machine will be provoked and act with "B" forever. That will lead to a worse human payoff $c < a$ afterwards.

21.3 The State Transfer Graph

In order to provide the theoretical proofs for the main results stated in the above section, we need to use the concept of State Transfer Graph (STG) together with some basic properties, as in the paper [25]. Throughout this section, the machine strategy $A = (a_1, a_2, a_3, a_4)$ is assumed to be known.

Given an initial state and a machine strategy, any human strategy $\{h(t)\}$ can lead to a realization of the states $\{s(1), s(2), ..., s(t), ...\}$. Hence, it also produces a sequence of human payoffs $\{p(s(1)), p(s(2)), ..., p(s(t)), ...\}$. Thus the Question 1) raised in Section 21.2 becomes to solve

$$\{h(t)\}_{t=1}^{\infty} = \arg\max P_{\infty}^{+}$$

among all possible human strategies.

In order to solve this problem, we need the definition of STG, and we refer to [24] for some standard concepts in graph theory, e.g. walk, path and cycle. We will only consider finite graphs (with finite vertices and finite edges) in the sequel.

Let $G = (V, E)$ be a directed graph with vertex set V and edge set E.

Definition 21.3.1. *A **walk** W is defined as an alternating sequence of vertices and edges, like* $v_0 e_1 v_1 e_2 ... v_{l-1} e_l v_l$, *abbreviated as* $v_0 v_1 ... v_{l-1} v_l$, *where* $e_i = \overline{v_{i-1} v_i}$ *is the edge from* v_{i-1} *to vi,* $1 \leq i \leq l$. *The total number of edges l is called the length of W.*

If $v_0 = v_l$, *then W is called closed, otherwise is called open.*

Definition 21.3.2. *A walk W,* $v_0 v_1 ... v_{l-1} v_l$, *is called a **path** (directed), if the vertices* $v_0, v_1, ... v_l$ *are distinct.*

Definition 21.3.3. [1] *A closed walk W:* $v_0 v_1 ... v_{l-1} v_l$, $v_0 = v_l$, $l \geq 1$, *is called a **cycle** if the vertices* $v_1, ..., v_l$ *are distinct.*

Definition 21.3.4. *A graph is called **strongly connected** if for any distinct vertices* v_i, v_j, *there exists a path starting from* v_i *and ending with* v_j.

Now, we are in a position to define the STG.

Note that any given machine strategy of k-memory, together with a human strategy, will determine an infinite Walk representing the state transfer process of the game.

Definition 21.3.5. *A directed graph with* 2^{2k} *vertices* $\{s_1, s_2, ...s_{2^{2k}}\}$ *is called the **State Transfer Graph** (STG), if it contains all the possible infinite walks corresponding to all possible human strategies, that equals to say, it contains all the possible one-step path or cycle in the walk.*

In the case of $k = 1$, for a machine strategy $A = (a_1, a_2, a_3, a_4)$, the STG is a directed graph with the vertices being the state $s(t) \in \{s_1, s_2, s_3, s_4\}$ with $s_i = i$.

An edge $\overline{s_i s_j}$ exists if $s(t+1) = s_j$ can be realized from $s(t) = s_i$ by choosing $h(t+1) = 0$ or 1. Since $s_i = i$, by (21.3) and (21.4), that means,

$$\textit{the edge } \overline{s_i s_j} \textit{ exists} \Leftrightarrow s_j = 2 \cdot a_i + 1 \textit{ or } s_j = 2 \cdot a_i + 2 \tag{21.8}$$

and the way to realize this transfer is taking human's action as $h = (s_j - 1) mod\ 2$ by (21.3).

By the definition above, one machine strategy leads to one STG, and vice versa.

Definition 21.3.6. *A state* s_j *is called **reachable** from the state* s_i, *if there exists a path (or cycle) starting from* s_i *and ending with* s_j. *All the vertices which are reachable from* s_i *constitute a set, called the **reachable set** of the state* s_i. *A STG is called **strongly connected** if any vertex* s_i *has all vertices in its reachable set.*

---

[1] The Definition 21.3.3 of cycle is a little different from [24]. We ignore the constraint that the length $l \geq 2$ and include 'loop' in the concept of "cycle".

Thus, the reachability of s_j from s_i means that there exists a finite number of human actions, such that the state $s(\cdot)$ can be transferred from s_i to s_j with the same number of steps.

Furthermore, we need to define the payoff of a walk on STG as follows:

Definition 21.3.7. *The averaged payoff of an open walk $W = v_0v_1...v_l$ on a STG, with $v_0 \neq v_l$, is defined as*

$$p_W \triangleq \frac{p(v_0)+p(v_1)+...+p(v_l)}{l+1}, \tag{21.9}$$

and the averaged payoff of a closed walk $W = v_0v_1...v_l$, with $v_0 = v_l$, is defined as

$$p_W \triangleq \frac{p(v_0)+p(v_1)+...+p(v_{l-1})}{l}. \tag{21.10}$$

Now, we can give some basic properties of STG below.

Lemma 21.3.1. *For a given STG, any closed walk can be divided into finite cycles, such that the edge set of the walk equals the union of the edges of these cycles. In addition, any open walk can be divided into finite cycles plus a path.*

Lemma 21.3.2. *Assume that a closed walk $W = v_0v_1...v_L$ with length L, can be partitioned into cycles $W_1, W_2, ..., W_m$, $m \geq 1$, with their respective lengths being $L_1, L_2, ..., L_m$. Then, p_W , the averaged payoff of W can be written as*

$$p_W = \sum_{j=1}^{m} \frac{L_j}{L} p_j, \tag{21.11}$$

where $p_1, p_2, ..., p_m$ are the averaged payoffs of the cycles $W_1, W_2, ..., W_m$, respectively.

By Theorem 21.2.1, the state of the repeated games will be periodic under the optimal human strategy. This enables us to find the optimal human strategy by searching on the STG, as will be illustrated in the example below. Similar to [25], we give an example for the Snowdrift game.

Example 21.3.1. Consider the "ALL A" strategy $A = (0,0,0,0)$ of the machine. Then the STG can be drawn as shown in Figure 21.4, in which $s_1(a,0)$ means that under the state s_1, the human gets his payoff vector $P(s_1) = (p(s_1), w(s_1)) = (a,0)$. The directed edge $\overline{s_1s_2}$ illustrates that if the human takes action D, he can transfer the state from s_1 to s_2 with payoff vector $(b,1)$. Others can be explained in the same way. Now we take the initial state as $s(0) = s_3 = (c,-1)$.Then the reachable set of s_3 is $\{s_1, s_2\}$, and we just need to search the cycle whose vertices are on this set.

Obviously, there are three possible cycles $W_1 = \{s_1\}$, $W_2 = \{s_2\}$, $W_3 = \{s_1, s_2\}$ and by (21.10), the averaged payoffs of the human are respectively $p_{W_1} = p(s_1) = a$, $p_{W_2} = p(s_2) = b$, $p_{W_3} = \frac{p(s_1)+p(s_2)}{2} = \frac{a+b}{2}$.

Obviously, the optimal payoff lies in the cycle $W_2 = \{s_2\}$. To induce the system state enters into this cycle, the human just take $h(1) = 1$. Then by taking $h(t) = 1, t \geq 2$, the optimal state sequence $s(t) = s_2, t \geq 1$ will be obtained from $s(0) = s_3$.

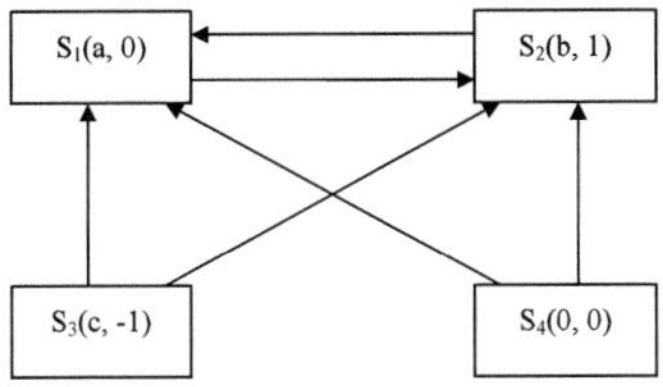

Fig. 21.4. STG of ALL A machine strategy $A = (0,0,0,0)$ in Snowdrift game

Remark 21.3.1. The search procedure above can be accomplished in the general case by an algorithm which is omitted here for brevity, and it can also be seen that for any given machine strategy with k-memory, there always exists a search method to find the optimal strategy of the human. Moreover, the optimal payoff remains the same when the initial state varies over a reachable set.

21.4 Proofs of the Main Results

First of all, it is not difficult to see that in the current general case, Theorem 21.2.1, Proposition 21.2.1 and Theorem 21.2.3 can be proven along the proof lines of those in [25], and so the details will be omitted here.

Remark 21.4.1. It is worth mentioning that the form of the averaged payoff criteria is important in Theorem 21.2.1. For other payoff criteria, similar results may not hold.

As for the proof of Theorem 21.2.2, the first conclusion on the Prisoners' Dilemma game can be seen in [25], and so we just need to prove the conclusions (2) and (3).

Proof of Theorem 21.2.2 (2).
The conclusion will be proven if we can find the required machine strategy. To this end, we just need take the "ALL B" strategy $(1,1,1,1)$ as the machine's strategy. Then starting from any initial state, to optimize his payoff value, the human has to take the action "A" always, which will lead to a payoff c for him while the machine will get b. Hence he will lose.

Proof of Theorem 21.2.2 (3).
For the game of the Battle of Sex, consider the following two cases:

Case 1: when $b > c$, the pure Nash equilibria of the game are the profile (A,A) and (B,B);

Case 2: when $b < c$, the pure Nash equilibrium of the game is the profile (A,B).

Then in Case 1, if the machine takes the "Always B" strategy $(1,1,1,1)$, then the optimal human strategy will always act "B" too. Thus the state will repeat the profile (B,B) and the human will get a payoff of b while the machine get a, which implies that the human will lose. In Case 2, similar to the proof of Theorem 21.2.2 in [25], we can prove that the human cannot lose to the machine in this case. This completes the proof of Theorem 21.2.2. □

Remark 21.4.2. Note that all the three games in Theorem 21.2.2 have some characters about "need for coordination", and the differences in the three assertions of Theorem 21.2.2 result from the different game structures. Specifically, the Snowdrift game has a more cruel assumption, i.e. the player must never choose "both B" which is the worst case for both, while the "A" player have to sacrifice in a (A,B) profile. For the Battle of Sex game, we can imagine that the parameters may measure whether the two players care more about the time they share or care more about their own interest. If they care more about the sharing time, i.e. when $b > c$, then the selfisher one can use this feature to win.

21.5 Extensions to 3×3 Matrix Games

In this section, we consider possible extensions of the results in the previous sections. Consider the 2-player 3-action games, in which there are 2 players while either one has three actions. The payoff matrix is then as in figure 21.5

		Player II		
		A	B	C
Player I	A	(a_{11}, b_{11})	(a_{12}, b_{12})	(a_{13}, b_{13})
	B	(a_{21}, b_{21})	(a_{22}, b_{22})	(a_{23}, b_{23})
	C	(a_{31}, b_{31})	(a_{32}, b_{32})	(a_{33}, b_{33})

Fig. 21.5. The payoff matrix of 2-player 3-action game

Similar to Section 21.2, we can formulate a repeated game and describe the corresponding dynamic rules by a STG. To this end, we need to define the system state first. For the 1-memory machine strategy, there are 3 actions for each player, which can be denoted as 0,1,2 like a ternary signal. So there are $3 \times 3 = 9$ 1-memory histories, and thus we can define the state as

$$s(t) = 3 \cdot m(t) + h(t) + 1, \tag{21.12}$$

and the machine strategy can be written as

$$m(t+1) = f(m(t), h(t)) = \sum_{i=1}^{9} a_i I_{\{s(t)=s_i\}} \tag{21.13}$$

Thus, the STG with have 9 vertices and can be formed and analyzed by similar methods as those in Section 21.3. It can be easily seen that Theorem 21.2.1 and Proposition 21.2.1 will hold true in this case since the proofs only use the finite state information. However, Theorem 21.2.2 must be checked for specific games and Theorem 21.2.3 must be modified for this kind of 2-player 3-action games. Also, extensions to 2-player n-action games can be carried out in a similar way.

A well-known example of 2-player 3-action games is the "Rock-Paper-Scissors" game whose payoff matrix can be specified by as in figure 21.6. This game is a zero-

Player I \ Player II	rock	paper	scissors
rock	(0, 0)	(-1, 1)	(1, -1)
paper	(1, -1)	(0,0)	(-1,-1)
scissors	(-1, 1)	(1, -1)	(0, 0)

Fig. 21.6. The payoff matrix of "Rock-Paper-Scissors" game

sum game, and the relationship between the optimality and the wining of the human is consistent. In fact, the human can select one from the three actions to beat his opponent, and the game is like history independent. So, once the machine's strategy is known, the human can always get his optimal payoff and win at the same time.

21.6 Concluding Remarks

In an attempt to study dynamical control systems which contain game-like mechanisms in the system structure, we have, in this paper, presented a preliminary investigation on optimization and identification problems for a specific non-equilibrium dynamic game where two heterogeneous agents, called "Human" and "Machine", play repeated games modeled by a generic 2×2 game. Some typical games including the Prisoner Dilemma game, Snowdrift game and the Battle of Sex game have been studied in certain detail. By using the concept and properties of the state transfer graph, we are able to establish some interesting theoretical results, which have not been observed in the traditional control framework. For example, we have shown that the optimal strategy of the game will be periodic after finite steps, and that optimizing one's payoff solely may lose to the opponent eventually. Possible extensions to more general game structures like 2-player 3-action games are also discussed. It goes without saying that there may be many implications and other extensions of these results. However, it would be more challenging to establish a mathematical theory for more complex systems, where many (possibly heterogeneous) agents interact with learning and adaptation, cooperation and competition, etc.

References

1. Astrom, K.J., Wittenmark, B.: Adaptive Control, 2nd edn. Addison-Wesley, Reading (1995)
2. Chen, H.F., Guo, L.: Identification and Stochastic Adaptive Control. Birkhäuser, Boston (1991)
3. Goodwin, G.C., Sin, K.S.: Adaptive Filtering, Prediction and Control. Prentice-Hall, Englewood Cliffs (1984)
4. Kumar, P.R., Varaiya, P.: Stochastic Systems: Estimation, Identification and Adaptive Control. Prentice Hall, Englewood Cliffs (1986)
5. Kristic, M., Kanellakopoulos, I., Kokotoric, P.: Nonlinear Adaptive Control Design. A Wiley-Interscience Publication, John Wiley & Sons, Chichester (1995)
6. Guo, L.: Adaptive Systems Theory: Some Basic Concepts, Methods andResults. Journal of Systems Science and Complexity 16, 293–306 (2003)
7. Holland, J.: Hidden Order: How Adaptation Builds Complexity. Addison-Wesley, Reading (1995)
8. Holland, J.: Studying Complex Adaptive Systems. Journal of System Science and Complexity 19, 1–8 (2006)
9. Basar, T., Olsder, G.J.: Dynamic Noncooperative Game Theory, the Society for Industrial Applied Mathematics. Academic Press, New York (1999)
10. Arthur, W.B., Durlauf, S.N., Lane, D.: The Economy As An Evolving Complex System II. Addison-Wesley, Reading (1997)
11. Weibull, J.W.: Evolutionary Game Theory. MIT Press, Cambridge (1995)
12. Hofbauer, J., Sigmund, K.: Evolutionary game dynamics. Bulletin of the American Mathematical Society 40, 479–519 (2003)
13. Fudenberg, D., Levine, D.K.: The Theory of Learning in Games. MIT Press, Cambridge (1998)
14. Fudenberg, D., Levine, D.K.: Learning and equilibrium (2008), Available: `http://www.dklevine.com/papers/annals38.pdf`
15. Kalai, E., Lehrer, E.: Rational learning leads to Nash equilibrium. Econometria 61, 1019–1045 (1993)
16. Kalai, E., Lehrer, E.: Subjective equilibrium in repeated games. Econometrica 61, 1231–1240 (1993)
17. Marden, J.R., Arslan, G., Shamma, J.S.: Joint strategy fictitious play with inertia for potential games. IEEE Trans. Automatic Control 54, 208–220 (2009)
18. Foster, D.P., Young, H.P.: Learning, hyperthesis testing ans Nash equilibrium. Games and Economic Behavior 45, 73–96 (2003)
19. Marden, J.R., Young, H.P., Arslan, G., Shamma, J.S.: Payoff based dynamics for multiplayer weakly acyclic games. In: Prodeedings of the 46th IEEE Conferrence on Decision and Control, New Orleans, USA, pp. 3422–3427 (2007)
20. Chang, Y.: No regrets about no-regret. Artificial Intelligence 171, 434–439 (2007)
21. Young, H.P.: The possible and the impossible in multi-agent learning. Artificial Intelligence 171, 429–433 (2007)
22. Axelrod, R.: The Evolution of Cooperation. Basic Books, New York (1984)
23. Puterman, M.: Markov Decision Processes:Discrete Stochastic Dynamic Programming. John Wiley & Sons, New York (1994)
24. B-Jensen, J., Gutin, G.: Digraphs: Theory, Algorithms and applications. Springer, London (2001)
25. Mu, Y.F., Guo, L.: Optimization and Identification in a Non-equilibrium Dynamic Game. In: Prodeedings of the 48th IEEE Conferrence on Decision and Control, Shanghai, China, December 16-18 (2009)

22

Rational Systems – Realization and Identification*

Jana Němcová and Jan H. van Schuppen

CWI, Science Park 123, 1098 XG Amsterdam, The Netherlands

Summary. In this expository paper we provide an overview of recent developments in realization theory and system identification for the class of rational systems. Rational systems are a sufficiently big class of systems to model various phenomena in engineering, physics, economy, and biology which still has a nice algebraic structure. By an algebraic approach we derive necessary and sufficient conditions for a response map to be realizable by a rational system. Further we characterize identifiability properties of rational systems with parameters. For the proofs we refer the reader to the corresponding papers.

22.1 Introduction

In the last few decades, control and system theory has been enriched by results obtained by the methods of commutative algebra and algebraic geometry. For the theory concerning linear systems see for example [7, 13, 14]. Polynomial systems are studied in [1, 2, 5], and others, and rational systems in algebraic-geometric framework are introduced in [4]. In this paper we present an algebraic approach motivated by [5, 4] to realization theory and system identification for the class of rational systems. The importance and usefulness of algebraic methods lies in their connection to computational algebra and consequently to the algorithms already implemented in many computer algebra systems like for example CoCoA [25], Macaulay 2 [11], Magma [6], Maxima [18], Reduce [10], Singular [26]. Many programmes can be found also in Maple, Mathematica, and Matlab.

Rational systems arise in several domains of the sciences like for example physics, engineering, and economics. Another area where rational systems are extensively used is systems biology. Biologists distinguish in a cell *metabolic networks* which handle the major material flows and the energy flow of a cell; *signaling networks* which convey signals from one location in a cell to another; and *genetic networks* which describe the process from the reading of DNA to the production of pro-

*This paper is dedicated to Christopher I. Byrnes and to Anders Lindquist for their contributions to control and system theory.

X. Hu et al. (Eds.): Three Decades of Progress in Control Sciences, pp. 327–341, 2010.

teins. For analysis and simulation purposes, mathematical models of these networks are needed and rational systems are widely used as such.

Realization theory of rational systems studies the problem of finding an initialized rational system such that it corresponds to an apriori given input-output or response map. The correspondence is given by getting the same output after applying the same input to the system and to the map. Further problems deal with minimality and canonicity of such systems, and with the development of algorithms related to these problems. The results can be applied to study control and observer synthesis, model reduction and for system identification of rational systems.

System identification of rational systems deals with the problem of obtaining rational systems as realistic models of observed phenomena. Very often these systems contain unknown parameters which have to be estimated to get a fully specified model. The uniqueness of parameter values determining the system modeling the phenomenon is referred as identifiability. In this paper we discuss the system identification procedure with the stress on identifiability and approximation steps.

The structure of the paper is as follows. The framework and motivation for rational systems are introduced in Section 22.2. Section 22.3 deals with realization theory of rational systems. System identification is discussed in Section 22.4. The last section provides an overview of some open problems for the class of rational systems.

22.2 Rational Systems

In this section we motivate the study of the class of rational systems by their application in biochemistry. Further we recall an algebraic framework for rational systems which we later use to solve the realization problem and derive the characterization of identifiability.

22.2.1 Biochemical Reaction Systems

The modeling of biochemical processes such as the glycolysis in *Trypanosoma brucei* or in Baker's Yeast (*Saccharomyces cerevisiae*) and ammonium stress response in *Escherichia coli* are part of the research area of biochemistry. Mathematical models of biochemical reaction networks are needed for providing the tools to analyze the reaction networks. The models allow (1) to evaluate the behaviour of a reaction network; (2) to determine the dynamic system properties of networks such as the existence of steady states, the uniqueness or the multiplicity of steady states, local or global asymptotic stability of steady states, periodic trajectories, the decomposition into slow and fast manifolds, etc.; and (3) to analyse control of such networks for rational drug design or for biotechnology.

Example 22.2.1. We derive a model of a reversible chemical reaction represented by the diagram

$$A_1 + A_2 \rightleftarrows A_3,$$

where A_1, A_2, A_3 denote the corresponding chemical species. The complexes of this reaction are $C_1 = A_1 + A_2$ and $C_2 = A_3$. The relation between complexes and species they are composed of is specified by the matrix

$$B = \begin{pmatrix} 1 & 0 \\ 1 & 0 \\ 0 & 1 \end{pmatrix} = \begin{pmatrix} B_1 & B_2 \end{pmatrix} \in \mathbb{N}^{3\times 2}.$$

In particular, if $B_{i,j}$ (the entry of the matrix B in i-th column and j-th row) equals $a \in \mathbb{N}$, then the i-th complex contains a units of the j-th species. The reaction network which in this case consists only of one reversible reaction (described be two irreversible reactions, one in either direction) is denoted by

$$\text{rnet} = \{(2,1),(1,2)\}.$$

The numbers 1 and 2 stand for the complexes C_1 and C_2. The rate of a reaction in a reaction network determines the speed with which a corresponding complex associates or dissociates. The rates of biochemical reactions can be modeled by different types of kinetics. In this example we consider the simplest one, so-called mass-action kinetics. Let x_1, x_2, x_3 denote the concentrations of the chemical species A_1, A_2, A_3 in the reaction system, respectively. The rates of the reactions are assumed to be proportional to the concentrations of species in the complexes which associate or dissociate. Therefore, the rate of the reaction $C_1 \to C_2$ is given as $r_{2,1}(x) = k_{2,1}x_1x_2$ where $k_{2,1} \in [0,\infty)$, and the rate of the reaction $C_2 \to C_1$ is $r_{1,2}(x) = k_{1,2}x_3$ where $k_{1,2} \in [0,\infty)$. Then, the reversible reaction $A_1 + A_2 \rightleftarrows A_3$ is modeled by the system of ordinary differential equations

$$\frac{dx(t)}{dt} = \sum_{(i,j)\in\{(2,1),(1,2)\}} (B_i - B_j) r_{i,j}(x(t)) = \begin{pmatrix} k_{1,2}x_3(t) - k_{2,1}x_1(t)x_2(t) \\ k_{1,2}x_3(t) - k_{2,1}x_1(t)x_2(t) \\ k_{2,1}x_1(t)x_2(t) - k_{1,2}x_3(t) \end{pmatrix}.$$

The dynamics of a general reaction system given by its reaction network rnet, chemical complexes and species is given as follows:

$$\frac{dx(t)}{dt} = \sum_{(i,j)\in\,\text{rnet}} (B(i) - B(j)) r_{i,j}(x(t)) u_{i,j}(t). \tag{22.1}$$

Here $u_{i,j}$ stands for an input influencing the corresponding reaction. If a reaction system is modeled by *mass-action kinetics* (as in the example above), then there exists a matrix $K \in \mathbb{R}_+^{n_c \times n_c}$ such that for all $(i,j) \in$ rnet it holds that

$$r_{i,j}(x) = K_{i,j} \prod_{s=1}^{n} x_s^{B_{s,j}}.$$

If the considered kinetics is *Michaelis-Menten*, then for all $(i,j) \in$ rnet it holds that

$$r_{i,j}(x) = \frac{p_{i,j}(x)}{q_{i,j}(x)},$$

where $p_{i,j}, q_{i,j}$ are polynomials and moreover $q_{i,j}$ is not constant. Note that in both cases the right-hand sides of the equations describing the dynamics of the reaction system are given as rational functions. Therefore, the systems modeling biochemical reactions by mass-action or Michaelis-Menten kinetics are rational systems.

22.2.2 Framework

To deal with rational systems we adopt the framework introduced by Z. Bartosiewicz in [4]. For the used terminology and basic facts of commutative algebra and algebraic geometry see [8, 16, 32].

A *real affine variety* X is a subset of $\mathbb{R}^n$ of zero points of finitely many polynomials with real coefficients in n variables, i.e. finitely many polynomials of $\mathbb{R}[X_1,\dots,X_n]$. We say that a variety is *irreducible* if it cannot be written as an union of two non-empty varieties which are its strict subvarieties. We consider the *Zariski topology* on $\mathbb{R}^n$ which is a topology with the closed sets defined as real affine varieties. By a *polynomial on a variety* X we mean a map $p : X \to \mathbb{R}$ for which there exists a polynomial $q \in \mathbb{R}[X_1,\dots,X_n]$ such that $p = q$ on X. We denote by A the algebra of all polynomials on X. It is a finitely generated algebra and, since X is irreducible, it is also an integral domain. Therefore, we can define the set Q of rational functions on X as the field of quotients of A. A *rational vector field* f on an irreducible real affine variety X is a $\mathbb{R}$-linear map $f : Q \longrightarrow Q$ such that $f(\varphi \cdot \psi) = f(\varphi)\cdot\psi + \varphi \cdot f(\psi)$ for $\varphi, \psi \in Q$. We say that f is defined at $x_0 \in X$ if $f(O_{x_0}) \subseteq O_{x_0}$ with $O_{x_0} = \{\varphi \in Q | \varphi = \frac{\varphi_n}{\varphi_d}, \varphi_n, \varphi_d \in A, \varphi_d(x_0) \neq 0\}$.

By a rational system we mean a dynamical system with inputs and outputs, with the dynamics defined by a family of rational vector fields, with the output function the components of which are rational functions, and with the specified initial state. The inputs to the system are assumed to be piecewise-constant functions with the values in an *input space* U which is an subset of $\mathbb{R}^m$. We denote the *space of input functions* by $\mathcal{U}_{pc}$. For every $u \in \mathcal{U}_{pc}$ there are $\alpha_1,\dots,\alpha_{n_u} \in U$ such that $u = (\alpha_1,t_1)(\alpha_2,t_2)\dots(\alpha_{n_u},t_{n_u})$. This means that for $t \in (\sum_{j=0}^{i} t_j, \sum_{j=0}^{i+1} t_j]$ with $t_0 = 0$ the input $u(t) = \alpha_{i+1} \in U$ for $i = 0,1,\dots,n_u-1$, and $u(0) = \alpha_1$. Every input function $u \in \mathcal{U}_{pc}$ has a time domain $[0,T_u]$ where $T_u = \sum_{j=1}^{n_u} t_j$ depends on u. The empty input e is such input that $T_e = 0$. Further we consider as an *output space* all $\mathbb{R}^r$.

Definition 22.2.1. *A* rational system Σ *is a quadruple* (X, f, h, x_0) *where*

(i) $X \subseteq \mathbb{R}^n$ *is an irreducible real affine variety,*
(ii) $f = \{f_\alpha | \alpha \in U\}$ *is a family of rational vector fields on* X*,*
(iii) $h : X \to \mathbb{R}^r$ *is an output map with rational components (*$h_j \in Q$ *for* $j = 1,\dots,r$*),*
(iv) $x_0 \in X$ *is the initial state such that all components of* h *and at least one of the vector fields* $f_\alpha, \alpha \in U$ *are defined at* x_0*.*

The *trajectory* of a rational system $\Sigma = (X, f = \{f_\alpha | \alpha \in U\}, h, x_0)$ corresponding to a constant input $u = (\alpha, T_u) \in \mathcal{U}_{pc}$ is the trajectory of the rational vector field f_α from x_0 at which f_α is defined, i.e. it is the map $x(\cdot; x_0, u) : [0,T_u] \to X$ for which $\frac{d}{dt}(\varphi \circ x)(t; x_0, u) = (f_\alpha \varphi)(x(t; x_0, u))$ and $x(0; x_0, u) = x_0$ for $t \in [0,T_u]$ and for

$\varphi \in A$. The trajectory of Σ corresponding to an input $u = (\alpha_1, t_1)\dots(\alpha_{n_u}, t_{n_u}) \in \mathscr{U}_{pc}$ with $T_u = \sum_{j=1}^{n_u} t_j$ is the map $x(\cdot; x_0, u) : [0, T_u] \to X$ such that $x(0; x_0, u) = x_0$, and $x(t; x_0, u) = x_{\alpha_i}(t - \sum_{j=0}^{i-1} t_j)$ for $t \in [\sum_{j=0}^{i-1} t_j, \sum_{j=0}^{i} t_j]$, $i = 1, \dots, n_u$ where $x_{\alpha_i} : [0, t_i] \to X$ is a trajectory of a vector field f_{α_i} from the initial state $x(\sum_{j=0}^{i-1} t_j; x_0, u) = x_{\alpha_{i-1}}(t_{i-1})$ for $i = 2, \dots, n_u$, and from the initial state x_0 for $i = 1$. Note that for any rational vector field f and any point x_0 at which f is defined there exists an unique trajectory of f from x_0 defined on the maximal interval $[0, T)$ (T may be infinite), see [4]. Because a trajectory of a rational system $\Sigma = (X, f, h, x_0)$ does not need to exist for every input $u \in \mathscr{U}_{pc}$, we define the set $\mathscr{U}_{pc}(\Sigma) = \{u \in \mathscr{U}_{pc} | x(\cdot; x_0, u) \text{ exists}\}$ of *admissible inputs for the system* Σ.

22.3 Realization of Rational Systems

Our approach to realization theory for rational systems is based on an algebraic approach to realization theory introduced in [3, 5] for the class of polynomial systems. The results we present in this section are derived in our papers [22, 21]. They are related to the solution of the immersion problem of a smooth system into a rational system presented in [4]. The problem of the existence of rational realizations is treated also in [30].

22.3.1 Response Maps

Since the realization problem deals with finding an internal representation of a phenomenon characterized externally, we first introduce the concept of external representations. Every phenomenon is characterized by the measurements of the inputs to the system and the corresponding outputs. The maps which describe the outputs immediately after applying finite parts of the inputs are called *response maps*. Hence, because we study realization problem for the class of rational systems and because the inputs for rational systems we consider are assumed to be piecewise-constant functions $\mathscr{U}_{pc}$, a response map φ is a map from $\mathscr{U}_{pc}$ to $\mathbb{R}^r$. To solve the realization problem for rational systems we consider arbitrary set $\widetilde{\mathscr{U}_{pc}} \subseteq \mathscr{U}_{pc}$ of *admissible inputs* to be a domain of response maps instead of $\mathscr{U}_{pc}$. Let us define $\widetilde{\mathscr{U}_{pc}}$ formally.

Definition 22.3.1. *A set* $\widetilde{\mathscr{U}_{pc}} \subseteq \mathscr{U}_{pc}$ *of input functions with the values in an input space* $U \subseteq \mathbb{R}^m$ *is called a* set of admissible inputs *if:*

(i) $\forall u \in \widetilde{\mathscr{U}_{pc}}\ \forall t \in [0, T_u] : u_{[0,t]} \in \widetilde{\mathscr{U}_{pc}}$,
(ii) $\forall u \in \widetilde{\mathscr{U}_{pc}}\ \forall \alpha \in U\ \exists t > 0 : (u)(\alpha, t) \in \widetilde{\mathscr{U}_{pc}}$,
(iii) $\forall u = (\alpha_1, t_1)\dots(\alpha_k, t_k) \in \widetilde{\mathscr{U}_{pc}}\ \exists \delta > 0\ \forall \overline{t_i} \in [0, t_i + \delta], i = 1, \dots, k :$

$$\overline{u} = (\alpha_1, \overline{t_1})\dots(\alpha_k, \overline{t_k}) \in \widetilde{\mathscr{U}_{pc}}.$$

The properties of a set $\widetilde{\mathscr{U}_{pc}}$ of admissible inputs with the values in $U \subseteq \mathbb{R}^m$ allow us to define derivations D_α, $\alpha \in U$ of real functions on $\widetilde{\mathscr{U}_{pc}}$. Consider a real function $\varphi : \widetilde{\mathscr{U}_{pc}} \to \mathbb{R}$. Then,

$$(D_\alpha \varphi)(u) = \frac{d}{dt}\varphi((u)(\alpha,t))|_{t=0+}$$

for $(u)(\alpha,t) \in \widetilde{\mathscr{U}_{pc}}$, where $t > 0$ is sufficiently small and $\alpha \in U$. Note that $(D_\alpha \varphi)(u)$ is well-defined if $\varphi((u)(\alpha,\hat{t})), \hat{t} \in [T_u, T_u + t]$ is differentiable at T_u+. To simplify the notation, the derivation $D_{\alpha_1} \dots D_{\alpha_i} \varphi$ can be rewritten as $D_\alpha \varphi$ where $\alpha = (\alpha_1, \dots, \alpha_i)$.

Definition 22.3.2. *Consider a set $\widetilde{\mathscr{U}_{pc}}$ of admissible inputs. Let $\varphi : \widetilde{\mathscr{U}_{pc}} \to \mathbb{R}$ be a real function such that for every input $u = (\alpha_1,t_1)\dots(\alpha_k,t_k) \in \widetilde{\mathscr{U}_{pc}}$ the function $\varphi_{\alpha_1,\dots,\alpha_k}(t_1,\dots,t_k) = \varphi((\alpha_1,t_1)\dots(\alpha_k,t_k))$ can be written in the form of convergent formal power series in k indeterminates. We denote the set of all such functions φ by $\mathscr{A}(\widetilde{\mathscr{U}_{pc}} \to \mathbb{R})$.*

From the two definitions above it follows that for any $f,g \in \mathscr{A}(\widetilde{\mathscr{U}_{pc}} \to \mathbb{R})$ it holds that if $fg = 0$ on $\widetilde{\mathscr{U}_{pc}}$, then $f = 0$ on $\widetilde{\mathscr{U}_{pc}}$ or $g = 0$ on $\widetilde{\mathscr{U}_{pc}}$. Thus, the set $\mathscr{A}(\widetilde{\mathscr{U}_{pc}} \to \mathbb{R})$ with $\widetilde{\mathscr{U}_{pc}}$ being a set of admissible inputs is an integral domain which implies the possibility to define the field $\mathscr{Q}(\widetilde{\mathscr{U}_{pc}} \to \mathbb{R})$ of the quotients of elements of $\mathscr{A}(\widetilde{\mathscr{U}_{pc}} \to \mathbb{R})$.

For well-definedness of the observation field of a response map (see Definition 22.3.4), which is one of the main algebraic objects used in the presented approach to realization theory of rational systems, we have to assume that the components of a response map generate an integral domain. Therefore, let us specify the response maps considered in this paper.

Definition 22.3.3. *Let $\widetilde{\mathscr{U}_{pc}}$ be a set of admissible inputs. A map $p : \widetilde{\mathscr{U}_{pc}} \to \mathbb{R}^r$ is called a* response map *if its components $p_i : \widetilde{\mathscr{U}_{pc}} \to \mathbb{R}$, $i = 1,\dots,r$ are such that $p_i \in \mathscr{A}(\widetilde{\mathscr{U}_{pc}} \to \mathbb{R})$.*

The following definition of the observation algebra and the observation field of a response map corresponds to the definition of the same objects for rational systems, see Definition 22.3.6.

Definition 22.3.4. *Let $\widetilde{\mathscr{U}_{pc}}$ be a set of admissible inputs and let $p : \widetilde{\mathscr{U}_{pc}} \to \mathbb{R}^r$ be a response map. The observation algebra $A_{obs}(p)$ of p is the smallest subalgebra of the algebra $\mathscr{A}(\widetilde{\mathscr{U}_{pc}} \to \mathbb{R})$ which contains the components $p_i, i = 1,\dots,r$ of p, and which is closed with respect to the derivations $D_\alpha, \alpha \in U$. The observation field $Q_{obs}(p)$ of p is the field of quotients of $A_{obs}(p)$.*

22.3.2 Problem Formulation

A rational system which for each input gives us the same output as a response map p is called a *rational realization of p* (a rational system *realizing* p). The realization problem for rational systems then deals with the existence of (canonical, minimal) rational realizations for a given response map and with the algorithms for constructing them.

Problem 22.3.1 (Existence of rational realizations). Let $\widetilde{\mathscr{U}_{pc}}$ be a set of admissible inputs. Consider a response map $p : \widetilde{\mathscr{U}_{pc}} \to \mathbb{R}^r$. The existence part of the realization problem for rational systems consists of determining a rational system $\Sigma = (X, f, h, x_0)$ such that

$$p(u) = h(x(T_u; x_0, u)) \text{ for all } u \in \widetilde{\mathscr{U}_{pc}} \text{ and } \widetilde{\mathscr{U}_{pc}} \subseteq \mathscr{U}_{pc}(\Sigma).$$

Let us introduce the concepts of algebraic reachability (Definition 22.3.5) and rational observability (Definition 22.3.6) of rational realizations. They are based on [4, Definition 3 and 4].

Definition 22.3.5. *Let $\Sigma = (X, f, h, x_0)$ be a rational realization of a response map $p : \widetilde{\mathscr{U}_{pc}} \to \mathbb{R}^r$ where $\widetilde{\mathscr{U}_{pc}}$ is a set of admissible inputs. If the reachable set*

$$\mathscr{R}(x_0) = \{x(T_u; x_0, u) \in X | u \in \widetilde{\mathscr{U}_{pc}} \subseteq \mathscr{U}_{pc}(\Sigma)\}$$

is dense in X in Zariski topology, then Σ is said to be algebraically reachable.

One can show that the closure of the reachable set $\mathscr{R}(x_0)$ in Zariski topology on X is an irreducible variety.

Definition 22.3.6. *Let $\Sigma = (X, f = \{f_\alpha | \alpha \in U\}, h, x_0)$ be a rational system and let Q denote the field of rational functions on X. The* observation algebra $A_{obs}(\Sigma)$ *of Σ is the smallest subalgebra of the field Q containing all components $h_i, i = 1, \dots, r$ of h, and closed with respect to the derivations given by rational vector fields $f_\alpha, \alpha \in U$. The* observation field $Q_{obs}(\Sigma)$ *of the system Σ is the field of quotients of $A_{obs}(\Sigma)$. The rational system Σ is called* rationally observable *if $Q_{obs}(\Sigma) = Q$.*

The irreducibility of the variety X implies that the observation algebra of Σ is an integral domain. Therefore, the observation field of Σ is well-defined. Further, $Q_{obs}(\Sigma)$ is closed with respect to the derivations given by rational vector fields $f_\alpha, \alpha \in U$.

Definition 22.3.7. *We call a rational realization of a response map* canonical *if it is both rationally observable and algebraically reachable.*

The dimension of a rational system Σ is given as the dimension of its state-space X. Because X is an irreducible real affine variety and because the dimension of an irreducible real affine variety X equals the maximal number of rational functions on X which are algebraically independent over $\mathbb{R}$, the dimension of a state-space X equals the transcendence degree (trdeg) of the field Q of all rational functions on X. Note that trdeg Q also corresponds to the dimension of the rational vector fields on X considered as a vector space over Q [15, Corollary to Theorem 6.1].

Definition 22.3.8. *We say that a rational realization $\Sigma = (X, f, h, x_0)$ of a response map p is* minimal *if for all rational realizations $\Sigma' = (X', f', h', x'_0)$ of p it holds that* $\dim X \leq \dim X'$.

One can prove that the dimension of a minimal rational realization Σ of a response map p equals the transcendence degree of the field $Q_{obs}(p)$.

Problem 22.3.2 (Canonical and minimal rational realizations). Consider a response map p. Does there exist a rational realization of p which is canonical and/or minimal? How to determine such realization from arbitrary rational realization of p?

The solution to Problem 22.3.2 is of practical relevance since it provides the realizations with minimal dimensions and the realizations with useful control theoretic properties. Obtaining the realizations in this form implies easier manipulation with the systems, faster predictions and validations. This problem is also closely related to the problem of the existence of algorithms and procedures for the construction of realizations with the desired properties.

Problem 22.3.3 (Algorithms). Let p be an arbitrary response map. Provide the algorithms for the construction of a rational realization of p, for the construction of a canonical and/or minimal rational realization of p, for the transformation of arbitrary realization of p to a realization of p which is canonical and/or minimal.

22.3.3 Rational Realizations

The following two theorems solve the existence parts of Problem 22.3.1 and Problem 22.3.2. Further, their proofs provide the procedures for constructing rational realizations of desired properties. Therefore, also Problem 22.3.3 is partly solved. Further research is needed for developing and implementing the corresponding algorithms by the means of existing computer algebra packages.

Theorem 22.3.1 (Existence of rational realizations). *A response map $p : \widetilde{\mathscr{U}_{pc}} \to \mathbb{R}^r$ has a rational realization if and only if $Q_{obs}(p)$ is finitely generated.*

Theorem 22.3.2 (Existence of canonical and minimal rational realizations). *Let p be a response map. The following statements are equivalent:*

(i) p has a rational realization,
(ii) p has a rationally observable rational realization,
(iii) p has a canonical rational realization,
(iv) p has a minimal rational realization.

nitial state $x_0' = x_0$, and by deriving the family f' of rational vector fields on X' and the output function h' in the following way. The output function h' is defined as The proof of Theorem 22.3.2(iii)⇒(iv) implies that a canonical rational realization Σ of a response map p is also a minimal realization of p. The converse is true only if the elements of $Q \setminus Q_{obs}(\Sigma)$ are not algebraic over $Q_{obs}(\Sigma)$.

Let us introduce the notion of birationally equivalent rational realizations by a slight modification of [4, Definition 8].

Definition 22.3.9. *We say that rational realizations $\Sigma = (X, f, h, x_0)$, $\Sigma' = \big(X', f', h', x_0'\big)$ of the same response map p with the same input space U and the same output space $\mathbb{R}^r$ are* birationally equivalent *if*

(i) the state-spaces X and X' are birationally equivalent (there exist rational mappings $\phi : X \to X'$, $\psi : X' \to X$ such that the equalities $\phi \circ \psi = id_{X'}$ and $\psi \circ \phi = id_X$ hold on Z-dense subsets of X' and X, respectively),

(ii) $h'\phi = h$,

(iii) $f_\alpha(\varphi \circ \phi) = (f'_\alpha \varphi) \circ \phi$ for all $\varphi \in Q'$, $\alpha \in U$,

(iv) ϕ is defined at x_0, and $\phi(x_0) = x'_0$.

Then, every rational realization of a response map which is birationally equivalent to a minimal rational realization of the same map, is itself minimal. On the other hand, all canonical rational realizations of the same response map are birationally equivalent. Therefore, minimal rational realizations are birationally equivalent if they are canonical.

22.4 Identification of Rational Systems

For the modeling of a particular biochemical phenomenon one can formulate a biochemical reaction system (22.1) which specifies the reaction network and the kinetics used. The class of selected systems then contains the systems which vary with the values of parameters in the system structure. In this section we derive for the class of rational systems the conditions under which the numerical values of parameters can be determined uniquely from the measurements characterizing the phenomenon. Further we discuss the way how to estimate these parameter values.

22.4.1 System Identification Procedure

The identification objectives are always twofold: (1) To obtain a realistic model which expresses as much as possible of the characteristics of a phenomenon to be modeled. (2) To strive for a system which is not too complex. Each rational system can be associated with a complexity measure, given for example as the dimension of a state-space or as the maximal degree of polynomials used in the system.

Let us recall the system identification procedure as it is described for example in [27, 28]. The procedure has the following steps.

1. *Modeling.* Formulate a physical model of a phenomenon or a model of the appropriate domain and, based on the physical model, formulate a mathematical model in the form of a control system. This system usually contains unknown parameters.
2. *Identifiability.* Determine whether the parametrization of the class of control systems selected in Step 1 is identifiable. Identifiability guarantees the uniqueness of parameter values for the considered model.
3. *Collection of data.* Design an experiment, carry out the experiment, collect the data in the form of a time series, and preprocess the time series.
4. *Approximation.* Select a system in the class of systems determined in Step 1 which best meets the observed time series according to an approximation criterion. The selection is carried out by estimating unknown parameters of the model.

5. *Evaluation and adjustment of the system class.* Compare the output of the system derived in Step 4 with the measured time series. If the comparison is not satisfactory, then adjust the system class chosen in Step 1 appropriately and continue with the subsequent steps of the procedure to derive another fully-determined system modeling the phenomenon. One may have to iterate this procedure till the comparison is satisfactory.

In the following two sections we discuss Step 2 and Step 4 in more detail.

22.4.2 Identifiability of Rational Systems

By choosing a model structure in the modeling step of system identification procedure we specify a system which is usually not fully determined, i.e. it contains unknown parameters. Depending on the modeling techniques, the parameters could have a physical or a biological meaning relevant for further investigation of the studied phenomenon. In this section we introduce the concept of parametrized systems within the class of rational systems and we derive necessary and sufficient conditions for the parametrizations of parametrized rational systems to be structurally identifiable.

The results presented in this section are derived in [19], an overview can be found in [20]. There are many approaches to study identifiability of parametrized systems, for example the approach based on a power series expansion of the output, differential algebra, generating series approach, and similarity transformation method. Our approach, which is related to similarity transformation or state isomorphism approach, strongly relies on the results of realization theory for rational systems presented in the previous section. For other approaches to identifiability of parametrized rational systems see [17, 9, 12, 31].

Throughout this section we assume that parameters take values in a set $P \subseteq \mathbb{R}^l$, $l \in \mathbb{N}$ which is an irreducible real affine variety. We refer such P as a *parameter set.*

Definition 22.4.1 (Parametrized systems). *By a* parametrized rational system $\Sigma(P)$ *we mean a family* $\{\Sigma(p) = (X^p, f^p, h^p, x_0^p) | p \in P\}$ *of rational systems where* $P \subseteq \mathbb{R}^l$ *is a parameter set. We assume that the systems* $\Sigma(p), p \in P$ *have the same input spaces* U *and the same output spaces* $\mathbb{R}^r$. *The map* $\mathscr{P} : P \to \Sigma(P)$ *defined as* $\mathscr{P}(p) = \Sigma(p)$ *for* $p \in P$ *is called the* parametrization *of* $\Sigma(P)$.

We say that a parametrized rational system $\Sigma(P)$ is *structurally reachable* (*structurally observable*) if there exists a variety $V \subsetneq P$ such that all rational systems $\Sigma(p)$ with $p \in P \setminus V$ are algebraically reachable (rationally observable). In the same way we define also *structural canonicity* of $\Sigma(P)$. It is easy to prove that $\Sigma(P)$ is structurally canonical if and only if it is structurally reachable and structurally observable.

Definition 22.4.2 (Identifiability). *Let* $P \subseteq \mathbb{R}^l$ *be a parameter set and let* $\widetilde{\mathscr{U}_{pc}}$ *be a set of admissible inputs. Let* $\Sigma(P)$ *be a parametrized rational system such that* $\widetilde{\mathscr{U}_{pc}} \subseteq \mathscr{U}_{pc}(\Sigma(p))$ *for all* $p \in P$. *We say that the parametrization* $\mathscr{P} : P \to \Sigma(P)$ *is* *(i)* globally identifiable *if the map*

$$p \mapsto h^p(x^p) = \{(u, h^p(x^p(T_u; x_0^p, u))) | u \in \widetilde{\mathcal{U}_{pc}}\}$$

is injective on P,

(ii) structurally identifiable *if the map*

$$p \mapsto h^p(x^p) = \{(u, h^p(x^p(T_u; x_0^p, u))) | u \in \widetilde{\mathcal{U}_{pc}}\}$$

is injective on $P \setminus S$ where S is a variety strictly contained in P.

Global identifiability of a parametrization of a parametrized system means that unknown parameters of the parametrized system can be determined uniquely from the measurements. Structural identifiability of a parametrization provides this uniqueness only on a Z-dense subset of a parameter set. Obviously, a globally identifiable parametrization of a parametrized system is structurally identifiable.

The following two theorems specify necessary and sufficient conditions for a parametrization of a parametrized rational system to be structurally identifiable.

Theorem 22.4.1 (Necessary condition for structural identifiability). *Let $P \subseteq \mathbb{R}^l$ be a parameter set and let $\Sigma(P)$ be a parametrized rational system with the parametrization $\mathscr{P} : P \to \Sigma(P)$. We assume that $\Sigma(P)$ is structurally canonical. Then the following statement holds.*

If the parametrization $\mathscr{P}$ is structurally identifiable, then there exists a variety $S \subsetneq P$ such that for any $p, p' \in P \setminus S$ any rational mapping relating the systems $\Sigma(p), \Sigma(p') \in \Sigma(P)$ as in Definition 22.3.9 is the identity.

We say that a parametrized rational system $\Sigma(P)$ is a *structured system* if all rational systems $\Sigma(p), \Sigma(p') \in \Sigma(P)$ are, after symbolic identification of parameter values p and p', birationally equivalent. Let us write all numerators and denominators of rational functions defining dynamics, output function, and initial state of a rational system $\Sigma(p) \in \Sigma(P)$ in the form of real polynomials in state variables with coefficients given as rational functions in parameters. If these functions generate the field of all rational functions on P, then we say that $\Sigma(p)$ *distinguishes parameters*. If this holds for all $p \in P \setminus D$ where $D \subsetneq P$ is a variety, then we say that $\Sigma(P)$ *structurally distinguishes parameters*.

Theorem 22.4.2 (Sufficient condition for structural identifiability). *Let $P \subseteq \mathbb{R}^l$ be a parameter set and let $\Sigma(P)$ be a structured rational system with the parametrization $\mathscr{P} : P \to \Sigma(P)$. We assume that $\Sigma(P)$ is structurally canonical and that it structurally distinguishes parameters. Then the following statement holds.*

If there exists a variety $S \subsetneq P$ such that for any $p, p' \in P \setminus S$ a rational mapping relating the systems $\Sigma(p), \Sigma(p') \in \Sigma(P)$ according to Definition 22.3.9 is the identity, then the parametrization $\mathscr{P}$ is structurally identifiable.

Let us discuss the main steps which have to be performed to check structural identifiability of a parametrization $\mathscr{P} : P \to \Sigma(P)$ by applying Theorem 22.4.2.

1. $\Sigma(P)$ *is a structured system.* To be able to apply Theorem 22.4.2 the considered parametrized system has to be structured. In most of applications the parametrized systems consist only of systems which all have the same state-spaces and which differ only by the values of parameters, i.e. they have the same structure. Because the parametrized systems having these properties are structured, in most of realistic examples the chosen model given as a parametrized system is structured.
2. $\Sigma(P)$ *is structurally canonical.* We need to verify whether $\Sigma(p)$ is algebraically reachable and rationally observable for almost all $p \in P$. We proceed to check these properties by various methods illustrated in [19]. The presence of parameters leads to constraints in the form of polynomial equations which then define a variety $RO \subsetneq P$ which has to be excluded.
3. $\Sigma(P)$ *structurally distinguishes parameters.* To check whether the systems $\Sigma(p)$ distinguish parameters for all $p \in P \setminus D$ where D is a strict subvariety of P, we check this property for a system of $\Sigma(P)$ with varying parameters. Then a variety D is derived as a by-product.
4. *Existence of a variety* $S \subsetneq P$. A variety $S \subsetneq P$ is in practice usually defined as $S = RO \cup D$. Here RO and D are the varieties determined in Step 2 and Step 3, respectively. Then all systems $\Sigma(p), p \in P \setminus S$ are canonical realizations of the same data and thus they are birationally equivalent. From Definition 22.3.9 we obtain the characterization of all isomorphisms ϕ relating the systems $\Sigma(p)$ and $\Sigma(p')$ for $p, p' \in P \setminus S$. Once this characterization implies that ϕ is the identity, the parametrization is structurally identifiable.

22.4.3 Approximation of Rational Systems

After the selection of a parametrized rational system, a check on the identifiability of its parametrization, and the collection of a time series, the next step of the system identification procedure is to estimate the parameter values of the parametrized system so that the corresponding system approximates the measured time series as well as possible. The approximation methods in system identification are generally distinguished into the *optimization approach* and the *algebraic system-theoretic approach*.

The optimization approach consists of infimizing an approximation criterion over a parameter set. For each value in the parameter set one computes the value of the approximation criterion by simulation of an observer or filter with the measured inputs and outputs, and a computation. The optimization approach is not guaranteed to work well since the approximation criterion is in general a nonconvex function of the parameters. Thus, any local optimization algorithm computes a local minimum of which there can be very many. Global optimization algorithms may help but so far the experience is not convincing and a convergence proof is also not known. Especially for rational systems, there is no proof of convergence for any optimization procedure for approximation and also, as the optimization approach requires the availability of an observer or filter, there is no observer known.

Based on the realization theory of Gaussian systems developed by P. Faurre with R.E. Kalman, H. Akaike, A. Lindquist, G. Picci, and others, there has been developed the *subspace identification algorithm* for Gaussian systems. In this algorithm, the infimization of an approximation criterion is achieved by algebraic means. It has been proven [24] that this procedure is the optimal solution of an approximation problem with the divergence criterion (Kullback-Leibler pseudo-distance) for finite-dimensional Gaussian random variables. The reader may find the algorithm described and many references in [29]. The algebraic system-theoretic approach to approximation of rational systems could be formulated in analogy with the subspace identification algorithm.

The easiest thing to do for the approximation problem is to apply a local linearization step followed by an application of the subspace identification algorithm to the linearized system, possibly transformed to a Gaussian system by addition of noise. This approach may not work in general since linearization does not preserve identifiability properties. Further, there is no guarantee that the estimate of the system for the approximation criterion is optimal. Another simple heuristic approach to the approximation problem for rational systems is to optimize the approximation criterion for one parameter at a time though in a global way, not a local way. Again, there is no guarantee that this will produce the optimal estimate, the procedure may even wander away from the optimum.

Most of rational systems arising in systems biology are of a too high state-space dimension and are often over-parametrized. Because of the wide-variety of time scales in most metabolic systems, the systems determined from a time series will have a much lower dimension than that of the system class. Most parameters in a rational system, even if identifiable, can probably be determined poorly and the corresponding terms in the numerator or denominator polynomial could therefore be eliminated from the representation of the system class. More experience with concrete examples is needed.

22.5 Concluding Remarks

We restricted our attention to rational systems with the state-spaces defined as irreducible real affine varieties. The generalization to reducible varieties is possible. Further, due to the applications in real life problems of biology and engineering, we have chosen to work with the field of real numbers. From the computational point of view, computable fields like the field of rational numbers could be considered.

Concerning realization theory for rational systems, smoothness, rationality, and other geometric properties of the possible state-spaces of rational realizations are of interest. Further, better insight to the characterization of birational equivalence classes of rational realizations can be given by the study of field isomorphisms. The application of the results of realization theory for rational systems to the problems of control and observer design and model reduction is to be carried out.

There are still many open problems concerning system identification for rational systems. One of them is the problem of determining the classes of inputs which are exciting the rational systems sufficiently to be able to determine their identifiability properties and consequently estimate the values of the parameters. For bilinear systems, the problem of characterizing sufficiently exciting inputs is considered in [23]. The problem of determining the numerical values of parameters from measurements is itself a major open problem. Further, structural indistinguishability which deals with the uniqueness of a model structure is of interest. In the case of rational systems it should be easily solvable by means of realization theory developed for this class of systems.

References

1. Baillieul, J.: The geometry of homogeneous polynomial dynamical systems. Nonlinear anal., Theory, Meth. and Appl. 4(5), 879–900 (1980)
2. Baillieul, J.: Controllability and observability of polynomial dynamical systems. Nonlinear Anal., Theory, Meth. and Appl. 5(5), 543–552 (1981)
3. Bartosiewicz, Z.: Realizations of polynomial systems. In: Fliess, M., Hazenwinkel, M. (eds.) Algebraic and Geometric Methods in Nonlinear Control Theory, pp. 45–54. D. Reidel Publishing Company, Dordrecht (1986)
4. Bartosiewicz, Z.: Rational systems and observation fields. Systems and Control Letters 9, 379–386 (1987)
5. Bartosiewicz, Z.: Minimal polynomial realizations. Mathematics of control, signals, and systems 1, 227–237 (1988)
6. Bosma, W., Cannon, J., Playoust, C.: The Magma algebra system. i. the user language. J. Symbolic Comput. 24(3-4), 235–265 (1997)
7. Byrnes, C.I., Falb, P.L.: Applications of algebraic geometry in system theory. American Journal of Mathematics 101(2), 337–363 (1979)
8. Cox, D., Little, J., O'Shea, D.: Ideals, varieties, and algorithms: An introduction to computational algebraic geometry and commutative algebra, 3rd edn. Springer, Heidelberg (2007)
9. Denis-Vidal, L., Joly-Blanchard, G., Noiret, C.: Some effective approaches to check identifiability of uncontrolled nonlinear systems. Mathematics and computers in simulation 57, 35–44 (2001)
10. REDUCE developers. REDUCE, Available at `http://reduce-algebra.com`
11. Eisenbud, D., Grayson, D.R., Stillman, M.E., Sturmfels, B. (eds.): Computations in algebraic geometry with Macaulay 2. Algorithms and Computations in Mathematics, vol. 8. Springer, Heidelberg (2001)
12. Evans, N.D., Chapman, M.J., Chappell, M.J., Godfrey, K.R.: Identifiability of uncontrolled nonlinear rational systems. Automatica 38, 1799–1805 (2002)
13. Falb, P.: Methods of algebraic geometry in control theory: Part 1, Scalar linear systems and affine algebraic geometry. Birkhäuser, Boston (1990)
14. Falb, P.: Methods of algebraic geometry in control theory: Part 2, Multivariable linear systems and projective algebraic geometry. Birkhäuser, Boston (1999)
15. Hermann, R.: Algebro-geometric and Lie-theoretic techniques in systems theory, Part A, Interdisciplinaty mathematics, Volume XIII. Math Sci Press, Brookline (1977)
16. Kunz, E.: Introduction to commutative algebra and algebraic geometry. Birkhäuser, Boston (1985)

17. Ljung, L., Glad, T.: On global identifiability for arbitrary model parametrizations. Automatica 30(2), 265–276 (1994)
18. Maxima.sourceforge.net. Maxima, a computer algebra system. version 5.18.1 (2009), Available at `http://maxima.sourceforge.net/`
19. Němcová, J.: Structural identifiability of polynomial and rational systems (submitted)
20. Němcová, J.: Structural and global identifiability of parametrized rational systems. In: Proceedings of 15th IFAC Symposium on System Identification, Saint-Malo, France (2009)
21. Němcová, J., van Schuppen, J.H.: Realization theory for rational systems: Minimal rational realizations. To appear in Acta Applicandae Mathematicae
22. Němcová, J., van Schuppen, J.H.: Realization theory for rational systems: The existence of rational realizations. To appear in SIAM Journal on Control and Optimization
23. Sontag, E.D., Wang, Y., Megretski, A.: Input classes for identification of bilinear systems. IEEE Transactions Autom. Control 54, 195–207 (2009)
24. Stoorvogel, A.A., van Schuppen, J.H.: Approximation problems with the divergence criterion for gaussian variables and processes. Systems and Control Letters 35, 207–218 (1998)
25. CoCoA Team: CoCoA: a system for doing Computations in Commutative Algebra. Available at `http://cocoa.dima.unige.it`
26. Singular team: Singular, a computer algebra system. Available at `http://www.singular.uni-kl.de/`
27. van den Hof, J.M.: System theory and system identification of compartmental systems. PhD thesis, Rijksuniversiteit Groningen, The Netherlands (1996)
28. van den Hof, J.M.: Structural identifiability of linear compartmental systems. IEEE Transactions Autom. Control 43(6), 800–818 (1998)
29. van Overschee, P., De Moor, B.L.R.: Subspace identification for linear systems. Kluwer Academic Publishers, Dordrecht (1996)
30. Wang, Y., Sontag, E.D.: Algebraic differential equations and rational control systems. SIAM J. Control Optim. 30(5), 1126–1149 (1992)
31. Xia, X., Moog, C.H.: Identifiability of nonlinear systems with application to HIV/AIDS models. IEEE Transactions Autom. Control 48(2), 330–336 (2003)
32. Zariski, O., Samuel, P.: Commutative algebra I, II. Springer, Heidelberg (1958)

23

Semi-supervised Regression and System Identification*,†

Henrik Ohlsson and Lennart Ljung

Division of Automatic Control, Department of Electrical Engineering,
Linköpings Universitet, SE-583 37 Linköping, Sweden

Summary. System Identification and Machine Learning are developing mostly as independent subjects, although the underlying problem is the same: To be able to associate "outputs" with "inputs". Particular areas in machine learning of substantial current interest are *manifold learning* and *unsupervised* and *semi-supervised regression*. We outline a general approach to semi-supervised regression, describe its links to Local Linear Embedding, and illustrate its use for various problems. In particular, we discuss how these techniques have a potential interest for the system identification world.

23.1 Introduction

A central problem in many scientific areas is to link certain observations to each other and build *models* for how they relate. In loose terms, the problem could be described as relating y to φ in

$$y = f(\varphi) \tag{23.1}$$

where φ is a vector of observed variables and y is a characteristic of interest.

In system identification φ could be observed past behavior of a dynamical system, and y the predicted next output. In classification problems φ would be the vector of features and y the class label. Following statistical nomenclature, we shall generally call φ the *regression vector* containing the *regressors* and following classification nomenclature we call y the corresponding *label*.

The information available could be a collection of labeled pairs

$$y(t) = f(\varphi(t)) + e(t), \quad t = 1, \ldots, N_l \tag{23.2}$$

where e accounts for possible errors in the measured labels. Constructing an estimate of the function f from labeled data $\{(y(t), \varphi(t)), t = 1, \ldots, N_l\}$ is a standard *regression problem* in statistics, see e.g. [11].

*This work was supported by the Strategic Research Center MOVIII, funded by the Swedish Foundation for Strategic Research, SSF, and CADICS, a Linnaeus center funded by the Swedish Research Council.

†Dedicated to Chris and Anders at the peak of their careers.

X. Hu et al. (Eds.): Three Decades of Progress in Control Sciences, pp. 343–360, 2010.

Fig. 23.1. Left side shows three regressors, two labeled, with the class label next to them, and one unlabeled regressor. Desiring an estimate of the label of the unlabeled regressor, having no further information, we would probably guess that it belongs to class B. Now, let us assume that we are provided the information that regressors are constrained to lay on the black areas (on the elliptic curve on which the labeled regressor of class A lay or on the elliptic filled area in which the labeled regressor belonging to B belongs) shown in the right part of the figure. What would the guess be now?

We shall in this contribution generally not seek explicit constructions of the estimate f, but be content by having a scheme that provides an estimate of $f(\varphi^*)$ for any given regressor φ^*. This approach has been termed *Model-on-Demand*, [23] or *Just-In-Time* modeling, [5].

The term *supervised learning* is also used for such algorithms, since the construction of f is "supervised" by the measured information in y. In contrast to this, *unsupervised learning* only has the information of the regressors $\{\varphi(t), t = 1, \ldots, N_u\}$. In unsupervised classification, e.g. [13], the classes are constructed by various clustering techniques. *Manifold learning*, e.g. [24, 21] deals with unsupervised techniques to construct a manifold in the regressor space that houses the observed regressors.

Semi-supervised algorithms are less common. In semi-supervised algorithms, both labeled and unlabeled regressors,

$$\{(y(t), \varphi(t)), t = 1, \ldots, N_l, \varphi(t), t = N_l + 1, \ldots, N_l + N_u\} \tag{23.3}$$

are used to construct f. This is particularly interesting if extra effort is required to measure the labels. Thus costly labeled regressors are supported by less costly unlabeled regressors to improve the result.

It is clear that unsupervised and semi-supervised algorithms are of interest only if the regressors have a pattern that is unknown *a priori*.

Semi-supervised learning is an active area within classification and machine learning (see [4, 28] and references therein). In classification, it is common to make the assumption that class labels do not change in areas with a high density of regressors. Figure 23.1 gives an illustration of this situation. To estimate the high density areas, unlabeled data are useful.

The main reason that semi-supervised algorithms are not often seen in regression and system identification may be that it is less clear when unlabeled regressors can be of use. We will try to bring some clarity to this through this chapter. Let us start directly by a pictorial example. Consider the 5 regressors shown in the left of Fig. 23.2. Four of the regressors are labeled and their labels are written out next to

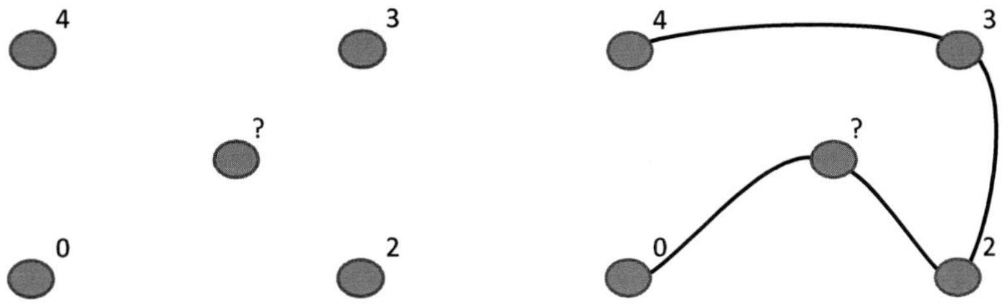

Fig. 23.2. The left side shows 5 regressors, four labeled and one unlabeled. Desiring an estimate of the label of the unlabeled regressor, we could simply weight together the two closest regressors' labels and get 2.5. Say now that the process that generated our regressors, traced out the path shown in the right part of the figure. Would we still guess 2.5?

them. One of the regressors is unlabeled. To estimate that label, we could compute the average of the two closest regressors' labels, which would give an estimate of 2.5. Let us now add the information that the regressors and the labels were sampled from an in time continuous process and that the value of the regressor was evolving along the curve shown in the right part of Fig. 23.2. Knowing this, a better estimate of the label would probably be 1. The knowledge of that the regressors are restricted to a certain region in the regressor space can hence make us reconsider our estimation strategy. Notice also that to estimate the region to which the regressors are restricted, both labeled and unlabeled regressors are useful.

Generally, regression problems having regressors constrained to rather limited regions in the regressor space may be suitable for a semi-supervised regression algorithm. It is also important that unlabeled regressors are available and comparably "cheap" to get as opposed to the labeled regressors.

The chapter is organized as follows: We start off by giving a background to semi-supervised learning and an overview of previous work, Sect. 23.2. We thereafter formalize the assumptions under which unlabeled data has potential to be useful, Sect. 23.3. A semi-supervised regression algorithm is described in Sect. 23.4 and exemplified in Sect. 23.5. In Sect. 23.6 we discuss the application to dynamical systems and we end by a conclusion in Sect. 23.7.

23.2 Background

Semi-supervised learning has been around since the 1970s (some earlier attempts exist). *Fisher's linear discriminant rule* was then discussed under the assumption that each of the class conditional densities was Gaussian. *Expectation maximization* was applied using both labeled and unlabeled regressors to find the parameters of the Gaussian densities [12]. During the 1990s the interest for semi-supervised learning increased, mainly due to its application to text classification, see e.g. [17]. The first usage of the word *semi-supervised* learning, as it is used today, was not until 1992 [14].

The boost in the area of manifold learning in the 1990s brought with it a number of semi-supervised methods. *Semi-supervised manifold learning* is a type of semi-supervised learning in which the map found by an unsupervised manifold learning algorithm is restricted by giving a number of labeled regressors as examples for how that map should be. Most of the algorithms are extensions of unsupervised manifold learning algorithms, see among others [2, 26, 16, 7, 6, 18, 27]. Another interesting contribution is the developments by Rahimi in [20]. A time series of regressors, some labeled and some unlabeled, are considered there. The series of labels best fitting the given labels and at the same time satisfying some temporal smoothness assumption is then computed.

Most of the references above are to semi-supervised classification algorithms. They are however relevant since most semi-supervised classification methods can, with minor modifications, be applied to regression problems. The modification or the application to regression problems are however almost never discussed or exemplified. For more historical notes on semi-supervised learning, see [4].

23.3 The Semi-supervised Smoothness Assumption

We are in regression interested in finding estimates for the conditional distribution $p(y|\varphi)$. For the unlabeled regressors to be useful, it is required that the regressor distribution $p(\varphi)$ brings information concerning the conditional $p(y|\varphi)$.

We saw from the pictorial example in Sect. 23.1 that one situation for which this is the case is when we make the assumption that the label changes continuously along high-density areas in the regressor space. This assumption is referred to as the *semi-supervised smoothness assumption* [4]:

Assumption 23.3.1 (Semi-supervised Smoothness). *If two regressors $\varphi(1)$, $\varphi(2)$ in a high-density region are close, then so should their labels.*

"High density region" is a somewhat loose term: In many cases it corresponds to a manifold in the regressor space, such that the regressors for the application in question are confined to this manifold. That two regressors are "close" then means that the distance between them along the manifold (the geodesic distance) is small.

In classification, this smoothness assumption is interpreted as that the class labels should be the same in the high-density regions. In regression, we interpret this as a slowly varying label along high-density regions. Note that in regression, it is common to assume that the label varies smoothly in the regressor space; the semi-supervised smoothness assumption is less conservative since it only assumes smoothness in the high-density regions in the regressor space. Two regressors could be close in the regressor space metric, but far apart along the high density region (the manifold): think of the region being a spiral in the regressor space.

23.4 Semi-supervised Regression: WDMR

Given a particular regressor φ^*, consider the problem of finding an estimate for $f(\varphi^*)$ given the measurements $\{(y(t), \varphi(t))\}_{t=1}^{N_l}$ generated by

$$y = f(\varphi) + e, \quad e \sim \mathcal{N}(0, \sigma). \tag{23.4}$$

This is a supervised regression problem. If unlabeled regressors $\{\varphi(t)\}_{t=N_l+1}^{N_l+N_u}$ are used as well, the regression becomes semi-supervised. Since we in the following will make no difference between the unlabeled regressor φ^* and $\{\varphi(t)\}_{t=N_l+1}^{N_l+N_u}$, we simply include φ^* in the set of unlabeled regressors to make the notation a bit less cluttered. We let $\hat{f}_t$ denote the estimates of $f(\varphi(t))$ and assume that $f: \mathscr{R}^{n_\varphi} \to \mathscr{R}$ for simplicity. In the following we will also need to introduce kernels as distance measure in the regressor space. To simplify the notation, we will use K_{ij} to denote a kernel $k(\cdot,\cdot)$ evaluated at the regressor pair $(\varphi(i), \varphi(j))$ i.e., $K_{ij} \triangleq k(\varphi(i), \varphi(j))$. A popular choice of kernel is the Gaussian kernel

$$K_{ij} = e^{-\|\varphi(i)-\varphi(j)\|^2/2\sigma^2}. \tag{23.5}$$

Since we will consider regressors constrained to certain regions of the regressor space (often manifolds), kernels constructed from manifold learning techniques, see Sect. 23.4.1, will be of particular interest. Notice however that we will allow us to use a kernel like

$$K_{ij} = \begin{cases} \frac{1}{K}, & \text{if } \varphi(j) \text{ is one of the } K \text{ closest} \\ & \text{neighbors of } \varphi(i), \\ 0, & \text{otherwise,} \end{cases} \tag{23.6}$$

and K_{ij} will therefore not necessarily be equal to K_{ji}. We will also always use the convention that $K_{ij} = 0$ if $i = j$.

Under the semi-supervised smoothness assumption, we would like the estimates belonging to two regressors which are close in a high-density region to have similar values. Using a kernel, we can express this as

$$\hat{f}_t = \sum_{i=1}^{N_l+N_u} K_{ti}\hat{f}_i, \quad t = 1 \ldots N_l + N_u \tag{23.7}$$

where K_{ti} is a kernel giving a measure of distance between $\varphi(t)$ and $\varphi(i)$, relevant to the assumed region.

So the sought estimates $\hat{f}_i$ should be such that they are smooth over the region. At the same time, for regressors with measured labels, the estimates should be close to those, meaning that

$$\sum_{t=1}^{N_l} (y(t) - \hat{f}_t)^2 \tag{23.8}$$

should be small. The two requirements (23.7) and (23.8) can be combined into a criterion

$$\lambda \sum_{i=1}^{N_l+N_u} (\hat{f}_i - \sum_{j=1}^{N_l+N_u} K_{ij}\hat{f}_j)^2 + (1-\lambda)\sum_{t=1}^{N_l}(y(t) - \hat{f}_t)^2 \tag{23.9}$$

to be minimized with respect to $\hat{f}_t$, $t = 1,\dots,N_l + N_u$. The scalar λ decides how trustworthy our labels are and is seen as a design parameter.

The criterion (23.9) can be given a Bayesian interpretation as a way to estimate $\hat{f}$ in (23.8) with a "smoothness prior" (23.7), with λ reflecting the confidence in the prior.

Introducing the notation

$$J \triangleq [I_{N_l \times N_l}\, 0_{N_l \times N_u}],$$
$$\mathbf{y} \triangleq [y(1)\, y(2) \dots y(N_l)]^T,$$
$$\hat{\mathbf{f}} \triangleq [\hat{f}_1\, \hat{f}_2 \dots \hat{f}_{N_l}\, \hat{f}_{N_l+1} \dots \hat{f}_{N_l+N_u}]^T,$$
$$K \triangleq \begin{bmatrix} K_{11} & K_{12} & 1ots & K_{1,N_l+N_u} \\ K_{21} & K_{22} & & K_{2,N_l+N_u} \\ \vdots & & \ddots & \vdots \\ K_{N_l+N_u,1} & K_{N_l+N_u,2} & \dots & K_{N_l+N_u,N_l+N_u} \end{bmatrix},$$

(23.9) can be written as

$$\lambda(\hat{\mathbf{f}} - K\hat{\mathbf{f}})^T(\hat{\mathbf{f}} - K\hat{\mathbf{f}}) - (1-\lambda)(\mathbf{y} - J\hat{\mathbf{f}})^T(\mathbf{y} - J\hat{\mathbf{f}}) \tag{23.10}$$

which expands into

$$\hat{\mathbf{f}}^T\left(\lambda(I - K - K^T + K^T K) - (1-\lambda)J^T J\right)\hat{\mathbf{f}} + 2(1-\lambda)\hat{\mathbf{f}}^T J^T \mathbf{y} + (1-\lambda)\mathbf{y}^T\mathbf{y}. \tag{23.11}$$

Setting the derivative with respect to $\hat{\mathbf{f}}$ to zero and solving gives the linear kernel smoother

$$\hat{\mathbf{f}} = (1-\lambda)\left(\lambda(I - K - K^T + K^T K) - (1-\lambda)J^T J\right)^{-1} J^T \mathbf{y}. \tag{23.12}$$

This regression procedure uses all regressors, both unlabeled and labeled, and is hence a semi-supervised regression algorithm. We call the kernel smoother *Weight Determination by Manifold Regularization* (WDMR, [18]). In this case the unlabeled regressors are used to get a better knowledge for what parts of the regressor space that the function f varies smoothly in.

Similar methods to the one presented here has also been discussed in [10, 26, 3, 2, 25]. [26] discusses manifold learning and construct a semi-supervised version of the manifold learning technique *Locally Linear Embedding* (LLE, [21]) which coincides with a particular choice of kernel in (23.9). More details about this kernel choice will be given in the next section. [10] studies graph based semi-supervised methods for classification and derives a similar objective function as (23.9). [3, 25] discuss a classification method called *label propagation* which is an iterative approach converging to (23.12). In [2], support vector machines is extended to work under the semi-supervised smoothness assumption.

23.4.1 LLE: A Way of Selecting the Kernel in WDMR

Local Linear Embedding, LLE, [21] is a technique to find lower dimensional manifolds to which an observed collection of regressors belong. A brief description of it is as follows:

Let $\{\varphi(i), i = 1,\ldots,N\}$ belong to $U \subset R^{n_\varphi}$ where U is an unknown manifold of dimension n_z. A coordinatization $z(i)$, $(z(i) \in R^{n_z})$ of U is then obtained by first minimizing the cost function

$$\varepsilon(l) = \sum_{i=1}^{N} \left\| \varphi(i) - \sum_{j=1}^{N} l_{ij}\varphi(j) \right\|^2 \tag{23.13a}$$

under the constraints

$$\begin{cases} \sum_{j=1}^{N} l_{ij} = 1, \\ l_{ij} = 0 \text{ if } \|\varphi(i) - \varphi(j)\| > C_i(K) \text{ or if } i = j. \end{cases} \tag{23.13b}$$

Here, $C_i(K)$ is chosen so that only K weights l_{ij} become nonzero for every i. K is a design variable. It is also common to add a regularization to (23.13a) not to get degenerate solutions.

Then for the determined l_{ij} find $z(i)$ by minimizing

$$\sum_{i=1}^{N} \left\| z(i) - \sum_{j=1}^{N} l_{ij} z(j) \right\|^2 \tag{23.14}$$

wrt $z(i) \in R^{n_z}$ under the constraint

$$\frac{1}{N}\sum_{i=1}^{N} z(i)z(i)^T = I_{n_z \times n_z}$$

$z(i)$ will then be the coordinate for $\varphi(i)$ in the lower dimensional manifold.

The link between WDMR and LLE is now clear: If we pick the kernel K_{ij} in (23.9) as l_{ij} from (23.13) and have no labeled regressors ($N_l = 0$) and add the constraint $\frac{1}{N_u}\hat{\mathbf{f}}^T\hat{\mathbf{f}} = I_{n_z \times n_z}$, minimization of the WDMR criterion (23.9) will yield $\hat{f}_i$ as the LLE coordinates $z(i)$.

In WDMR with labeled regressors, the addition of the criterion (23.8) will replace the constraint $\frac{1}{N_u}\hat{\mathbf{f}}^T\hat{\mathbf{f}} = I_{n_z \times n_z}$ as an anchor to prevent a trivial zero solution. Thus WDMR is a natural semi-supervised version of LLE, [18].

23.4.2 A Comparison with K Nearest Neighbor Averages: K-NN

It is interesting to notice the difference between using the kernel given in (23.6) and

$$K_{ij} = \begin{cases} \frac{1}{K}, & \text{if } \varphi(j) \text{ is one of the } K \text{ closest} \\ & \underline{\text{labeled}} \text{ neighbors of } \varphi(i), \\ 0, & \text{otherwise.} \end{cases} \tag{23.15}$$

To illustrate the difference, let us return to the pictorial example discussed in Fig. 23.2. We now add 5 unlabeled regressors to the 5 previously considered. Hence we have 10 regressors, 4 labeled and 6 unlabeled, and we desire an estimate of the label marked with a question mark in Fig. 23.3. The left part of Fig. 23.3 shows how WDMR solves the estimation problem if the kernel in (23.15) is used. Since the kernel will cause the searched label to be similar to the label of the K closest labeled regressors, the result will be similar to using the algorithm *K-nearest neighbor average* (K-NN, see e.g. [11]). In the right part of Fig. 23.3, WDRM with the kernel given in (23.6) is used. This kernel grants estimates of the K closest regressors (labeled or not) to be similar. Since the closest regressors, to the regressor for which we search the label, are unlabeled, information is propagated from the labeled regressors towards the one for which we search a label along the chain of unlabeled regressors. The shaded regions in both the left and right part of the figure symbolize the way information is propagated using the different choices of kernels. In the left part of the figure we will therefore obtain an estimate equal to 2.5 while in the right we get an estimate equal to 1.

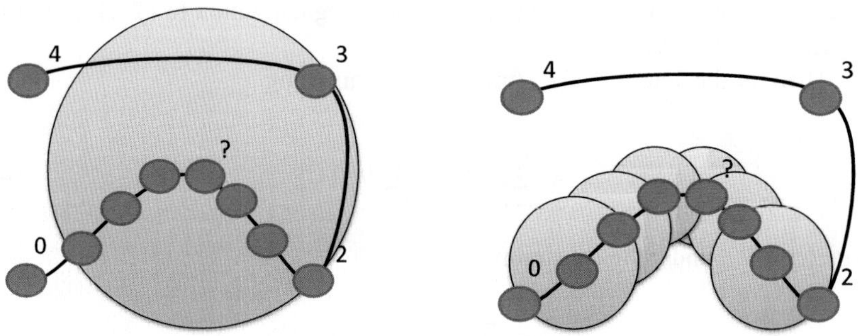

Fig. 23.3. An illustration of the difference of using the kernel given in (23.15) (left part of the figure) and (23.6) (right part of the figure).

23.5 Examples

We give in the following two examples of regression problem for which the semi-supervised smoothness assumption is motivated. Estimates are computed using WDMR and comparisons to conventional supervised regression methods are given.

23.5.1 fMRI

Functional Magnetic Resonance Imaging, fMRI is a technique to measure brain activity. The fMRI measurements give a measure of the degree of oxygenation in the blood, it measures the *Blood Oxygenation Level Dependent* (BOLD) response. The degree of oxygenation reflects the neural activity in the brain and fMRI is therefore an indirect measure of brain activity.

Measurements of brain activity can with fMRI be acquired as often as once a second and are given as an array, each element giving a scalar measure of the average activity in a small volume element of the brain. These volume elements are commonly called *voxels* (short for *volume pixel*) and they can be as small as one cubic millimeter. The fMRI measurements are heavily affected by noise.

In this example, we consider measurements from an $8 \times 8 \times 2$ array covering parts of the visual cortex gathered with a sampling period of 2 seconds. To remove noise, data was prefiltered by applying a spatial and temporal filter with a Gaussian kernel. The filtered fMRI measurements at each time t were vectorized into the regression vector $\varphi(t)$. fMRI data was acquired during 240 seconds (giving 120 samples, since the sampling period was 2 seconds) from a subject that was instructed to look away from a flashing checkerboard covering 30% of the field of view. The flashing checkerboard moved around and caused the subject to look to the left, right, up and down. The direction in which the person was looking was seen as the label. The label was chosen to 0 when the subject was looking to the right, $\pi/2$ when looking up, π when looking to the left and $-\pi/2$ when looking down.

The direction in which the person was looking is described by its angle, a scalar. The fMRI data should hence be constrained to a one-dimensional closed manifold residing in the 128 dimensional regressor space (since the regressors can be parameterized by the angle). If we assume that the semi-supervised smoothness assumption holds, WDMR therefore seems like a good choice.

The 120 labeled regressors were separated into two sets, a training set consisting of 80 labeled regressors and a test set consisting of 40 labeled regressors. The training set was further divided into an estimation set and a validation set, both of the same size. The estimation set and the regressors of the validation set were used in WDMR. The estimated labels of the validation regressors were compared to the measured labels and used to determine the design parameters. λ in (23.9) was chosen as 0.8 and K (using the kernel determined by LLE, see (23.13)) as 6. The tuned WDMR regression algorithm was then used to predict the direction in which the person was looking. The result from applying WDMR to the 40 regressors of the test set are shown in Fig. 23.4.

The result is satisfactory but it is not clear to what extent the one-dimensional manifold has been found. The number of unlabeled regressors used are rather low and it is therefore not surprising that K-NN can be shown to do almost as good as WDMR in this example. One would expect that adding more unlabeled regressors would improve the result obtained by WDMR. The estimates of K-NN would however stay unchanged since K-NN is a supervised method and therefore not affected by unlabeled data.

23.5.2 Climate Reconstruction

There exist a number of climate recorders in nature from which the past temperature can be extracted. However, only a few natural archives are able to record climate

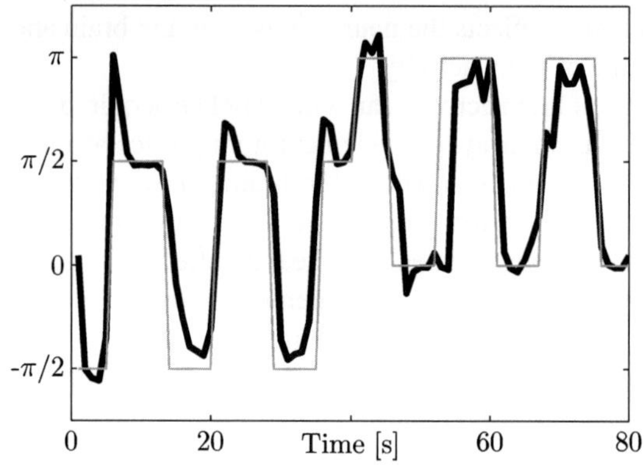

Fig. 23.4. WDMR applied to brain activity measurements (fMRI) of the visual cortex in order to tell in what direction the subject in the MR scanner was looking. Thin gray line shows the direction in which the subject was looking and thick black line, the estimated direction by WDMR.

fluctuations with high enough resolution so that the seasonal variations can be reconstructed. One such archive is a bivalve shell. The chemical composition of a shell of a bivalve depends on a number of chemical and physical parameters of the water in which the shell was composed. Of these parameters, the water temperature is probably the most important one. It should therefore be possible to estimate the water temperature for the time the shell was built, from measurements of the shell's chemical composition. This would e.g. give climatologists the ability to estimate past water temperatures by analyzing ancient shells.

In this example, we used 10 shells grown in Belgium. Since the temperature in the water had been monitored for these shells, this data set provides excellent means to test the ability to predict water temperature from chemical composition measurements. For these shells, the chemical composition measurements had been taken along the growth axis of the shells and paired up with temperature measurements. Between 30 and 52 measurement were provided from each shell, corresponding to a time period of a couple of months. The 10 shells were divided into an estimation set and a validation set. The estimation set consisted of 6 shells (a total of 238 labeled regressors) grown in Terneuzen in Belgium. Measurements from five of these shells are shown in Fig. 23.5. The figure shows measurements of the relative concentrations of Sr/Ca, Mg/Ca and Ba/Ca (Pb/Ca is also measured but not shown in the figure). The line shown between measurements connects the measurements coming from a shell and gives the chronological order of the measurements (two in time following measurements are connected by a line).

As seen in the figure, measurements are highly restricted to a small region in the measurement space. Also, the water temperature (gray level coded in Fig. 23.5) varies smoothly in the high-density regions. This together with that it is a biological process generating data, motivates the semi-supervised smoothness assumption

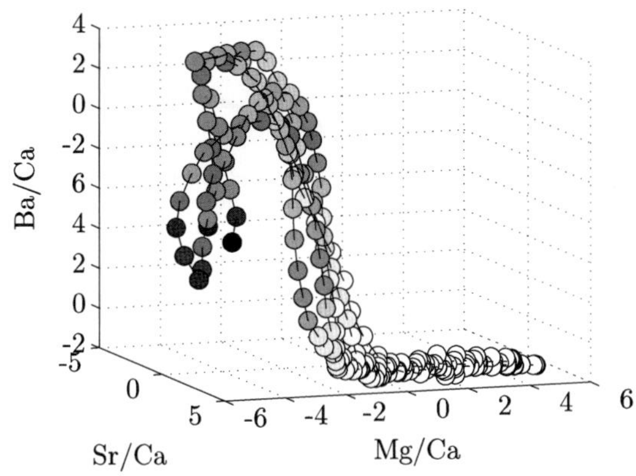

Fig. 23.5. A plot of the Sr/Ca, Mg/Ca and Ba/Ca concentration ratio measurements from five shells. Lines connects measurements (ordered chronologically) coming from the same shell. The temperatures associated with the measurements were color coded and are shown as different gray scales on the measurement points.

when trying to estimate water temperature (labels) from chemical composition measurements (4-dimensional regressors).

The four shells in the validation set came from four different sites (Terneuzen, Breskens, Ossenisse, Knokke) and from different time periods. The estimated temperatures for the validation data obtained by using WDMR with the kernel determined by LLE (see (23.13)) are shown in Fig. 23.6. For comparison purpose, it could be mentioned that K-NN had a *Mean Absolute Error* (MAE) nearly twice as high as WDMR.

A more detailed discussion of this exampled is presented in [1]. The data sets used were provided by Vander Putten and colleagues [19] and Gillikin and colleagues [8, 9].

23.6 Dynamical Systems

23.6.1 Analysis of a Circadian Clock

The circadian rhythmic living among people and animals are kept by robustly coupled chemical processes in cells of the suprachiasmatic nucleus (SCN) in the brain. The whole system is affected by light and goes under the name *biological clock*. The biological clock synchronizes the periodic behavior of many chemical processes in the body and is crucial for the survival of most species.

The chemical processes cause protein and messenger RNA (mRNA) concentrations in the cells of the SCN to fluctuate. The "freerunning" rhythm (no external input) of the fluctuations is however not the same as the light/dark cycle and environmental cues, such as light, cause it to synchronize with the environmental rhythm. We use the nonlinear biological clock model by [22]

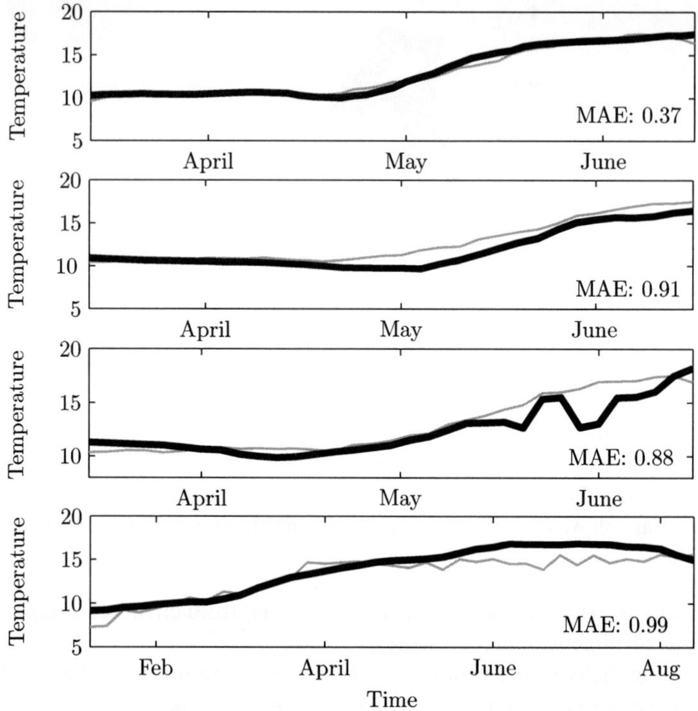

Fig. 23.6. Water temperature estimations using WDMR for validation data (thick line) and measured temperature (thin line). From top to bottom figure: Terneuzen, Breskens, Ossenisse, Knokke.

$$\frac{dM(t)}{dt} = \frac{r_M(t)}{1+P(t)^2} - 0.21M(t), \tag{23.16}$$

$$\frac{dP(t)}{dt} = M(t-4)^3 - 0.21P(t), \tag{23.17}$$

to generate simulated data and simulate the affect of the light cue by letting the mRNA production rate r_M vary periodically. M and P are the relative concentrations of mRNA and the protein. Figure 23.7 shows the (periodic) response of P to the (periodic) stimuli r_M. We see r_M as input and P as output and we want to predict P from measured $r_M(t)$. The measurements of P are rather costly in real applications, while $r_M(t)$ can be inferred from simple measurements of the light.

We seek to describe the output P as a nonlinear FIR (NFIR) model from two previous inputs $[r_M(t) \quad r_M(t-4)]^T$ (the regression vector) and collect 230 measurements of this regression vector. Only 6 out of these are labeled by the corresponding $P(t)$. We thus have a situation (23.3) with $N_l = 6$ and $N_u = 224$. Applying the WDMR algorithm (23.12) with $\lambda = 0.5$ and the kernel defined in (23.6) (with $K = 4$) gives an estimate of P corresponding to all the 230 time points. This estimate is shown in Fig. 23.8 together with the true values.

We note that the estimate is quite good, despite the very small number of labeled measurements. In this case the two-dimensional regression vector is confined to a 1-dimensional manifold (this follows since r_M is periodic: one full period will create a track in the regressor space that can be parameterized by the scalar time variable over one period). This means that this application can make full use of the dimension reduction that is inherent in WDMR. On the other hand, the model is tailored to the specific choice of input. (This, by the way, is true for any non-linear identification method)

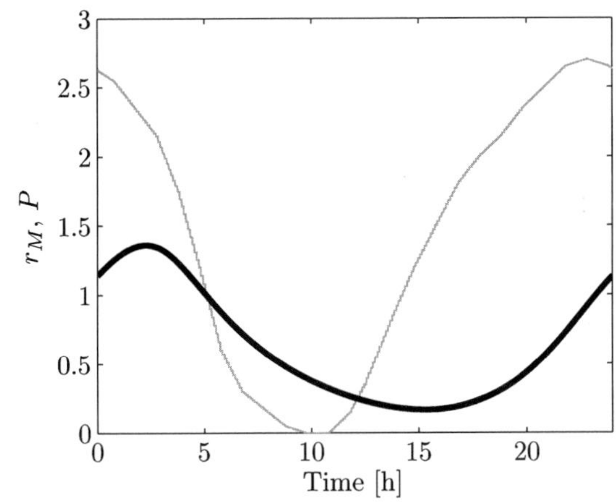

Fig. 23.7. The circadian clock is affected by light. This is in Example 23.6.1 modeled by letting r_M vary in a periodic manner. One period of r_M (thin gray line) and a period of P (thick black line) are shown in the figure. The synchronization between r_M and P is characteristic for a circadian clock and crucial for surviving.

Let us compare with the estimates obtained by K-NN, using the K-NN kernel given in (23.15). The dashed line in Fig. 23.8 shows the estimated protein levels using the K-NN (using only the labeled regressors). Since using only one neighbor ($K = 1$) gave the best result in K-NN, only this result is shown.

The result shown in Fig. 23.8 confirms the previous discussion around the pictorial example, see Fig. 23.3. K-NN average together the Euclidean closest regressors' labels while WDMR search for labeled regressors along the manifold and then assumes a slowly varying function along the manifold.

23.6.2 The Narendra–Li System

Let us now consider a standard test example from [15], "the Narendra-Li example":

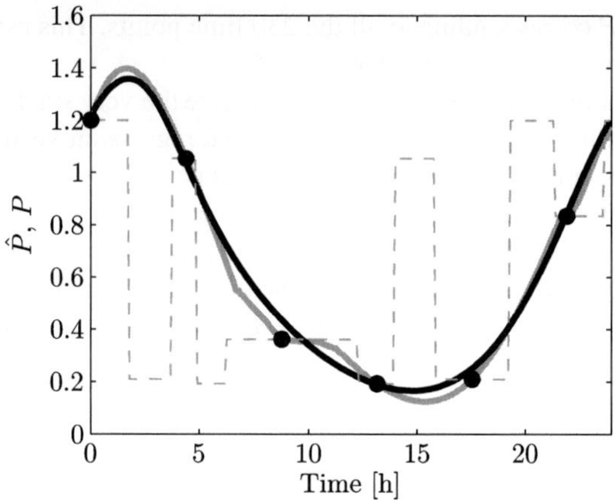

Fig. 23.8. Estimated relative protein concentration by K-NN ($K = 1$ gave the best result and therefore shown) and WDMR using the K-nearest neighbor kernel ($K = 4$ gave the best result and therefore shown). K-NN: dashed gray line; true P: solid black line; WDMR: solid gray line; estimation data: filled circles.

$$x_1(t+1) = \left(\frac{x_1(t)}{1+x_1^2(t)} + 1\right)\sin(x_2(t)) \tag{23.18a}$$

$$x_2(t+1) = x_2(t)\cos(x_2(t)) + x_1(t)\exp\left(-\frac{x_1^2(t)+x_2^2(t)}{8}\right) + \frac{u^3(t)}{1+u^2(t)+0.5\cos(x_1(t)+x_2(t))} \tag{23.18b}$$

$$y(t) = \frac{x_1(t)}{1+0.5\sin(x_2(t))} + \frac{x_2(t)}{1+0.5\sin(x_1(t))} + e(t) \tag{23.18c}$$

This dynamical system was simulated with 2000 samples using a random binary input, giving input output data $\{y(t), u(t), t = 1,\ldots,2000\}$. A separate set of 50 validation data were also generated with a sinusoidal input. The chosen regression vector was

$$\varphi(t) = \left[y(t-1)\ y(t-2)\ y(t-3)\ u(t-1)\ u(t-2)\ u(t-3)\right]^T \tag{23.19}$$

A standard Sigmoidal Neural Network model with one hidden layer with 18 (gave the best result) units was applied to this data set, and a corresponding NLARX model $y(t) = f(\varphi(t))$ was constructed. The prediction performance for the validation data is illustrated in Fig. 23.9. As a numerical measure of how good the prediction is, the "fit" is shown in the figure. The fit is the relative norm of the difference between the curves, expressed in %. 100% is thus a perfect fit.

The semi-supervised algorithm, WDMR (23.12) with a kernel determined from LLE (as described in (23.13)) was also applied to these data. Then unlabeled regression vectors from the validation data was appended to the estimation data. The resulting prediction performance is also shown in Fig. 23.9.

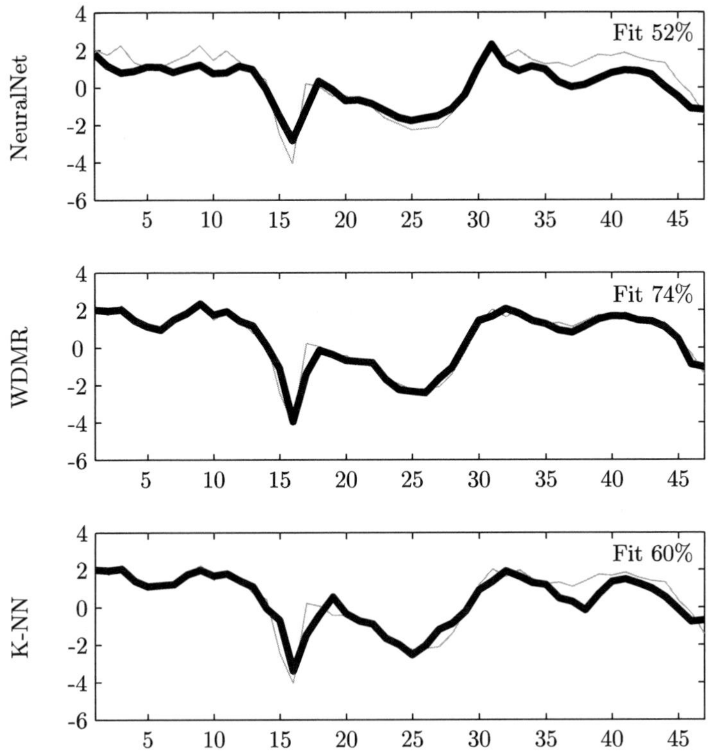

Fig. 23.9. One step ahead prediction for the models of the Narendra-Li example. Top: Neural Network (18 units); middle: WDMR; bottom: K-Nearest Neighbor (K=15). Thin line: true validation outputs; thick line: model output.

We see that WDMR gives a significantly better model than the standard neural network technique. In this case it is not clear that the regressors are constrained to a manifold. Therefore *semi*-supervised aspect is not so pronounced, and anyway the (validation) set of unlabeled regressors is quite small in comparison to the labeled (estimation) ones. WDMR in this case can be seen as a kernel method, and the message is perhaps that the neural network machinery is too heavy an artillery for this application. For comparison we also computed a K-nearest neighbor model for the same data. Experiments showed that $K = 15$ neighbors gave the best prediction fit to validation data, and the result is also depicted in Fig. 23.9. It is better than a Neural Network, but worse than WDMR.

In System Identification it is common that the regression vector contains old outputs as in (23.19). Then it is not so natural to think of "unlabeled" regressor sets, since they would contain outputs = "labels", for other regressors. But WDMR still provides a good algorithms as we saw in the example.

Also, one may discuss how common it is in system identification that the regressors are constrained to a manifold. The input signal part of the regression vector should according to identification theory be "persistently exciting" which is precisely the opposite of being constrained. However, in many biological applications and in DAE (differential algebraic equation) modeling such structural constraints are frequently occurring.

Anyway, even in the absence of manifold constraints it may be a good idea to require smoothness in dense regressor regions as in (23.18). The Narendra-Li example showed the benefits of WDMR also in this more general context.

23.7 Conclusion

The purpose of this contribution was to explore what current techniques typical in machine learning has to offer for system identification problems. We outlined the ideas behind semi-supervised learning, that even regressors without corresponding outputs can improve the model fit, due to inherent constraints in the regressor space. We described a particular method, WDMR, which we believe to be novel, how to use both labeled and unlabeled regressors in regression problems. The usefulness of this method was illustrated on a number of examples, including some problems of a traditional non-linear system identification character. Even though WDMR compared favorably to more conventional methods for these problems, further analysis and comparisons must be made before a full evaluation of this approach can be made.

References

1. Bauwens, M., Ohlsson, H., Barbé, K., Beelaerts, V., Dehairs, F., Schoukens, J.: On climate reconstruction using bivalve shells: Three methods to interpret the chemical signature of a shell. In: 7th IFAC Symposium on Modelling and Control in Biomedical Systems (April 2009)
2. Belkin, M., Niyogi, P., Sindhwani, V.: Manifold regularization: A geometric framework for learning from labeled and unlabeled examples. J. Mach. Learn. Res. 7, 2399–2434 (2006)
3. Bengio, Y., Delalleau, O., Le Roux, N.: Label propagation and quadratic criterion. In: Chapelle, O., Schölkopf, B., Zien, A. (eds.) Semi-Supervised Learning, pp. 193–216. MIT Press, Cambridge (2006)
4. Chapelle, O., Schölkopf, B., Zien, A. (eds.): Semi-Supervised Learning. MIT Press, Cambridge (2006)
5. Cybenko, G.: Just-in-time learning and estimation. In: Bittanti, S., Picci, G. (eds.) Identification, Adaptation, Learning. The Science of Learning Models from data. NATO ASI Series, pp. 423–434. Springer, Heidelberg (1996)
6. de Ridder, D., Duin, R.: Locally linear embedding for classification. Tech Report, PH-2002-01, Pattern Recognition Group, Dept. of Imaging Science & Technology, Delft University of Technology, Delft, The Netherlands (2002)
7. de Ridder, D., Kouropteva, O., Okun, O., Pietikäinen, M.: Supervised locally linear embedding. In: Kaynak, O., Alpaydın, E., Oja, E., Xu, L. (eds.) ICANN 2003 and ICONIP 2003. LNCS, vol. 2714, pp. 333–341. Springer, Heidelberg (2003)

8. Gillikin, D.P., Dehairs, F., Lorrain, A., Steenmans, D., Baeyens, W., André, L.: Barium uptake into the shells of the common mussel (Mytilus edulis) and the potential for estuarine paleo-chemistry reconstruction. Geochimica et Cosmochimica Acta 70(2), 395–407 (2006)
9. Gillikin, D.P., Lorrain, A., Bouillon, S., Willenz, P., Dehairs, F.: Stable carbon isotopic composition of Mytilus edulis shells: relation to metabolism, salinity, $\delta^{13}C_{DIC}$ and phytoplankton. Organic Geochemistry 37(10), 1371–1382 (2006)
10. Goldberg, A.B., Zhu, X.: Seeing stars when there aren't many stars: Graph-based semi-supervised learning for sentiment categorization. In: HLT-NAACL 2006 Workshop on Textgraphs: Graph-based Algorithms for Natural Language Processing (2006)
11. Hastie, T., Tibshirani, R., Friedman, J.: The Elements of Statistical Learning: Data Mining, Inference, and Prediction. Springer Series in Statistics. Springer, New York (2001)
12. Hosmer Jr., D.W.: A comparison of iterative maximum likelihood estimates of the parameters of a mixture of two normal distributions under three different types of sample. Biometrics 29(4), 761–770 (1973)
13. Kohonen, T.: Self-Organizing Maps. Springer Series in Information Sciences, vol. 30. Springer, Berlin (1995)
14. Merz, C.J., St. Clair, D.C., Bond, W.E.: Semi-supervised adaptive resonance theory (smart2). In: International Joint Conference on Neural Networks, IJCNN, Jun. 1992, vol. 3, pp. 851–856 (1992)
15. Narendra, K.S., Li, S.-M.: Neural networks in control systems. In: Smolensky, P., Mozer, M.C., Rumelhard, D.E. (eds.) Mathematical Perspectives on Neural Networks, pp. 347–394. Lawrence Erlbaum Associates, Mahwah (1996)
16. Navaratnam, R., Fitzgibbon, A.W., Cipolla, R.: The joint manifold model for semi-supervised multi-valued regression. In: IEEE 11th International Conference on Computer Vision, ICCV 2007, October 2007, pp. 1–8 (2007)
17. Nigam, K., McCallum, A., Thrun, S., Mitchell, T.: Learning to classify text from labeled and unlabeled documents. In: AAAI '98/IAAI '98: Proceedings of the fifteenth national/tenth conference on Artificial intelligence/Innovative applications of artificial intelligence, pp. 792–799. AAAI Press, Menlo Park (1998)
18. Ohlsson, H., Roll, J., Ljung, L.: Manifold-constrained regressors in system identification. In: Proc. 47st IEEE Conference on Decision and Control, December 2008, pp. 1364–1369 (2008)
19. Putten, E.V., Dehairs, F., André, L., Baeyens, W.: Quantitative in situ microanalysis of minor and trace elements in biogenic calcite using infrared laser ablation - inductively coupled plasma mass spectrometry: a critical evaluation. Analytica Chimica Acta 378(1-3), 261–272 (1999)
20. Rahimi, A., Recht, B., Darrell, T.: Learning to transform time series with a few examples. IEEE Transactions on Pattern Analysis and Machine Intelligence 29(10), 1759–1775 (2007)
21. Roweis, S.T., Saul, L.K.: Nonlinear Dimensionality Reduction by Locally Linear Embedding. Science 290(5500), 2323–2326 (2000)
22. olde Scheper, T., Klinkenberg, D., Pennartz, C., van Pelt, J.: A Mathematical Model for the Intracellular Circadian Rhythm Generator. J. Neurosci. 19(1), 40–47 (1999)
23. Stenman, A.: Model on Demand: Algorithms, Analysis and Applications. Linköping Studies in science and Technology. Thesis No 571, Linköping University, SE-581 83 Linköping, Sweden (April 1999)
24. Tenenbaum, J.B., de Silva, V., Langford, J.C.: A global geometric framework for nonlinear dimensionality reduction. Science 290(5500), 2319–2323 (2000)

25. Wang, F., Zhang, C.: Label propagation through linear neighborhoods. IEEE Transactions on Knowledge and Data Engineering 20(1), 55–67 (2008)
26. Yang, X., Fu, H., Zha, H., Barlow, J.: Semi-supervised nonlinear dimensionality reduction. In: ICML '06: Proceedings of the 23rd international conference on Machine learning, pp. 1065–1072. ACM, New York (2006)
27. Zhao, L., Zhang, Z.: Supervised locally linear embedding with probability-based distance for classification. Comput. Math. Appl. 57(6), 919–926 (2009)
28. Zhu, X.: Semi-supervised learning literature survey. Technical Report 1530, Computer Sciences, University of Wisconsin-Madison (2005)

24

Path Integrals and Bézoutians for a Class of Infinite-Dimensional Systems*

Yutaka Yamamoto[1] and Jan C. Willems[2]

[1] Department of AACDS, Kyoto University, Kyoto 606-8501, Japan
[2] SISTA, Department of Electrical Engineering, K.U. Leuven, B-3001 Leuven, Belgium

Summary. There is an effective way of constructing a Lyapunov function without recourse to a state space construction. This is based upon an integral of special type called a path integral, and this approach is particularly suited for behavior theory. The theory successfully exhibits a deep connection between Lyapunov theory and Bézoutians. This paper extends the theory to a class of distributed parameter systems called *pseudorational*. A new construction of Lyapunov functions via an infinite-dimensional version of Bézoutians is presented. An example is given to illustrate the theory.

24.1 Introduction

It is our pleasure to dedicate this article to Chris Byrnes and Anders Lindquist on this special occasion. Their work has been a source of inspiration for us both, and their contributions to system and control theory are to be noted in many respects. In this article, we will deal with one of the classical aspects of control theory, in particular, stability theory and present results Lyapunov theory and Bézoutians.

It is well known and generally appreciated that Lyapunov theory plays a key role in stability theory of dynamical systems. The notion of Lyapunov functions defined on a state space is a central tool in both linear and nonlinear system theory.

It is perhaps less appreciated that there is an effective way of constructing a Lyapunov function and discussing stability without recourse to a state space formalism.

*This research is supported in part by the JSPS Grant-in-Aid for Scientific Research (B) No. 18360203, and Grant-in-Aid for Exploratory Research No. 1765138. The SISTA--SMC research program is supported by the Research Council KUL: GOA AMBioRICS, CoE EF/05/006 Optimization in Engineering (OPTEC), IOF-SCORES4CHEM, several PhD/postdoc and fellow grants; by the Flemish Government: FWO: PhD/postdoc grants, projects G.0452.04 (new quantum algorithms), G.0499.04 (Statistics), G.0211.05 (Nonlinear), G.0226.06 (cooperative systems and optimization), G.0321.06 (Tensors), G.0302.07 (SVM/Kernel, research communities (ICCoS, ANMMM, MLDM)); and IWT: PhD Grants, McKnow-E, Eureka-Flite; by the Belgian Federal Science Policy Office: IUAP P6/04 (DYSCO, Dynamical systems, control and optimization, 2007-2011); and by the EU: ERNSI.

X. Hu et al. (Eds.): Three Decades of Progress in Control Sciences, pp. 361–374, 2010.

This approach is based upon an integral of special type, called *path integral*. Given a dynamical system and the trajectories associated with it, an integral is said to be a path integral if its value is independent of the trajectory except that it depends only on the value of the integrand and its derivatives at the end points of integration.

This leads to an elegant theory for constructing Lyapunov functions for linear systems. This method was developed in late 60s by R.W. Brockett [1]. Recently, new light has been shed on this approach in the behavioral context [4, 5].

This approach provides a basis-free approach for the general theory of stability and the construction of Lyapunov functions, but was restricted to finite-dimensional systems. In [12] the behavioral approach to linear systems has been extended to a class of infinite-dimensional systems, in the context of pseudorational transfer functions. This setting provides a suitable framework for generalizing path integrals and related Lyapunov theory. This is the subject of the present article.

Let W be a transfer function, and A be its associated impulse response. Roughly speaking, W or A is said to be *pseudorational*, if A is expressible as the ratio of two distributions with compact support, with respect to convolution. To be precise, $A = p^{-1} * q$ for some distributions with compact support p, q, and the inverse is taken with respect to convolution.

Due to the compactness of the support of p, this allows for a bounded-time construction for a standard state space construction, and the fractional representation structure is particularly amenable to behavior theory. A typical example is $W(s) = 1/(se^s - 1)$. Two specific features are particularly relevant. One, the spectrum of the system is given by the zeros of $\hat{p}(s)$, i.e., the Laplace transform of p, and two, the stability is determined by the location of the spectrum. None of these properties hold generally for infinite-dimensional systems, and it is these properties that make the generalization of Bézoutians a rather pleasant and fruitful task.

We proceed as follows: After fixing notation, we give generalized notions of quadratic differential forms, and introduce path integrals. We then introduce pseudorational behaviors and path integrals along behaviors. We then discuss the relationships among stability, Lyapunov functions and Bézoutians.

24.2 Notation and Nomenclature

$\mathscr{C}^\infty(\mathbb{R},\mathbb{R})$ ($\mathscr{C}^\infty$ for short) is the space of C^∞ functions on $(-\infty,\infty)$. Similarly for $\mathscr{C}^\infty(\mathbb{R},\mathbb{R}^{\mathtt{q}})$ with higher dimensional codomains. $\mathscr{D}(\mathbb{R},\mathbb{R}^{\mathtt{q}})$ denote the space of $\mathbb{R}^{\mathtt{q}}$-valued C^∞ functions having compact support in $(-\infty,\infty)$. $\mathscr{D}'(\mathbb{R},\mathbb{R}^{\mathtt{q}})$ is its dual, the space of distributions. $\mathscr{D}'_+(\mathbb{R},\mathbb{R}^{\mathtt{q}})$ is the subspace of $\mathscr{D}'$ with support bounded on the left. $\mathscr{E}'(\mathbb{R},\mathbb{R}^{\mathtt{q}})$ denotes the space of distributions with compact support in $(-\infty,\infty)$. $\mathscr{E}'(\mathbb{R},\mathbb{R}^{\mathtt{q}})$ is a convolution algebra and acts on $\mathscr{C}^\infty(\mathbb{R},\mathbb{R})$ by the action: $p* : \mathscr{C}^\infty(\mathbb{R},\mathbb{R}) \to \mathscr{C}^\infty(\mathbb{R},\mathbb{R}) : w \mapsto p * w$. $\mathscr{C}^\infty(\mathbb{R},\mathbb{R})$ is a module over $\mathscr{E}'$ via this action. Similarly, $\mathscr{E}'(\mathbb{R}^2,\mathbb{R}^{\mathtt{q}})$ denotes the space of distributions in two variables having compact support in $\mathbb{R}^2$. For simplicity of notation, we may drop the range space $\mathbb{R}^{\mathtt{q}}$ and write $\mathscr{E}'(\mathbb{R})$, etc., when no confusion is likely,

A distribution α is said to be of *order at most m* if it can be extended as a continuous linear functional on the space of m-times continuously differentiable functions. Such a distribution is said to be of *finite order*. The largest number m, if one exists, is called the *order* of α ([2, 3]). The delta distribution δ_a ($a \in \mathbb{R}$) is of order zero, while its derivative δ_a' is of order one, etc. A distribution with compact support is known to be always of finite order ([2, 3]).

The Laplace transform of $p \in \mathscr{E}'(\mathbb{R}, \mathbb{R}^{\mathrm{q}})$ is defined by

$$\mathscr{L}[p](\zeta) = \hat{p}(\zeta) := \langle p, e^{-\zeta t} \rangle_t \tag{24.1}$$

where the action is taken with respect to t. Likewise, for $p \in \mathscr{E}'(\mathbb{R}^2, \mathbb{R}^{\mathrm{q}})$, its Laplace transform is defined by

$$\hat{p}(\zeta, \eta) := \langle p, e^{-(\zeta s + \eta t)} \rangle_{s,t} \tag{24.2}$$

where the distribution action is taken with respect to two variables s and t. For example, $\mathscr{L}[\delta_s'' \otimes \delta_t'] = \zeta^2 \cdot \eta$.

By the well-known Paley-Wiener theorem [2, 3], $\hat{p}(\zeta)$ is an entire function of exponential type satisfying the *Paley-Wiener estimate*

$$|\hat{p}(\zeta)| \le C(1 + |\zeta|)^r e^{a|\mathrm{Re}\,\zeta|} \tag{24.3}$$

for some $C, a \ge 0$ and a nonnegative integer r.

Likewise, for $p \in \mathscr{E}'(\mathbb{R}^2, \mathbb{R}^{\mathrm{q}})$, there exist $C, a \ge 0$ and a nonnegative integer r such that its Laplace transform

$$|\hat{p}(\zeta, \eta)| \le C(1 + |\zeta| + |\eta|)^r e^{a(|\mathrm{Re}\,\zeta| + |\mathrm{Re}\,\eta|)}. \tag{24.4}$$

This is also a sufficient condition for a function $\hat{p}(\cdot, \cdot)$ to be the Laplace transform of a distribution in $\mathscr{E}'(\mathbb{R}^2, \mathbb{R}^{\mathrm{q}})$. We denote by $\mathscr{PW}$ the class of functions satisfying the estimate above for some C, a, m. In other words, $\mathscr{PW} = \mathscr{L}[\mathscr{E}']$.

Other spaces, such as L^2, L^2_{loc} are all standard. For a vector space X, X^{n} and $X^{\mathrm{n} \times \mathrm{m}}$ denote, respectively, the spaces of n products of X and the space of $\mathrm{n} \times \mathrm{m}$ matrices with entries in X. When a specific dimension is immaterial, we will simply write $X^\bullet$ $X^{\bullet \times \bullet}$.

24.3 Quadratic Differential Forms

First consider the symmetric two-variable polynomial matrix $\Phi = \Phi^* \in \mathbb{R}^{q \times q}[\zeta, \eta]$, where $\Phi^*[\zeta, \eta] := \Phi^T[\eta, \zeta]$, with coefficient matrices as $\Phi(\zeta, \eta) = \sum_{k,\ell} \Phi_{k,\ell} \zeta^k \eta^\ell$. The *quadratic differential form* (QDF for short) $Q_\Phi : (\mathscr{C}^\infty)^q \to (\mathscr{C}^\infty)^q$ is defined by

$$Q_\Phi(w) := \sum_{k,\ell} \left(\frac{d^k}{dt^k} \overline{w} \right)^T \Phi_{k,\ell} \left(\frac{d^\ell}{dt^\ell} w \right).$$

For example, $\Phi = (\zeta + \eta)/2$ yields the QDF $Q_\Phi = w(dw/dt)$. [To be precise, $[(\overline{w}(dw/dt) + (d\overline{w}/dt)w)]/2$ but when everything is real valued, we consider real-valued forms only, and consider QDFs with $\overline{w} = w$.]

Observing this example, we notice that we can view Φ as the Laplace transform of two-variable distributions $(\delta_s' \otimes \delta_t + \delta_s \otimes \delta_t')/2$ where δ_s' denotes the derivative of the delta distribution in the variable s, and likewise for δ_t', δ_s, δ_t, etc.; $\alpha_s \otimes \beta_t$ denotes the tensor product of two distributions α and β. (In fact, $\mathscr{L}[\delta_s'] = \zeta$, and $\mathscr{L}[\delta_t'] = \eta$.)

We can easily extend the definition above to tensor products of distributions in variables s and t, and then to distributions $\Phi \in \mathscr{E}'(\mathbb{R}^2)$. Indeed, if $\Phi = \alpha_s \otimes \beta_t$, $\alpha, \beta \in \mathscr{E}'(\mathbb{R})$

$$Q_\Phi(w) = (\overline{w} * \alpha) \cdot (\beta * w),$$

and extend linearly for the elements of form $\sum_{k,\ell} \alpha_s^i \otimes \beta_t^j$. Since $\mathscr{E}'(\mathbb{R}) \otimes \mathscr{E}'(\mathbb{R})$ is dense in $\mathscr{E}'(\mathbb{R}^2)$ (cf., [3]), we can extend this definition to the whole of $\mathscr{E}'(\mathbb{R}^2)$. Finally, for the matrix case, we apply the definition above to each entries.

In short, given $\Phi \in \mathscr{E}'(\mathbb{R}^2, \mathbb{R}^{\mathtt{q}})$,

$$\Phi(v, w) = \overline{v}_s * \Phi * w_t \tag{24.5}$$

where the convolution from the left is taken with respect to the variable s while that on the right is taken with respect to t. For example, $\overline{v} * (\sum \alpha_k \otimes \beta_\ell) * w = \sum_{k,\ell} (\overline{v} * \alpha_k)_s (\beta_\ell * w)_t$. This gives a bilinear mapping from $(\mathscr{C}^\infty)$ to $(\mathscr{C}^\infty)$. Then the *quadratic differential form* Q_Φ associated with Φ is defined by

$$Q_\Phi(w) := \Phi(w, w) = \overline{w}_s * \Phi * w_t|_{s=t}. \tag{24.6}$$

Given $\Phi \in \mathscr{E}'(\mathbb{R}^2)^{\mathtt{q} \times \mathtt{q}}$ such that $\Phi^* = \Phi$, we define the *quadratic differential form* $Q_\Phi : (\mathscr{C}^\infty)^{\mathtt{q}} \to (\mathscr{C}^\infty)^{\mathtt{q}}$ associated with Φ by

$$Q_\Phi(w) := \Phi(w, w) = (\overline{w}_s * \Phi * w_t)|_{s=t} \tag{24.7}$$

as a function of a single variable $t \in \mathbb{R}$.

Example 24.3.1. Define $\Phi := (1/2)[\delta_s' \otimes \delta_t + \delta_s \otimes \delta_t']$. Then $\Phi(v, w) = (1/2)[(dv/ds)(s) \cdot w(t) + v(s) \cdot (dw/dt)(t)]$ and $Q_\Phi(w) = (1/2)[(dw/dt)(t) \cdot w(t) + w(t) \cdot (dw/dt)(t)]$.

Example 24.3.2. For $\Phi := \delta''_{-1} \otimes \delta'_{-1}$,

$$Q_\Phi(w) = \Phi(w, w) = \frac{d^2 w}{dt^2}(t+1) \cdot \frac{dw}{dt}(t+1).$$

Basic Operations on $\mathscr{E}'(\mathbb{R}^2, \mathbb{R}^{\mathtt{q}})$ and $\mathscr{PW}$

We generalize some fundamental operations polynomials matrices to the present context following [4]. Let $P \in (\mathscr{E}'(\mathbb{R}^2))^{\mathtt{n}_1 \times \mathtt{n}_2}$. Define $P^\sim \in (\mathscr{E}')^{n_2 \times n_1}$ by

$$P^\sim := (\overline{\check{P}})^T \tag{24.8}$$

where $\check{\alpha}$ is defined by

$$\langle \check{\alpha}, \phi \rangle := \langle \alpha, \phi(-\cdot) \rangle, \alpha \in \mathscr{E}', \phi \in \mathscr{C}^\infty(\mathbb{R}, \mathbb{R}).$$

Hence for $\hat{P} \in (\mathscr{PW})^{n_1 \times n_2}$, $\hat{P}^{\sim}(\zeta) = (\check{P}^T)^{\wedge} = (P^{\sim})^{\wedge}(\zeta) = \hat{P}^T(-\zeta)$.

For $\hat{P} \in \mathscr{PW}^{\bullet\times\bullet}[\zeta, \eta]$, $\hat{P}^*(\zeta, \eta) := \hat{P}^T(\eta, \zeta)$. Also,

$$\hat{P}^\bullet(\zeta, \eta) := (\zeta + \eta)\hat{P}(\zeta, \eta). \tag{24.9}$$

In the (s,t)-domain, this corresponds to

$$P^\bullet = (\delta_s' * P) + (\delta_t' * P) = \left(\frac{\partial}{\partial s} + \frac{\partial}{\partial t}\right). \tag{24.10}$$

The operator $\partial : \mathscr{PW} \to \mathscr{PW}$ is defined by

$$\partial \hat{P}(\xi) := \hat{P}(-\xi, \xi). \tag{24.11}$$

For an element P of type $P = \alpha_s \otimes \beta_t$, this means

$$\partial P = \check{\alpha}_t \otimes \beta_t.$$

The formula for the general case is obtained by extending this linearly.

We note the following lemma for the expression $\hat{\Phi}(\zeta, \eta)/(\zeta + \eta)$ to belong to the class $\mathscr{PW}$:

Lemma 24.3.1. *Let $f \in (\mathscr{PW})^{\bullet\times\bullet}$. $f(\zeta, \eta)/(\zeta + \eta)$ belongs to the class $\mathscr{PW}$ if and only if $\partial f = 0$, i.e., $f(-\xi, \xi) = 0$.*

Proof. Omitted. See [13]. □

The following lemma is a direct consequence of the definition of $\Psi^\bullet$:

Lemma 24.3.2. *For $\Psi \in \mathscr{E}'(\mathbb{R}^2, \mathbb{R}^q)^{\bullet\times\bullet}$,*

$$\frac{d}{dt} Q_\Psi = Q_{\Psi^\bullet}.$$

Proof. Consider $\alpha_s \otimes \beta_t$, and consider the action $w \mapsto (w * \alpha) \cdot (\beta * w)$. According to (24.10), differentiation of this yields $(w * (d\alpha/ds)) \cdot (\beta * w) + (w * \alpha) \cdot ((d\beta/dt) * w)|_{s=t} = (w * \delta_s' * \alpha \cdot (\beta * w) + (w * \alpha) \cdot ((\delta_t' * \beta) * w)|_{s=t} = Q_{\Psi^\bullet}(w)$. Extend linearly and then also extend continuously to complete the proof. □

24.4 Path Integrals

The integral

$$\int_{t_1}^{t_2} Q_\Phi(w) dt \tag{24.12}$$

(or briefly $\int Q_\Phi$) is said to be *independent of path*, or simply a *path integral* if it depends only on the values taken on by w and its derivatives at end points t_1 and t_2 (*but not on the intermediate trajectories between them*).

The following theorem gives equivalent conditions for Φ to give rise to a path integral.

Theorem 24.4.1. *Let $\Phi \in \mathscr{E}'(\mathbb{R}^2)^{\mathtt{q}\times\mathtt{q}}$, and Q_Φ the quadratic differential form associated with Φ. The following conditions are equivalent:*

1. *$\int Q_\Phi$ is a path integral;*
2. *$\partial\Phi = 0$;*
3. *$\int_{-\infty}^{\infty} Q_\Phi(w)dt = 0$ for all $w \in \mathscr{D}(\mathbb{R}, \mathbb{R}^{\mathtt{q}})$;*
4. *the expression $\hat{\Phi}(\zeta,\eta)/(\zeta+\eta)$ belongs to the class $\mathscr{PW}$.*
5. *there exists a two-variable matrix $\Psi \in \mathscr{E}'(\mathbb{R}^2)^{\mathtt{q}\times\mathtt{q}}$ that defines a Hermitian bilinear form on $(\mathscr{C}^\infty)^{\mathtt{q}} \otimes (\mathscr{C}^\infty)^{\mathtt{q}}$ such that*

$$\frac{d}{dt} Q_\Psi(w) = Q_\Phi(w) \tag{24.13}$$

for all $w \in \mathscr{C}^\infty(\mathbb{R}, \mathbb{R}^{\mathtt{q}})$.

Proof. This is essentially the same as that given in [4, 5], and hence omitted. The key facts are Parseval's identity, Lemmas 24.3.1 and 24.3.2. □

24.5 Pseudorational Behaviors

Let us review some basic facts on pseudorational behaviors [12].

Definition 24.5.1. Let R be a $\mathtt{p}\times\mathtt{w}$ matrix ($\mathtt{w} \geq \mathtt{p}$) with entries in $\mathscr{E}'$. It is said to be *pseudorational* if there exists a $\mathtt{p}\times\mathtt{p}$ submatrix P such that

1. $P^{-1} \in \mathscr{D}'_+(\mathbb{R})$ exists with respect to convolution;
2. $\mathrm{ord}(\det P^{-1}) = -\mathrm{ord}(\det P)$, where $\mathrm{ord}\,\psi$ denotes the order of a distribution ψ [2, 3] (for a definition, see the Appendix).

Definition 24.5.2. Let R be pseudorational as defined above. The *behavior* $\mathscr{B}$ defined by R is given by

$$\mathscr{B} := \{w \in \mathscr{C}^\infty(\mathbb{R}, \mathbb{R}^{\mathtt{q}}) : R * w = 0\} \tag{24.14}$$

The convolution $R * w$ is taken in the sense of distributions. Since R has compact support, this convolution is always well defined [2].

Remark 24.5.1. We here took $\mathscr{C}^\infty(\mathbb{R}, \mathbb{R}^{\mathtt{q}})$ as the signal space in place of $L^2_{loc}(\mathbb{R}, \mathbb{R}^{\mathtt{q}})$ in [12], but the basic structure remains intact.

A state space formalism is possible for this class and it yields various nice properties as follows:

Suppose, without loss of generality, that R is partitioned as $R = \begin{bmatrix} P & Q \end{bmatrix}$ such that P satisfies the invertibility condition of Definition 24.5.1, i.e., we consider the kernel representation

$$P * y + Q * u = 0 \tag{24.15}$$

where $w := \begin{bmatrix} y & u \end{bmatrix}^T$ is partitioned conformably with the sizes of P and Q.

A nice consequence of pseudorationality is that this space X is always a closed subspace of the following more tractable space X^P:

$$X^P := \{x \in (L^2_{[0,\infty)})^{\mathrm{p}} \mid P * x|_{[0,\infty)} = 0\}, \tag{24.16}$$

and it is possible to give a realization using X^P as a state space. The state transition is generated by the left shift semigroup:

$$(\sigma_\tau x)(t) := x(t+\tau)$$

and its infinitesimal generator A determines the spectrum of the system ([6]). We have the following facts concerning the spectrum, stability, and coprimeness of the representation $\begin{bmatrix} P & Q \end{bmatrix}$ ([6, 8, 9, 10]):

Theorem 24.5.1.

1. *The spectrum $\sigma(A)$ is given by*

$$\sigma(A) = \{\lambda \mid \det \hat{P}(\lambda) = 0\}. \tag{24.17}$$

   Furthermore, every $\lambda \in \sigma(A)$ is an eigenvalue with finite multiplicity. The corresponding eigenfunction for $\lambda \in \sigma(A)$ is given by $e^{\lambda t} v$ where $\hat{P}(\lambda)v = 0$. Similarly for generalized eigenfunctions such as $te^{\lambda t} v'$.
2. *The semigroup σ_t is exponentially stable, i.e., satisfies for some $C, \beta > 0$*

$$\|\sigma_t\| \le Ce^{-\beta t}, \quad t \ge 0,$$

   if and only if there exists $\rho > 0$ such that

$$\sup\{\mathrm{Re}\,\lambda : \det \hat{P}(\lambda) = 0\} \le -\rho.$$

24.6 Path Integrals along a Behavior

Generalizing the results of Section 24.4 on path integrals in the unconstrained case, we now study path integrals *along a behavior $\mathscr{B}$*.

Definition 24.6.1. Let $\mathscr{B}$ be the behavior (24.14) with pseudorational R. The integral $\int Q_\Phi$ is said to be *independent of path* or a *path integral along $\mathscr{B}$* if the path independence condition holds for all $w_1, w_2 \in \mathscr{B}$.

Let $\mathscr{B}$ be as above, i.e.,

$$\mathscr{B} := \ker R = \{w \in \mathscr{C}^\infty(\mathbb{R}, \mathbb{R}^{\mathrm{q}}) : R * w = 0\}. \tag{24.18}$$

We assume that $\mathscr{B}$ also admits an *image representation*, i.e., there exists M with entries in $\mathscr{E}'(\mathbb{R}, \mathbb{R}^{\mathrm{q}})$ such that

$$\mathscr{B} = \mathrm{im}\ M = \{w = M * \varphi \in \mathscr{C}^\infty(\mathbb{R}, \mathbb{R}^{\mathrm{q}}) : \varphi \in \mathscr{C}^\infty(\mathbb{R}, \mathbb{R}^{\mathrm{q}})\}. \tag{24.19}$$

This implies that $\mathscr{B}$ is controllable [12]. In fact, for a polynomial R, controllability of $\mathscr{B}$ is also sufficient for the existence of an image representation, but in the present situation, it is not fully known. A partial necessary and sufficient result for the scalar case is given in [12].

We then have the following theorem.

Theorem 24.6.1. *Let $\mathscr{B}$ be a behavior defined by a pseudorational R, and suppose that $\mathscr{B}$ admits an image representation (24.18) for some M. Let $\Phi \in \mathscr{E}'(\mathbb{R}^2)^{q\times q}$, and Q_Φ the quadratic differential form associated with Φ. Then the following conditions are equivalent:*

1. *$\int Q_\Phi$ is a path integral along $\mathscr{B}$;*
2. *there exists $\Psi = \Psi^* \in \mathscr{PW}^{q\times q}[\zeta,\eta]$ such that*

$$\frac{d}{dt}Q_\Psi(w) = Q_\Phi(w) \tag{24.20}$$

   for all $w \in \mathscr{B}$;
3. *$\int Q_{\Phi'}$ is a path integral where Φ' is defined by $\Phi'(\zeta,\eta) := M^T(\zeta)\Phi(\zeta,\eta)M(\eta)$;*
4. *$\partial\Phi' = 0$;*
5. *there exists $\Psi' = (\Psi')^\bullet = \mathscr{PW}^{q\times q}[\zeta,\eta]$ such that*

$$\frac{d}{dt}Q_{\Psi'}(\ell) = Q_{\Phi'}(\ell)$$

   for all $\ell \in \mathscr{C}^\infty$, i.e., $\Psi'^\bullet = \Phi'$.

Proof. The equivalence of 3, 4 and 5 is a direct consequence of the image representation $\mathscr{B} = M * \mathscr{C}^\infty$ and Theorem 24.4.1. The crux here is that the image representation reduces these statements on $w \in \mathscr{B}$ to the unconstrained ℓ via $w = M * \ell$. The equivalence of 2 and 5 is also an easy consequence of the image representation: for every $w \in \mathscr{B}$ there exists $\ell \in \mathscr{C}^\infty$ such that $w = M * \ell$.

Now the implications $2 \Rightarrow 1$ and $1 \Rightarrow 4$ are obvious. □

We also have the following proposition:

Proposition 24.6.1. *Let $\mathscr{B}$ be as above, admitting an image representation $\mathscr{B} = \mathrm{im} M*$. Suppose that the extended Lyapunov equation*

$$X^* * R + R^* * X = \partial\Phi \tag{24.21}$$

has a solution $X \in \mathscr{E}'(\mathbb{R}^2)^{q\times q}$. Then $\int Q_\Phi$ is a path integral.

Proof. Omitted. See [13]. □

24.7 Stability

Let $R \in (\mathscr{E}'(\mathbb{R}))^{\mathrm{q}\times\mathrm{q}}$, and consider the autonomous behavior $\mathscr{B} = \{w : R * w = 0\}$.

The following lemma claims that for pseudorational R, the stability is determined by location of the zeros of $\det \hat{R}$.

Lemma 24.7.1. *The behavior $\mathscr{B}$ is exponentially stable if and only if*

$$\sup\{\mathrm{Re}\,\lambda : \det \hat{R}(\lambda) = 0\} < 0. \tag{24.22}$$

Proof. (Outline) Without loss of generality, we can shift R to left so that $\mathrm{supp} R \subset (-\infty, 0]$. Consider $\overline{R} := \begin{bmatrix} R & I \end{bmatrix}$, and define

$$\overline{\mathscr{B}} := \{\begin{bmatrix} y & u \end{bmatrix}^T : \overline{R} * \begin{bmatrix} y & u \end{bmatrix}^T = 0\}.$$

Then $\mathscr{B} \subset \pi_1(\overline{\mathscr{B}})$, where π_1 denotes the projection to the first component. Hence $\mathscr{B}$ is asymptotically stable if every element of $\pi_1(\overline{\mathscr{B}})$ decays to zero asymptotically. Now note that $\overline{\mathscr{B}}$ is trivially controllable, every trajectory $w \in \mathscr{B}$ can be concatenated with zero trajectory as

$$w'(t) = \begin{cases} w(t), & t \geq 0 \\ 0, & t \leq -T \end{cases}$$

for some $T > 0$. Then $\pi_1(w')$ clearly belongs to X^R because $R * w = 0$. According to Theorem 24.5.1, $w(t)$ goes to zero as $t \to \infty$, and this decay is exponential. This proves the claim. □

24.8 Lyapunov Stability

A characteristic feature in stability for the class of pseudorational transfer functions is that asymptotic stability is determined by the location of poles, i.e., zeros of $\det \hat{R}(\zeta)$. Indeed, as we have seen in Lemma 24.7.1, the behavior

$$\mathscr{B} = \{w : R * w = 0\},$$

is exponentially stable if and only if $\sup\{\mathrm{Re}\,\lambda : \det \hat{R}(\lambda) = 0\} < 0$, and this is determined how each characteristic solution $e^{\lambda t} a$, $a \in \mathbb{C}^{\mathrm{q}}$ ($\det \hat{R}(\lambda) = 0$), behaves. This plays a crucial role in discussing stability in the Lyapunov theory. We start with the following lemma which tells us how $p \in \mathscr{E}'(\mathbb{R}, \mathbb{R}^{\mathrm{q}})$ acts on $e^{\lambda t}$ via convolution:

Lemma 24.8.1. *For $p \in \mathscr{E}'(\mathbb{R}, \mathbb{R}^{\mathrm{q}})$, $p * e^{\lambda t} = \hat{p}(\lambda) e^{\lambda t}$.*

Proof. This is obvious for elements of type $\sum \alpha_i \delta_{t_i}$. Since such elements form a dense subspace of $\mathscr{E}'$ ([2]), the result readily follows. □

We now give some preliminary notions on positivity (resp. negativity).

Definition 24.8.1. The QDF Q_Φ induced by Φ is said to be *nonnegative* (denoted $Q_\Phi \geq 0$) if $Q_\Phi(w) \geq 0$ for all $w \in \mathscr{C}^\infty(\mathbb{R}, \mathbb{R}^{\mathtt{q}})$, and *positive* (denoted $Q_\Phi(w) > 0$) if it is nonnegative and $Q_\Phi(w) = 0$ implies $w = 0$.

Let $\mathscr{B} = \{w : R * w = 0\}$ be a pseudorational behavior. The QDF Q_Φ induced by Φ is said to be *$\mathscr{B}$-nonnegative* (denoted $Q_\Phi \overset{\mathscr{B}}{\geq} 0$) if $Q_\Phi(w) \geq 0$ for all $w \in \mathscr{B}$, and *$\mathscr{B}$-positive* (denoted $Q_\Phi(w) \overset{\mathscr{B}}{>} 0$) if it is $\mathscr{B}$-nonnegative and if $Q_\Phi(w)$ and $w \in \mathscr{B}$ imply $w = 0$. $\mathscr{B}$-nonpositivity and $\mathscr{B}$-negativity are defined if the respective conditions hold for $-Q_\Phi$.

We say that Q_Φ *weakly strictly positive along* $\mathscr{B}$ if

- Q_Φ is $\mathscr{B}$-positive; and
- for every $\gamma > 0$ there exists c_γ such that $\overline{a}^T \hat{\Phi}(\overline{\lambda}, \lambda) a \geq c_\gamma \|a\|^2$ for all λ with $\hat{p}(\lambda) = 0$, $\mathrm{Re}\,\lambda \geq -\gamma$ and $a \in \mathbb{C}^q$.

Similarly, for *weakly strict negativity along* $\mathscr{B}$.

For a polynomial $\hat{\Phi}$, $\mathscr{B}$-positivity clearly implies the second condition. However, for pseudorational behaviors, this may not be true. Note that we require the above estimate only for the eigenvalues λ, whence the term "weakly".

The following theorem is a consequence of Lemma 24.7.1 that asserts asymptotic stability can be concluded from the location of the spectrum.

Theorem 24.8.1. *Let $\mathscr{B} = \{w : R * w = 0\}$ be a pseudorational behavior. Then $\mathscr{B}$ is asymptotically stable if there exists $\Psi = \Psi^* \in \mathscr{E}'(\mathbb{R}^2)^{\mathtt{q}\times\mathtt{q}}$ whose elements are measures (i.e., distributions of order 0) such that Q_Ψ is weakly strictly positive along $\mathscr{B}$ and $\Psi^\bullet$ weakly strictly negative along $\mathscr{B}$.*

Proof. Let $\exp_\lambda : \mathbb{R} \to \mathbb{C} : t \mapsto e^{\lambda t}$ denote the exponential function with exponent parameter λ. Lemma 24.7.1 implies that we can deduce stability of $\mathscr{B}$ if there exists $c > 0$ such that $a\exp_\lambda(\cdot) \in \mathscr{B}$, $a \neq 0$ implies $\mathrm{Re}\,\lambda \leq -c < 0$. Now take any $\gamma > 0$ and consider $a\exp_\lambda(\cdot) \in \mathscr{B}$ with $\mathrm{Re}\,\lambda \geq -\gamma$. Then

$$Q_\Psi(a\exp_\lambda) = [\overline{a}^T \hat{\Psi}(\overline{\lambda}, \lambda) a](\exp_{2\mathrm{Re}\,\lambda}(\cdot)),$$

and

$$Q_{\Psi^\bullet}(a\exp_\lambda) = (2\mathrm{Re}\,\lambda)[\overline{a}^T \hat{\Psi}(\overline{\lambda}, \lambda) a](\exp_{2\mathrm{Re}\,\lambda}(\cdot)).$$

Hence the weak strict positivity of $Q_\Psi(w)$ implies $a^T \hat{\Psi}(\overline{\lambda}, \lambda) a \geq c_\gamma \|a\|^2 \geq 0$. Also since the elements of $\hat{\Psi}$ are measures, $a^T \hat{\Psi}(\overline{\lambda}, \lambda) a \leq \beta \|a\|^2$. On the other hand, weak strict negativity of $Q_{\Psi^\bullet}$ implies

$$Q_{\Psi^\bullet}(a\exp_\lambda(\cdot)) \leq -\rho \|a\|^2.$$

Combining these, we obtain

$$(2\mathrm{Re}\,\lambda) \cdot c \|a\|^2 \leq -\rho \|a\|^2$$

and hence $\mathrm{Re}\,\lambda \leq -\rho/(2c) < 0$ for such λ. Since other λ's satisfying $\hat{p}(\lambda) = 0$ satisfy $\mathrm{Re}\,\lambda < -\gamma$, this yields exponential stability of $\mathscr{B}$. □

Remark 24.8.1. In the theorem above, the condition that the elements of Ψ be measures is necessary to guarantee the boundedness of $\Psi(\lambda,\lambda)$. However, for the single variable case, one can reduce the general case to this case. See the next section.

Proposition 24.8.1. *Under the hypotheses of Theorem 24.8.1,*

$$Q_\Psi(w)(0) = -\int_0^\infty Q_{\Psi^\bullet}(w)dt \tag{24.23}$$

Proof. Note that

$$Q_\Psi(w)(t) - Q_\Psi(w)(0) = \int_0^t Q_{\Psi^\bullet}(w)dt.$$

By Theorem 24.8.1, $Q_\Psi(w)(t) \to 0$ as $t \to \infty$, the result follows. □

24.9 The Bézoutian

We have seen that exponential stability can be deduced from the existence of a suitable positive definite quadratic form Ψ that works as a Lyapunov function. The question then hinges upon how one can find such a Ψ. The objective of this section is to show that for the single-variable case, the Bézoutian gives a universal construction for obtaining a Lyapunov function.

In this section we confine ourselves to the case $q = 1$, that is, given $p \in \mathscr{E}'(\mathbb{R},\mathbb{R}^q)$, we consider the behavior

$$\mathscr{B} = \{w : p * w = 0\}.$$

Define the Bézoutian $b(\zeta,\eta)$ by

$$b(\zeta,\eta) := \frac{p(\zeta)p(\eta) - p(-\zeta)p(-\eta)}{\zeta+\eta}. \tag{24.24}$$

Note that this expression belongs to the class $\mathscr{PW}[\zeta,\eta]$, and hence its inverse Laplace transform is a distribution having compact support. Let us further assume that p is a measure, i.e., distribution of order 0. If not, $\hat{p}(s)$ possess (stable) zeros, and we can reduce $\hat{p}(s)$ to a measure by extracting such zeros. For details, see [7].

Our main result is the following theorem:

Theorem 24.9.1. *Suppose that $p \in \mathscr{E}'$ is a measure. The following conditions are equivalent:*

1. *$\mathscr{B} = \{w : p * w = 0\}$ is exponentially stable;*
2. *there exists $\rho > 0$ such that $\sup\{\lambda : \hat{p}(\lambda) = 0\} \le -\rho$;*
3. *$Q_b \ge 0$ and the pair $(p, p^{\sim})$ is coprime in the following sense: there exists $\phi,\psi \subset \mathscr{E}'$ such that*
$$p * \phi + p^{\sim} * \psi = \delta \tag{24.25}$$
4. *Q_b is weakly strictly positive definite, and $Q_{\overset{\bullet}{b}}$ is weakly strictly negative definite.*

Proof. The equivalence of 1 and 2 are already shown.

Note first that for $w \in \mathscr{B}$, we have

$$\frac{d}{dt}Q_b(w) = |p * w|^2 - |p^{\sim} * w|^2 = -|p^{\sim} * w|^2 \tag{24.26}$$

because $p * w = 0$.

1 ⇒ 3 Since $\mathscr{B}$ is asymptotically stable, we have from (24.26)

$$Q_b(w)(0) = \int_0^\infty |p^{\sim} * w)|^2 dt \geq 0.$$

Now exponential stability implies that $\sup\{\lambda : \hat{p}(\lambda) = 0\} \leq -\rho$ for some $\rho > 0$ and also

$$\left|\frac{1}{\hat{p}(\zeta)}\right| \leq C, \quad |\zeta| \geq 0. \tag{24.27}$$

This implies that for λ_n, $n = 1, 2, \ldots$ with $\hat{p}(\lambda_n) = 0$, $|\hat{p}^{\sim}(\lambda_n)| = |\hat{p}(-\lambda_n)| \geq C$. Then by the coprimeness condition [12, Theorem 4.1], $(p, p^{\sim})$ satisfies the Bézout identity (24.25).

3 ⇒ 1 and 4 By (24.26), we have for $w \in \mathscr{B}$,

$$\frac{d}{dt}Q_b(w) \leq 0.$$

We show that $(d/dt)Q_b(w) < 0$. Suppose that $(d/dt)Q_b(w) = 0$ for some w, i.e., $p^{\sim} * w = 0$ according to (24.26). Then $w \in \mathscr{B} \cap \mathscr{B}_{p^{\sim}}$, where

$$\mathscr{B}_{p^{\sim}} := \{w \in \mathscr{C}^\infty(\mathbb{R}, \mathbb{R}) : p^{\sim} * w = 0\}.$$

Since $(p, p^{\sim})$ satisfies (24.25), $\mathscr{B} \cap \mathscr{B}_{p^{\sim}} = 0$ because for $w \in \mathscr{B} \cap \mathscr{B}_{p^{\sim}}$

$$w = (\phi * p + \psi * p^{\sim}) * w = 0.$$

Hence $(d/dt)Q_b(w) < 0$. Again by [12, Theorem 4.1] and (24.25) there exists $c > 0$ such that

$$|\hat{p}^{\sim}(\lambda_n)| \geq c > 0$$

for all λ_n with $\hat{p}(\lambda_n) = 0$. Then

$$-|\hat{p}^{\sim}(\lambda_n)|^2 = -|\hat{p}(-\lambda_n)|^2 \leq -c^2. \tag{24.28}$$

Hence $Q_{\overset{\bullet}{b}}$ is weakly strictly negative definite. Furthermore,

$$Q_b(\exp_{\lambda_n}(\cdot)) = \frac{-\hat{p}(-\lambda_n)\hat{p}(-\overline{\lambda}_n)}{2\,\mathrm{Re}\,\lambda_n}\exp_{2\,\mathrm{Re}\,\lambda_n}(\cdot)$$

Now take any $\gamma > 0$, and suppose $\mathrm{Re}\,\lambda_n \geq -\gamma$. Then by (24.28)

$$\frac{-\hat{p}(-\lambda_n)\hat{p}(-\overline{\lambda}_n)}{2\,\mathrm{Re}\,\lambda_n} \geq \frac{|\hat{p}(-\lambda_n)|^2}{2\gamma} \geq \frac{c^2}{2\gamma} > 0.$$

Hence Q_b is weakly strictly positive definite. Hence by Theorem 24.8.1, $\mathscr{B}$ is asymptotically stable. This proof also shows that 3 implies 4.

4 ⇒ 1 This is already proved in Theorem 24.8.1. □

Remark 24.9.1. Condition 4 above may appear too strong, given its counterpart in the finite-dimensional case. Indeed, in the finite-dimensional context, one needs to require only the $\mathscr{B}$-positivity of Q_b, and coprimeness of $(p, p^{\sim})$ follows (and stability also). In the present context, however, there can be a case in which there are infinitely many λ_n's that approach the imaginary axis as $n \to \infty$, and this situation is not well controlled by the positivity of Q_b. An exception is the case of retarded delay systems, or its generalized version of class $\mathscr{R}$ where it is guaranteed that there are always only a finite number of poles to the right of any vertical axis parallel to the imaginary axis, as we see below.

Corollary 24.9.1. *Let p be pseudorational, and suppose that p belong to the class* $\mathbb{R}$ *as defined in [11]. Then* $\mathscr{B}$ *is exponentially stable if* Q_b *is* $\mathscr{B}$*-positive.*

This is obvious since there are only finitely many zeros of $\hat{p}(\zeta)$ in $\{\zeta : -\rho < \operatorname{Re}\zeta < 0\}$ for arbitrary ρ. A simplified proof of Theorem 24.8.1 without requiring uniformity works, just as in the finite-dimensional case. Note that we do not have to require weak strict positivity.

Example 24.9.1. Let $p := \delta_{-1} - \alpha\delta$, with $\alpha \in \mathbb{R}$, $|\alpha| < 1$. Then $\hat{p}(\zeta)e^{\zeta} - \alpha$. An easy calculation yields

$$\begin{aligned} b(\zeta,\eta) &:= \frac{\hat{p}(\zeta)\hat{p}(\eta) - \hat{p}(-\zeta)\hat{p}(-\eta)}{\zeta+\eta} \\ &= \frac{e^{\zeta+\eta} - e^{-\zeta-\eta} - \alpha(e^{\zeta} - e^{-\zeta} + e^{\eta} - e^{-\eta})}{\zeta+\eta}. \end{aligned}$$

Clearly the numerator vanishes for $\zeta + \eta = 0$. Let λ be a zero of $\hat{p}(\zeta)$. Then for $\zeta = \lambda$ and $\eta = \overline{\lambda}$, the numerator becomes

$$-\hat{p}(-\lambda)\hat{p}(-\overline{\lambda}) = -\alpha^2 + (2\operatorname{Re} e^{\lambda}) - e^{2\operatorname{Re}\lambda}.$$

It is easy to see that this is negative, and hence b is weakly strictly positive along $\mathscr{B}$ if and only if $\operatorname{Re}\lambda < 0$.

References

1. Brockett, R.W.: Finite Dimensional Linear Systems. Wiley, New York (1970)
2. Schwartz, L.: Théorie des Distribution. Hermann, Paris (1966)
3. Treves, F.: Topological Vector Spaces, Distributions and Kernels. Academic Press, London (1967)
4. Willems, J.C.: Path integrals and stability. In: Baillieul, J., Willems, J.C. (eds.) Mathematical Control Theory, Festschrift on the occasion of 60th birthday of Roger Brockett, pp. 1–32. Springer, Heidelberg (1999)
5. Willems, J.C., Trentelman, H.L.: On quadratic differential forms. SIAM J. Control & Optimization 36, 1703–1749 (1998)

6. Yamamoto, Y.: Pseudo-rational input/output maps and their realizations: a fractional representation approach to infinite-dimensional systems. SIAM J. Control & Optimiz. 26, 1415–1430 (1988)
7. Yamamoto, Y., Hara, S.: Relationships between Internal and External Stability for Infinite-Dimensional Systems with Applications to a Servo Problem. IEEE Transactions on Automatic Control 33(11), 1044–1052 (1988)
8. Yamamoto, Y.: Reachability of a class of infinite-dimensional linear systems: an external approach with applications to general neutral systems. SIAM J. Control & Optimiz. 27, 217–234 (1989)
9. Yamamoto, Y.: Equivalence of internal and external stability for a class of distributed systems. Math. Control, Signals and Systems 4, 391–409 (1991)
10. Yamamoto, Y.: Pseudorational transfer functions—A survey of a class of infinite-dimensional systems. In: Proc. 46th IEEE CDC 2007, New Orleans, pp. 848–853 (2007)
11. Yamamoto, Y., Hara, S.: Internal and external stability and robust stability condition for a class of infinite-dimensional systems. Automatica 28, 81–93 (1992)
12. Yamamoto, Y., Willems, J.C.: Behavioral controllability and coprimeness for a class of infinite-dimensional systems. In: Proc. 47th IEEE CDC 2008, Cancun, pp. 1513–1518 (2008)
13. Yamamoto, Y., Willems, J.C.: Path integrals and Bézoutians for pseudorational transfer functions. Submitted to 47th IEEE CDC (2009)